National Center for Construction Education and Research

Electronic Systems Technician Level Two

PEARSON
Prentice
Hall

Upper Saddle River, New Jersey
Columbus, Ohio

contren®
Learning Series

National Center for Construction Education and Research

President: Don Whyte
Director of Curriculum Revision and Development: Daniele Dixon
Electronic Systems Technician Project Manager: Daniele Dixon
Production Manager: Jessica Martin
Production Maintenance Supervisor: Debie Ness
Editor: Brendan Coote
Desktop Publishers: Laura Parker, Debie Ness

NCCER would like to acknowledge the contract service provider for this curriculum:
Topaz Publications, Liverpool, New York.

This information is general in nature and intended for training purposes only. Actual performance of activities described in this manual requires compliance with all applicable operating, service, maintenance, and safety procedures under the direction of qualified personnel. References in this manual to patented or proprietary devices do not constitute a recommendation of their use.

10 9 8 7 6 5 4 3 2 1
ISBN 0-13-109199-9

Preface

This volume was developed by the National Center for Construction Education and Research (NCCER) in response to the training needs of the construction, maintenance, and pipeline industries. It is one of many in NCCER's *Contren® Learning Series*. The program, covering training for close to 40 construction and maintenance areas, and including skills assessments, safety training, and management education, was developed over a period of years by industry and education specialists.

NCCER also maintains a National Registry that provides transcripts, certificates, and wallet cards to individuals who have successfully completed modules of NCCER's *Contren® Learning Series*, when the training program is delivered by an NCCER Accredited training Sponsor.

The NCCER is a not-for-profit 501(c)(3) education foundation established in 1995 by the world's largest and most progressive construction companies and national construction associations. It was founded to address the severe workforce shortage facing the industry and to develop a standardized training process and curricula. Today, NCCER is supported by hundreds of leading construction and maintenance companies, manufacturers, and national associations, including the following partnering organizations:

PARTNERING ASSOCIATIONS

- American Fire Sprinkler Association
- American Petroleum Institute
- American Society for Training & Development
- American Welding Society
- Associated Builders & Contractors, Inc.
- Association for Career and Technical Education
- Associated General Contractors of America
- Carolinas AGC, Inc.
- Citizens Development Corps
- Construction Industry Institute
- Construction Users Roundtable
- Design-Build Institute of America

- Electronic Systems Industry Consortium
- Merit Contractors Association of Canada
- Metal Building Manufacturers Association
- National Association of Minority Contractors
- National Association of State Supervisors for Trade and Industrial Education
- National Association of Women in Construction
- National Insulation Association
- National Ready Mixed Concrete Association
- National Systems Contractors Association
- National Utility Contractors Association
- National Technical Honor Society
- North American Crane Bureau
- North American Technician Excellence
- Painting & Decorating Contractors of America
- Portland Cement Association
- SkillsUSA
- Steel Erectors Association of America
- Texas Gulf Coast Chapter ABC
- U.S. Army Corps of Engineers
- University of Florida
- Women Construction Owners & Executives, USA
- Youth Training and Development Consortium

Some features of NCCER's *Contren® Learning Series* are:

- An industry-proven record of success
- Curricula developed by the industry for the industry
- National standardization providing portability of learned job skills and educational credits
- Credentials for individuals through NCCER's National Registry
- Compliance with Apprenticeship, Training, Employer, and Labor Services (ATELS) requirements for related classroom training (CFR 29:29)
- Well-illustrated, up-to-date, and practical information

Acknowledgments

Special thanks to the National Systems Contractors Association (NSCA) for their continued support.

This curriculum was revised as a result of the farsightedness and leadership of the following sponsors:

BICI
Copp Systems Integrator
Electronic Systems Industry
  Consortium

Lincoln Technical Institute
Pratt Landry Associates
Simplex Grinnell
Sound Com Corporation

This curriculum would not exist were it not for the dedication and unselfish energy of those volunteers who served on the Authoring Team. A sincere thanks is extended to:

Stephen Clare
Mark Curry
Brendan Dillon
Mike Friedman
Dick Fyten

Larry Garter
Bruce Nardone
Joe Jones
Ken Nieto
Paul Salyers

Contents

DC Circuits

COURSE MAP

This course map shows all of the modules in the second level of the *Electronic Systems Technician* curriculum. The suggested training order begins at the bottom and proceeds up. Skill levels increase as you advance on the course map. The local Training Program Sponsor may adjust the training order.

ELECTRONIC SYSTEMS TECHNICIAN LEVEL TWO

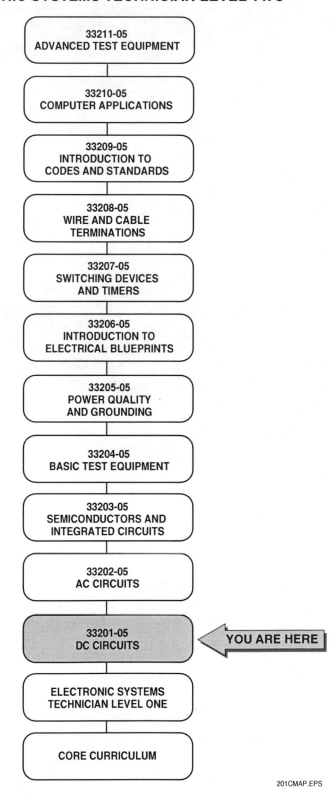

33211-05
ADVANCED TEST EQUIPMENT

33210-05
COMPUTER APPLICATIONS

33209-05
INTRODUCTION TO
CODES AND STANDARDS

33208-05
WIRE AND CABLE
TERMINATIONS

33207-05
SWITCHING DEVICES
AND TIMERS

33206-05
INTRODUCTION TO
ELECTRICAL BLUEPRINTS

33205-05
POWER QUALITY
AND GROUNDING

33204-05
BASIC TEST EQUIPMENT

33203-05
SEMICONDUCTORS AND
INTEGRATED CIRCUITS

33202-05
AC CIRCUITS

33201-05
DC CIRCUITS ◁ YOU ARE HERE

ELECTRONIC SYSTEMS
TECHNICIAN LEVEL ONE

CORE CURRICULUM

201CMAP.EPS

MODULE 33201-05 CONTENTS

Figures

Table

DC Circuits

Objectives

When you have completed this module, you will be able to do the following:

1. Define voltage and identify the ways in which it can be produced.
2. Explain the difference between conductors and insulators.
3. Explain how voltage, current, and resistance are related to each other.
4. Using the formula for Ohm's law, calculate an unknown value.
5. Describe the function of resistors and explain their color codes.
6. Explain the different types of meters used to measure voltage, current, and resistance.
7. Using the power formula, calculate the amount of power used by a circuit.
8. Explain the basic characteristics of series, parallel, and series-parallel circuits.
9. Using Kirchhoff's laws, calculate the voltage drop and current in series, parallel, and series-parallel circuits.
10. Find the total amount of resistance in series, parallel, and series-parallel circuits.

Prerequisites

Before you begin this module, it is recommended that you successfully complete *Core Curriculum* and *Electronic Systems Technician Level One*.

Required Trainee Materials

1. Paper and pencil
2. Appropriate personal protective equipment
3. Scientific calculator

1.0.0 ◆ INTRODUCTION

As an electronic systems technician, you must work with a force that cannot be seen. However, electricity is there on the job, every day of the year. It is necessary that you understand the forces of electricity so that you will be safe on the job. The first step is a basic understanding of the principles of electricity.

The relationships among **current**, **voltage**, **resistance**, and **power** in a basic direct current (DC) **series circuit** are common to all types of electrical **circuits**. This module provides a general introduction to the electrical concepts used in **Ohm's law**. It also presents the opportunity to practice applying these basic concepts to DC circuits. In this way, you can prepare for further study in electrical and electronics theory and maintenance techniques. By practicing these techniques for all combinations of DC circuits, you will be prepared to work on any DC circuits you might encounter.

2.0.0 ◆ CONDUCTORS AND INSULATORS

In order to understand the behavior of conductors and insulators, we must first examine their elemental structures. The most basic component of an element is the **atom**.

2.1.0 The Atom

The atom is the smallest part of an element that enters into a chemical change, but it does so in the form of a charged particle. These charged particles are called ions, and are of two types: positive and negative. A positive ion may be defined as an atom that has become positively charged. A negative ion is an atom that has become negatively charged.

One of the properties of charged ions is that ions of the same charge tend to repel one another, whereas ions of unlike charge will attract one another. The term charge is defined here as a quantity of electricity that is either positive or negative.

The structure of an atom is best explained by a detailed analysis of the simplest of all atoms, that of the element hydrogen. The hydrogen atom in *Figure 1* is composed of a **nucleus** containing one **proton** and a single orbiting **electron**. As the electron revolves around the nucleus, it is held in this orbit by two counteracting forces. One of these forces is called centrifugal force, which is the force that tends to cause the electron to fly outward as it travels around its circular orbit. The second force acting on the electron is electrostatic force. This force tends to pull the electron toward the nucleus and is generated by the mutual attraction between the positive nucleus and the negative electron. At some given radius, the two forces will balance each other, providing a stable path for the electron.

Basically, an atom contains three types of subatomic particles that are of interest in electricity: electrons, protons, and **neutrons**.

- A proton (+) repels another proton (+).
- An electron (−) repels another electron (−).
- A proton (+) attracts an electron (−).

The protons and neutrons are located in the center, or nucleus, of the atom, and the electrons travel about the nucleus in orbits.

Because protons are relatively heavy, the repulsive force they exert on one another in the nucleus of an atom has little effect.

The attracting and repelling forces on charged materials occur because of the electrostatic lines of force that exist around the charged materials. In a negatively charged object, the lines of force of the excess electrons add to produce an electrostatic field that has lines of force coming into the object from all directions. In a positively charged object, the lines of force of the excess protons add to produce an electrostatic field that has lines of force going out of the object in all directions. The electrostatic fields either aid or oppose each other to attract or repel.

2.1.1 The Nucleus

The nucleus is the central part of the atom. It is made up of heavy particles called protons and neutrons. The proton is a charged particle containing the smallest known unit of positive electricity. The neutron has no electrical charge. The number of protons in the nucleus determines how the atom of one element differs from the atom of another element.

Although a neutron is actually a particle by itself, it is generally thought of as an electron and proton combined and is electrically neutral. For this reason, neutrons are not considered important to the electrical nature of atoms.

2.1.2 Electrical Charges

The negative charge of an electron is equal but opposite to the positive charge of a proton. The charges of an electron and a proton are called electrostatic

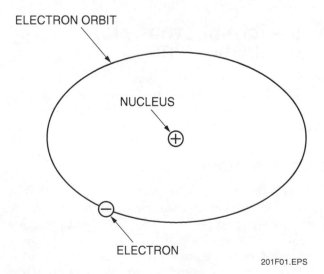

ELECTRON ORBIT

NUCLEUS

⊕

⊖

ELECTRON

201F01.EPS

Figure 1 ◆ Hydrogen atom.

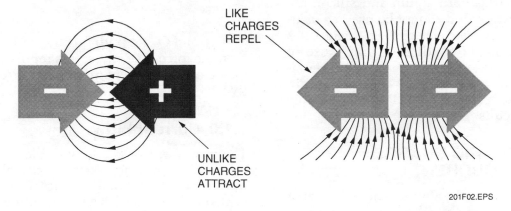

LIKE CHARGES REPEL

UNLIKE CHARGES ATTRACT

201F02.EPS

Figure 2 ◆ Law of electrical charges.

charges. The lines of force associated with each particle produce electrostatic fields. Because of the way these fields act together, charged particles can attract or repel one another. The Law of Electrical Charges states that particles with like charges repel each other, and those with unlike charges attract each other. This is shown in *Figure 2*.

2.2.0 Conductors and Insulators

The difference between atoms, with respect to chemical activity and stability, depends on the number and position of the electrons included within the atom. In general, the electrons reside in groups of orbits called shells. The shells are arranged in steps that correspond to fixed energy levels.

The number of electrons in the outermost shell determines the valence of an atom. For this reason, the outer shell of an atom is called the **valence shell**, and the electrons contained in this shell are called valence electrons (*Figure 3*). The valence of

an atom determines its ability to gain or lose an electron, which in turn determines the chemical and electrical properties of the atom. An atom that is lacking only one or two electrons from its outer shell will easily gain electrons to complete its shell, but a large amount of energy is required to free any of its electrons. An atom having a relatively small number of electrons in its outer shell in comparison to the number of electrons required to fill the shell will easily lose these valence electrons.

Valence electrons are what we are most concerned with in electricity. These are the electrons that are easiest to break loose from their parent atom. Normally, a **conductor** has three or fewer valence electrons, an **insulator** has five or more valence electrons, and semiconductors usually have four valence electrons.

All the elements of which **matter** is made may be placed into one of three categories: conductors, insulators, and semiconductors.

Conductors, for example, are elements such as copper and silver that will conduct a flow of electricity very readily. Because of their good conducting abilities, they are formed into wire and used to transfer electrical energy from one point to another.

Insulators, on the other hand, do not conduct electricity to any great degree and are used to prevent the flow of electricity. Compounds such as porcelain and plastic are good insulators.

VALENCE (OUTER) SHELL

VALENCE ELECTRON

201F03.EPS

Figure 3 ◆ Valence shell and electrons.

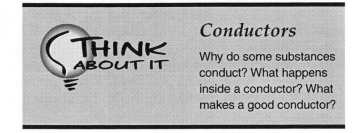

Conductors

Why do some substances conduct? What happens inside a conductor? What makes a good conductor?

Materials such as germanium and silicon are not good conductors but cannot be used as insulators either, since their electrical characteristics fall between those of conductors and those of insulators. These in-between materials are classified as semiconductors. As you will learn later in your training, semiconductors play a crucial role in electronic circuits.

3.0.0 ◆ ELECTRIC CHARGE AND CURRENT

An electric charge has the ability to do the work of moving another charge by attraction or repulsion. The ability of a charge to do work is called its potential. When one charge is different from another, there must be a difference in potential between them. The sum of the difference of potential of all the charges in the electrostatic field is referred to as electromotive force (emf) or voltage. Voltage is frequently represented by the letter E.

Electric charge is measured in **coulombs**. An electron has 1.6×10^{-19} coulombs of charge. Therefore, it takes 6.25×10^{18} electrons to make up one coulomb of charge, as shown below.

$$\frac{1}{1.6 \times 10^{-19}} = 6.25 \times 10^{18} \text{ electrons}$$

If two particles, one having charge Q_1 and the other charge Q_2, are a distance (d) apart, then the force between them is given by Coulomb's law, which states that the force is directly proportional to the product of the two charges and a constant (k) with a value of 10^9, and inversely proportional to the square of the distance between them:

$$\text{Force} = \frac{k \times Q_1 \times Q_2}{d^2}$$

If Q_1 and Q_2 are both positive or both negative, then the force is positive; it is repulsive. If Q_1 and Q_2 are of opposite charges, then the force is negative; it is attractive.

3.1.0 Current Flow

The movement of the flow of electrons is called current. To produce current, the electrons are moved by a potential difference. Current is represented by the letter I. The basic unit in which current is measured is the **ampere (A)**, also called the amp. One ampere of current is defined as the movement of one coulomb past any point of a conductor during one second of time. One coulomb is equal to 6.25×10^{18} electrons; therefore, one ampere is equal to 6.25×10^{18} electrons moving past any point of a conductor during one second of time.

The definition of current can be expressed as an equation:

$$I = \frac{Q}{T}$$

Where:

I = current (amperes)

Q = charge (coulombs)

T = time (seconds)

Charge differs from current in that Q is an accumulation of charge, while I measures the intensity of moving charges.

Units of Electricity and Volta

A disagreement with a fellow scientist over the twitching of a frog's leg eventually led 18th-century physicist Alessandro Volta to theorize that when certain objects and chemicals come into contact with each other, they produce an electric current. Believing that electricity came from contact between metals only, Volta coined the term metallic electricity. To demonstrate his theory, Volta placed two discs, one of silver and the other of zinc, into a weak acidic solution. When he linked the discs together with wire, electricity flowed through the wire. Thus, Volta introduced the world to the battery, also known as the Voltaic pile. Volta needed a term to measure the strength of the electric push or the flowing charge; the volt is that measure.

Law of Electrical Force and de Coulomb

In the 18th century, a French physicist named Charles de Coulomb was concerned with how electric charges behaved. He watched the repelling forces that electric charges exerted by measuring the twist in a wire. An object's weight acted as a turning force to twist the wire, and the amount of twist was proportional to the object's weight. After many experiments with opposing forces, de Coulomb proposed the Inverse Square Law, later known as the Law of Electrical Force.

In a conductor, such as copper wire, the free electrons are charges that can be forced to move with relative ease by a potential difference. If a potential difference is connected across two ends of a copper wire, as shown in *Figure 4*, the applied voltage forces the free electrons to move. This current is a flow of electrons from the point of negative charge (–) at one end of the wire, moving through the wire to the positive charge (+), at the other end. The direction of the electron flow is from the negative side of the **battery**, through the wire, and back to the positive side of the battery. The direction of current flow is therefore dependent on the polarity of the voltage source; that is, from a point of negative potential to a point of positive potential.

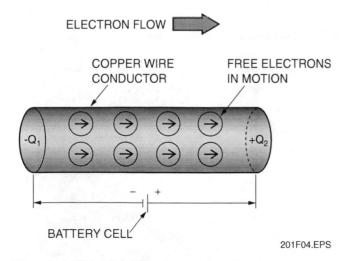

ELECTRON FLOW

COPPER WIRE CONDUCTOR

FREE ELECTRONS IN MOTION

$-Q_1$

$+Q_2$

– +

BATTERY CELL

201F04.EPS

Figure 4 ◆ Potential difference causing electric current.

3.2.0 Voltage

The force that causes electrons to move is called voltage, potential difference, or electromotive force (emf). One **volt (V)** is the potential difference between two points for which one coulomb of electricity will do one **joule (J)** of work. A battery is one of several means of creating voltage. It chemically creates a large reserve of free electrons at the negative (–) terminal. The positive (+) terminal has electrons chemically removed and will therefore accept them if an external path is provided from the negative (–) terminal. When a battery is no longer able to chemically deposit electrons at the negative (–) terminal, it is said to be dead, or in need of recharging. Batteries are normally rated in volts. Large batteries are also rated in ampere-hours, where one ampere-hour is a current of one amp supplied for one hour.

4.0.0 ◆ RESISTANCE

Resistance is directly related to the ability of a material to conduct electricity. Conductors have very low resistance; insulators have very high resistance.

4.1.0 Characteristics of Resistance

Resistance can be defined as the opposition to current flow. To add resistance to a circuit, electrical components called **resistors** are used. A resistor is a device whose resistance to current flow is a known, specified value. Resistance is measured in **ohms (Ω)** and is represented by the symbol R in equations. One ohm is defined as the amount of

Joule's Law

While other scientists of the 19th century were experimenting with batteries, cells, and circuits, James Joule was theorizing about the relationship between heat and energy. He discovered, contrary to popular belief, that work does not just move heat from one place to another; work, in fact, generates heat. Furthermore, he demonstrated that over time, a relationship exists between the temperature of water and electric current. These ideas formed the basis for the concept of energy. In his honor, the modern unit of energy was named the joule.

The Visual Language of Electricity

Learning to read circuit diagrams is like learning to read a book; first you learn to read the letters, then you learn to read the words, and before you know it, you are reading without paying attention to the individual letters anymore. Circuits are the same way; you will struggle at first with the individual pieces, but before you know it, you will be reading a circuit without even thinking about it. Studying the following table will help you to understand the fundamental language of electricity.

What Is Measured	Unit of Measurement	Ohm's Law Symbol	Symbol Used in Calculations
Amount of current	Amp	A	I
Electrical power	Watt	W	P
Force of current	Volt	V	E
Resistance to current	Ohm	Ω	R

resistance that will limit the current in a conductor to one ampere when the voltage applied to the conductor is one volt.

The resistance of a wire is proportional to the length of the wire, inversely proportional to the cross-sectional area of the wire, and dependent upon the kind of material from which the wire is made. The relationship for finding the resistance of a wire is as follows:

$$R = \rho \frac{L}{A}$$

Where:

R = resistance (ohms)

L = length of wire (feet)

A = area of wire (circular mils, CM, or cm²)

ρ = specific resistance (ohm-CM/ft or microhm-CM)

A mil equals 0.001 inch; a circular mil is the cross-sectional area of a wire that is one mil in diameter.

The specific resistance is a constant that depends on the material of which the wire is made. *Table 1* shows the properties of various wire conductors.

Table 1 shows that at 75°F, a one-mil diameter, pure annealed copper wire that is one foot long has a resistance of 10.351 ohms. A one-mil diameter, one-foot-long aluminum wire has a resistance of 16.758 ohms at 75°F. Temperature is important in determining the resistance of a wire. The hotter a wire, the greater its resistance.

Table 1 Conductor Properties

Metal	Specific Resistance (Resistance of 1 CM/ft in ohms)	
	32°F or 0°C	75°F or 23.8°C
Silver, pure annealed	8.831	9.674
Copper, pure annealed	9.39	10.351
Copper, annealed	9.59	10.505
Copper, hard-drawn	9.81	10.745
Gold	13.216	14.404
Aluminum	15.219	16.758
Zinc	34.595	37.957
Iron	54.529	62.643

Voltage Matters

Standard household voltage is different the world over, from 100V in Japan to 600V in Bombay, India. Many countries have no standard voltage; for example, France varies from 110V to 360V. If you were to plug a 120V hair dryer into England's 240V standard system, you would burn out the dryer. Use basic electric theory to explain exactly what would happen to destroy the hair dryer.

4.2.0 Ohm's Law

Ohm's law defines the relationship between current, voltage, and resistance. There are three ways to express Ohm's law mathematically.

- The current in a circuit is equal to the voltage applied to the circuit divided by the resistance of the circuit:

$$I = \frac{E}{R}$$

- The resistance of a circuit is equal to the voltage applied to the circuit divided by the current in the circuit:

$$R = \frac{E}{I}$$

- The applied voltage to a circuit is equal to the product of the current and the resistance of the circuit:

$$E = I \times R = IR$$

Where:

I = current (amperes)

R = resistance (ohms)

E = voltage or emf (volts)

If any two of the quantities E, I, or R are known, the third can be calculated.

The Ohm's law equations can be memorized and practiced effectively by using an Ohm's law circle, as shown in *Figure 5*. To find the equation for E, I, or R when two quantities are known, cover the unknown third quantity. The other two quantities in the circle will indicate how the covered quantity may be found.

Example 1: Find I when E = 120V and R = 30Ω.

$$I = \frac{E}{R}$$

$$I = \frac{120V}{30\Omega}$$

$$I = 4A$$

This formula shows that in a DC circuit, current (I) is directly proportional to voltage (E) and inversely proportional to resistance (R).

Example 2: Find R when E = 240V and I = 20A.

$$R = \frac{E}{I}$$

$$R = \frac{240V}{20A}$$

$$R = 12\Omega$$

Example 3: Find E when I = 15A and R = 8Ω.

$$E = I \times R$$

$$E = 15A \times 8\Omega$$

$$E = 120V$$

5.0.0 ◆ SCHEMATIC REPRESENTATION OF CIRCUIT ELEMENTS

A simple electric circuit is shown in both pictorial and schematic forms in *Figure 6*. The schematic diagram is a shorthand way to draw an electric circuit, and circuits are usually represented in this way. In addition to the connecting wire, three components are shown symbolically: the battery, the switch, and the lamp. Note the positive (+) and negative (–) markings in both the pictorial

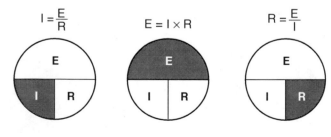

	LETTER SYMBOL	UNIT OF MEASUREMENT
CURRENT	I	AMPERES (A)
RESISTANCE	R	OHMS (Ω)
VOLTAGE	E	VOLTS (V)

201F05.EPS

Figure 5 ◆ Ohm's law circle.

Using Your Intuition

Learning the meanings of various electrical symbols may seem overwhelming, but if you take a moment to study *Figure 7*, you will see that most of them are intuitive—that is, they are shaped in a symbolic way to represent the actual object. For example, the battery shows + and –, just like an actual battery. The motor has two arms that suggest a spinning rotor. The transformer shows two coils. The resistor has a jagged edge to suggest pulling or resistance. Connected wires have a black dot that reminds you of solder. Unconnected wires simply cross. The fuse stretches out in both directions as though to provide extra slack in the line. The circuit breaker shows a line with a break in it. The capacitor shows a gap. The variable resistor has an arrow like a swinging compass needle. As you learn to read schematics, take the time to make mental connections between the symbol and the object it represents.

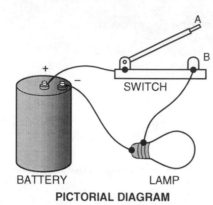

PICTORIAL DIAGRAM   SCHEMATIC DIAGRAM

201F06.EPS

Figure 6 ◆ Simple electrical symbols.

Ammeter	—(A)—	Motor (DC)	—(B)—
Battery	—+‖⊢—	Resistor (fixed)	—⋁⋁⋁—
Capacitor (fixed)	—⊣(—	Resistor (variable)	—⋀⋀⋀—
Capacitor (variable)	—⫫—	Rheostat	
Circuit breaker	—⌒—	Switch	—o⸝ o—
Crystal	—⊣▯⊢—	Semiconductor diode	—+▶⊢ -
Fuse	—⌇—	Transformer (general)	
Generator (AC)	—(∼)—	Transformer (iron core)	
Generator (DC)	(G)	Transistor (NPN)	
Ground	⏚ or ⎓	Transistor (PNP)	
Inductor (air core)	⌒⌒⌒	Voltmeter	—(V)—
Inductor (iron core)	⌒⌒⌒	Wattmeter	—(W)—
Inductor (tapped)		Wires (connected)	—•—
Lamp		Wires (unconnected)	—┼—
Motor (AC)	(◎)	Zener diode	—▶⌐

201F07.EPS

Figure 7 ◆ Standard schematic symbols.

and schematic representations of the battery. The schematic components represent the pictorial components in a simplified manner. A schematic diagram is one that shows, by means of graphic symbols, the electrical connections and functions of the different parts of a circuit.

The standard graphic symbols for commonly used electrical and electronic components are shown in *Figure 7*.

6.0.0 ◆ RESISTORS

The function of a resistor is to offer a particular resistance to current flow. For a given current and known resistance, the change in voltage across the component, or voltage drop, can be predicted using Ohm's law. Voltage drop refers to a specific amount of voltage used, or developed, by that component. An example is a very basic circuit of a 10V battery and a single resistor in a series circuit. The voltage drop across that resistor is 10V because it is the only component in the circuit, and all voltage must be dropped across that resistor. Similarly, for a given applied voltage, the current that flows may be predetermined by selection of the resistor value. The required power dissipation largely dictates the construction and physical size of a resistor.

The two most common types of electronic resistors are wire-wound and carbon composition construction. A typical wire-wound resistor consists of a length of nickel wire wound on a ceramic tube and covered with porcelain. Low-resistance connecting wires are provided, and the resistance value is usually printed on the side of the component. *Figure 8* illustrates the construction of typical resistors. Carbon composition resistors are constructed by molding mixtures of powdered carbon and insulating materials into a cylindrical shape. An outer sheath of insulating material affords mechanical and electrical protection, and copper connecting wires are provided at each end. Carbon composition resistors are smaller and less expensive than the wire-wound type. However, the wire-wound type is the more rugged of the two and is able to survive much larger power dissipations than the carbon composition type.

Most resistors have standard fixed values, so they can be termed fixed resistors. Variable resistors, also known as adjustable resistors, are used a great deal in electronics. Two common symbols for a variable resistor are shown in *Figure 9*.

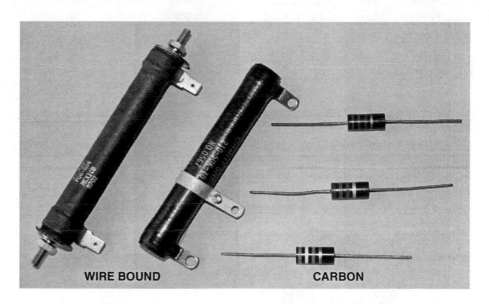

WIRE BOUND   **CARBON**

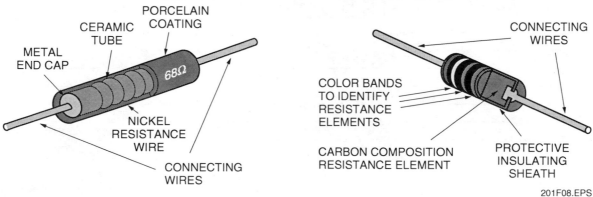

201F08.EPS

Figure 8 ◆ Common resistors.

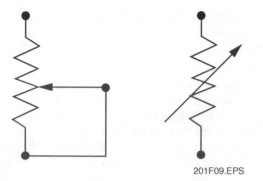

Figure 9 ◆ Symbols used for variable resistors.

A variable resistor consists of a coil of closely wound, insulated resistance wire formed into a partial circle. The coil has a low-resistance terminal at each end, and a third terminal is connected to a movable contact with a shaft adjustment facility. The movable contact can be set to any point on a connecting track that extends over one (uninsulated) edge of the coil.

Using the adjustable contact, the resistance from either end terminal to the center terminal may be adjusted from zero to the maximum coil resistance.

Another type of variable resistor is known as a decade resistance box. This is a laboratory component that contains precise values of switched series-connected resistors.

6.1.0 Resistor Color Codes

Because carbon composition resistors are physically small (some are less than 1 cm in length), it is not convenient to print the resistance value on the side. Instead, a color code in the form of colored bands is employed to identify the resistance value and tolerance. The color code is illustrated in *Figure 10*. Starting from one end of the resistor, the first two bands identify the first and second digits of the resistance value, and the third band indicates the number of zeros. An exception to this is when the third band is either silver or gold, which indicates a 0.01 or 0.1 multiplier, respectively. The fourth band represents the tolerance. A silver band represents a ±10 percent tolerance, while a gold band is ±5 percent. Precision resistors use red to indicate a tolerance of ±2 percent and brown for ±1 percent. Where no fourth band is present, the resistor tolerance is ±20 percent.

We can put this information to practical use by determining the range of values for the carbon resistor in *Figure 11*.

The color code for this resistor is as follows:

- Brown = 1, black = 0, red = 2, gold = a tolerance of ±5 percent
- First digit = 1, second digit = 0, number of zeros (2) = 1,000Ω

Since this resistor has a value of 1,000Ω ±5 percent, the resistor can range in value from 950Ω to 1,050Ω.

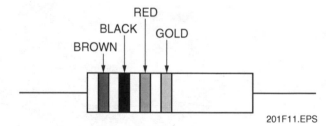

Figure 11 ◆ Sample color codes on a fixed resistor.

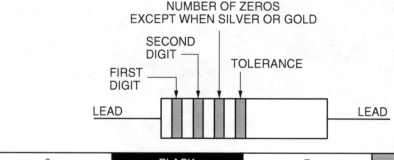

0	BLACK	7	VIOLET
1	BROWN	8	GREY
2	RED	9	WHITE
3	ORANGE	0.1	GOLD
4	YELLOW	0.01	SILVER
5	GREEN	5%	GOLD – TOLERANCE
6	BLUE	10%	SILVER – TOLERANCE

Figure 10 ◆ Resistor color codes.

7.0.0 ◆ MEASURING VOLTAGE, CURRENT, AND RESISTANCE

Working with electricity requires making accurate measurements. This section will discuss the basic meters used to measure voltage, current, and resistance: the **voltmeter**, **ammeter**, and **ohmmeter**.

WARNING!

Only qualified individuals may use these meters. Consult your company's safety policy for applicable rules.

7.1.0 Basic Meter Operation

When troubleshooting or testing equipment, you will need various meters to check for proper circuit voltages, currents, and resistances and to determine if the wiring is defective. Meters are used in repairing, maintaining, and troubleshooting electrical circuits and equipment. The best and most expensive measuring instrument is of no use to you unless you know what you are measuring and what each reading indicates. Remember that the purpose of a meter is to measure quantities existing within a circuit. For this reason, when the meter is connected to a circuit, it must not change the condition of the circuit.

The three basic electrical quantities discussed in this section are current, voltage, and resistance. It is really current that causes the meter to respond even when voltage or resistance is being measured. In a basic meter, the measurement of current can be calibrated to indicate almost any electrical quantity based on the principle of Ohm's law. The amount of current that flows through a meter is determined by the voltage applied to the meter and the resistance of the meter, as stated by $I = E \div R$.

For a given meter resistance, different values of applied voltage will cause specific values of current to flow. Although the meter actually measures current, the meter scale can be calibrated in units of voltage. Similarly, for a given applied voltage, different values of resistance will cause specific values of current to flow. Therefore, the meter scale can also be calibrated in units of resistance rather than current. The same holds true for power, since power is proportional to current, as stated by $P = EI$. It is on this principle that the meter was developed and its construction allows for the measurement of various parameters by actually measuring current.

You must understand the purpose and function of each individual piece of test equipment and any limitations associated with it. It is also extremely important that you understand how to use each piece of equipment safely. If you understand the capabilities of the test equipment, you can better use the equipment, better understand the indications on the equipment, and know what substitute or backup meters can be used.

7.2.0 Voltmeter

A simple voltmeter consists of the meter movement in series with the internal resistance of the voltmeter itself. For example, a meter with a 50-microamp (µA) meter movement and a 1,000Ω internal resistance can be used to measure voltages directly up to 0.05V. (The prefix **micro** means one-millionth.) When the meter is placed across the voltage source (*Figure 12*), a current determined by the internal resistance of the meter flows through the meter movement. A voltmeter's internal resistance is typically high to minimize meter loading effects on the source.

To measure larger voltages, a multiplier resistor is used. This increased series resistance limits the current that can flow through the meter movement, thus extending the range of the meter.

To avoid damage to the meter movement, the following precautions should be taken when using a voltmeter:

- Always set the full-scale voltage of the meter to be larger than the expected voltage to be measured.
- Always ensure that the internal resistance of the voltmeter is much greater than the resistance of the component to be measured. This means that the current it takes to drive the voltmeter (about 50 microamps) should be a negligible fraction of the current flowing through the circuit element being measured.
- If you are unsure of the level of the voltage to be measured, take a reading at the highest range of the voltmeter and progressively lower the range until the reading is obtained.

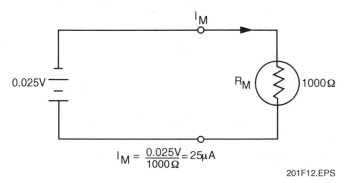

Figure 12 ◆ Simple voltmeter.

In most commercial voltmeters, the internal resistance is expressed by the ohms-per-volt rating of the meter. A typical meter has a rating of 20,000 ohms-per-volt with a 50-microamp movement. This quantity tells what the internal resistance of the meter is on any particular full-scale setting. In general, the meter's internal resistance is the ohms-per-volt rating multiplied by the full-scale voltage. The higher the ohms-per-volt rating, the higher the internal resistance of the meter, and the smaller the effect of the meter on the circuit.

7.3.0 Ammeter

A current meter, usually called an ammeter, is used by placing the meter in series with the wire through which the current is flowing. This method of connection is shown in *Figure 13*. Notice how the magnitude of load current will flow through the ammeter. Because of this, an ammeter's internal resistance must be low to minimize the circuit-loading effects as seen by the source. Also, high current magnitudes flowing through an ammeter can damage it. For this reason, ammeter shunts are employed to reduce the ammeter circuit current to a fraction of the current flowing through the load.

To avoid damage to the meter movement, take the following precautions when measuring current with an ammeter:

- Always check the polarity of the ammeter. Make certain that the meter is connected to the circuit so that electrons flow into the negative lead and out of the positive lead. It is easy to tell which is the positive lead because it is normally red. The negative lead is usually black.
- Always set the full-scale deflection of the meter to be larger than the expected current. To be safe, set the full-scale current several times larger than the expected current, and then slowly increase the meter sensitivity to the appropriate scale.

- Always connect the ammeter in series with the circuit element through which the current to be measured is flowing. Never connect the ammeter in parallel. When an ammeter is connected across a constant-potential source of appreciable voltage, the low internal resistance of the meter bypasses the circuit resistance. This results in the application of the source voltage directly to the meter terminals. The resulting excess current will burn out the meter coil.

7.4.0 Ohmmeter

An ohmmeter is used to measure resistance and check continuity. The deflection of the pointer of an ohmmeter is controlled by the amount of battery current passing through the coil. Current flow depends on the applied voltage and the circuit resistance. By applying a constant source voltage to the circuit under test, the resultant current flow depends only on circuit resistance. This magnitude of current will create meter movement. By knowing the relationship between current and resistance, an ohmmeter's scale can be calibrated to indicate circuit resistance based on the magnitude of current for a constant source voltage. A simple ohmmeter circuit is shown in *Figure 14*.

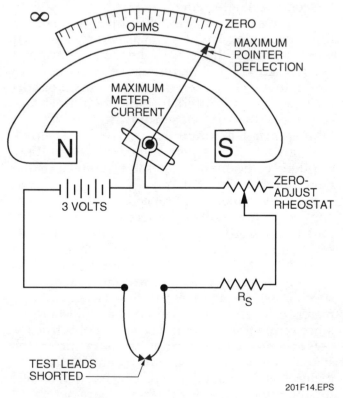

Figure 14 ◆ Simple ohmmeter circuit.

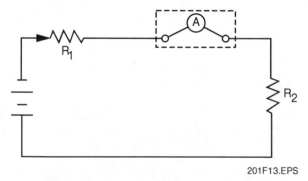

Figure 13 ◆ Ammeter connection.

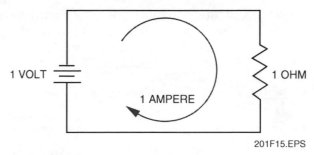

Figure 15 ◆ One watt.

201F15.EPS

8.0.0 ◆ ELECTRICAL POWER

Power is defined as the rate of doing work. This is equivalent to the rate at which energy is used or dissipated. Electrons passing through a resistance dissipate energy in the form of heat. In electrical circuits, power is measured in units called **watts (W)**. The power in watts equals the rate of energy conversion. One watt of power equals the work done in one second by one volt of potential difference in moving one coulomb of charge. One coulomb per second is an ampere; therefore, power in watts equals the product of amperes times volts.

The work done in an electrical circuit can be useful work or it can be wasted work. In both cases, the rate at which the work is done is still measured in power. The turning of an electric motor is useful work. On the other hand, the heating of wires or resistors in a circuit is wasted work, since no useful function is performed by the heat.

The unit of electrical work is the joule, the amount of work done by one coulomb flowing through a potential difference of one volt. Thus, if five coulombs flow through a potential difference of one volt, five joules of work are done. The time it takes these coulombs to flow through the potential difference has no bearing on the amount of work done.

It is more convenient when working with circuits to think of amperes of current rather than coulombs. As previously discussed, one ampere equals one coulomb passing a point in one second. Using amperes, one joule of work is done in one second when one ampere moves through a potential difference of one volt. This rate of one joule of work in one second is the basic unit of power, and is called a watt. Therefore, a watt is the power used when one ampere of current flows through a potential difference of one volt, as shown in *Figure 15*.

8.1.0 Power Equation

When one ampere flows through a difference of two volts, two watts must be used. In other words, the number of watts used is equal to the number of amperes of current times the potential difference. This is expressed in equation form as:

$$P = I \times E \text{ or } P = IE$$

Where:

P = power used in watts

I = current in amperes

E = potential difference in volts

The equation is sometimes called Ohm's law for power because it is similar to Ohm's law. This equation is used to find the power consumed in a circuit or load when the values of current and voltage are known. The second form of the equation is used to find the voltage when the power and current are known:

$$E = \frac{P}{I}$$

The third form of the equation is used to find the current when the power and voltage are known:

$$I - \frac{P}{E}$$

Using these three equations, the power, voltage, or current in a circuit can be calculated whenever any two of the values are already known.

Example 1: Calculate the power in a circuit where the source of 100V produces 2A in a 50Ω resistance.

P = IE

P = 2 × 100

P = 200W

This means the source generates 200W of power while the resistance dissipates 200W in the form of heat.

Example 2: Calculate the source voltage in a circuit that consumes 1,200W at a current of 5A.

$$E = \frac{P}{I}$$

$$E = \frac{1,200}{5}$$

E = 240V

Example 3: Calculate the current in a circuit that consumes 600W with a source voltage of 120V.

$$I = \frac{P}{E}$$

$$I = \frac{600}{120}$$

$$I = 5A$$

Components that use the power dissipated in their resistance are generally rated in terms of power. The power is rated at normal operating voltage, which is usually 120V. For instance, an appliance that draws 5A at 120V would dissipate 600W. The rating for the appliance would then be 600W/120V.

To calculate I or R for components rated in terms of power at a specified voltage, it may be convenient to use the power formula in different forms. There are actually three basic power formulas, but each can be rearranged into three other forms for a total of nine combinations:

$$P = IE \qquad P = I^2R \qquad P = \frac{E^2}{R}$$

$$I = \frac{P}{E} \qquad R = \frac{P}{I^2} \qquad R = \frac{E^2}{P}$$

$$E = \frac{P}{I} \qquad I = \sqrt{\frac{P}{R}} \qquad E = \sqrt{PR}$$

Note that all of these formulas are based on Ohm's law (E = IR) and the power formula (P = IE). *Figure 16* shows all of the applicable power, voltage, resistance, and current equations.

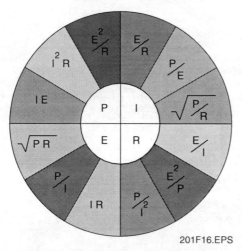

201F16.EPS

Figure 16 ◆ Expanded Ohm's law circle.

8.2.0 Power Rating of Resistors

If too much current flows through a resistor, the heat caused by the current will damage or destroy the resistor. This heat is caused by I^2R heating, which is power loss expressed in watts. Therefore, every resistor is given a wattage, or power rating, to show how much I^2R heating it can take before it burns out. This means that a resistor with a power rating of one watt will burn out if it is used in a circuit where the current causes it to dissipate heat at a rate greater than one watt.

If the power rating of a resistor is known, the maximum current it can carry is found by using an equation derived from $P = I^2R$:

$P = I^2R$ becomes $I^2 = P \div R$, which becomes $I = \sqrt{P \div R}$

Using this equation, find the maximum current that can be carried by a 1Ω resistor with a power rating of 4W:

$$I = \sqrt{P \div R} = \sqrt{4 \div 1} = 2 \text{ amperes}$$

If such a resistor conducts more than 2 amperes, it will dissipate more than its rated power and burn out.

Power ratings assigned by resistor manufacturers are usually based on the resistors being mounted in an open location where there is free air circulation, and where the temperature is not higher than 104°F (40°C). Therefore, if a resistor is mounted in a small, crowded, enclosed space, or where the temperature is higher than 104°F, there is a good chance it will burn out even before its power rating is exceeded. Also, some resistors are designed to be attached to a chassis or frame that will carry away the heat.

9.0.0 ◆ RESISTIVE CIRCUITS

A resistive circuit is simply a circuit that includes a resistor, such as a lamp or other device. Resistive circuits can be wired in series, in parallel, or as a combination (series-parallel).

9.1.0 Resistances in Series

A series circuit is a circuit in which there is only one path for current flow. In the series circuit shown in *Figure 17*, the current (I) is the same in all

Power Ratings

Notice the common electrical devices in the building you're in. What is their wattage rating? How much current do they draw? How would you test their voltage or amperage?

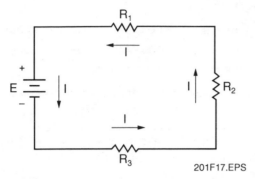

201F17.EPS

Figure 17 ◆ Series circuit.

parts of the circuit. This means that the current flowing through R_1 is the same as the current flowing through R_2 and R_3, and it is also the same as the current supplied by the battery.

When resistances are connected in series as in this example, the total resistance in the circuit is equal to the sum of the resistances of all the parts of the circuit:

$$R_T = R_1 + R_2 + R_3$$

Where:

R_T = total resistance
$R_1 + R_2 + R_3$ = resistances in series

Example 1: The circuit shown in *Figure 18A* has 50Ω, 75Ω, and 100Ω resistors in series. Find the total resistance of the circuit.

Add the values of the three resistors in series:

$$R_T = R_1 + R_2 + R_3 = 50 + 75 + 100 = 225\Omega$$

Example 2: The circuit shown in *Figure 18B* has three lamps connected in series with the resistances shown. Find the total resistance of the circuit.

Add the values of the three lamp resistances in series:

$$R_T = R_1 + R_2 + R_3 = 20 + 40 + 60 = 120\Omega$$

9.2.0 Resistances in Parallel

A **parallel circuit** contains two or more parallel paths through which current can flow. The total resistance in a parallel resistive circuit is given by this formula:

$$R_T = \cfrac{1}{\cfrac{1}{R_1} + \cfrac{1}{R_2} + \cfrac{1}{R_3} + \cfrac{1}{R_n}}$$

Where:

R_T = total resistance in parallel
R_1, R_2, R_3, and R_n = branch resistances

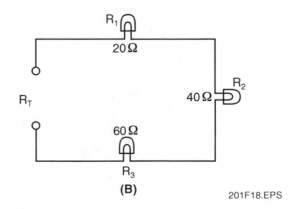

(A)   (B)

201F18.EPS

Figure 18 ◆ Total resistance.

Example 1: Find the total resistance of the 2Ω, 4Ω, and 8Ω resistors in parallel shown in *Figure 19*.

Write the formula for the three resistances in parallel:

$$R_T = \cfrac{1}{\cfrac{1}{R_1} + \cfrac{1}{R_2} + \cfrac{1}{R_3}}$$

Substitute the resistance values:

$$R_T = \cfrac{1}{\cfrac{1}{2} + \cfrac{1}{4} + \cfrac{1}{8}}$$

$$R_T = \frac{1}{0.5 + 0.25 + 0.125}$$

$$R_T = \frac{1}{0.875}$$

$$R_T = 1.14\Omega$$

Note that when resistances are connected in parallel, the total resistance is always less than the resistance of any single branch. In this case:

$$R_T = 1.14\Omega < R_1 = 2\Omega, R_2 = 4\Omega, \text{ and } R_3 = 8\Omega$$

Example 2: Add a fourth parallel resistor of 2Ω to the circuit in *Figure 19*. What is the new total resistance, and what is the net effect of adding another resistance in parallel?

Write the formula for four resistances in parallel:

$$R_T = \cfrac{1}{\cfrac{1}{R_1} + \cfrac{1}{R_2} + \cfrac{1}{R_3} + \cfrac{1}{R_4}}$$

Substitute values:

$$R_T = \cfrac{1}{\cfrac{1}{2} + \cfrac{1}{4} + \cfrac{1}{8} + \cfrac{1}{2}}$$

$$R_T = \frac{1}{0.5 + 0.25 + 0.125 + 0.5}$$

$$R_T = \frac{1}{1.375}$$

$$R_T = 0.73\Omega$$

The net effect of adding another resistance in parallel is a reduction of the total resistance from 1.14Ω to 0.73Ω.

9.2.1 Simplified Formulas

The total resistance of equal resistors in parallel is equal to the resistance of one equal resistor divided by the number of equal resistors:

$$R_T = \frac{R}{N}$$

Where:

R_T = total resistance of equal resistors in parallel

R = resistance of one of the equal resistors

N = number of equal resistors

If two resistors with the same resistance are connected in parallel, the equivalent resistance is half of that value, as shown in *Figure 20*.

The two 200Ω resistors in parallel are the equivalent of one 100Ω resistor; the two 100Ω resistors are the equivalent of one 50Ω resistor; and the two 50Ω resistors are the equivalent of one 25Ω resistor.

When any two unequal resistors are in parallel, it is often easier to calculate the total resistance by multiplying the two resistances and then dividing the product by the sum of the resistances:

$$R_T = \frac{R_1 \times R_2}{R_1 + R_2}$$

Where:

R_T = total resistance of unequal resistors in parallel

R_1, R_2 = two unequal resistors in parallel

Example 1: Find the total resistance of a 6Ω (R_1) resistor and an 18Ω (R_2) resistor in parallel:

$$R_T = \frac{R_1 \times R_2}{R_1 + R_2}$$

$$R_T = \frac{6 \times 18}{6 + 18}$$

$$R_T = \frac{108}{24}$$

$$R_T = 4.5\Omega$$

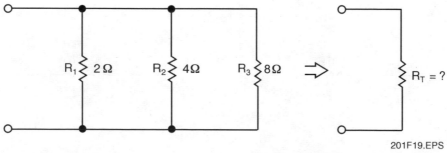

201F19.EPS

Figure 19 ◆ Parallel branch.

Parallel Circuits

An interesting fact about circuits is the drop in resistance in a parallel circuit as more resistors are added. But this fact does not mean that you can add an endless number of devices, such as lamps, in a parallel circuit. Why not?

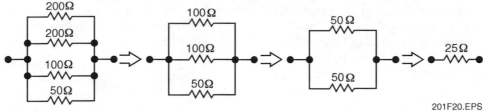

Figure 20 ◆ Equal resistances in a parallel circuit.

Example 2: Find the total resistance of a 100Ω (R_1) resistor and a 150Ω (R_2) resistor in parallel:

$$R_T = \frac{R_1 \times R_2}{R_1 + R_2}$$

$$R_T = \frac{100 \times 150}{100 + 150}$$

$$R_T = \frac{15,000}{250}$$

$$R_T = 60Ω$$

9.3.0 Series-Parallel Circuits

To find current, voltage, and resistance in series circuits and parallel circuits is fairly easy. When working with either type, use only the rules that apply to that type. In a **series-parallel circuit**, some parts of the circuit are series connected and other parts are parallel connected. Thus, in some parts the rules for series circuits apply, and in other parts, the rules for parallel circuits apply. To analyze or solve a problem involving a series-parallel circuit, it is necessary to recognize which parts of the circuit are series connected and which parts are parallel connected. This is obvious if the circuit is simple. Many times, however, the circuit must be redrawn, putting it into a form that is easier to recognize.

In a series circuit, the current is the same at all points. In a parallel circuit, there are one or more points where the current divides and flows in separate branches. In a series-parallel circuit, there are both separate branches and series loads. The easiest way to find out whether a circuit is a series, parallel, or series-parallel circuit is to start at the negative terminal of the power source and trace the path of current through the circuit back to the positive terminal of the power source. If the current does not divide anywhere, it is a series circuit. If the current divides into separate branches, but there are no series loads, it is a parallel circuit. If the current divides into separate branches and there are also series loads, it is a series-parallel circuit. *Figure 21* shows electric lamps connected in series, parallel, and series-parallel circuits.

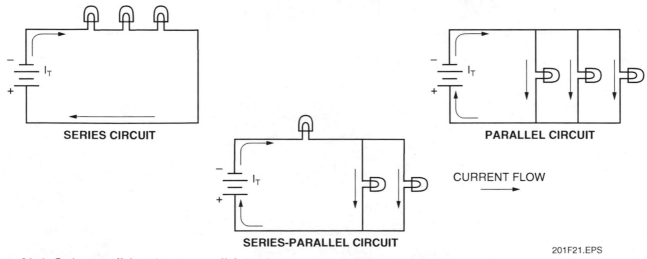

SERIES CIRCUIT

SERIES-PARALLEL CIRCUIT

PARALLEL CIRCUIT

CURRENT FLOW

Figure 21 ◆ Series, parallel, and series-parallel circuits.

Series-Parallel Circuits

Explain Figure 22. Which resistors are in series and which are in parallel?

After determining that a circuit is series-parallel, redraw the circuit so that the branches and the series loads are more easily recognized. This is especially helpful when computing the total resistance of the circuit. *Figure 22* shows resistors connected in a series-parallel circuit and the equivalent circuit redrawn to simplify it.

9.3.1 Reducing Series-Parallel Circuits

Very often, all that is known about a series-parallel circuit is the applied voltage and the values of the individual resistances. To find the voltage drop across any of the loads or the current in any circuit branch, the total circuit current must usually be known. To find the total current, the total resistance of the circuit must be known. To find the total resistance, reduce the circuit to its simplest form, which is usually one resistance that forms a series circuit with the voltage source. This simple series circuit has the equivalent resistance of the series-parallel circuit it was derived from, and also has the same total current. There are four basic steps in reducing a series-parallel circuit:

• If necessary, redraw the circuit so that all parallel combinations of resistances and series resistances are easily recognized.
• For each parallel combination of resistances, calculate its effective resistance.
• Replace each of the parallel combinations with one resistance whose value is equal to the effective resistance of that combination. This provides a circuit with all series loads.
• Find the total resistance of this circuit by adding the resistances of all the series loads.

Examine the series-parallel circuit shown in *Figure 23*, and reduce it to an equivalent series circuit.

In this circuit, resistors R_2 and R_3 are connected in parallel, but resistor R_1 is in series with both the battery and the parallel combination of R_2 and R_3. The current I_T leaving the negative terminal of the voltage source travels through resistor R_1 before it is divided at the junction of resistors R_1, R_2, and R_3 (Point A) to go through the two branches formed by resistors R_2 and R_3.

Given the information in *Figure 23*, calculate the resistance of R_2 and R_3 in parallel and the total resistance of the circuit, R_T.

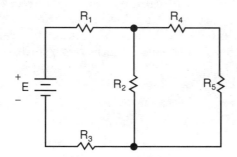
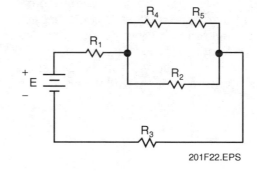

201F22.EPS

Figure 22 ◆ Redrawing a series-parallel circuit.

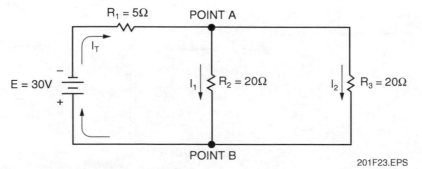

201F23.EPS

Figure 23 ◆ Reducing a series-parallel circuit.

The total resistance of the circuit is the sum of R_1 and the equivalent resistance of R_2 and R_3 in parallel. To find R_T, first find the resistance of R_2 and R_3 in parallel. Because the two resistances have the same value of 20Ω, the resulting equivalent resistance is 10Ω. We can then calculate the total resistance:

$R_T = 15\Omega$ $(5\Omega + 10\Omega = 15\Omega)$

9.4.0 Applying Ohm's Law

In resistive circuits, unknown circuit parameters can be found by using Ohm's law and the techniques for determining equivalent resistance. The calculation varies depending on whether it is a series circuit, a parallel circuit, or a series-parallel circuit.

9.4.1 Voltage and Current in Series Circuits

Ohm's law may be applied to an entire series circuit or to the individual parts of the circuit. When it is used on a particular part of a circuit, the voltage across that part is equal to the current in that part multiplied by the resistance of that part.

For example, given the information in *Figure 24*, calculate the total resistance (R_T) and the total current (I_T).

To find R_T:

$R_T = R_1 + R_2 + R_3$

$R_T = 20 + 50 + 120$

$R_T = 190\Omega$

To find I_T using Ohm's law:

$I_T = \dfrac{E_T}{R_T}$

$I_T = \dfrac{95}{190}$

$I_T = 0.5A$

Find the voltage across each resistor. In a series circuit, the current is the same; that is, $I = 0.5A$ through each resistor:

$E_1 = IR_1 = 0.5(20) = 10V$

$E_2 = IR_2 = 0.5(50) = 25V$

$E_3 = IR_3 = 0.5(120) = 60V$

The voltages E_1, E_2, and E_3 found for *Figure 24* are known as voltage drops or IR drops. Their effect is to reduce the voltage that is available to be applied across the rest of the components in the circuit. The sum of the voltage drops in any series circuit is always equal to the voltage that is applied to the circuit. The total voltage (E_T) is the same as the applied voltage and can be verified in this example:

$E_T = 10 + 25 + 60 = 95V$

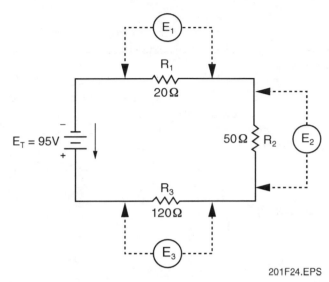

201F24.EPS

Figure 24 ◆ Calculating voltage drops.

9.4.2 Voltage and Current in Parallel Circuits

A parallel circuit is a circuit in which two or more components are connected across the same voltage source, as illustrated in *Figure 25*. The resistors R_1, R_2, and R_3 are in parallel with each other and with the battery. Each parallel path is then a branch with its own individual current. When the total current I_T leaves the voltage source E, part I_1 of the current I_T will flow through R_1, part I_2 will flow through R_2, and the remainder I_3 will flow through R_3. The branch currents I_1, I_2, and I_3 can be different. However, if a voltmeter is connected across R_1, R_2, and R_3, the respective voltages E_1, E_2, and E_3 will be equal to the source voltage E.

The total current I_T is equal to the sum of all branch currents.

Voltage Drops

Calculating voltage drops is not just a schoolroom exercise. It is important to know the voltage drop when sizing circuit components. What would happen if you sized a component without accounting for a substantial voltage drop in the circuit?

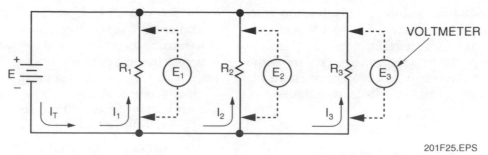

Figure 25 ◆ Parallel circuit.

201F25.EPS

This formula applies for any number of parallel branches whether the resistances are equal or unequal.

Using Ohm's law, each branch current equals the applied voltage divided by the resistance between the two points where the voltage is applied. Hence, for each branch in *Figure 25* we have the following equations:

Branch 1: $I_1 = \dfrac{E_1}{R_1} = \dfrac{E}{R_1}$

Branch 2: $I_2 = \dfrac{E_2}{R_2} = \dfrac{E}{R_2}$

Branch 3: $I_3 = \dfrac{E_3}{R_3} = \dfrac{E}{R_3}$

With the same applied voltage, any branch that has less resistance allows more current through it than a branch with higher resistance.

Example 1: The two branches R_1 and R_2, shown in *Figure 26A*, across a 110V power line draw a total line current of 20A. Branch R_1 takes 12A. What is the current I_2 in branch R_2?

Transpose to find I_2 and then substitute given values:

$I_T = I_1 + I_2$
$I_2 = I_T - I_1$
$I_2 = 20 - 12 = 8A$

Example 2: As shown in *Figure 26B*, the two branches R_1 and R_2 across a 240V power line draw a total line current of 35A. Branch R_2 takes 20A. What is the current I_1 in branch R_1?

Transpose to find I_1 and then substitute given values:

$I_T = I_1 + I_2$
$I_1 = I_T - I_2$
$I_1 = 35 - 20 = 15A$

9.4.3 Voltage and Current in Series-Parallel Circuits

Series-parallel circuits combine the elements and characteristics of both the series and parallel configurations. By properly applying the equations and methods previously discussed, the values of individual components of the circuit can be determined. *Figure 27* shows a simple series-parallel circuit with a 1.5V battery.

The current and voltage associated with each component can be determined by first simplifying the circuit to find the total current and then working across the individual components.

This circuit can be broken into two components: the series resistances R_1 and R_2, and the parallel resistances R_3 and R_4.

R_1 and R_2 can be added together to form the equivalent series resistance R_{1+2}:

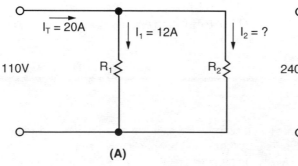

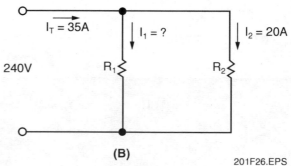

(A)   (B)

201F26.EPS

Figure 26 ◆ Solving for an unknown current.

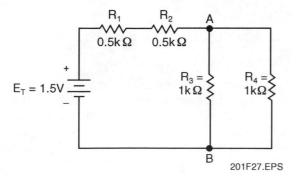

Figure 27 ◆ Series-parallel circuit.

$$R_{1+2} = R_1 + R_2$$
$$R_{1+2} = 0.5k\Omega + 0.5k\Omega$$
$$R_{1+2} = 1k\Omega$$

R_3 and R_4 can be totaled using either the general reciprocal formula or, since there are two resistances in parallel, the product-over-sum method. Both methods are shown below.

$$R_{3+4} = \frac{1}{\frac{1}{R_3} + \frac{1}{R_4}} = \frac{1}{\frac{1}{1k\Omega} + \frac{1}{1k\Omega}}$$

$$R_{3+4} = \frac{1}{\frac{2}{1,000\Omega}} = \frac{1}{0.002} = 500\Omega$$

and

$$R_{3+4} = \frac{R_1 \times R_2}{R_1 + R_2} = \frac{1k\Omega \times 1k\Omega}{1k\Omega + 1k\Omega}$$

$$R_{3+4} = \frac{1,000,000\Omega}{R_1 + R_2} = 500\Omega$$

The equivalent circuit containing the R_{1+2} resistance of $1k\Omega$ and the R_{3+4} resistance of 500Ω is shown in *Figure 28*.

Using the Ohm's law relationship that total current equals voltage divided by circuit resistance, the circuit current can be determined. First, however, total circuit resistance must be found. Since

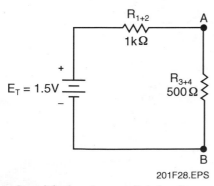

Figure 28 ◆ Simplified series-parallel circuit.

the simplified circuit consists of two resistances in series, they are simply added together to obtain total resistance.

$$R_T = R_{1+2} + R_{3+4}$$
$$R_T = 1k\Omega + 500\Omega$$
$$R_T = 1.5k\Omega$$

Applying this to the current/voltage equation:

$$I_T = \frac{E_T}{R_T}$$

$$I_T = \frac{1.5V}{1.5k\Omega}$$

$$I_T = 1mA \text{ of } 0.001A$$

Now that the total current is known, voltage drops across individual components can be determined:

$$E_{R1} = I_T R_1 = 1mA \times 0.5k\Omega = 0.5V$$
$$E_{R2} = I_T R_2 = 1mA \times 0.5k\Omega = 0.5V$$

Since the total voltage equals the sum of all voltage drops, the voltage drop from A to B can be determined by subtraction:

$$E_T = E_{R1} + E_{R2} + E_{A+B}$$
$$E_T - E_{R1} - E_{R2} = E_{A+B}$$
$$1.5V - 0.5V - 0.5V = E_{A+B} = 0.5V$$

Since R_3 and R_4 are in parallel, some of the total current must pass through each resistor. R_3 and R_4 are equal, so the same current should flow through each branch. Using the relationship:

$$I = \frac{E}{R}$$

$$I_{R3} = \frac{E_{R3}}{R_3} \qquad\qquad I_{R4} = \frac{E_{R4}}{R_4}$$

$$I_{R3} = \frac{0.5V}{1k\Omega} \qquad\qquad I_{R4} = \frac{0.5V}{1k\Omega}$$

$$I_{R3} = 0.5mA \qquad\qquad I_{R4} = 0.5mA$$

$$0.5mA + 0.5mA = 1mA$$

Therefore, the total current for the circuit passes through R_1 and R_2 and is evenly divided between R_3 and R_4.

10.0.0 ◆ KIRCHHOFF'S LAWS

Kirchhoff's laws provide a simple, practical method of solving for unknown parameters in a circuit. Kirchhoff's laws include **Kirchhoff's current law** and **Kirchhoff's voltage law**.

10.1.0 Kirchhoff's Current Law

In its most general form, Kirchhoff's current law, which is also called Kirchhoff's first law, states that at any point in a circuit, the total current entering that point must equal the total current leaving that point. For parallel circuits, this implies that the current in a parallel circuit is equal to the sum of the currents in each branch.

When using Kirchhoff's laws to solve circuits, it is necessary to adopt conventions that determine the algebraic signs for current and voltage terms. A convenient system for current is to consider all current flowing into a branch point as positive, and all current directed away from that point as negative.

As an example, in *Figure 29*, the currents can be written as:

$$I_A + I_B - I_C = 0$$

or

$$5A + 3A - 8A = 0$$

Currents I_A and I_B are positive terms because these currents flow into P, but I_C, because it flows out of P, is negative.

For a circuit application, refer to Point C at the top of the diagram in *Figure 30*. The 6A I_T into Point C divides into the 2A I_3 and 4A I_{4+5} which are directed out. Note that I_{4+5} is the current through R_4 and R_5. The algebraic equation is:

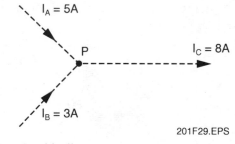

$$I_T - I_3 - I_{4+5} = 0$$

Substituting the values for each current:

$$6A - 2A - 4A = 0$$

For the opposite direction, refer to Point D at the bottom of *Figure 30*. Here, the branch currents into Point D combine to equal the mainline current I^T returning to the voltage source. Now, I_T is directed out from Point D, with I_3 and I_{4+5} directed in. The algebraic equation is:

$$I_3 + I_{4+5} - I_T = 0$$

$$2A + 4A - 6A = 0$$

Note that at either Point C or Point D, the sum of the 2A and 4A branch currents must equal the 6A total line current. Therefore, Kirchhoff's current law can also be stated as:

$$I_{IN} = I_{OUT}$$

For *Figure 30*, the equations for current can be written as shown below.

At Point C: $6A = 2A + 4A$

At Point D: $2A + 4A = 6A$

Kirchhoff's current law is really the basis for the practical rule in parallel circuits that the total line current must equal the sum of the branch currents.

10.2.0 Kirchhoff's Voltage Law

Kirchhoff's voltage law states that the algebraic sum of the voltages around any closed path is zero.

Referring to *Figure 31*, the sum of the voltage drops around the circuit must equal the voltage applied to the circuit:

$$E_A = E_1 + E_2 + E_3$$

Where:

E_A = voltage applied to the circuit

E_1, E_2, and E_3 = voltage drops in the circuit

201F29.EPS

Figure 29 ◆ Kirchhoff's current law.

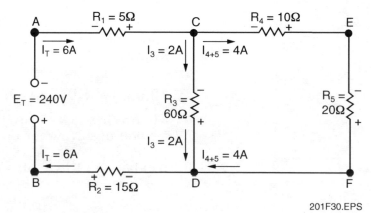

201F30.EPS

Figure 30 ◆ Application of Kirchhoff's current law.

ELECTRONIC SYSTEMS TECHNICIAN LEVEL TWO — TRAINEE MODULE 33201-05

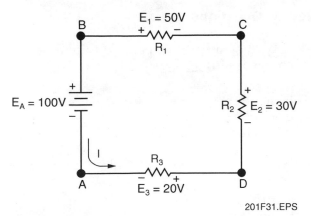

Figure 31 ◆ Kirchhoff's voltage law.

201F31.EPS

Another way of stating this law is that the algebraic sum of the voltage rises and voltage drops must be equal to zero. A voltage source is considered a voltage rise; a voltage across a resistor is a voltage drop. For convenience in labeling, letter subscripts are shown for voltage sources and numerical subscripts are used for voltage drops. This form of the law can be written by transposing the right members to the left side:

Voltage applied − sum of voltage drops = 0

Substitute letters:

$E_A − E_1 − E_2 − E_3 = 0$
$E_A − (E_1 + E_2 + E_3) = 0$

10.3.0 Loop Equations

Any closed path is called a loop. A loop equation specifies the voltages around the loop. Refer to *Figure 32*.

Consider the inside loop A, C, D, B, A, including the voltage drops E_1, E_3, and E_2, and the source E_T. In a clockwise direction, starting at Point A, the algebraic sum of the voltages is:

$− E_1 − E_3 − E_2 + E_T = 0$

or

$− 30V − 120V − 90V + 240V = 0$

Voltages E_1, E_3, and E_2 have a negative value, because there is a decrease in voltage seen across each of the resistors in a clockwise direction. However, the source E_T is a positive term because an increase in voltage is seen in that same direction.

For the opposite direction, going counterclockwise in the same loop from Point A, E_T is negative while E_1, E_2, and E_3 have positive values. Therefore:

$− E_T + E_2 + E_3 + E_1 = 0$

or

$− 240V + 90V + 120V + 30V = 0$

When the negative term is transposed, the equation becomes:

$240V = 90V + 120V + 30V$

In this form, the loop equation shows that Kirchhoff's voltage law is really the basis for the practical rule in series circuits that the sum of the voltage drops must equal the applied voltage.

For example, determine the voltage E_B for the circuit shown in *Figure 33*. The direction of the current flow is shown by the arrow. First mark the polarity of the voltage drops across the resistors, and trace the circuit in the direction of the current flow starting at Point Λ. Then write the voltage equation around the circuit:

$− E_3 − E_B − E_2 − E_1 + E_A = 0$

Solve for E_B:

$E_B = E_A − E_3 − E_2 − E_1$
$E_B = 15V − 2V − 6V − 3V$
$E_B = 4V$

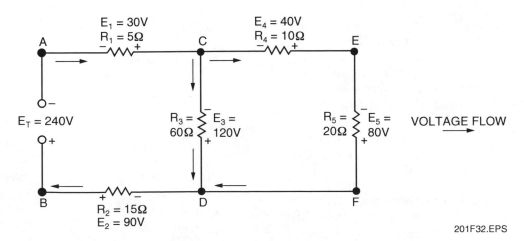

Figure 32 ◆ Loop equation.

201F32.EPS

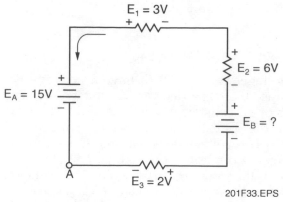

Figure 33 ◆ Applying Kirchhoff's voltage law.

201F33.EPS

Since E_B was found to be positive, the assumed direction of current is in fact the actual direction of current. In its most general form, Kirchhoff's voltage law (also called Kirchhoff's second law) states that the algebraic sum of all the potential differences in a closed loop is equal to zero. A closed loop means any completely closed path consisting of wire, resistors, batteries, or other components. For series circuits, this implies that the sum of the voltage drops around the circuit is equal to the applied voltage. For parallel circuits, this implies that the voltage drops across all branches are equal.

Parallel Circuit Exercise

Draw four 60W lamps in parallel with a 120V power source. What is the amperage in the circuit? What would happen to the amperage if we doubled the voltage?

Summary

The relationships among current, voltage, resistance, and power in Ohm's law are the same for both DC series and DC parallel circuits. Understanding and being able to apply these concepts is necessary for effective circuit analysis and troubleshooting. DC series-parallel circuits also have these fundamental relationships. Since DC series-parallel circuits are a combination of simple series and parallel circuits, Kirchhoff's voltage and current laws will apply. Calculating I, E, R, and P for series-parallel circuits is no more difficult than calculating these values for simple series or parallel circuits. However, for series-parallel circuits, these calculations require more careful circuit analysis in order to use Ohm's law correctly.

Review Questions

1. A type of subatomic particle with a negative charge is a(n) _____.
    a. proton
    b. neutron
    c. electron
    d. nucleus

2. A type of subatomic particle with a positive charge is a(n) _____.
    a. proton
    b. neutron
    c. electron
    d. nucleus

3. Which of the following is an insulator?
    a. Silver
    b. Copper
    c. Silicon
    d. Porcelain

4. An electron has _____ coulombs of charge.
    a. 1.6×10^{-9}
    b. 1.6×10^{9}
    c. 1.6×10^{19}
    d. 1.6×10^{-19}

5. The quantity that Ohm's law does *not* express a relationship for in an electrical circuit is _____.
    a. charge
    b. resistance
    c. voltage
    d. current

6. The color band that represents tolerance on a resistor is the _____ band.
    a. first
    b. second
    c. third
    d. fourth

7. A resistor with a color code of red/red/orange indicates a value of _____.
    a. 66 ohms
    b. 220 ohms
    c. 223 ohms
    d. 22,000 ohms

8. An ammeter is placed in _____ with the circuit being tested.
    a. parallel
    b. series

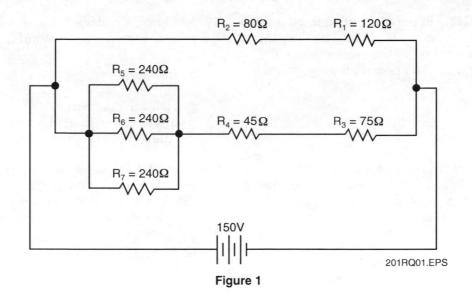

Figure 1

9. The basic unit of power is the _____ .

   a. volt
   b. ampere
   c. coulomb
   d. watt

10. The power in a circuit with 120 volts and 5 amps is _____.

   a. ¼₄ watt
   b. 24 watts
   c. 600 watts
   d. 6,000 watts

11. The formula for calculating the total resistance in a series circuit with three resistors is _____.

   a. $R_T = R_1 + R_2 + R_3$
   b. $R_T = R_1 - R_2 - R_3$
   c. $R_T = R_1 \times R_2 \times R_3$
   d. $R_T = \dfrac{1}{\dfrac{1}{R_1} + \dfrac{1}{R_2} + \dfrac{1}{R_3}}$

12. Find the total resistance in a series circuit with three resistances of 10Ω, 20Ω, and 30Ω.

   a. 1Ω
   b. 15Ω
   c. 20Ω
   d. 60Ω

13. The formula for calculating the total resistance in a parallel circuit with three resistors is _____.

   a. $R_T = R_1 + R_2 + R_3$
   b. $R_T = R_1 - R_2 - R_3$
   c. $R_T = R_1 \times R_2 \times R_3$
   d. $R_T = \dfrac{1}{\dfrac{1}{R_1} + \dfrac{1}{R_2} + \dfrac{1}{R_3}}$

14. The approximate total resistance in *Figure 1* is _____.

   a. 100Ω
   b. 129Ω
   c. 157Ω
   d. 1,035Ω

15. In a parallel circuit, the voltage across each path is equal to the _____.

   a. total circuit resistance times path current
   b. source voltage minus path voltage
   c. path resistance times total current
   d. source voltage

16. The value for total current in *Figure 2* is _____ amps.

   a. 1.25
   b. 2.50
   c. 5
   d. 10

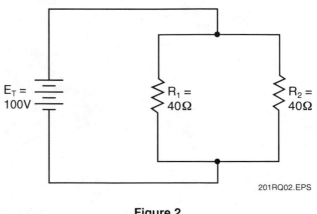

Figure 2

17. A resistor of 32Ω is in parallel with a resistor of 36Ω, and a 54Ω resistor is in series with the pair. When 350V is applied to the combination, the current through the 54Ω resistor is _____ amps.

    a. 2.87
    b. 3.26
    c. 4.93
    d. 5.86

18. A 242Ω resistor is in parallel with a 180Ω resistor, and a 420Ω resistor is in series with the combination. A current of 22mA flows through the 242Ω resistor. The current through the 180Ω resistor is _____ mA.

    a. 19.8
    b. 29.6
    c. 36.4
    d. 40.2

19. Kirchhoff's voltage law states that the algebraic sum of the voltages around any closed path is _____.

    a. infinity
    b. zero
    c. twice the current
    d. always less than the individual voltages due to voltage drop

20. Two 24Ω resistors are in parallel, and a 42Ω resistor is in series with the combination. When 78V is applied to the three resistors, the voltage drop across the 42Ω resistor is about _____ volts.

    a. 50
    b. 55
    c. 61
    d. 65

Trade Terms Introduced in This Module

Ammeter: An instrument for measuring electrical current.

Ampere (A): A unit of electrical current. For example, one volt across one ohm of resistance causes a current flow of one ampere.

Atom: The smallest particle to which an element may be divided and still retain the properties of the element.

Battery: A DC voltage source consisting of two or more cells that convert chemical energy into electrical energy.

Circuit: A complete path for current flow.

Conductor: A material that offers very little resistance to current flow.

Coulomb: An electrical charge equal to 6.25×10^{18} electrons or 6,250,000,000,000,000,000 electrons. A coulomb is the common unit of quantity used for specifying the size of a given charge.

Current: The movement, or flow, of electrons in a circuit. Current (I) is measured in amperes.

Electron: A negatively charged particle that orbits the nucleus of an atom.

Insulator: A material that offers resistance to current flow.

Joule (J): A unit of measurement that represents one newton-meter (Nm), which is a unit of measure for doing work.

Kilo: A prefix used to indicate one thousand; for example, one kilowatt is equal to one thousand watts.

Kirchhoff's current law: The statement that the total amount of current flowing through a parallel circuit is equal to the sum of the amounts of current flowing through each current path.

Kirchhoff's voltage law: The statement that the sum of all the voltage drops in a circuit is equal to the source voltage of the circuit.

Matter: Any substance that has mass and occupies space.

Mega: A prefix used to indicate one million; for example, one megawatt is equal to one million watts.

Micro: A prefix used to indicate one-millionth; for example, one microwatt is equal to one-millionth of a watt.

Neutrons: Electrically neutral particles (neither positive nor negative) that have the same mass as a proton and are found in the nucleus of an atom.

Nucleus: The center of an atom. It contains the protons and neutrons of the atom.

Ohm (Ω): The basic unit of measurement for resistance.

Ohmmeter: An instrument used for measuring resistance.

Ohm's law: A statement of the relationships among current, voltage, and resistance in an electrical circuit: current (I) equals voltage (E) divided by resistance (R). Generally expressed as a mathematical formula: $I = E \div R$.

Parallel circuits: Circuits containing two or more parallel paths through which current can flow.

Power: The rate of doing work or the rate at which energy is used or dissipated. Electrical power is the rate of doing electrical work. Electrical power is measured in watts.

Protons: The smallest positively charged particles of an atom. Protons are contained in the nucleus of an atom.

Resistance: An electrical property that opposes the flow of current through a circuit. Resistance (R) is measured in ohms.

Resistor: Any device in a circuit that resists the flow of electrons.

Schematic: A type of drawing in which symbols are used to represent the components in a system.

Series circuit: A circuit with only one path for current flow.

Series-parallel circuits: Circuits that contain both series and parallel current paths.

Valence shell: The outermost ring of electrons that orbit about the nucleus of an atom.

Volt (V): The unit of measurement for voltage (electromotive force). One volt is equivalent to the force required to produce a current of one ampere through a resistance of one ohm.

Voltage: The driving force that makes current flow in a circuit. Voltage (E) is also referred to as electromotive force or potential.

Voltage drop: The change in voltage across a component that is caused by the current flowing through it and amount of resistance opposing it.

Voltmeter: An instrument for measuring voltage. The resistance of the voltmeter is fixed. When the voltmeter is connected to a circuit, the current passing through the meter will be directly proportional to the voltage at the connection points.

Watt (W): The basic unit of measurement for electrical power.

Additional Resources

This module is intended to be a thorough resource for task training. The following reference works are suggested for further study. These are optional materials for continued education rather than for task training.

Electronics Fundamentals: Circuits, Devices, and Applications, 2000. Thomas L. Floyd. New York, NY: Prentice Hall.

Principles of Electric Circuits: Conventional Current Version, 2002. Thomas L. Floyd. New York, NY: Prentice Hall.

Figure Credits

Topaz Publications, Inc.   201F08B

CONTREN® LEARNING SERIES — USER UPDATE

NCCER makes every effort to keep these textbooks up-to-date and free of technical errors. We appreciate your help in this process. If you have an idea for improving this textbook, or if you find an error, a typographical mistake, or an inaccuracy in NCCER's Contren® textbooks, please write us, using this form or a photocopy. Be sure to include the exact module number, page number, a detailed description, and the correction, if applicable. Your input will be brought to the attention of the Technical Review Committee. Thank you for your assistance.

Instructors – If you found that additional materials were necessary in order to teach this module effectively, please let us know so that we may include them in the Equipment/Materials list in the Annotated Instructor's Guide.

Write: Product Development and Revision
   National Center for Construction Education and Research
   P.O. Box 141104, Gainesville, FL 32614-1104

Fax: 352-334-0932

E-mail: curriculum@nccer.org

Craft _____ Module Name _____

Copyright Date _____ Module Number _____ Page Number(s) _____

Description _____
_____
_____
_____
_____

(Optional) Correction _____
_____
_____

(Optional) Your Name and Address _____
_____
_____

AC Circuits

COURSE MAP

This course map shows all of the modules in the second level of the *Electronic Systems Technician* curriculum. The suggested training order begins at the bottom and proceeds up. Skill levels increase as you advance on the course map. The local Training Program Sponsor may adjust the training order.

ELECTRONIC SYSTEMS TECHNICIAN LEVEL TWO

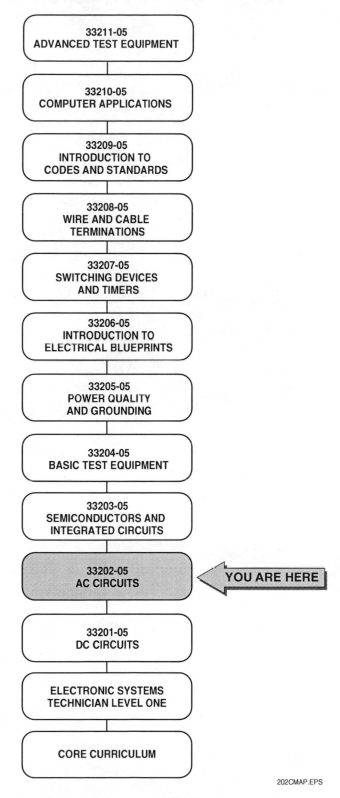

33211-05
ADVANCED TEST EQUIPMENT

33210-05
COMPUTER APPLICATIONS

33209-05
INTRODUCTION TO
CODES AND STANDARDS

33208-05
WIRE AND CABLE
TERMINATIONS

33207-05
SWITCHING DEVICES
AND TIMERS

33206-05
INTRODUCTION TO
ELECTRICAL BLUEPRINTS

33205-05
POWER QUALITY
AND GROUNDING

33204-05
BASIC TEST EQUIPMENT

33203-05
SEMICONDUCTORS AND
INTEGRATED CIRCUITS

33202-05
AC CIRCUITS ◄ YOU ARE HERE

33201-05
DC CIRCUITS

ELECTRONIC SYSTEMS
TECHNICIAN LEVEL ONE

CORE CURRICULUM

202CMAP.EPS

Figures

AC Circuits

Objectives

When you have completed this module, you will be able to do the following:

1. Calculate the peak and effective voltage or current values for an AC waveform.
2. Calculate the phase relationship between two AC waveforms.
3. Describe the voltage and current phase relationship in a resistive AC circuit.
4. Describe the voltage and current transients that occur in an inductive circuit.
5. Define inductive reactance and state how it is affected by frequency.
6. Describe the voltage and current transients that occur in a capacitive circuit.
7. Define capacitive reactance and state how it is affected by frequency.
8. Explain the relationship between voltage and current in the following types of AC circuits:
   - RL
   - RC
   - LC
   - RLC
9. Describe the effect that resonant frequency has on impedance and current flow in a series or parallel resonant circuit.
10. Define bandwidth and describe how it is affected by resistance in a series or parallel resonant circuit.
11. Explain the following terms as they relate to AC circuits:
    - True power
    - Apparent power
    - Reactive power
    - Power factor
12. Explain basic transformer action.

Prerequisites

Before you begin this module, it is recommended that you successfully complete *Core Curriculum*; *Electronic Systems Technician Level One*; and *Electronic Systems Technician Level Two*, Module 33201-05.

Required Trainee Materials

1. Pencil and paper
2. Appropriate personal protective equipment
3. Scientific calculator

1.0.0 ◆ INTRODUCTION

Alternating current (AC) and its associated voltage reverse between positive and negative polarities and vary in amplitude with time. One complete waveform or cycle includes a complete set of variations, with two alternations in polarity. Many sources of voltage change direction with time and produce a resultant waveform. The most common AC waveform is the sine wave.

In this module, we will cover the characteristics of the sine wave and other waveforms; the difference between AC and DC; the effects of capacitance and inductance on AC circuits; and the uses of transformers in AC circuits. Because of the alternating nature of AC, the components of an AC circuit react differently than those of a DC circuit. It is important to understand these differences, because they provide the foundation for understanding audio, video, and radio frequency systems.

2.0.0 ◆ SINE WAVE GENERATION

To understand how the alternating current sine wave is generated, some of the basic principles learned in magnetism should be reviewed. Two

principles form the basis of all electromagnetic phenomena:

- An electric current in a conductor creates a magnetic field that surrounds the conductor.
- Relative motion between a conductor and a magnetic field, when at least one component of that relative motion is in a direction that is perpendicular to the direction of the field, creates a voltage in the conductor.

Figure 1 shows how these principles are applied to generate an AC waveform in a simple one-loop rotary generator. The conductor loop rotates through the magnetic field to generate the induced AC voltage across its open terminals. The magnetic flux shown here is vertical.

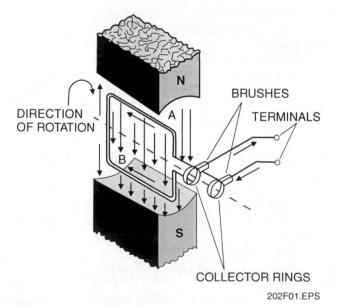

202F01.EPS

Figure 1 ◆ Conductor moving across a magnetic field.

There are several factors affecting the magnitude of voltage developed by a conductor through a magnetic field. They are the strength of the magnetic field, the length of the conductor, and the rate at which the conductor cuts directly across or perpendicular to the magnetic field.

Assuming that the strength of the magnetic field and the length of the conductor making the loop are both constant, the voltage produced will vary depending on the rate at which the loop cuts directly across the magnetic field.

The rate at which the conductor cuts the magnetic field depends on two things: the speed of the generator in revolutions per minute (rpm) and the angle at which the conductor is traveling through the field. If the generator is operated at a constant rpm, the voltage produced at any moment will depend on the angle at which the conductor is cutting the field at that instant.

In *Figure 2*, the magnetic field is shown as parallel lines called lines of flux. These lines always go from the north to south poles in a generator. The motion of the conductor is shown by the large arrow.

Assuming the speed of the conductor is constant, as the angle between the flux and the conductor motion increases, the number of flux lines cut in a given time (the rate) increases. When the conductor is moving parallel to the lines of flux (angle of 0 degrees), it is not cutting any of them, and the voltage will be zero.

The angle between the lines of flux and the motion of the conductor is called θ (theta). The magnitude of the voltage produced will be proportional to the sine of the angle. Sine is a trigonometric function. Each angle has a sine value that never changes.

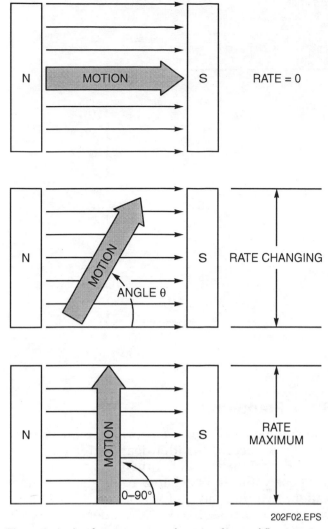

Figure 2 ♦ Angle versus rate of cutting lines of flux.

The sine of 0 degrees is 0. It increases to a maximum of 1 at 90 degrees. From 90 degrees to 180 degrees, the sine decreases back to 0. From 180 degrees to 270 degrees, the sine decreases to –1. Then from 270 degrees to 360 degrees (back to 0 degrees), the sine increases to its original 0.

Because voltage is proportional to the sine of the angle, as the loop goes 360 degrees around the circle the voltage will increase from 0 to its maximum at 90 degrees, back to 0 at 180 degrees, down to its maximum negative value at 270 degrees, and back up to 0 at 360 degrees, as shown in *Figure 3*.

Notice that at 180 degrees the polarity reverses. This is because the conductor has turned completely around and is now cutting the lines of flux in the opposite direction. This can be shown using the left-hand rule for generators. The curve shown in *Figure 3* is called a sine wave because its shape is generated by the trigonometric function sine. The value of voltage at any point along the sine wave can be calculated if the angle and the maximum obtainable voltage (E_{max}) are known. The formula used is as follows:

$$E = E_{max} \sin \theta$$

Where:

E = voltage induced

E_{max} = maximum induced voltage

θ = angle at which the voltage is induced

Using the above formula, the values of voltage anywhere along the sine wave in *Figure 4* can be calculated.

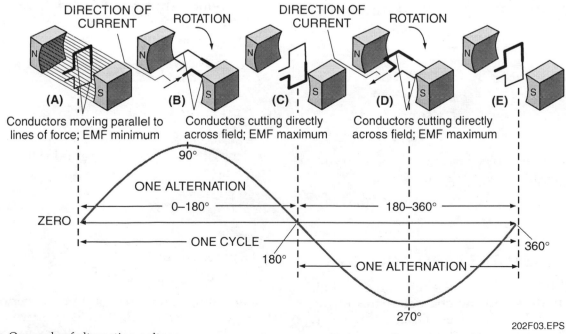

Figure 3 ♦ One cycle of alternating voltage.

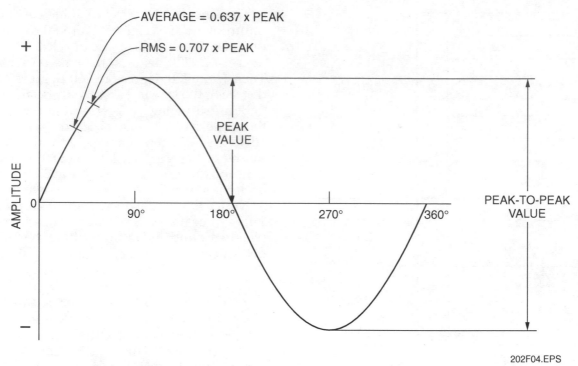

AVERAGE = 0.637 x PEAK

RMS = 0.707 x PEAK

Figure 4 ◆ Amplitude values for a sine wave.

Sine values can be found using either a scientific calculator or trigonometric tables. With an E_{max} of 10 volts (V), the following values are calculated as examples:

θ = 0°, sine = 0.0	θ = 45°, sine = 0.707
E = E_{max}sine θ	E = E_{max}sine θ
E = (10V)(0)	E = (10V)(0.707)
E = 0V	E = 7.07V
θ = 90°, sine = 1.0	θ = 135°, sine = 0.707
E = E_{max}sine θ	E = E_{max}sine θ
E = (10V)(1.0)	E = (10V)(0.707)
E = 10V	E = 7.07V
θ = 180°, sine = 0	θ = 225°, sine = –0.707
E = E_{max}sine θ	E = E_{max}sine θ
E = (10V)(0)	E = (10V)(–0.707)
E = 0V	E = –7.07V
θ = 270°, sine = –1.0	θ = 315°, sine = –0.707
E = E_{max}sine θ	E = E_{max}sine θ
E = (10V)(–1.0)	E = (10V)(–0.707)
E = –10V	E = –7.07V

3.0.0 ◆ SINE WAVE TERMINOLOGY

In order to fully understand alternating current, you must understand the characteristics of the sine wave. The key characteristics are frequency, wavelength, and voltage values.

3.1.0 Frequency

The **frequency** of a waveform is the number of times per second an identical pattern repeats itself. Each time the waveform changes from zero to a peak value and back to zero is called an alternation. Two alternations form one cycle. The number of cycles per second is the frequency. The unit of frequency is **hertz (Hz)**. One hertz equals one cycle per second (cps).

For example, let us determine the frequency of the waveform shown in *Figure 5*. In one-half second, the basic sine wave is repeated five times. Therefore, the frequency (f) is:

$$f = \frac{5 \text{ cycles}}{0.5 \text{ second}} = 10 \text{ cycles per second (Hz)}$$

3.1.1 Period

The period of a waveform is the time (t) required to complete one cycle. The period is the inverse of frequency:

$$t = \frac{1}{f}$$

Where:

t = period (seconds)

f = frequency (Hz or cps)

For example, let us determine the period of the waveform in *Figure 5*. If there are five cycles in one-half second, then the frequency for one cycle is 10 cps (5 ÷ 0.5 = 10). Therefore, the period is:

THINK ABOUT IT

Frequency

The frequency of the utility power generated in the United States is normally 60Hz. In some European countries and elsewhere, utility power is often generated at a frequency of 50Hz. Which of these frequencies (60Hz or 50Hz) has the shortest period?

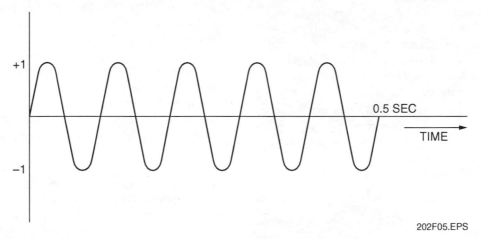

202F05.EPS

Figure 5 ◆ Frequency measurement.

$$t = \frac{1}{cps}$$

$$t = \frac{1}{10} = 0.1 \text{ second}$$

3.2.0 Wavelength

The wavelength or λ (lambda) is the distance traveled by a waveform during one period. Since electricity travels at the speed of light (186,000 miles/second or 300,000,000 meters/second), the wavelength of electrical waveforms equals the product of the period and the speed of light (c):

$$\lambda = tc$$

or

$$\lambda = \frac{c}{f}$$

Where:

λ = wavelength (meters)

t = period (seconds)

c = speed of light (meters/second)

f = frequency (Hz or cps)

3.3.0 Peak Value

The peak value is the maximum value of voltage (V_M) or current (I_M). For example, specifying that a sine wave has a **peak voltage** of 170V applies to either the positive or the negative peak. To include both peak amplitudes, the peak-to-peak (p–p) value may be specified. The peak-to-peak value is 340V, or double the peak value of 170V, because the positive and negative peaks are symmetrical. However, the two opposite peak values cannot occur at the same time. Also, in some waveforms, the two peaks are not equal. The positive peak value and peak-to-peak value of a sine wave are shown in *Figure 4*.

3.4.0 Average Value

The average value is calculated from all the values in a sine wave for one alternation or half cycle. The half cycle is used for the average because over a full cycle the average value is zero, which is useless for comparison purposes. If the sine values for all angles up to 180 degrees in one alternation are added and then divided by the number of values, this average equals 0.637.

Since the peak value of the sine is 1 and the average equals 0.637, the average value can be calculated as follows:

Average value = 0.637 × peak value

For example, with a peak of 170V, the average value is 0.637 × 170V, which equals approximately 108V. *Figure 4* shows where the average value would fall on a sine wave.

Left-Hand Rule for Generators

Hand rules for generators and motors give direction to the basic principles of induction. For a generator, if you move a conductor through a magnetic field made up of flux lines, you will induce an EMF, which drives current through a conductor. The left-hand rule for generators will help you determine which direction the current will flow in the conductor. It states that if you hold the thumb, first, and middle fingers of the left hand at right angles to one another with the first finger pointing in the flux direction (from the north pole to the south pole), and the thumb pointing in the direction of motion of the conductor, the middle finger will point in the direction of the induced voltage (EMF). The polarity of the EMF determines the direction in which current will flow as a result of this induced EMF. The left-hand rule for generators is also called Fleming's first rule.

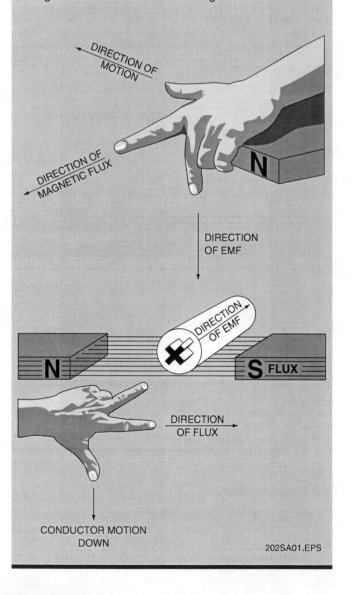

202SA01.EPS

3.5.0 Root-Mean-Square or Effective Value

Meters used in AC circuits indicate a value called the effective value. The effective value is the value of the AC current or voltage wave that indicates the same energy transfer as an equivalent direct current (DC) or voltage.

The direct comparison between DC and AC is in the heating effect of the two currents. Heat produced by current is a function of current amplitude only and is independent of current direction. Thus, heat is produced by both alternations of the AC wave, although the current changes direction during each alternation.

In a DC circuit, the current maintains a steady amplitude. Therefore, the heat produced is steady and is equal to I^2R. In an AC circuit, the current is continuously changing; periodically high, periodically low, and periodically zero. To produce the same amount of heat from AC as from an equivalent amount of DC, the instantaneous value of the AC must at times exceed the DC value.

By averaging the heating effects of all the instantaneous values during one cycle of alternating current, it is possible to find the average heat produced by the AC current during the cycle. The amount of DC required to produce that heat will be equal to the effective value of the AC.

The most common method of specifying the amount of a sine wave of voltage or current is by stating its value at 45 degrees, which is 70.7 percent of the peak. This is its **root-mean-square (rms)** value. Therefore:

Value of rms = 0.707 × peak value

For example, with a peak of 170V, the rms value is 0.707 × 170, or approximately 120V. This is the voltage of the commercial AC power line, which is always given in rms value.

4.0.0 ◆ AC PHASE RELATIONSHIPS

In AC systems, phase is involved in two ways: the location of a point on a voltage or current wave with respect to the starting point of the wave or with respect to some corresponding point on the same wave. In the case of two waves of the same frequency, it is the time at which an event of one takes place with respect to a similar event of the other.

Often, the event is the starting of the waves at zero or the points at which the waves reach their maximum values. When two waves are compared in this manner, there is a phase lead or lag of one with respect to the other unless they are alternating in unison, in which case they are said to be in phase.

RMS Amplitude

The root-mean-square (rms) value, also called the effective value, is the value assigned to an alternating voltage or current that results in the same power dissipation in a given resistance as DC voltage or current of the same numerical value. This is illustrated below. As shown, 120 (peak) VAC will not produce the same light (350 lumens versus 500 lumens) as 120VDC from a 60W lamp. In order to produce the same light (500 lumens), 120V rms must be applied to the lamp. This requires that the applied sinusoidal AC waveform have a peak voltage of about 170V (170V × 0.707V = 120V).

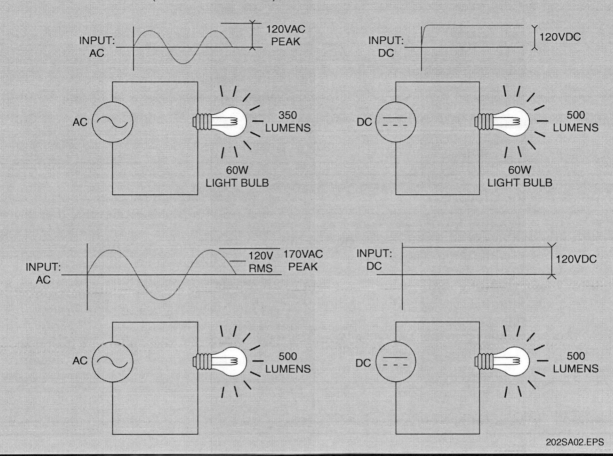

202SA02.EPS

4.1.0 Phase Angle

Suppose that a generator started its cycle at 90 degrees where maximum voltage output is produced instead of starting at the point of zero output. The two output voltage waves are shown in *Figure 6*. Each is the same waveform of alternating voltage, but wave B starts at the maximum value while wave A starts at zero. The complete cycle of wave B through 360 degrees takes it back to the maximum value from which it started.

Wave A starts and finishes its cycle at zero. With respect to time, wave B is ahead of wave A in its values of generated voltage. The amount it leads in time equals one quarter revolution, which is 90 degrees. This angular difference is the phase angle between waves B and A. Wave B leads wave A by the phase angle of 90 degrees.

The 90-degree phase angle between waves B and A is maintained throughout the complete cycle and in all successive cycles as long as they both have the same frequency. At any instant in time, wave B has the value that A will have 90 degrees later. For instance, at 180 degrees, wave A is at zero, but B is already at its negative maximum value, the point where wave A will be later at 270 degrees.

To compare the phase angle between two waves, both waves must have the same frequency. Otherwise, the relative phase keeps changing. Both waves must also have sine wave variations, because this is the only kind of waveform that is measured in angular units of time. The amplitudes can be different for the two waves. The phases of two voltages, two currents, or a current with a voltage can be compared.

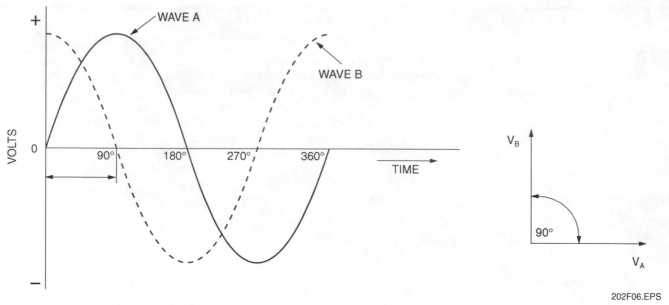

Figure 6 ◆ Voltage waveforms 90 degrees out of phase.

4.2.0 Phase Angle Diagrams

To compare AC phases, it is much more convenient to use vector diagrams corresponding to the voltage and current waveforms, as shown in *Figure 6*. V_A and V_B represent the vector quantities corresponding to the generator voltage.

A vector is a quantity that has magnitude and direction. The length of the arrow indicates the magnitude of the alternating voltage in rms, peak, or any AC value as long as the same measure is used for all the vectors. The angle of the arrow with respect to the horizontal axis indicates the phase angle.

In *Figure 6*, the vector V_A represents the voltage wave A, with a phase angle of 0 degrees. This angle can be considered as the plane of the loop in the rotary generator where it starts with zero output voltage. The vector V_B is vertical to show the phase angle of 90 degrees for this voltage wave, corresponding to the vertical generator loop at the start of its cycle. The angle between the two vectors is the phase angle.

The symbol for a phase angle is θ (theta). In *Figure 7*, θ = 0 degrees. *Figure 7* shows the waveforms and phasor diagram of two waves that are in phase but have different amplitudes.

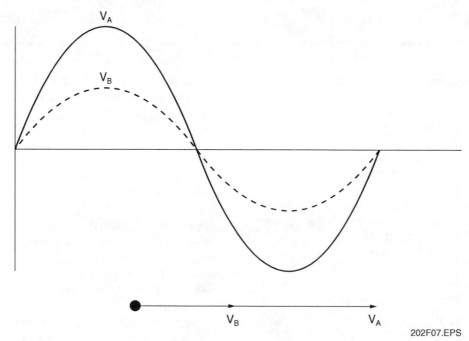

Figure 7 ◆ Waves in phase.

5.0.0 ◆ NONSINUSOIDAL WAVEFORMS

The sine wave is the basic waveform for AC variations for several reasons. This waveform is produced by a rotary generator, as the output is proportional to the angle of rotation. Because of its derivation from circular motion, any sine wave can be analyzed in angular measure, either in degrees from 0 to 360 degrees, or in **radians** from 0 to 2π radians.

In many electronic applications, however, other waveshapes are important. Any waveform that is not a sine (or cosine) wave is a nonsinusoidal waveform. Common examples are the square wave and the sawtooth wave in *Figure 8*.

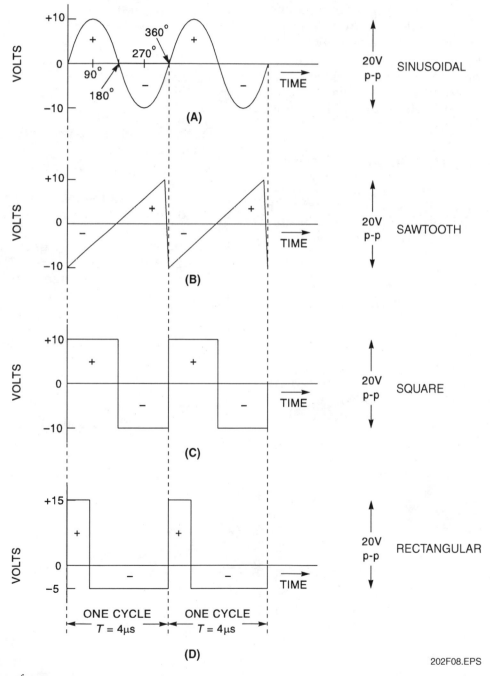

202F08.EPS

Figure 8 ◆ AC waveforms.

With nonsinusoidal waveforms for either voltage or current, there are important differences and similarities to consider. Note the following comparisons with sine waves:

- In all cases, the cycle is measured between two points having the same amplitude and varying in the same direction. The period is the time for one cycle.
- Peak amplitude is measured from the zero axis to the maximum positive or negative value. However, peak-to-peak amplitude is better for measuring nonsinusoidal waveshapes because they can have asymmetrical peaks, as with the rectangular wave in *Figure 8.*
- The rms value 0.707 of peak applies only to sine waves, as this factor is derived from the sine values in the angular measure used only for the sine waveform.
- Phase angles apply only to sine waves, as angular measure is used only for sine waves. Note that the phase angle is indicated only on the sine wave of *Figure 8.*

6.0.0 ◆ RESISTANCE IN AC CIRCUITS

An AC circuit has an AC voltage source. Note the circular symbol with the sine wave inside it shown in *Figure 9.* It is used for any source of sine wave alternating voltage. This voltage connected across an external load resistance produces alternating current of the same waveform, frequency, and phase as the applied voltage.

According to Ohm's law, current (I) equals voltage (E) divided by resistance (R). When E is an rms value, I is also an rms value. For any instantaneous value of E during the cycle, the value of I is for the corresponding instant of time.

In an AC circuit with only resistance, the current variations are in phase with the applied voltage, as shown in *Figure 9.* This in-phase relationship between E and I means that such an AC circuit can be analyzed by the same methods used for DC circuits since there is not a phase angle to consider. Components that have only resistance include resistors, the filaments for incandescent light bulbs, and vacuum tube heaters.

In purely resistive AC circuits, the voltage, current, and resistance are related by Ohm's law because the voltage and current are in phase:

$$I = \frac{E}{R}$$

Unless otherwise noted, the calculations in AC circuits are generally in rms values. For example, in *Figure 9,* the 120V applied across the 10Ω resistance R_L produces an rms current of 12A. This is determined as follows:

$$I = \frac{E}{R_L}$$

$$I = \frac{120V}{10\Omega}$$

$$I = 12A$$

Furthermore, the rms power (true power) dissipation is I²R or:

$$P = (12A)^2 \times 10\Omega = 1,440W$$

Figure 10 shows the relationship between voltage and current in purely resistive AC circuits. The voltage and current are in phase, their cycles begin and end at the same time, and their peaks occur at the same time.

The value of the voltage shown in *Figure 10* depends on the applied voltage to the circuit. The value of the current depends on the applied voltage and amount of resistance. If resistance is changed, it will affect only the magnitude of the current.

The total resistance in any AC circuit, whether it is a series, parallel, or series-parallel circuit, is calculated using the same rules that you learned and applied to DC circuits with resistance.

Power computations are discussed later in this module.

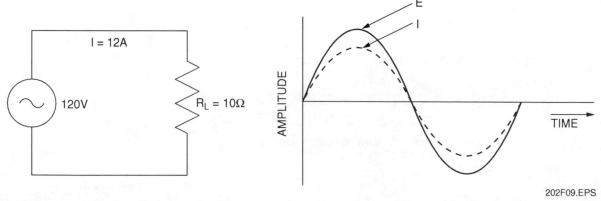

202F09.EPS

Figure 9 ◆ Resistive AC circuit.

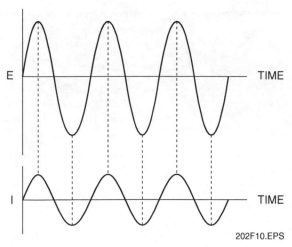

Figure 10 ◆ Voltage and current in a resistive AC circuit.

7.0.0 ◆ INDUCTANCE IN AC CIRCUITS

Inductance is the characteristic of an electrical circuit that opposes the change of current flow. It is the result of the expanding and collapsing field caused by the changing current. This moving flux cuts across the conductor that is providing the current, producing induced voltage in the wire itself. Furthermore, any other conductor in the field, whether carrying current or not, is also cut by the varying flux and has induced voltage. This induced current opposes the current flow that generated it.

In DC circuits, a change must be initiated in the circuit to cause inductance. The current must change to provide motion of the flux. A steady DC of 10A cannot produce any induced voltage as long as the current value is constant. A current of 1A changing to 2A does induce voltage. Also, the faster the current changes, the higher the induced voltage becomes, because when the flux moves at a higher speed it can induce more voltage.

However, in an AC circuit the current is continuously changing and producing induced voltage. Lower frequencies of AC require more inductance to produce the same amount of induced voltage as a higher frequency current. The current can have any waveform as long as the amplitude is changing.

The ability of a conductor to induce voltage in itself when the current changes is its **self-inductance** or simply inductance. The symbol for inductance is L and its unit is the henry (H). One henry is the amount of inductance that allows one volt to be induced when the current changes at the rate of one ampere per second.

7.1.0 Factors Affecting Inductance

An inductor is a coil of wire that may be wound on a core of metal or paper, or it may be self supporting. It may consist of turns of wire placed side by side to form a layer of wire over the core or coil form. The inductance of a coil or inductor depends on its physical construction. The following factors affecting inductance are shown in *Figure 11*:

- *Number of turns* – The greater the number of turns, the greater the inductance. In addition, the spacing of the turns on a coil also affects inductance. A coil that has widely spaced turns has a lower inductance than one with the same number of more closely spaced turns. The reason for this higher inductance is that the closely wound turns produce a more concentrated magnetic field, causing the coil to exhibit a greater inductance.

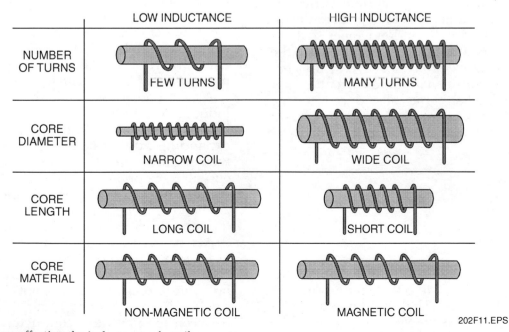

Figure 11 ◆ Factors affecting the inductance of a coil.

- *Coil diameter* – The inductance increases directly as the cross-sectional area of the coil increases.
- *Length of the coil* – When the length of the coil is decreased, the turn spacing is decreased, increasing the inductance of the coil.
- *Core material* – The core of the coil can be either a magnetic material (such as iron) or a non-magnetic material (such as paper or air). Coils wound on a magnetic core produce a stronger magnetic field than those with non-magnetic cores, giving them higher values of inductance. Air-core coils are used where small values of inductance are required.
- *Winding the coil in layers* – The more layers used to form a coil, the greater the effect the magnetic field has on the conductor. Layering a coil can increase the inductance.

7.2.0 Voltage and Current in an Inductive AC Circuit

The self-induced voltage across an inductance L is produced by a change in current with respect to time ($\Delta i / \Delta t$) and can be stated as follows:

$$V_L = L \frac{\Delta i}{\Delta t}$$

Where:

Δ = change

V_L = volts

L = henrys

$\Delta i / \Delta t$ = amperes per second

This gives the voltage in terms of how much magnetic flux is cut per second. When the magnetic flux associated with the current varies the same as I, this formula gives the same results for calculating induced voltage. Remember that the induced voltage across the coil is actually the result of inducing electrons to move in the conductor, so there is also an induced current.

For example, what is the self-induced voltage V_L across a 4h inductance produced by a current change of 12A per second?

$$V_L = L \frac{\Delta i}{\Delta t}$$

$$V_L = 4h \frac{12A}{1}$$

$$V_L = 4 \times 12$$

$$V_L = 48V$$

The current through a 200 microhenry (µh) inductor changes from 0 to 200 milliamps (mA) in 2 microseconds (µsec). (The prefix **micro** means one-millionth). What is the V_L?

$$V_L = L \frac{\Delta i}{\Delta t}$$

$$V_L = (200 \times 10^{-6}) \frac{200 \times 10^{-3}}{2 \times 10^{-6}}$$

$$V_L = 20V$$

The induced voltage is an actual voltage that can be measured, although V_L is produced only while the current is changing. When $\Delta i / \Delta t$ is present for only a short time, V_L is in the form of a voltage pulse. With a sine wave current that is always changing, V_L is a sinusoidal voltage that is 90 degrees out of phase with I_L.

The current that flows in an inductor is induced by the changing magnetic field that surrounds the inductor. This changing magnetic field is produced by an AC voltage source that is applied to the inductor. The magnitude and polarity of the induced current depend on the field strength, direction, and rate at which the field cuts the inductor windings. The overall effect is that the current is out of phase and lags the applied voltage by 90 degrees.

At 270 degrees in *Figure 12*, the applied electromotive force (EMF) is zero, but it is increasing in the positive direction at its greatest rate of change. Likewise, electron flow due to the applied EMF is also increasing at its greatest rate. As the electron flow increases, it produces a magnetic field that is building with it. The lines of flux cut the conductor as they move outward from it with the expanding field.

As the lines of flux cut the conductor, they induce a current into it. The induced current is at its maximum value because the lines of flux are expanding outward through the conductor at their greatest rate. The direction of the induced current is in opposition to the force that generated it. Therefore, at 270 degrees the applied voltage is zero and is increasing to a positive value, while the current is at its maximum negative value.

At 0 degrees in *Figure 12*, the applied voltage is at its maximum positive value, but its rate of change is zero. Therefore, the field it produces is no longer expanding and is not cutting the conductor. Because there is no relative motion between the field and conductor, no current is

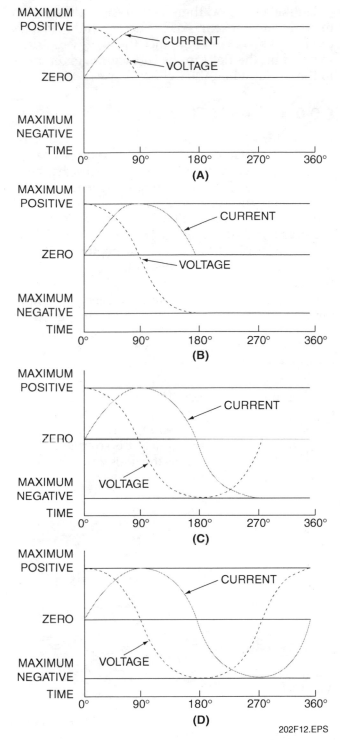

Figure 12 ◆ Inductor voltage and current relationship.

induced. Therefore, at 0 degrees voltage is at its maximum positive value, while current is zero.

At 90 degrees in *Figure 12*, voltage is once again zero, but this time it is decreasing toward negative at its greatest rate of change. Because the applied voltage is decreasing, the magnetic field is collapsing inward on the conductor. This has the effect of reversing the direction of motion between the field and conductor that existed at 0 degrees.

Therefore, the current will flow in a direction opposite of what it was at 0 degrees. Also, because the applied voltage is decreasing at its greatest rate, the field is collapsing at its greatest rate. This causes the flux to cut the conductor at the greatest rate, causing the induced current magnitude to be maximum. At 90 degrees, the applied voltage is zero decreasing toward negative, while the current is maximum positive.

At 180 degrees in *Figure 12*, the applied voltage is at its maximum negative value, but just as at 0 degrees, its rate of change is zero. At 180 degrees, therefore, current will be zero. This explanation shows that the voltage peaks positive first, then 90 degrees later the current peaks positive. Current thus lags the applied voltage in an inductor by 90 degrees. This can easily be remembered using this phrase: ELI the ICE man. ELI represents voltage (E), inductance (L), and current (I). In an inductor, the voltage leads the current just like the letter E leads or comes before the letter I. The word ICE will be explained in the section on **capacitance**.

7.3.0 Inductive Reactance

The opposing force that an inductor presents to the flow of alternating current cannot be called resistance since it is not the result of friction within a conductor. The name given to this force is inductive **reactance** because it is the reaction of the inductor to alternating current. Inductive reactance is measured in ohms and its symbol is X_L.

Remember that the induced voltage in a conductor is proportional to the rate at which magnetic lines of force cut the conductor. The greater the rate or higher the frequency, the greater the counter-electromotive force (CEMF). Also, the induced voltage increases with an increase in

ELI in ELI the ICE Man

Remembering the phrase "ELI" as in "ELI the ICE man" is an easy way to remember the phase relationships that always exist between voltage and current in an inductive circuit. An inductive circuit is a circuit where there is more inductive reactance than capacitive reactance. The L in ELI indicates inductance. The E (voltage) is stated before the I (current) in ELI, meaning that the voltage leads the current in an inductive circuit.

inductance; the more turns, the greater the CEMF. Reactance then increases with an increase of frequency and with an increase in inductance. The formula for inductive reactance is as follows:

$$X_L = 2\pi fL$$

Where:

X_L = inductive reactance in ohms

2π = a constant in which the Greek letter pi (π) represents 3.14 and 2 × pi = 6.28

f = frequency of the alternating current in hertz

L = inductance in henrys

For example, if f is equal to 60Hz and L is equal to 20h, find X_L:

$$X_L = 2\pi fL$$
$$X_L = 6.28 \times 60Hz \times 20h$$
$$X_L = 7,536\Omega$$

Once calculated, the value of X_L is used like resistance in a form of Ohm's law:

$$I = \frac{E}{X_L}$$

Where:

I = effective current (amps)

E = effective voltage (volts)

X_L = inductive reactance (ohms)

Unlike a resistor, there is no power dissipation in an ideal inductor. An inductor limits current, but it uses no net energy since the energy required to build up the field in the inductor is given back to the circuit when the field collapses.

8.0.0 ◆ CAPACITANCE

A capacitor stores an electric charge in a dielectric material. Capacitance is the ability to store a charge. In storing a charge, a capacitor opposes a change in voltage. *Figure 13* shows a simple capacitor in a circuit, schematic representations of two types of capacitors, and a photo of common capacitors.

Figure 14A shows a capacitor in a DC circuit. When voltage is applied, the capacitor begins to charge (*Figure 14B*), which continues until the potential difference across the capacitor is equal to the applied voltage. This charging current is transient, or temporary, flowing only until the capacitor is charged to the applied voltage. Then there is no current in the circuit. *Figure 14C* shows this with the voltage across the capacitor equal to the battery voltage, or 10V.

The capacitor can be discharged by connecting a conducting path across the dielectric. The stored charge across the dielectric provides the potential difference to produce a discharge current, as shown in *Figure 14D*. Once the capacitor is completely discharged, the voltage across it equals zero, and there is no discharge current.

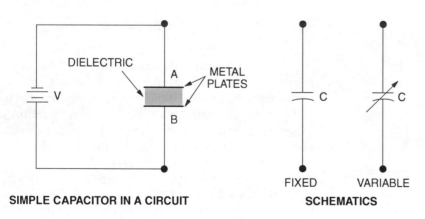 

SIMPLE CAPACITOR IN A CIRCUIT   **SCHEMATICS**   **COMMON CAPACITORS**

202F13.EPS

Figure 13 ◆ Capacitors.

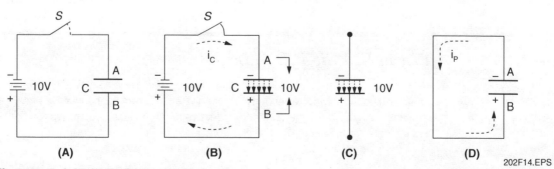

(A)   (B)   (C)   (D)

202F14.EPS

Figure 14 ◆ Charging and discharging capacitor.

In a capacitive circuit, the charge and discharge current must always be in opposite directions. Current flows in one direction to charge the capacitor and in the opposite direction when the capacitor is allowed to discharge.

Current will flow in a capacitive circuit with AC voltage applied because of the capacitor charge and discharge current. There is no current through the dielectric, which is an insulator. While the capacitor is being charged by increasing applied voltage, the charging current flows in one direction to the plates. While the capacitor is discharging as the applied voltage decreases, the discharge current flows in the reverse direction. With alternating voltage applied, the capacitor alternately charges and discharges.

First, the capacitor is charged in one polarity, and then it discharges; next, the capacitor is charged in the opposite polarity, and then it discharges again. The cycles of charge and discharge current provide alternating current in the circuit at the same frequency as the applied voltage. The amount of capacitance in the circuit will determine how much current is allowed to flow.

Capacitance is measured in farads (F), where one farad is the capacitance when one coulomb is stored in the dielectric with a potential difference of one volt. Smaller values are measured in microfarads (μF). A small capacitance will allow less charge and discharge current to flow than a larger capacitance. The smaller capacitor has more opposition to alternating current, because less current flows with the same applied voltage.

In summary, capacitance exhibits the following characteristics:

- DC is blocked by a capacitor. Once charged, no current will flow in the circuit.
- AC flows in a capacitive circuit with AC voltage applied.
- A smaller capacitance allows less current.

8.1.0 Factors Affecting Capacitance

A capacitor consists of two conductors separated by an insulating material called a dielectric. There are many types and sizes of capacitors with different dielectric materials. The capacitance of a capacitor is determined by three factors:

- *Area of the plates* – The initial charge displacement on a set of capacitor plates is related to the number of free electrons in each plate. Larger plates produce a greater capacitance than smaller ones. Therefore, the capacitance varies directly with the area of the plates. For example, if the area of the plates is doubled, the capacitance is doubled. If the size of the plates is reduced by 50 percent, the capacitance is also reduced by 50 percent.

- *Distance between plates* – As two capacitor plates are brought closer together, more electrons move away from the positively charged plate and into the negatively charged plate. This is because the mutual attraction between the opposite charges on the plates increases as we move the plates closer together. This added movement of charge increases the capacitance of the capacitor. In a capacitor composed of two plates of equal area, the capacitance varies inversely with the distance between the plates. For example, if the distance between the plates is decreased by one-half, the capacitance is doubled. If the distance between the plates is doubled, the capacitance is one-half as great.

- *Dielectric permittivity* – The dielectric is the material between the capacitor plates in which the electric field appears. Relative permittivity expresses the ratio of the electric field strength in a dielectric to that in a vacuum. Permittivity has nothing to do with the dielectric strength of the medium or the breakdown voltage. An insulating material that will withstand a higher applied voltage than some other substance does not always have a higher dielectric permittivity. Many insulating materials have a greater dielectric permittivity than air. For a given applied voltage, a greater attraction exists between the opposite charges on the capacitor plates, and an electric field can be set up more easily than when the dielectric is air. The capacitance of the capacitor is increased when the permittivity of the dielectric is increased if all the other parameters remain unchanged.

Capacitance

The concept of capacitance, like many electrical quantities, is often hard to visualize or understand. A comparison with a balloon may help to make this concept clearer. Electrical capacitance has a charging effect similar to blowing up a balloon and holding it closed. The expansion capacity of the balloon can be changed by changing the thickness of the balloon walls. A balloon with thick walls will expand less (have less capacity) than one with thin walls. This is like a small 10μF capacitor that has less capacity and will charge less than a larger 100μF capacitor.

8.2.0 Calculating Equivalent Capacitance

Connecting capacitors in parallel is equivalent to adding the plate areas. Therefore, the total capacitance is the sum of the individual capacitances, as shown in *Figure 15*.

A 10µF capacitor in parallel with a 5µF capacitor, for example, provides a 15µF capacitance for the parallel combination. The voltage is the same across the parallel capacitors. Note that adding parallel capacitance is opposite to the case of inductances in parallel and resistances in parallel.

Connecting capacitances in series is equivalent to increasing the thickness of the dielectric. Therefore, the combined capacitance is less than the smallest individual value. The combined equivalent capacitance is calculated by the reciprocal formula, as shown in *Figure 16*.

Capacitors connected in series are combined like resistors in parallel. Any of the shortcut calculations for the reciprocal formula apply. For example, the combined capacitance of two equal capacitances of 10µF in series is 5µF.

Capacitors are used in series to provide a higher voltage breakdown rating for the combination. For instance, each of three equal capacitances in series has one-third the applied voltage.

In series, the voltage across each capacitor is inversely proportional to its capacitance. The smaller capacitance has the larger proportion of the

applied voltage. The reason is that the series capacitances all have the same charge because they are in one current path. With equal charge, a smaller capacitance has a greater potential difference.

8.3.0 Capacitor Specifications

In addition to its capacitance rating, the capacitor is rated by its operating voltage and leakage resistance. These factors must be considered in the selection of a capacitor.

8.3.1 Voltage Rating

This rating specifies the maximum potential difference that can be applied across the plates without puncturing the dielectric. Usually, the voltage rating is for temperatures up to about 60°C. High temperatures result in a lower voltage rating. Voltage ratings for general-purpose paper, mica, and ceramic capacitors are typically 200V to 500V. Ceramic capacitors with ratings of 1 to 5kV are also available.

Electrolytic capacitors are commonly used in 25V, 150V, and 450V ratings. In addition, 6V and 10V electrolytic capacitors are often used in transistor circuits. For applications where a lower voltage rating is permissible, more capacitance can be obtained in a smaller physical size.

The potential difference across the capacitor depends on the applied voltage and is not necessarily equal to the voltage rating. A voltage rating higher than the potential difference applied across the capacitor provides a safety factor for long life in service. With electrolytic capacitors, however, the actual capacitor voltage should be close to the rated voltage to produce the oxide film that provides the specified capacitance.

The voltage ratings are for applied DC voltage. The breakdown rating is lower for AC voltage because of the internal heat produced by continuous charge and discharge.

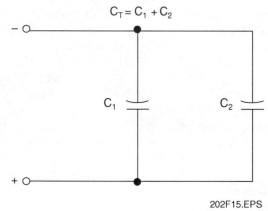

$$C_T = C_1 + C_2$$

202F15.EPS

Figure 15 ◆ Capacitors in parallel.

$$C_T = \cfrac{1}{\cfrac{1}{C_1} + \cfrac{1}{C_2}}$$

C_1 C_2

202F16.EPS

Figure 16 ◆ Capacitors in series.

8.3.2 Leak Resistance

Consider a capacitor charged by a DC voltage source. After the charging voltage is removed, a perfect capacitor would keep its charge indefinitely. After a long period of time, however, the charge will be neutralized by a small leakage current through the dielectric and across the insulated case between terminals, because there is no perfect insulator. For paper, ceramic, and mica capacitors, the leakage current is very slight, or inversely, the leakage resistance is very high. For paper, ceramic, or mica capacitors, leakage resistance is 100MΩ or more. However, electrolytic capacitors may have a leakage resistance of 0.5MΩ or less.

8.4.0 Voltage and Current in a Capacitive AC Circuit

In a capacitive circuit driven by an AC voltage source, the voltage is continuously changing. Thus, the charge on the capacitor is also continuously changing. The four parts of *Figure 17* show the variation of the alternating voltage and current in a capacitive circuit for each quarter of one cycle.

The solid line represents the voltage across the capacitor, and the dotted line represents the current. The line running through the center is the zero or reference point for both the voltage and the current. The bottom line marks off the time of the cycle in terms of electric degrees. Assume that the AC voltage has been acting on the capacitor for some time before the time represented by the starting point of the sine wave.

At the beginning of the first quarter-cycle (0 to 90 degrees), the voltage has just passed through zero and is increasing in the positive direction. Since the zero point is the steepest part of the sine wave, the voltage is changing at its greatest rate.

The charge on a capacitor varies directly with the voltage; therefore, the charge on the capacitor is also changing at its greatest rate at the beginning of the first quarter-cycle. In other words, the greatest number of electrons are moving off one plate and onto the other plate. Thus, the capacitor current is at its maximum value.

As the voltage proceeds toward maximum at 90 degrees, its rate of change becomes lower and lower, making the current decrease toward zero. At 90 degrees, the voltage across the capacitor is maximum, the capacitor is fully charged, and there is no further movement of electrons from plate to plate. That is why the current at 90 degrees is zero.

At the end of the first quarter-cycle, the alternating voltage stops increasing in the positive direction and starts to decrease. It is still a positive voltage; but to the capacitor, the decrease in voltage

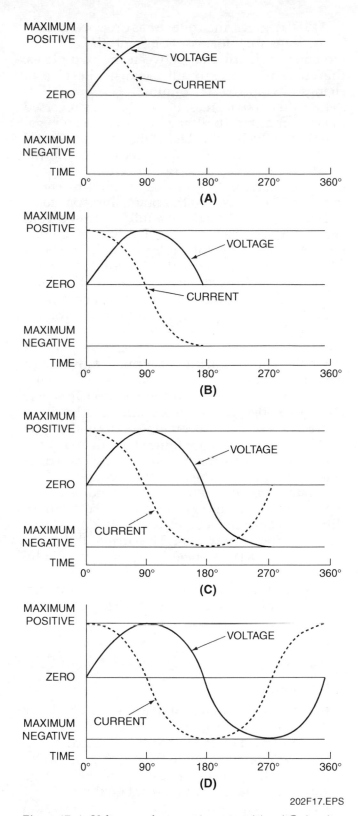

202F17.EPS

Figure 17 ◆ Voltage and current in a capacitive AC circuit.

means that the plate that has just accumulated an excess of electrons must lose some electrons. The current flow must reverse its direction. The second part of the figure shows the current curve to be below the zero line (negative current direction) during the second quarter-cycle (90 to 180 degrees).

At 180 degrees, the voltage has dropped to zero. This means that, for a brief instant, the electrons are equally distributed between the two plates; the current is maximum because the rate of change of voltage is maximum.

Just after 180 degrees, the voltage has reversed polarity and starts building to its maximum negative peak, which is reached at the end of the third quarter-cycle (180 to 270 degrees). During the third quarter-cycle, the rate of voltage change gradually decreases as the charge builds to a maximum at 270 degrees. At this point, the capacitor is fully charged and carries the full impressed voltage. Because the capacitor is fully charged, there is no further exchange of electrons and the current flow is zero at this point. The conditions are exactly the same as at the end of the first quarter-cycle (90 degrees), but the polarity is reversed.

Just after 270 degrees, the impressed voltage once again starts to decrease, and the capacitor must lose electrons from the negative plate. It must discharge, starting at a minimum rate of flow and rising to a maximum. This discharging action continues through the last quarter-cycle (270 to 360 degrees) until the impressed voltage has reached zero. The beginning of the entire cycle is 360 degrees, and everything starts over again.

In *Figure 17*, note that the current always arrives at a certain point in the cycle 90 degrees ahead of the voltage because of the charging and discharging action. This voltage-current phase relationship in a capacitive circuit is exactly opposite to that in an inductive circuit. The current through a capacitor leads the voltage across the capacitor by 90 degrees. A convenient way to remember this is the phrase ELI the ICE man (ELI refers to inductors, as previously explained). ICE pertains to capacitors as follows:

I = current

C = capacitor

E = voltage

In capacitors (C), current (I) leads voltage (E) by 90 degrees.

It is important to realize that the current and voltage are both going through their individual cycles at the same time during the period the AC voltage is impressed. The current does not go through part of its cycle (charging or discharging) and then stop and wait for the voltage to catch up. The amplitude and polarity of the voltage and the amplitude and direction of the current are continually changing.

Their positions, with respect to each other and to the zero line at any electrical instant or any degree between 0 and 360 degrees, can be seen by reading upward from the time-degree line. The current swing from the positive peak at 0 degrees to the negative peak at 180 degrees is not a measure of the number of electrons or the charge on the plates. It is a picture of the direction and strength of the current in relation to the polarity and strength of the voltage appearing across the plates.

8.5.0 Capacitive Reactance

Capacitors offer a very real opposition to current flow. This opposition arises from the fact that, at a given voltage and frequency, the number of electrons that go back and forth from plate to plate is limited by the storage ability or the capacitance of the capacitor. As the capacitance is increased, a greater number of electrons changes plates every cycle. Since current is a measure of the number of electrons passing a given point in a given time, the current is increased.

Increasing the frequency will also decrease the opposition offered by a capacitor. This occurs because the number of electrons that the capacitor is capable of handling at a given voltage will change plates more often. As a result, more electrons will pass a given point in a given time (greater current flow). The opposition that a capacitor offers to AC is therefore inversely proportional to frequency and capacitance. This opposition is called capacitive reactance. Capacitive reactance decreases with increasing frequency or, for a given frequency, the capacitive reactance decreases with increasing capacitance. The symbol for capacitive reactance is X_C. The formula is:

$$X_c = \frac{1}{2\pi fC}$$

Where:

X_c = capacitive reactance in ohms

f = frequency in hertz

C = capacitance in farads

2π = 6.28 (2 × 3.14)

For example, what is the capacitive reactance of a 0.05µF capacitor in a circuit whose frequency is 1 megahertz?

$$X_c = \frac{1}{2\pi fC}$$

$$X_c = \frac{1}{6.28(10^6 \text{ hertz})(5 \times 10^{-8} \text{ farads})}$$

$$X_c = \frac{1}{6.28(5 \times 10^{-2})}$$

$$X_c = \frac{1}{31.4 \times 10^{-2}}$$

$$X_c = \frac{1}{0.314} = 3.18 \text{ ohms}$$

Frequency and Capacitive Reactance

A variable capacitor is used in the tuner of an AM radio to tune the radio to the desired station. Will its capacitive reactance value be higher or lower when it is tuned to the low end of the frequency band (550kHz) than it would be when tuned to the high end of the band (1,440kHz)?

The capacitive reactance of a 0.05μF capacitor operated at a frequency of 1 megahertz is 3.18 ohms. Suppose this same capacitor is operated at a lower frequency of 1,500 hertz instead of 1 megahertz. What is the capacitive reactance now? Substituting where $1,500 = 1.5 \times 10^3$ hertz:

$$X_c = \frac{1}{2\pi fC}$$

$$X_c = \frac{1}{6.28(1.5 \times 10^3 \text{ hertz})(5 \times 10^{-8} \text{ farads})}$$

$$X_c = \frac{1}{6.28(7.5 \times 10^{-5})}$$

$$X_c = \frac{1}{47.1 \times 10^{-5}}$$

$$X_c = \frac{1}{0.000471} = 2,123 \text{ ohms}$$

Note a very interesting point from these two examples. As frequency is decreased from 1 megahertz to 1,500 hertz, the capacitive reactance increases from 3.18 ohms to 2,123 ohms. Capacitive reactance increases as the frequency decreases.

9.0.0 ◆ RL, RC, LC, AND RLC CIRCUITS

AC circuits often contain inductors, capacitors, and/or resistors connected in series or parallel combinations. When this is done, it is important to determine the resulting phase relationship between the applied voltage and the current in the circuit. The simplest method of combining factors that have different phase relationships is vector addition with the trigonometric functions. Each quantity is represented as a vector, and the resultant vector and phase angle are then calculated.

In purely resistive circuits, the voltage and current are in phase. In inductive circuits, the voltage leads the current by 90 degrees. In capacitive circuits, the current leads the voltage by 90 degrees. *Figure 18* shows the phase relationships of these components used in AC circuits. Recall that these characteristics are summarized by the phrase ELI the ICE man.

ELI = E Leads I (inductive)

ICE = I Capacitive (leads) E

The **impedance** Z of a circuit is defined as the total opposition to current flow. The magnitude of the impedance Z is given by the following equation in a series circuit:

$$Z = \sqrt{R^2 + X^2}$$

Where:

Z = impedance (ohms)

R = resistance (ohms)

X = net reactance (ohms)

The current through a resistance is always in phase with the voltage applied to it; thus resistance is shown along the 0-degree axis. The voltage across an inductor leads the current by 90 degrees; thus inductive reactance is shown along the 90-degree axis. The voltage across a capacitor lags the current by 90 degrees; thus capacitive reactance is shown along the 90-degree axis. The net reactance is the algebraic difference between the inductive reactance and the capacitive reactance:

X = net reactance (ohms)

X_L = inductive reactance (ohms)

X_C = capacitive reactance (ohms)

The impedance Z is the vector sum of the resistance R and the net reactance X. The angle, called the phase angle, gives the phase relationship between the applied voltage and current.

9.1.0 RL Circuits

RL circuits combine resistors and inductors in a series, parallel, or series-parallel configuration. In a pure inductive circuit, the current lags the voltage by an angle of 90 degrees. In a circuit containing both resistance and inductance, the current will lag the voltage by some angle between 0 and 90 degrees.

9.1.1 Series RL Circuit

Figure 19 shows a series RL circuit. Since it is a series circuit, the current is the same in all portions of the loop. Using the values shown, the circuit will be analyzed for unknown values such as X_L, Z, I, E_L, and E_R.

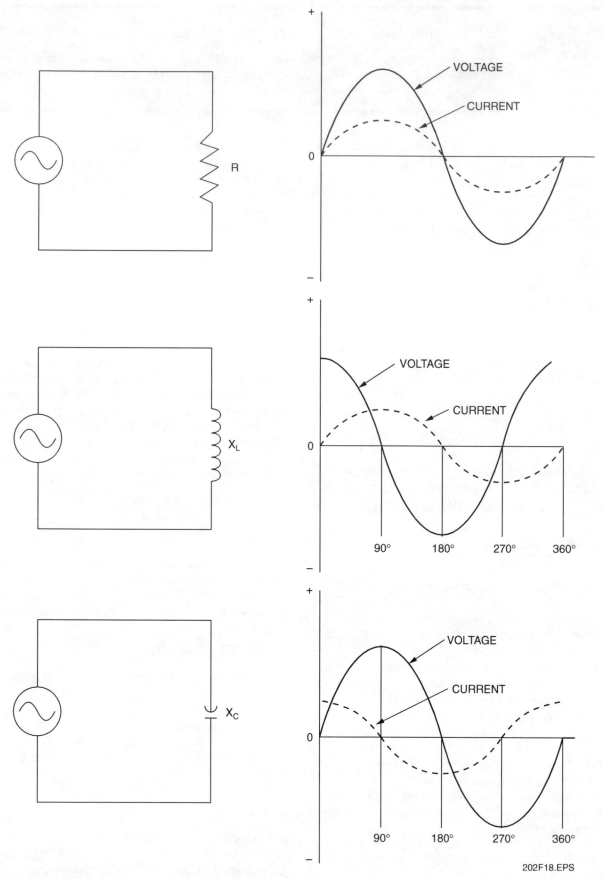

Figure 18 ◆ Summary of AC circuit phase relationships.

202F18.EPS

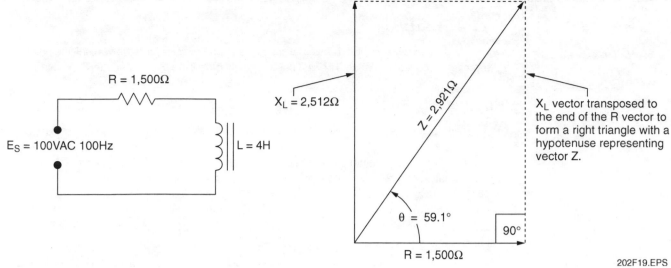

R = 1,500Ω

E_S = 100VAC 100Hz   L = 4H

X_L = 2,512Ω

Z = 2,921Ω

X_L vector transposed to the end of the R vector to form a right triangle with a hypotenuse representing vector Z.

θ = 59.1°

90°

R = 1,500Ω

202F19.EPS

Figure 19 ◆ Series RL circuit and vector diagram.

The solution would be worked as follows:

Step 1 Compute the value of X_L.

$X_L = 2\pi fL$

$X_L = 6.28 \times 100 \times 4 = 2{,}512$ ohms

Step 2 Draw vectors R and X_L as shown in *Figure 19*. R is drawn horizontally because the circuit current and voltage across R are in phase. It therefore becomes the reference line from which other angles are measured. X_L is drawn upward at 90 degrees from R because voltage across X_L leads circuit current through R.

NOTE

Rectangular coordinates and the Pythagorean theorem for right triangles are used for the solution of the following problem and the problems in the remaining sections of this module.

Step 3 Compute the value of circuit impedance Z, which is equal to the vector sum (hypotenuse) of vectors of X_L and R that form the right triangle shown in *Figure 19*.

$Z = \sqrt{X_L{}^2 + R^2}$

$Z = \sqrt{2{,}512\Omega^2 + 1{,}500\Omega^2}$

$Z = \sqrt{6{,}310{,}144 + 2{,}250{,}000}$

$Z = \sqrt{8{,}560{,}144}$

$Z = 2{,}926\Omega$ (rounded)

Step 4 Compute the circuit current using Ohm's law for AC circuits.

$I = \dfrac{E}{Z}$

$I = \dfrac{100}{2{,}926\Omega} = 0.034A$ (rounded)

Step 5 Compute voltage drops in the circuit.

$E_L = IX_L = 0.034 \times 2{,}512 = 85$ volts

$E_n = IR = 0.034 \times 1{,}500 = 51$ volts

Note that the voltage drops across the resistor and inductor do not equal the supply voltage because they must be added vectorially. This can be done using the same method as used to determine Z except that voltage (E_Z) is substituted for Z and 85 and 51 volts for the terms X_L and R.

Alternately, the voltage as well as its phase angle can be determined using trigonometry as shown below.

Determine the phase angle of the voltage vector E_Z:

Phase angle of E_Z is the arc tangent of $E_L \div E_R$

$\tan = E_L \div E_R$

$\tan = 85 \div 51 = 1.667$ (rounded)

arctan $1.667 = 59.04°$

Determine the voltage vector E_Z:

$\cos 59.04° = E_R \div E_Z$

Solving for E_Z yields:

$E_Z = E_R \div \cos 59.04°$

$E_Z = 51 \div 0.5144 = 99.14$ volts (rounded)

In this inductive circuit, the current lags the applied voltage by an angle equal to 59.04°.

Figure 20 shows another series RL circuit, its associated waveforms, and vector diagrams. This circuit is used to summarize the characteristics of a series RL circuit:

- The current I flows through all the series components.
- The voltage across X_L, labeled V_L, can be considered an IX_L voltage drop, just as V_R is used for an IR voltage drop.
- The current I through X_L must lag V_L by 90 degrees, as this is the angle between current through an inductance and its self-induced voltage.
- The current I through R and its IR voltage drop have the same phase. There is no reactance to sine

wave current in any resistance. Therefore, I and IR have the same phase, or this phase angle is 0 degrees.

- V_T is the vector sum of the two out-of-phase voltages V_R and V_L.
- Circuit current I lags V_T by the phase angle.
- Circuit impedance is the vector sum of R and X_L.

In a series circuit, the higher the value of X_L compared with R, the more inductive the circuit is. This means there is more voltage drop across the inductive reactance, and the phase angle increases toward 90 degrees. The series current lags the applied generator voltage.

Several combinations of X_L and R in series are listed in *Table 1* with their resultant impedance and

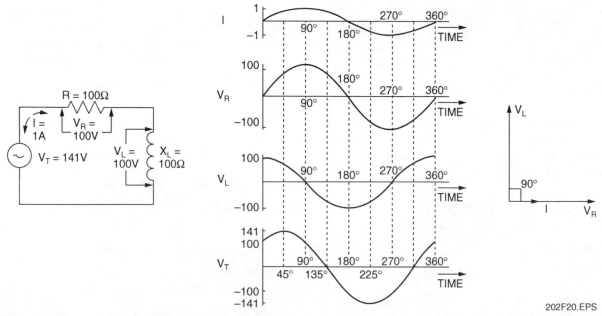

202F20.EPS

Figure 20 ◆ Series RL circuit with waveforms and vector diagram.

Table 1 Series R and X_L Combinations

R (Ω)	X_L (Ω)	Z (Ω) (Approx.)	Phase Angle (θ) (in degrees)
1	10	$\sqrt{101} = 10$	84.3
10	10	$\sqrt{200} = 14$	45.0
10	1	$\sqrt{101} = 10$	5.7

phase angle. Note that a ratio of 10:1 or more for X_L/R means that the circuit is practically all inductive. The phase angle of 84.3 degrees is only slightly less than 90 degrees for the ratio of 10:1, and the total impedance Z is approximately equal to X_L. The voltage drop across X_L in the series circuit will be equal to the applied voltage, with almost none across R.

At the opposite extreme, when R is 10 times as large as X_L, the series circuit is mainly resistive. The phase angle of 5.7 degrees means the current has almost the same phase as the applied voltage, the total impedance Z is approximately equal to R, and the voltage drop across R is practically equal to the applied voltage, with almost none across X_L.

9.1.2 Parallel RL Circuit

In a parallel RL circuit, the resistance and inductance are connected in parallel across a voltage source. Such a circuit thus has a resistive branch and an inductive branch.

The 90-degree phase angle must be considered for each of the branch currents, instead of voltage drops in a series circuit. Remember that any series circuit has different voltage drops, but one common current. A parallel circuit has different branch currents, but one common voltage.

In the parallel circuit in *Figure 21*, the applied voltage V_A is the same across X_L, R, and the generator, since they are all in parallel. There cannot be any phase difference between these voltages. Each branch, however, has its individual current. For the resistive branch $I_R = V_A \div R$; in the inductive branch $I_L = V_A \div X_L$.

The resistive branch current I_R has the same phase as the generator voltage V_A. The inductive branch current I_L lags V_A, however, because the current in an inductance lags the voltage across it by 90 degrees.

The total line current, therefore, consists of I_R and I_L, which are 90 degrees out of phase with each other. The phasor sum of I_R and I_L equals the total line current I_T. These phase relations are shown by the waveforms and vectors in *Figure 21*. I_T will lag V_A by some phase angle that results from the vector addition of I_R and I_L.

The impedance of a parallel RL circuit is the total opposition to current flow by the R of the resistive branch and the X_L of the inductive branch. Since X_L and R are vector quantities, they must be added vectorially.

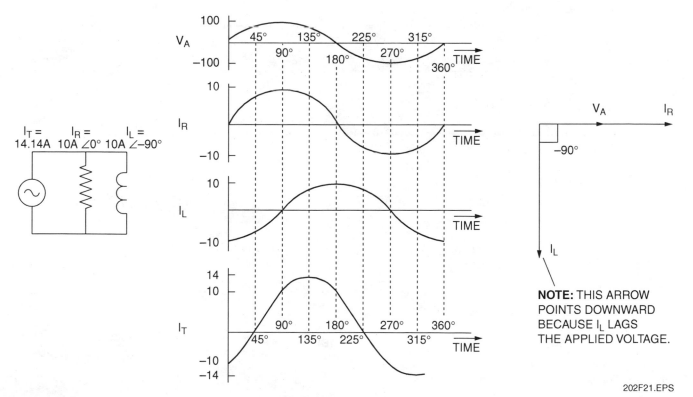

Figure 21 ◆ Parallel RL circuit with waveforms and vector diagram.

202F21.EPS

If the line current and the applied voltage are known, Z can also be calculated by the following equation:

$$Z = \frac{V_A}{I_{Line}}$$

The Z of a parallel RL circuit is always less than the R or X_L of any one branch. The branch of a parallel RL circuit that offers the most opposition to current flow has the lesser effect on the phase angle of the current.

Several combinations of X_L and R in parallel are listed in Table 2. When X_L is 10 times R, the parallel circuit is practically resistive because there is little inductive current in the line. The small value of I_L results from the high X_L. The total impedance of the parallel circuit is approximately equal to the resistance then, since the high value of X_L in a parallel branch has little effect. The phase angle of –5.7 degrees is practically 0 degrees because almost all the line current is resistive.

As X_L becomes smaller, it provides more inductive current in the main line. When X_L is ⅒R, practically all the line current is the I_L component. Then, the parallel circuit is practically all inductive, with a total impedance practically equal to X_L. The phase angle of –84.3 degrees is almost –90 degrees because the line current is mostly inductive. Note that these conditions are opposite from the case of X_L and R in series.

9.2.0 RC Circuits

In a circuit containing resistance only, the current and voltage are in phase. In a circuit of pure capacitance, the current leads the voltage by an angle of 90 degrees. In a circuit that has both resistance and capacitance, the current will lead the voltage by some angle between 0 and 90 degrees.

9.2.1 Series RC Circuit

Figure 22A shows a series RC circuit with resistance R in series with capacitive reactance X_C. Current I is the same in X_C and R since they are in series. Each has its own series voltage drop, equal to IR for the resistance and IX_C for the reactance.

In *Figure 22B*, the current phasor is shown horizontal as the reference phase, because I is the same throughout the series circuit. The resistive voltage drop IR has the same phase as I. The capacitor voltage IX_C must be 90 degrees clockwise from I and IR, as the capacitive voltage lags. Note that the IX_C phasor is downward, exactly opposite from an IX_L phasor, because of the opposite phase angle.

If the capacitive reactance alone is considered, its voltage drop lags the series current I by 90 degrees. The IR voltage has the same phase as I, however, because resistance provides no phase shift. Therefore, R and X_C combined in series must be added by

Table 2 Parallel R and X_L Combinations

R (Ω)	X_L (Ω)	I_R (A)	I_L (A)	I_T (A) (Approx.)	$Z_T = V_A/I_T$ (Ω)	Phase Angle (θ)ᵢ
1	10	10	1	$\sqrt{101}$ = 10	1	–5.7 degrees
10	10	1	1	$\sqrt{2}$ = 1.4	7.07	–45.0 degrees
10	1	1	10	$\sqrt{101}$ = 10	1	–84.3 degrees

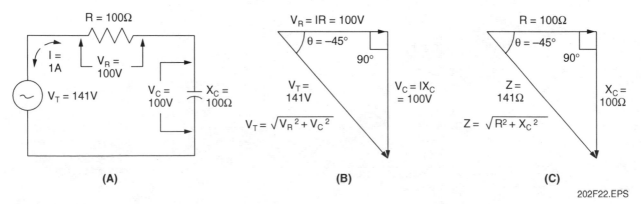

(A)   (B)   (C)

202F22.EPS

Figure 22 ◆ Series RC circuit with vector diagrams.

vectors because they are 90 degrees out of phase with each other, as shown in *Figure 22C*. As with inductive reactance, θ (theta) is the phase angle between the generator voltage and its series current. As shown in *Figure 22B* and *22C*, θ can be calculated from the voltage or impedance triangle.

With series X_C the phase angle is negative, clockwise from the zero reference angle of I because the X_C voltage lags its current. To indicate the negative phase angle, this 90-degree phasor points downward from the horizontal reference, instead of upward as with the series inductive reactance. In series, the higher the X_C compared with R, the more capacitive the circuit. There is more voltage drop across the capacitive reactance, and the phase angle increases toward –90 degrees. The series X_C always makes the current lead the applied voltage. With all X_C and no R, the entire applied voltage is across X_C and equals –90 degrees. Several combinations of X_C and R in series are listed in *Table 3*.

Table 3 Series R and X_C Combinations

R (Ω)	X_C (Ω)	Z (Ω) (Approx.)	Phase Angle (θ)$_Z$ (in degrees)
1	10	$\sqrt{101}$ = 10	84.3
10	10	$\sqrt{200}$ = 14	45.0
10	1	$\sqrt{101}$ = 10	5.7

9.2.2 Parallel RC Circuit

In a parallel RC circuit, as shown in *Figure 23A*, a capacitive branch as well as a resistive branch are connected across a voltage source. The current that leaves the voltage source divides among the branches, so there are different currents in each branch. The current is therefore not a common quantity, as it is in the series RC circuit.

In a parallel RC circuit, the applied voltage is directly across each branch. Therefore, the branch voltages are equal in value to the applied voltage and all voltages are in phase. Since the voltage is common throughout the parallel RC circuit, it serves as the common quantity in any vector representation of parallel RC circuits. This means the reference vector will have the same phase relationship or direction as the circuit voltage. Note in *Figure 23B* that V_A and I_R are both shown as the 0-degree reference.

Current within an individual branch of a RC parallel circuit is dependent on the voltage across the branch and on the R or X_C contained in the branch. The current in the resistive branch is in phase with the branch voltage, which is the applied voltage. The current in the capacitive branch leads V_A by 90 degrees. Since the branch voltages are the same, I_C leads I_R by 90 degrees, as shown in *Figure 23B*. Since the branch currents are out of phase, they have to be added vectorially to find the line current.

The phase angle (θ) is 45 degrees because R and X_C are equal, resulting in equal branch currents. The phase angle is between the total current I_T and the generator voltage V_A. However, the phase of V_A is the same as the phase of I_R. Therefore, θ is also between I_T and I_R.

The impedance of a parallel RC circuit represents the total opposition to current flow offered by the resistance and capacitive reactance of the circuit. The equation for calculating the impedance of a parallel RC circuit is:

$$Z = \frac{RX_C}{\sqrt{I_R^2 + I_C^2}}$$

or

$$Z = \frac{V_A}{I_T}$$

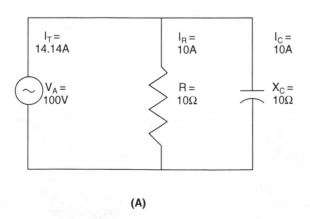
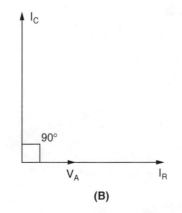
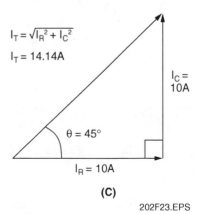

Figure 23 ◆ Parallel RC circuit with vector diagrams.

202F23.EPS

For the example shown in *Figure 23*, Z is:

$$Z = \frac{V_A}{I_T}$$

$$Z = \frac{100}{14.14A} = 7.07\Omega$$

This is the opposition in ohms across the generator. This Z of 7.07Ω is equal to the resistance of 10Ω in parallel with the reactance of 10Ω. Notice that the impedance of equal values of R and X_C is not one-half, but equals 70.7 percent of either one.

When X_C is high relative to R, the parallel circuit is practically resistive because there is little leading capacitive current in the main line. The small value of I_C results from the high reactance of shunt X_C. The total impedance of the parallel circuit is approximately equal to the resistance, since the high value of X_C in a parallel branch has little effect.

As X_C becomes smaller, it provides more leading capacitive current in the main line. When X_C is very small relative to R, practically all the line current is the I_C component. The parallel circuit is practically all capacitive, with a total impedance practically equal to X_C.

The characteristics of different circuit arrangements are shown in *Table 4*.

9.3.0 LC Circuits

An LC circuit consists of an inductance and a capacitance connected in series or in parallel with a voltage source. There is no resistor physically in an LC circuit, but every circuit contains some resistance. Since the circuit resistance of the wiring and voltage source is usually so small, it has little or no effect on circuit operation.

In a circuit with both X_L and X_C, the opposite phase angles enable one to cancel the effect of the other. For X_L and X_C in series, the net reactance is the difference between the two series reactances, resulting in less reactance than either one. In parallel circuits, the I_L and I_C branch currents cancel each other out. The net line current is then the difference between the two branch currents, resulting in less total line current than either branch current.

9.3.1 Series LC Circuit

As in all series circuits, the current in a series LC circuit is the same at all points. Therefore, the current in the inductor is the same as, and in phase with, the current in the capacitor. Because of this, on the vector diagram for a series LC circuit, the direction of the current vector is the reference or in the 0-degree direction, as shown in *Figure 24*.

When there is current flow in a series LC circuit, the voltage drops across the inductor and capacitor depend on the circuit current and the values of X_L and X_C. The voltage drop across the inductor leads the circuit current by 90 degrees, and the voltage drop across the capacitor lags the circuit current by 90 degrees. Using Kirchhoff's voltage law, the source voltage equals the sum of the voltage drops across the inductor and capacitor, with respect to the polarity of each.

Since the current through both is the same, the voltage across the inductor leads that across the capacitor by 180 degrees. The method used to add

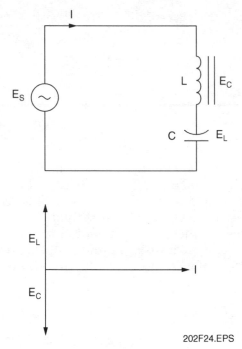

202F24.EPS

Figure 24 ◆ Series LC circuit with vector diagram.

Table 4 Parallel R and X_C Combinations

R (Ω)	X_C (Ω)	I_R (A)	I_C (A)	I_T (A) (Approx.)	Z_T (Ω) (Approx.)	Phase Angle (θ)$_i$
1	10	10	1	$\sqrt{101}$ = 10	1	5.7 degrees
10	10	1	1	$\sqrt{2}$ = 1.4	7.07	45.0 degrees
10	1	1	10	$\sqrt{101}$ = 10	1	84.3 degrees

the two voltage vectors is to subtract the smaller vector from the larger, and assign the resultant the direction of the larger. When applied to a series LC circuit, this means the applied voltage is equal to the difference between the voltage drops (E_L and E_C), with the phase angle between the applied voltage (E_T) and the circuit current determined by the larger voltage drop.

In a series LC circuit, one or both of the voltage drops are always greater than the applied voltage. Remember that although one or both of the voltage drops are greater than the applied voltage, they are 180 degrees out of phase. One of them effectively cancels a portion of the other so that the total voltage drop is always equal to the applied voltage.

Recall that X_L is 180 degrees out of phase with X_C. The impedance is then the vector sum of the two reactances. The reactances are 180 degrees apart, so their vector sum is found by subtracting the smaller one from the larger.

Unlike RL and RC circuits, the impedance in an LC circuit is either purely inductive or purely capacitive.

9.3.2 Parallel LC Circuit

In a parallel LC circuit there is an inductance and a capacitance connected in parallel across a voltage source. *Figure 25* shows a parallel LC circuit with its vector diagram.

As in any parallel circuit, the voltage across the branches is the same as the applied voltage. Since they are actually the same voltage, the branch voltages and applied voltage are in phase. Because of this, the voltage is used as the 0-degree phase reference and the phases of the other circuit quantities are expressed in relation to the voltage.

The currents in the branches of a parallel LC circuit are both out of phase with the circuit voltage. The current in the inductive branch (I_L) lags the voltage by 90 degrees, while the current in the capacitive branch (I_C) leads the voltage by 90 degrees. Since the voltage is the same for both branches, currents I_L and I_C are therefore 180 degrees out of phase. The amplitudes of the branch currents depend on the value of the reactance in the respective branches.

With the branch currents being 180 degrees out of phase, the line current is equal to their vector sum. This vector addition is done by subtracting the smaller branch current from the larger.

The line current for a parallel LC circuit, therefore, has the phase characteristics of the larger branch current. Thus, if the inductive branch current is the larger, the line current is inductive and lags the applied voltage by 90 degrees; if the capacitive branch current is the larger, the line current is capacitive, and leads the applied voltage by 90 degrees.

The line current in a parallel LC circuit is always less than one of the branch currents and sometimes less than both. The reason that the line current is less than the branch currents is because the two branch currents are 180 degrees out of phase. As a result of the phase difference, some cancellation takes place between the two currents when they combine to produce the line current. The impedance of a parallel LC circuit can be found using the following equations:

$$Z = \frac{X_L \times X_C}{X_L - X_C} \quad \text{(for } X_L \text{ larger than } X_C)$$

or

$$Z = \frac{X_L \times X_C}{X_C - X_L} \quad \text{(for } X_C \text{ larger than } X_L)$$

When using these equations, the impedance will have the phase characteristics of the smaller reactance.

9.4.0 RLC Circuits

Circuits in which the inductance, capacitance, and resistance are all connected in series are called series RLC circuits. The fundamental properties of series RLC circuits are similar to those for series LC circuits. The differences are caused by the

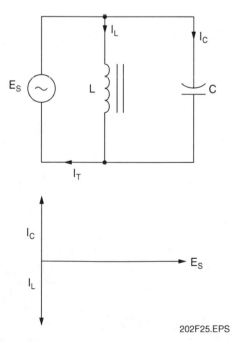

202F25.EPS

Figure 25 ◆ Parallel LC circuit with vector diagram.

effects of the resistance. Any practical series LC circuit contains some resistance. When the resistance is very small compared to the circuit reactance, it has almost no effect on the circuit and can be considered as zero. When the resistance is appreciable, though, it has a significant effect on the circuit operation and therefore must be considered in any circuit analysis.

9.4.1 Series RLC Circuit

In a series RLC circuit, the same current flows through each component. The phase relationships between the voltage drops are the same as they were in series RC, RL, and LC circuits. The voltage drops across the inductance and capacitance are 180 degrees out of phase. With current the same throughout the circuit as a reference, the inductive voltage drop (E_L) leads the resistive voltage drop (E_R) by 90 degrees, and the capacitive voltage drop (E_C) lags the resistive voltage drop by 90 degrees.

Figure 26 shows a series RLC circuit and the vector diagram used to determine the applied voltage. The vector sum of the three voltage drops is equal to the applied voltage. However, to calculate this vector sum, a combination of the methods learned for LC, RL, and RC circuits must be used. First, calculate the combined voltage drop of the two reactances. This value is designated E_X and is found as in pure LC circuits by subtracting the smaller reactive voltage drop from the larger. This is shown in *Figure 26* as E_X. The result of this calculation is the net reactive voltage drop and is either inductive or capacitive,

depending on which of the individual voltage drops is larger. In *Figure 26*, the net reactive voltage drop is inductive since E_L is greater than E_C. Once the net reactive voltage drop is known, it is added vectorially to the voltage drop across the resistance.

The angle between the applied voltage E_A and the voltage across the resistance E_R is the same as the phase angle between E_A and the circuit current. The reason for this is that E_R and I are in phase.

The impedance of a series RLC circuit is the vector sum of the inductive reactance, the capacitive reactance, and the resistance. This is done using the same method as for voltage drop calculations.

When X_L is greater than X_C, the net reactance is inductive, and the circuit acts essentially as an RL circuit. Similarly, when X_C is greater than X_L, the net reactance is capacitive, and the circuit acts as an RC circuit.

The same current flows in every part of a series RLC circuit. The current always leads the voltage across the capacitance by 90 degrees and is in phase with the voltage across the resistance. The phase relationship between the current and the applied voltage, however, depends on the circuit impedance. If the impedance is inductive (X_L greater than X_C), the current is inductive and lags the applied voltage by some phase angle less than 90 degrees. If the impedance is capacitive (X_C greater than X_L), the current is capacitive, and leads the applied voltage by some phase angle also less than 90 degrees. The angle of the lead or lag is determined by the relative values of the net reactance and the resistance.

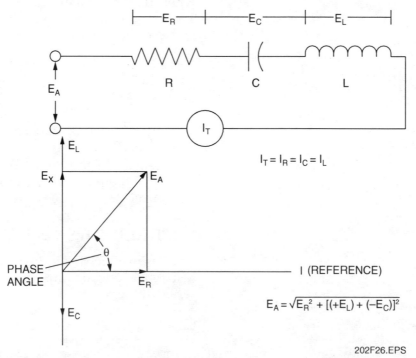

Figure 26 ◆ Series RLC circuit and vector diagram.

$$E_A = \sqrt{E_R{}^2 + [(+E_L) + (-E_C)]^2}$$

$$I_T = I_R = I_C = I_L$$

202F26.EPS

The greater the value of X or the smaller the value of R, the larger the phase angle, and the more reactive (or less resistive) the current. Similarly, the smaller the value of X or the larger the value of R, the more resistive (or less reactive) the current. If either R or X is 10 or more times greater than the other, the circuit will essentially act as though it is purely resistive or reactive, as the case may be.

9.4.2 Series Resonance

Recall that the formula for inductive reactance is as follows:

$$X_L = 2\pi fL$$

Also recall that the formula for capacitive reactance is as follows:

$$X_C = \frac{1}{2\pi fc}$$

With X_L and X_C being frequency sensitive, any change in frequency will affect the operating characteristics of any reactive circuit.

The effects of frequency variations on the input voltage of a series RLC circuit are shown in *Figure 27* and summarized as follows:

- As frequency is increased, X_L will become larger and X_C will become smaller. As a result, the circuit becomes even more inductive, θ increases, and the voltage across L will increase.

- As frequency decreases, X_L will become smaller and X_C larger, and θ will decrease toward zero.
- At a certain frequency, X_L will equal X_C. This is referred to as **resonance**.
- A further decrease in frequency will make X_C larger than X_L. The circuit will become capacitive and will increase in a negative direction.

The resonant frequency at which $X_L = X_C$ can be calculated as follows:

$$X_L = X_C$$

$$X_L = 2\pi fL$$

$$X_C = \frac{1}{2\pi fc}$$

Therefore, at resonance:

$$2\pi fL = \frac{1}{2\pi fc}$$

or

$$\text{resonant frequency} = f_r = \frac{1}{2\pi\sqrt{LC}}$$

Figure 27 shows a vector diagram of the resistance and reactance at a frequency when $X_L = X_C$, or resonance. Since X_L and X_C are equal and 180 degrees out of phase, the algebraic sum of X_L and X_C is zero. The only opposition to current flow in the circuit is R. At series resonance:

$$Z = R$$

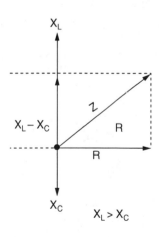
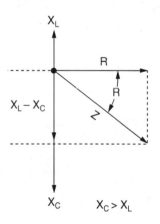
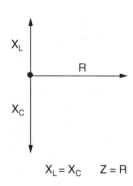

202F27.EPS

Figure 27 ◆ Effects of frequency variations on an RLC circuit.

AC Circuits

In a simple series circuit comprised of an ON/OFF switch, small lamp, motor, and capacitor, which components insert resistance, inductive reactance, and capacitive reactance into the circuit?

At frequencies above resonance, X_L is greater than X_C. The circuit becomes inductive and Z increases. At frequencies below resonance, X_C is greater than X_L. The circuit becomes capacitive and Z increases. The point of lowest impedance of the circuit is at resonance. These characteristics are shown in *Figure 28*.

The impedance of a series RLC circuit is minimum at resonance, so the current must therefore be maximum. The further the frequency is from the resonant frequency, the greater the impedance, and the smaller the current becomes. At any frequency, the current can be calculated from Ohm's law for AC circuits using the equation $I = E \div Z$. Since at the resonant frequency the impedance equals the resistance, the equation for current at resonance becomes $I = E \div R$.

The letter Q is used to designate the quality of a tuned circuit. It is an indication of its maximum response as well as its ability to respond within a band of frequencies.

To secure maximum currents and response, the resistance must be kept at a low value. At resonance, R is the only resistance in the circuit. The Q of a circuit is the relationship of the reactance of the circuit to its resistance:

$$Q = \frac{X_L}{R} = \frac{X_C}{R}$$

The bandwidth or bandpass of a tuned circuit is defined as those frequency limits above and below resonant frequency where the response of the circuit will drop to 0.707 of its peak response. If current or voltage drops to 0.707 of its peak value, the power drops to 50 percent.

Bandwidth is frequency above and below resonance where power drops to one-half of its peak value. These are called the half-power points.

If the frequency of an RLC circuit is varied and the values of current at the different frequencies are plotted on a graph, the result is a curve known as the resonance curve of the circuit, as shown in *Figure 29*.

Actually, *Figure 29* shows two curves: one with a high resistance that results in a lower current flow, low Q, and wide bandwidth; and the second with a low resistance that results in high current flow at resonance, high Q, and low bandwidth.

9.4.3 Parallel RLC Circuit

A parallel RLC circuit is a parallel LC circuit with an added parallel branch of resistance. The solution of a parallel circuit involves the solution of a parallel LC circuit, and then the solution of either a parallel RL circuit or a parallel RC circuit. The reason for this is that a parallel combination of L and C appears to the source as a pure L or a pure C. So by solving the LC portion of a parallel RLC circuit first, the circuit is reduced to an equivalent RL or RC circuit.

The distribution of the voltage in a parallel RLC circuit is no different from what it is in a parallel LC circuit or in any parallel circuit. The branch voltages are all equal and in phase, since they are the same as the applied voltage. The resistance is simply another branch across which the applied voltage appears. Because the voltages throughout the circuit are the same, the applied voltage is again used as the θ phase reference.

Figure 30 shows the current relationship in a parallel RLC circuit.

The three branch currents in a parallel RLC circuit are an inductive current I_L, a capacitive current I_C, and a resistive current I_R. Each is independent of the other, and depends only on the applied voltage and the branch resistance or reactance.

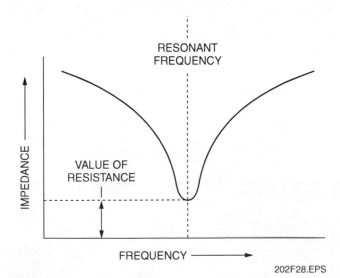

Figure 28 ◆ Frequency-impedance curve.

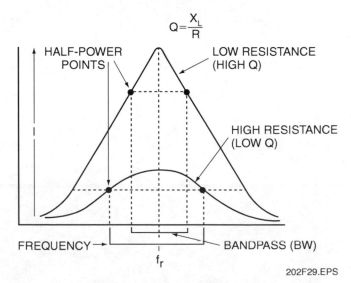

Figure 29 ◆ Typical series resonance curve.

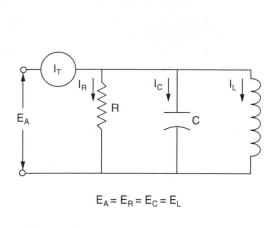

$E_A = E_R = E_C = E_L$

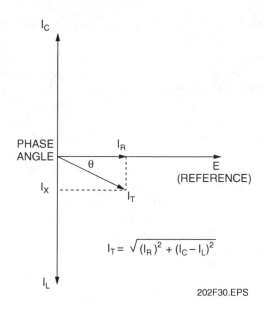

$$I_T = \sqrt{(I_R)^2 + (I_C - I_L)^2}$$

202F30.EPS

Figure 30 ◆ Parallel RLC circuit and vector diagram.

The three branch currents all have different phases with respect to the branch voltages. I_L lags the voltage by 90 degrees, I_C leads the voltage by 90 degrees, and I_R is in phase with the voltage. Since the voltages are the same, I_L and I_C are 180 degrees out of phase with each other, and both are 90 degrees out of phase with I_R. Because I_R is in phase with the voltage, it has the same zero-reference direction as the voltage. So I_C leads I_R by 90 degrees, and I_L lags I_R by 90 degrees.

The line current (I_T), or total current, is the vector sum of the three branch currents, and can be calculated by adding I_L, I_C, and I_R vectorially. Whether the line current leads or lags the applied voltage depends on which of the reactive branch currents (I_L or I_C) is the larger. If I_L is larger, I_T lags the applied voltage. If I_C is larger, I_T leads the applied voltage.

To determine the impedance of a parallel RLC circuit, first determine the net reactance X of the inductive and capacitive branches. Then use X to determine the impedance Z, the same as in a parallel RL or RC circuit.

Whenever Z is inductive, the line current will lag the applied voltage. Similarly, when Z is capacitive, the line current will lead the applied voltage.

9.4.4 Parallel Resonance

A parallel resonant circuit is a circuit in which the voltage source is in parallel with L and C. The characteristics of parallel resonance are quite different from those of series resonance. However, the frequency at which parallel resonance takes place is identical to the frequency at which series resonance takes place. Therefore, parallel resonance uses the same formula as series resonance:

$$f_r = \frac{1}{2\pi\sqrt{LC}}$$

The properties of a parallel resonant circuit are based on the action that takes place between the parallel inductance and capacitance, which is often called a tank circuit because it has the ability to store electrical energy.

The action of a tank circuit is one of interchange of energy between the inductance and capacitance. If a voltage is momentarily applied across the tank circuit, C charges to this voltage. When the applied voltage is removed, C discharges through L, and a magnetic field is built up around L by the discharge current. When C has discharged, the field around L collapses, and in doing so induces a current that is in the same direction as the current that created the field. This current, therefore, charges C in the opposite direction. When the field around L has collapsed, C again discharges, but this time in the direction opposite to before. The discharge current again causes a magnetic field around L, which, when it collapses, charges C in the same direction in which it was initially charged.

This interchange of energy and the circulating current it produces would continue indefinitely, producing a series of sine waves, if this were an ideal tank circuit with no resistance. However, since some resistance is always present, the circulating current gradually diminishes as the resistance dissipates the energy in the circuit in the form of heat. This causes the sine wave current to be damped out. If a voltage were again momentarily applied across the circuit, the interchange of energy and accompanying circulating current would begin again.

At resonance, X_L equals X_C, so the two currents I_L and I_C are also equal. Because the two currents in a parallel LC circuit are 180 degrees out of phase, the line current, which is their vector sum, must be zero. Thus, the only current is the circulating current in the tank circuit. No line current flows, therefore the circuit has infinite impedance as far as the voltage source is concerned.

These two conditions of zero line current and infinite impedance are characteristic of ideal parallel resonant circuits at resonance. In practical circuits that contain some resistance, the theoretical conditions of zero line current and infinite impedance are not realized. Instead, practical parallel resonant circuits have minimum line current and maximum impedance at resonance. This is the exact opposite of series resonant circuits, which have maximum current and minimum impedance at resonance.

In the ideal parallel resonant circuit at resonance, the branch currents I_L and I_C are equal, so the line current is zero and the circuit impedance is infinite. Above and below the resonant frequency, one of the reactances X_L or X_C is larger than the other. The two branch currents are therefore unequal, and the line current, which equals their vector sum (or arithmetic difference), has some value greater than zero. Since line current flows, the circuit impedance is no longer infinite. The further the frequency is from the resonant frequency, the greater the difference between the values of the reactances. As a result, the line current is larger and the circuit impedance is smaller.

The principal effect of the resistance in a parallel resonant circuit is that it causes the current in the inductive branch to lag the applied voltage by a phase angle of less than 90 degrees, instead of exactly 90 degrees as in the case of the ideal circuit. As a result, the two branch currents are not 180 degrees out of phase. For simplicity, resonance can still be considered as occurring when X_L equals X_C, but now when the two branch currents are added vectorially, their sum is not zero. This means that at resonance, some line current flows. Since there is line current, the impedance cannot be infinite, as it is in the ideal circuit. Thus at resonance, practical parallel resonant circuits have minimum line current and maximum resistance, instead of zero line current and infinite impedance, as do ideal circuits.

For parallel resonance, Q also measures the quality of a circuit. In parallel resonant circuits, Q depends on circuit resistance. The Q of a parallel resonant circuit is defined as follows:

$$Q = \frac{X_L}{R} \text{ or } \frac{X_C}{R}$$

Recognize this as the same equation used for the Q of a series resonant circuit. As a result, resistance has the same effect on the Q of a parallel resonant circuit as it does on a series resonant circuit. The lower the resistance, the higher the Q of the circuit and the narrower its bandpass. Conversely, the greater the resistance, the lower the Q and the wider the bandpass.

Recall that for every series resonant circuit there is a range of frequencies above and below the resonant frequency at which, for practical purposes, the circuit can be considered as being at resonance. This range of frequencies is called the bandwidth, and consists of all the frequencies at which the circuit current was 0.707 or more times its value at resonance. Parallel resonant circuits also have a bandwidth, but it is defined in terms of the frequency-vs.-impedance curve, and consists of all the frequencies that produce a circuit impedance 0.707 or more times the impedance at resonance. *Figure 31* shows the bandpass or bandwidth as all the frequencies between F_1 and F_2.

Circuit resistance affects the width and steepness of the frequency-impedance curve. Therefore, resistance affects the circuit bandpass. A low resistive circuit causes a steep curve and narrow bandpass. A high resistive circuit causes a flatter frequency-impedance curve and therefore, a wide bandpass.

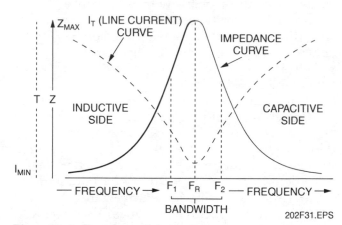

Figure 31 ◆ Tuned parallel circuit curves.

10.0.0 ◆ POWER IN AC CIRCUITS

In DC circuits, the power consumed is the sum of all the I^2R heating in the resistors. It is also equal to the power produced by the source, which is the product of the source voltage and current. In AC circuits containing only resistors, the same relationship also holds true.

10.1.0 True Power

The power consumed by resistance is true power and is measured in watts. True power is the product of the resistor current squared and the resistance:

ELECTRONIC SYSTEMS TECHNICIAN LEVEL TWO — TRAINEE MODULE 33202-05

The Importance of Resonance

A common use of RLC circuits is in resonant filter applications. Because RLC circuits provide a means for frequency selectivity, resonance in electrical circuits is very important to many types of electrical and electronic systems. Series and/or parallel RLC resonant filters are widely used to filter out or reduce the effects of unwanted harmonic frequencies and/or unwanted electromagnetic interference (EMI) signals present in power distribution circuits. Resonant filters are also used to increase the efficiency or power factor of a circuit or system.

In communications, the ability of a radio or television receiver to select a certain frequency transmitted by a certain station, and eliminate frequencies from other stations, is made possible by the use of series and/or parallel RLC resonant circuits.

$$P_T = I^2R$$

This formula applies because current and voltage have the same phase across a resistance.

To find the corresponding value of power as a product of voltage and current, this product must be multiplied by the cosine of the phase angle θ:

$$P_T = I^2R \ or \ P_T = EI \times \cos\theta$$

Where E and I are in rms values to calculate the true power in watts, multiplying I by the cosine of the phase angle provides the resistive component for true power equal to I^2R.

For example, a series RL circuit has 2A through a 100Ω resistor in series with the X_L of 173Ω. Therefore:

$$P_T = I^2R$$
$$P_T = 4 \times 100$$
$$P_T = 400W$$

Furthermore, in this circuit the phase angle is 60 degrees with a cosine of 0.5. The applied voltage is 400V. Therefore:

$$P_T = EI \times \cos\theta$$
$$P_T = 400 \times 2 \times 0.5$$
$$P_T = 400W$$

In both cases, the true power is the same (400W) because this is the amount of power supplied by the generator and dissipated in the resistance. Either formula can be used for calculating the true power.

10.2.0 Apparent Power

In ideal AC circuits containing resistors, capacitors, and inductors, the only mechanism for power consumption is $I^2_{eff}R$ heating in the resistors. Inductors and capacitors consume no power. The only function of inductors and capacitors is to store and release energy. However, because of the phase shifts that are introduced by these elements, the power consumed by the resistors is not equal to the product of the source voltage and current. The product of the source voltage and current is called apparent power and has units of volt-amperes (VA).

The apparent power is the product of the source voltage and the total current. Therefore, apparent power is actual power delivered by the source. The formula for apparent power is:

$$P_A = (E_A)(I)$$

Figure 32 shows a series RL circuit and its associated vector diagram.

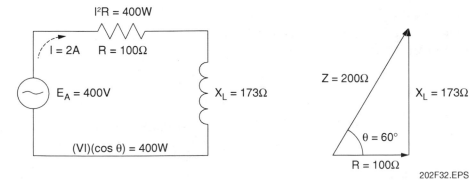

Figure 32 ◆ Power calculations in an AC circuit.

This circuit is used to calculate the apparent power and compare it to the circuit's true power:

$P_A = (EA)(I)$

$P_A = (400V)(2A)$

$P_A = 800VA$

$P_T = EI \times \cos \theta$

$\theta = \dfrac{R}{X_L} = \dfrac{173}{100} = 60°$

$P_T = (400V)(2A)(\cos 60°)$

$P_T = (400V)(2A)(0.5)$

$P_T = 400W$

Note that the apparent power formula is the product of EI alone without considering the cosine of the phase angle.

10.3.0 Reactive Power

Reactive power is that portion of the apparent power that is caused by inductors and capacitors in the circuit. Inductance and capacitance are always present in real AC circuits. No work is performed by reactive power; the power is stored in the inductors and capacitors, then returned to the circuit. Therefore, reactive power is always 90 degrees out of phase with true power. The units for reactive power are volt-amperes-reactive (VARs).

In general, for any phase angle θ between E and I, multiplying EI by sine θ gives the vertical component at 90 degrees for the value of the VARs. In *Figure 32*, the value of sine 60 degrees is 800 × 0.866 = 692.8 VARs.

Note that the factor sine q for the volt-amperes-reactive (VARs) gives the vertical or reactive component of the apparent power EI. However, multiplying EI by cosine q as the power factor gives the horizontal or resistive component for the real power.

10.4.0 Power Factor

Because it indicates the resistive component, cosine q is the power factor (pf) of the circuit, converting the EI product to real power. For series circuits, use the following formula:

$pf = \cos \theta = \dfrac{R}{Z}$

For parallel circuits, use this formula:

$pf = \cos \theta = \dfrac{I_R}{I_T}$

In *Figure 32* as an example of a series circuit, R and Z are used for the calculations:

$pf = \cos \theta = \dfrac{R}{Z}$

$pf = \dfrac{100\Omega}{200\Omega} = 0.5$

The power factor is not an angular measure but a numerical ratio with a value between 0 and 1, equal to the cosine of the phase angle. With all resistance and zero reactance, R and Z are the same for a series circuit of I_R and I_T and are the same for a parallel circuit. The ratio is 1. Therefore, unity power factor means a resistive circuit. At the opposite extreme, all reactance with zero resistance makes the power factor zero, meaning that the circuit is all reactive.

The power factor gives the relationship between apparent power and true power. The power factor can thus be defined as the ratio of true power to apparent power:

$pf = \dfrac{P_T}{P_A}$

For example, calculate the power factor of the circuit shown in *Figure 33*.

The true power is the product of the resistor current squared and the resistance:

$P_T = I^2R$

$P_T = 10A^2 \times 10\Omega$

$P_T = 1,000W$

The apparent power is the product of the source voltage and total current:

$P_A = I_TE$

$P_A = 10.2A \times 100V$

$P_A = 1,020VA$

Calculating total current:

$I_T = \sqrt{I_R^2 + (I_C - I_L)^2}$

$I_T = \sqrt{10A^2 + (4A - 2A)^2}$

$I_T = 10.2A$

The power factor is the ratio of true power to apparent power:

$pf = \dfrac{P_T}{P_A}$

$pf = \dfrac{1,000}{1,020}$

$pf = 0.98$

As illustrated in the previous example, the power factor is determined by the system load. If the load contained only resistance, the apparent power would equal the true power and the power factor would be at its maximum value of one.

Managing Power Factor

In hot weather, everyone turns on air conditioners. Air conditioners have compressors and fans that are driven by motors. These motors are inductive loads. The addition of so many inductive loads in the system can result in a poor system power factor and excessive power losses.

Banks of capacitors are automatically switched into the system so that their capacitive reactance cancels out as near as possible the inductive reactance presented by the motors. Remember that inductors shift the current and voltage away from each other in one direction (ELI), and capacitors shift them away from each other in the opposite direction (ICE). This maintains the system power factor at an acceptable value and keeps system losses to a minimum.

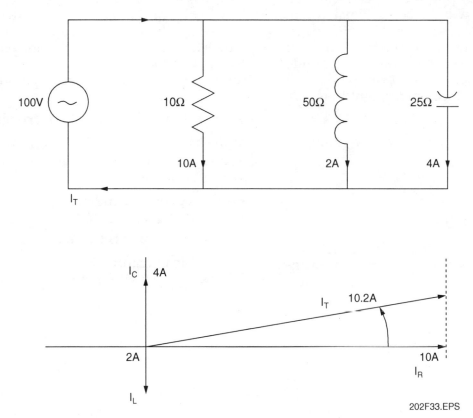

202F33.EPS

Figure 33 ◆ RLC circuit calculation.

Purely resistive circuits have a power factor of unity or one. If the load is more inductive than capacitive, the apparent power will lag the true power and the power factor will be lagging. If the load is more capacitive than inductive, the apparent power will lead the true power and the power factor will be leading. If there is any reactive load on the system, the apparent power will be greater than the true power and the power factor will be less than one.

10.5.0 Power Triangle

The phase relationships among the three types of AC power are easily visualized on the power triangle shown in *Figure 34*. The true power (W) is the horizontal leg, the apparent power (VA) is the hypotenuse, and the cosine of the phase angle between them is the power factor. The vertical leg of the triangle is the reactive power and has units of volt-amperes-reactive (VARs).

AC CIRCUITS

2.35

As illustrated on the power triangle (*Figure 34*), the apparent power will always be greater than the true power or reactive power. Also, the apparent power is the result of the vector addition of true and reactive power. The power magnitude relationships shown in *Figure 34* can be derived from the Pythagorean theorem for right triangles:

$$c^2 = a^2 + b^2$$

Therefore, c also equals the square root of $a^2 + b^2$, as shown below:

$$c = \sqrt{a^2 + b^2}$$

11.0.0 ◆ TRANSFORMERS

A transformer is a device that transfers electrical energy from one circuit to another by electromagnetic induction (transformer action). The electrical energy is always transferred without a change in frequency, but the transfer may involve changes in the magnitudes of voltage and current. Because a transformer works on the principle of electromagnetic induction, it must be used with an input source voltage that varies in amplitude.

11.1.0 Transformer Construction

Figure 35 shows the basic components of a transformer. In its most basic form, a transformer consists of the following:

- Primary coil or winding
- Secondary coil or winding
- Core that supports the coils or windings

A simple transformer action is shown in *Figure 36*. The primary winding is connected to a 60Hz

$$P_A = \sqrt{(P_T{}^2) + (P_{RX}{}^2)}$$

$$P_T = \sqrt{(P_A{}^2) - (P_{RX}{}^2)}$$

$$P_{RX} = \sqrt{(P_A{}^2) - (P_T{}^2)}$$

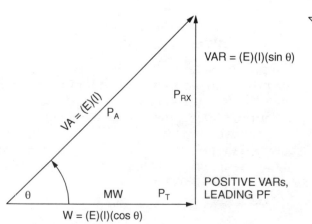

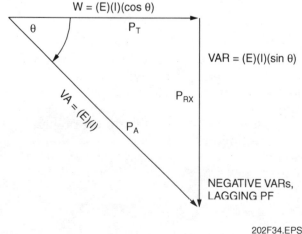

202F34.EPS

Figure 34 ◆ Power triangle.

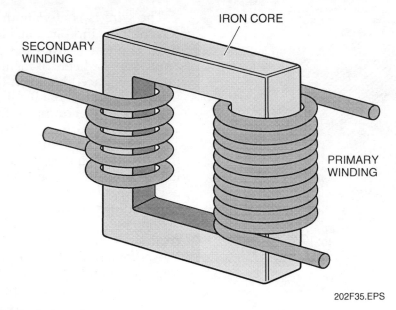

Figure 35 ◆ Basic components of a transformer.

AC voltage source. The magnetic field or flux builds up (expands) and collapses (contracts) around the primary winding. The expanding and contracting magnetic field around the primary winding cuts the secondary winding and induces an alternating voltage into the winding. This voltage causes AC to flow through the load. The voltage may be stepped up or down depending on the design of the primary and secondary windings.

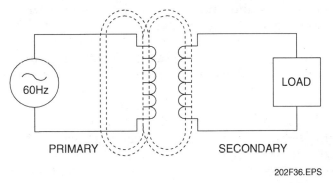

Figure 36 ◆ Transformer action.

11.1.1 Core Characteristics

Commonly used core materials are air, soft iron, and steel. Each of these materials is suitable for particular applications and unsuitable for others. Generally, air-core transformers are used when the voltage source has a high frequency (above 20kHz). Iron-core transformers are generally used when the source frequency is low (below 20kHz). A soft-iron transformer is very useful where the transformer must be physically small yet efficient. The iron-core transformer provides better power transfer than the air-core transformer. Laminated sheets of steel are often used in a transformer to reduce one type of power loss known as eddy currents. These are undesirable currents, induced into the core, which circulate around the core. Laminating the core reduces these currents to smaller levels. These steel laminations are insulated with a nonconducting material, such as varnish, and then formed into a core as shown in *Figure 37.* It takes about 50 such laminations to make a core one-inch thick. The most efficient

Transformers

Transformers are essential to all electrical systems and all types of electronic equipment. They are especially crucial to the operation of AC high-voltage power distribution systems. Transformers are used to both step up voltage and step down voltage throughout the distribution process. For example, a typical power generation plant might generate AC power at 13,800V, step it up to 230,000V for distribution over long transmission lines, step it down to 13,800V again at substations located at different points for local distribution, and finally step it down again to 240V and 120V for lighting and local power use.

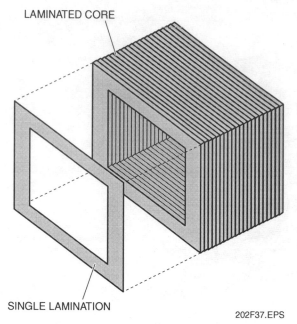

Figure 37 ◆ Steel laminated core.

transformer core is one that offers the best path for the most lines of flux, with the least loss in magnetic and electrical energy.

11.1.2 Transformer Windings

A transformer consists of two coils called windings, which are wrapped around a core. The transformer operates when a source of AC voltage is connected to one of the windings and a load device is connected to the other. The winding that is connected to the source is called the primary winding. The winding that is connected to the load is called the secondary winding. *Figure 38* shows a cutaway view of a typical transformer.

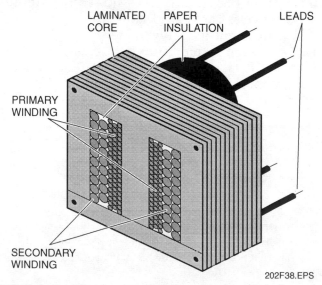

Figure 38 ◆ Cutaway view of a transformer core.

The wire is coated with varnish so that each turn of the winding is insulated from every other turn. In a transformer designed for high-voltage applications, sheets of insulating material such as paper are placed between the layers of windings to provide additional insulation.

When the primary winding is completely wound, it is wrapped in insulating paper or cloth. The secondary winding is then wound on top of the primary winding. After the secondary winding is complete, it too is covered with insulating paper. Next, the core is inserted into and around the windings as shown.

Sometimes, terminals may be provided on the enclosure for connections to the windings. *Figure 38* shows four leads, two from the primary and two from the secondary. These leads must be connected to the source and load, respectively.

11.2.0 Operating Characteristics

Regardless of type, most transformers operate in the same way. This section covers the no-load operation and phase relationships in a simple transformer.

11.2.1 Energized with No Load

A no-load condition is said to exist when a voltage is applied to the primary, but no load is connected to the secondary. Assume the output of the secondary is connected to a load by an open switch. Because of the open switch, there is no current flowing in the secondary winding. With the switch open and an AC voltage applied to the primary, there is, however, a very small amount of current, called exciting current, flowing in the primary. Essentially, what this current does is excite the coil of the primary to create a magnetic field. The amount of exciting current is determined by three factors: the amount of voltage applied (E_A); the resistance (R) of the primary coil's wire and core losses; and the X_L, which is dependent on the frequency of the exciting current. These factors are all controlled by transformer design.

This very small amount of exciting current serves two functions:

- Most of the exciting energy is used to support the magnetic field of the primary.
- A small amount of energy is used to overcome the resistance of the wire and core. This is dissipated in the form of heat (power loss).

Exciting current will flow in the primary winding at all times to maintain this magnetic field, but no transfer of energy will take place as long as the secondary circuit is open.

11.2.2 Phase Relationship

The secondary voltage of a simple transformer may be either in phase or out of phase with the primary voltage. This depends on the direction in which the windings are wound and the arrangement of the connection to the external circuit (load). Simply, this means that the two voltages may rise and fall together, or one may rise while the other is falling. Transformers in which the secondary voltage is in phase with the primary are referred to as like-wound transformers, while those in which the voltages are 180 degrees out of phase are called unlike-wound transformers.

Dots are used to indicate points on a transformer schematic symbol that have the same instantaneous polarity (points that are in phase). The use of phase-indicating dots is illustrated in *Figure 39*. In the first part of the figure, both the primary and secondary windings are wound from top to bottom in a clockwise direction, as viewed from above the windings. When constructed in this manner, the top lead of the primary and the top lead of the secondary have the same polarity. This is indicated by the dots on the transformer symbol.

The second part of the figure illustrates a transformer in which the primary and secondary are wound in opposite directions. As viewed from above the windings, the primary is wound in a clockwise direction from top to bottom, while the secondary is wound in a counterclockwise direction. Notice that the top leads of the primary and secondary have opposite polarities. This is indicated by the dots being placed on opposite ends of the transformer symbol. Thus, the polarity of voltage at the terminals of the transformer secondary depends on the direction in which the secondary is wound with respect to the primary.

11.3.0 Turns and Voltage Ratios

To understand how a transformer can be used to step up or step down voltage, the concept of turns ratio must be understood. The total voltage induced into the secondary winding of a transformer is determined mainly by the ratio of the number of turns in the primary to the number of turns in the secondary, and by the amount of voltage applied to the primary. Therefore, to set up a formula:

$$\text{Turns ratio} = \frac{\text{number of turns in the primary}}{\text{number of turns in the secondary}}$$

The first transformer in *Figure 40* shows a transformer whose primary consists of 10 turns of wire, and whose secondary consists of a single turn of wire. As lines of flux generated by the primary expand and collapse, they cut both the 10 turns of

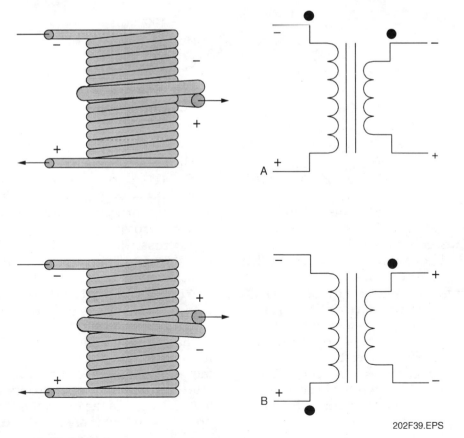

202F39.EPS

Figure 39 ◆ Transformer winding polarity.

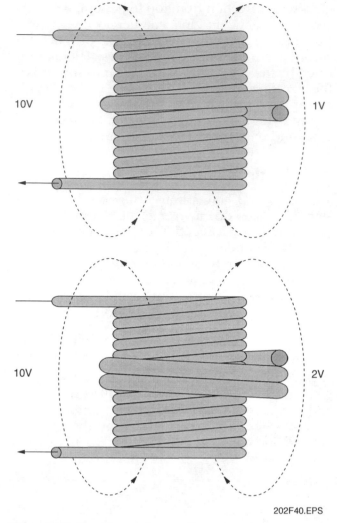

202F40.EPS

Figure 40 ◆ Transformer turns ratio.

the primary and the single turn of the secondary. Since the length of the wire in the secondary is approximately the same as the length of the wire in each turn of the primary, the EMF induced into the secondary will be the same as the EMF induced into each turn of the primary.

This means that if the voltage applied to the primary winding is 10 volts, the EMF in the primary is almost 10 volts. Thus, each turn in the primary will have an induced EMF of approximately one-tenth of the total applied voltage, or one volt. Since the same flux lines cut the turns in both the secondary and the primary, each turn will have an EMF of one volt induced into it. The first transformer in *Figure 40* has only one turn in the secondary, thus, the EMF across the secondary is one volt.

The second transformer represented in *Figure 40* has a 10-turn primary and a two-turn secondary. Since the flux induces one volt per turn, the total voltage across the secondary is two volts. Notice

that the volts per turn are the same for both primary and secondary windings. Since the EMF in the primary is equal (or almost) to the applied voltage, a proportion may be set up to express the value of the voltage induced in terms of the voltage applied to the primary and the number of turns in each winding. This proportion also shows the relationship between the number of turns in each winding and the voltage across each winding, and is expressed by the equation:

$$\frac{E_S}{E_P} = \frac{N_S}{N_P}$$

Where:

N_P = number of turns in the primary

E_P = voltage applied to the primary

E_S = voltage induced in the secondary

N_S = number of turns in the secondary

The equation shows that the ratio of secondary voltage to primary voltage is equal to the ratio of secondary turns to primary turns. The equation can be written as:

$$E_P N_S = E_S N_P$$

For example, a transformer has 100 turns in the primary, 50 turns in the secondary, and 120VAC applied to the primary (E_P). What is the voltage across the secondary (E_S)?

N_P = 100 turns

N_S = 50 turns

E_P = 120VAC

$$\frac{E_S}{E_P} = \frac{N_S}{N_P} \; or \; E_S = \frac{E_P N_S}{N_P}$$

$$E_S = \frac{120V \times 50 \text{ turns}}{100 \text{ turns}} = 60VAC$$

The transformers in *Figure 40* have fewer turns in the secondary than in the primary. As a result, there is less voltage across the secondary than across the primary. A transformer in which the voltage across the secondary is less than the voltage across the primary is called a step-down transformer. The ratio of a 10-to-1 step-down transformer is written as 10:1.

A transformer that has fewer turns in the primary than in the secondary will produce a greater voltage across the secondary than the voltage applied to the primary. A transformer in which the voltage across the secondary is greater than the voltage applied to the primary is called a step-up transformer. The ratio of a 1-to-4 step-up transformer should be written 1:4. Notice in the two ratios that the value of the primary winding is always stated first.

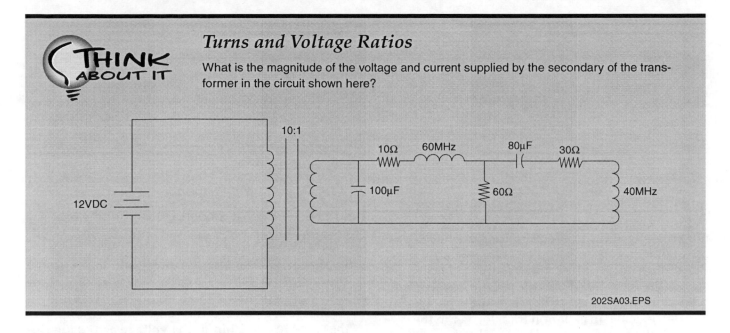

Turns and Voltage Ratios

What is the magnitude of the voltage and current supplied by the secondary of the transformer in the circuit shown here?

202SA03.EPS

11.4.0 Types of Transformers

Transformers are widely used to permit the use of trip coils and instruments of moderate current and voltage capacities and to measure the characteristics of high-voltage and high-current circuits. Since secondary voltage and current are directly related to primary voltage and current, measurements can be made under the low-voltage or low-current conditions of the secondary circuit and still determine primary characteristics. Tripping transformers and instrument transformers are examples of this use of transformers.

A transformer's primary or secondary coils can be tapped to permit multiple input and output voltages. *Figure 41* shows several tapped transformers. The center-tapped transformer is particularly important because it can be used in conjunction with other components to convert an AC input to a DC output.

11.4.1 Isolation Transformer

Isolation transformers are wound so that their primary and secondary voltages are equal. Their purpose is to electrically isolate a piece of electrical equipment from the power distribution system.

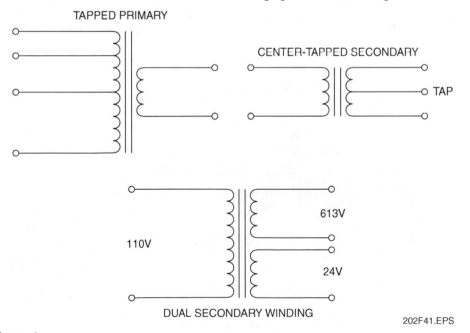

Figure 41 ◆ Tapped transformers.

Many pieces of electronic equipment use the metal chassis on which the components are mounted as part of the circuit (*Figure 42*). Personnel working with this equipment may accidentally come in contact with the chassis, completing the circuit to ground, and receive a shock, as shown in *Figure 42A*. If the resistances of their body and the ground path are low, the shock can be fatal. Placing an isolation transformer in the circuit, as shown in *Figure 42B*, breaks the ground current path that includes the worker. Current can no longer flow from the power supply through the chassis and worker to ground; however, the equipment is still supplied with the normal operating voltage and current.

11.4.2 Autotransformer

In a transformer, it is not necessary for the primary and secondary to be separate and distinct windings. *Figure 43* is a schematic diagram of what is known as an autotransformer. Note that a single coil of wire is tapped to produce what is electrically both a primary and a secondary winding.

The voltage across the secondary winding has the same relationship to the voltage across the primary that it would have if they were two distinct windings. The movable tap in the secondary is used to select a value of output voltage either higher or lower than E_P, within the range of the transformer. When the tap is at Point A, E_S is less than E_P; when the tap is at Point B, E_S is greater than E_P.

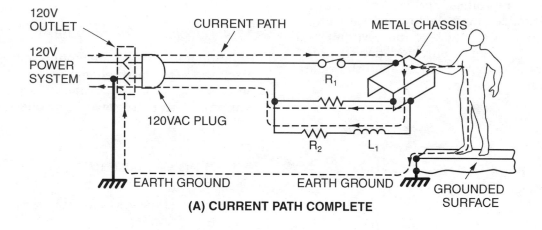

(A) CURRENT PATH COMPLETE

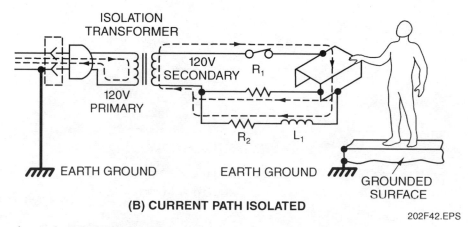

(B) CURRENT PATH ISOLATED

202F42.EPS

Figure 42 ◆ Importance of an isolation transformer.

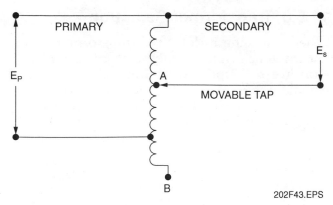

Figure 43 ◆ Autotransformer schematic diagram.

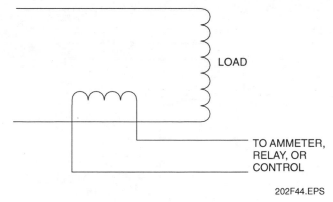

Figure 44 ◆ Current transformer schematic diagram.

Autotransformers rely on self-induction to induce their secondary voltage. The term autotransformer can be broken down into two words: auto, meaning self; and transformer, meaning to change potential. The autotransformer is made of one winding that acts as both a primary and a secondary winding. It may be used as either a step-up or step-down transformer. Some common uses of autotransformers are as variable AC voltage supplies and fluorescent light ballast transformers, and to reduce the line voltage for various types of low-voltage motor starters.

11.4.3 Current Transformer

A current transformer differs from other transformers in that the primary is a conductor to the load and the secondary is a coil wrapped around the wire to the load. Just as any ammeter is connected in line with a circuit, the current transformer is connected in series with the current to be measured. *Figure 44* is a diagram of a current transformer.

WARNING!
Do not open a current transformer under load. Although the use of a current transformer completely isolates the secondary and the related ammeter from the high-voltage lines, the secondary of a current transformer should never be left open circuited. To do so may result in dangerously high voltage being induced in the secondary.

Since current transformers are series transformers, the usual voltage and current relationships do not apply. Current transformers vary considerably in rated primary current, but are usually designed with ampere-turn ratios such that the secondary delivers five amperes at full primary load.

Current transformers are generally constructed with only a few turns or no turns in the primary. The voltage in the secondary is induced by the changing magnetic field that exists around a single conductor. The secondary is wound on a circular core, and the large conductor that makes up the primary passes through the hole in its center. Because the primary has few or no turns, the secondary must have many turns (providing a high turns ratio) in order to produce a usable voltage. The advantage of this is that you get an output off the secondary proportional to the current flowing through the primary, without an appreciable voltage drop across the primary. This is because the primary voltage equals the current times the impedance. The impedance is kept near zero by using no or very few primary turns. The disadvantage is that you cannot open the secondary circuit with the primary energized. To do so would cause the secondary current to drop rapidly to zero. This would cause the magnetic field generated by the secondary current to collapse rapidly. The rapid collapse of the secondary field through the many turns of the secondary winding would induce a dangerously high voltage in the secondary, creating an equipment and personnel hazard.

Because the output of current transformers is proportional to the current in the primary, they are most often used to power current-sensing meters and relays. This allows the instruments to respond to primary current without having to handle extreme magnitudes of current.

11.4.4 Potential Transformer

The primary of a potential transformer is connected across or in parallel with the voltage to be measured, just as a voltmeter is connected across a circuit. *Figure 45* shows the schematic diagram for a potential transformer.

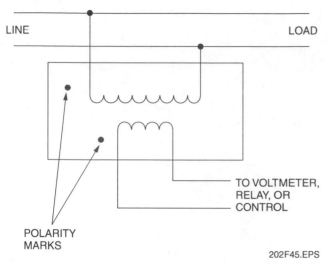

LINE

LOAD

TO VOLTMETER,
RELAY, OR
CONTROL

POLARITY
MARKS

202F45.EPS

Figure 45 ◆ Potential transformer.

Potential transformers are basically the same as other single-phase transformers. Although primary voltage ratings vary widely according to the specific application, secondary voltage ratings are usually 120V, a convenient voltage for meters and relays.

Because the output of potential transformers is proportional to the phase-to-phase voltage of the primary, they are often used to power voltage-sensing meters and relays. This allows the instruments to respond to primary voltage while having to handle only 120V. Also, potential transformers are essentially single-phase, step-down transformers. Therefore, power to operate low-voltage auxiliary equipment associated with high-voltage switchgear can be supplied off the high-voltage lines that the equipment serves via potential transformers.

11.5.0 Transformer Selection

When replacing a defective transformer that is known to be original equipment, try to replace it with an exact duplicate. If it cannot be determined whether the transformer is original equipment or not, make sure that the replacement transformer will safely and efficiency handle the primary voltage and current, and support the voltage and current demands of the load.

Depending on its use, additional load demand characteristics may have to be considered when selecting a replacement control transformer. Three important characteristics are total steady-state VA (sometimes referred to as sealed VA), total inrush VA, and inrush load power factor. These three factors usually apply to transformer applications associated with motor control circuitry and electromagnetic control devices such as relays.

The total steady-state VA requirement is the rating that provides sufficient volt-amperes to the load in order to hold the load in a continuous energized state.

Electromagnetic control devices such as magnetic motor starter or relay coils may take anywhere from 20 to 60 milliseconds to become totally energized. The coil may draw up to ten times the normal operating current during this time. This is referred to as the inrush VA and the transformer must be rated to handle these inrush current levels.

The true inrush load power factor cannot be determined without applying complex vector analysis to control transformer load components. Therefore, in determining the inrush load power

AC Power Provides Many Benefits, but Working with It Can Be Dangerous

Working with AC power can be dangerous unless proper safety methods and procedures are followed. The National Institute for Occupational Safety and Health (NIOSH) investigated 224 incidents of electrocutions that resulted in occupational fatalities. One hundred twenty-one of the victims were employed in the construction industry. Two hundred twenty-one of the incidents (99 percent) involved AC. Of the 221 AC electrocutions, 74 (33 percent) involved AC voltages less than 600V and 147 (66 percent) involved 600V or more. Forty of the lower-voltage electrocutions involved 120/240V.

Factors relating to the causes of these electrocutions included the lack of enforcement of existing employer policies including the use of personal protective equipment and the lack of supervisory intervention when existing policies were being violated. Of the 224 victims, 194 (80 percent) had some type of electrical safety training. Thirty-nine victims had no training at all. It is notable that 100 of the victims had been on the job less than one year.

Never assume that you are safe when working at lower voltages. All voltage levels must be considered potentially lethal. Also, safety training does no good if you don't put it into practice every day. Always put safety first.

factor, it is a safe assumption to apply a power factor of forty percent. This means that from the transformer's available inrush power, only 40 percent will be usable inrush power for the load. In other words, the transformer's total inrush VA rating must be approximately 160 percent inrush requirements of the loads if the total load consists of electromagnetic devices. To simplify calculations, most replacement transformers have their inrush VA rating calculated for a power factor of 40 percent.

If it is necessary to replace an isolation transformer, make sure to select a transformer designed for isolation. Its source and load impedance characteristics must match those of the original transformer. The replacement should also be designed to handle the signal levels and frequency range.

 NOTE
Although an isolation transformer may have a 1:1 turns ratio, it should not be reversed in the circuit because a change in operating performance could result.

Summary

The process by which current is produced electromagnetically is called induction. As the conductor moves across the magnetic field, it cuts the lines of force, and electrons within the conductor flow, creating an electromotive force (EMF).

Inductive and capacitance components offer impedance to current flow that is called reactance, rather than resistance. Inductive and capacitance reactances are out of phase with each other and with any DC resistive components in the circuit. Therefore, the total impedance to current flow must take the phase shift into account. In addition, reactive components do not consume power, thus the true power consumed by an AC circuit cannot be determined in the same way it is in a DC circuit.

Transformers are key components in AC circuits. They are used to provide circuit isolation and to increase (step up) or decrease (step down) AC voltage. The ratio of wire turns between the transformer primary and secondary determines the amount the voltage will be stepped up or down.

Power is the amount of energy consumed by a circuit. In a DC circuit, a decrease in voltage occurs as the conductor intersects the magnetic field at an angle less than 90 degrees. The greatest current is produced when the conductor intersects the magnetic field at right angles (perpendicular) to the flux lines.

Review Questions

1. An electric current always produces _____.
   a. mutual inductance
   b. a magnetic field
   c. capacitive reactance
   d. high voltage

2. The number of cycles that an alternating electric current undergoes per second is known as _____.
   a. amperage
   b. frequency
   c. voltage
   d. resistance

3. If the frequency of an AC current is 30 cycles per one-half second, the hertz rating of the circuit is _____ hertz.
   a. 15
   b. 30
   c. 60
   d. 90

4. The period of a waveform is the _____ of frequency.
   a. inverse
   b. square root
   c. sine
   d. square

5. What is the peak voltage in a 120VAC circuit?
   a. 117 volts
   b. 120 volts
   c. 150 volts
   d. 170 volts

6. When the current increases in an AC circuit, what role does inductance play?
   a. It increases the current.
   b. It plays no role at all.
   c. It causes the overcurrent protection to open.
   d. It reduces the current.

7. The inductive reactance of a 15-henry inductor in a 60Hz circuit is _____.
   a. 2,132Ω
   b. 4,712Ω
   c. 5,652Ω
   d. 6,937Ω

8. The total capacitance of a series circuit containing a 60µF capacitor and a 40µF capacitor is approximately _____.
   a. 24µF
   b. 32µF
   c. 40µF
   d. 62µF

9. The total capacitance of a 30-microfarad capacitor in parallel with a 20-microfarad capacitor is _____ microfarads.
   a. 20
   b. 30
   c. 50
   d. 600

10. The opposition to current flow offered by the capacitance of a circuit is known as _____.
    a. mutual inductance
    b. pure resistance
    c. inductive reactance
    d. capacitive reactance

11. Which of the following conditions exist in a circuit of pure resistance?
    a. The voltage and current are in phase.
    b. The voltage and current are 90 degrees out of phase.
    c. The voltage and current are 120 degrees out of phase.
    d. The voltage and current are 180 degrees out of phase.

12. Which of the following conditions exist in a circuit of pure inductance?
    a. The voltage and current are in phase.
    b. The voltage and current are 90 degrees out of phase.
    c. The voltage and current are 120 degrees out of phase.
    d. The voltage and current are 180 degrees out of phase.

13. The total opposition to current flow in an AC circuit is known as _____.
    a. resistance
    b. capacitive reactance
    c. inductive reactance
    d. impedance

14. The formula $\sqrt{R^2 + X^2}$ is used to solve for the _____.
    a. total impedance of an AC circuit
    b. frequency of a parallel RC circuit
    c. total current in an AC circuit
    d. power factor of an AC circuit

15. Which of the following best describes true power?
    a. The power consumed by the combination of the resistive and reactive components of a circuit
    b. The power consumed by the resistive components of a circuit
    c. The power consumed by the inductors and capacitors in a circuit
    d. The product of source voltage and current in an AC circuit

16. A power factor is not an angular measure, but a numerical ratio with a value between 0 and 1, equal to the _____ of the phase angle.
    a. sine
    b. tangent
    c. cosine
    d. cotangent

17. The two windings of a conventional transformer are known as the _____ windings.
    a. mutual and inductive
    b. high- and low-voltage
    c. primary and secondary
    d. step-up and step-down

18. A transformer has 300 turns on the primary, 100 turns on the secondary and 120VAC applied to the primary. The voltage across the secondary is _____.
    a. 33V
    b. 40V
    c. 90V
    d. 120V

19. A transformer with a primary of 200 turns and a secondary of 400 turns has a _____.
    a. step-up ratio of 4:2
    b. step-down ratio of 2:4
    c. step-up ratio of 1:2
    d. step-down ratio of 2:1

20. Inrush current is a factor in selecting transformers for _____ circuits.
    a. RC
    b. resistive
    c. capacitive
    d. inductive

Trade Terms Introduced in This Module

Capacitance: The storage of electricity in a capacitor; capacitance produces an opposition to voltage change. The unit of measurement for capacitance is the farad (F) or microfarad (µF).

Frequency: The number of cycles an alternating electric current, sound wave, or vibrating object undergoes per second.

Hertz (Hz): A unit of frequency; one hertz equals one cycle per second.

Impedance: The opposition to current flow in an AC circuit; impedance includes resistance (R), capacitive reactance (X_C), and inductive reactance (X_L). Impedance is measured in ohms.

Inductance: The creation of a voltage due to a time-varying current; also, the opposition to current change, causing current changes to lag behind voltage changes. The unit of measure for inductance is the henry (H).

Micro: Prefix designating one-millionth of a unit. For example, one microfarad is one-millionth of a farad.

Peak voltage: The peak value of a sinusoidally varying (cyclical) voltage or current is equal to the root-mean-square (rms) value multiplied by the square root of two (1.414). AC voltages are usually expressed as rms values; that is, 120 volts, 208 volts, 240 volts, 277 volts, 480 volts, etc., are all rms values. The peak voltage, however, differs. For example, the peak value of 120 volts (rms) is actually $120 \times 1.414 = 169.68$ volts.

Radian: An angle at the center of a circle, subtending (opposite to) an arc of the circle that is equal in length to the radius.

Reactance: The imaginary part of impedance. Also, the opposition to alternating current (AC) due to capacitance (X_C) and/or inductance (X_L).

Resonance: A condition reached in an electrical circuit when the inductive reactance neutralizes the capacitance reactance, leaving ohmic resistance as the only opposition to the flow of current.

Root-mean-square (rms): The square root of the average of the square of the function taken throughout the period. The rms value of a sinusoidally varying voltage or current is the effective value of the voltage or current.

Self-inductance: A magnetic field induced in the conductor carrying the current.

Additional Resources

This module is intended to be a thorough resource for task training. The following reference works are suggested for further study. These are optional materials for continued education rather than for task training.

Introduction to Electric Circuits, 2004. Richard C. Dorf, James A. Svoboda. Hoboken, NJ: John Wiley & Sons.

Principles of Electric Circuits: Conventional Current Version, 2002. Thomas L. Floyd. New York: Prentice Hall.

Figure Credits

Topaz Publications, Inc.   202F13C

NCCER makes every effort to keep these textbooks up-to-date and free of technical errors. We appreciate your help in this process. If you have an idea for improving this textbook, or if you find an error, a typographical mistake, or an inaccuracy in NCCER's Contren® textbooks, please write us, using this form or a photocopy. Be sure to include the exact module number, page number, a detailed description, and the correction, if applicable. Your input will be brought to the attention of the Technical Review Committee. Thank you for your assistance.

Instructors – If you found that additional materials were necessary in order to teach this module effectively, please let us know so that we may include them in the Equipment/Materials list in the Annotated Instructor's Guide.

Write: Product Development and Revision
National Center for Construction Education and Research
P.O. Box 141104, Gainesville, FL 32614-1104

Fax: 352-334-0932

E-mail: curriculum@nccer.org

Craft _____ Module Name _____

Copyright Date _____ Module Number _____ Page Number(s) _____

Description _____

(Optional) Correction _____

(Optional) Your Name and Address _____

Semiconductors and Integrated Circuits

COURSE MAP

This course map shows all of the modules in the second level of the *Electronic Systems Technician* curriculum. The suggested training order begins at the bottom and proceeds up. Skill levels increase as you advance on the course map. The local Training Program Sponsor may adjust the training order.

ELECTRONIC SYSTEMS TECHNICIAN LEVEL TWO

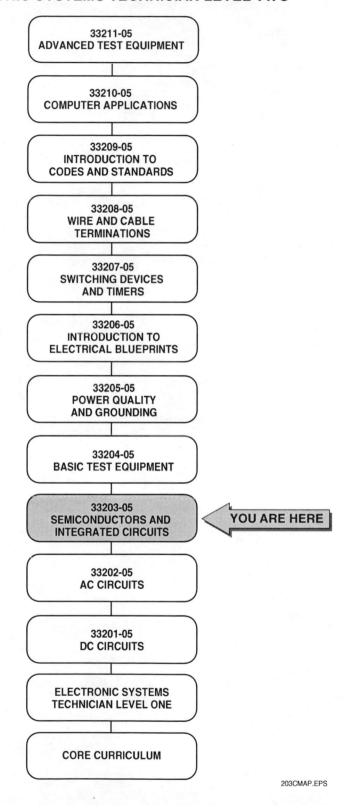

33211-05
ADVANCED TEST EQUIPMENT

33210-05
COMPUTER APPLICATIONS

33209-05
INTRODUCTION TO
CODES AND STANDARDS

33208-05
WIRE AND CABLE
TERMINATIONS

33207-05
SWITCHING DEVICES
AND TIMERS

33206-05
INTRODUCTION TO
ELECTRICAL BLUEPRINTS

33205-05
POWER QUALITY
AND GROUNDING

33204-05
BASIC TEST EQUIPMENT

33203-05
SEMICONDUCTORS AND
INTEGRATED CIRCUITS   YOU ARE HERE

33202-05
AC CIRCUITS

33201-05
DC CIRCUITS

ELECTRONIC SYSTEMS
TECHNICIAN LEVEL ONE

CORE CURRICULUM

203CMAP.EPS

Figures

Tables

Semiconductors and Integrated Circuits

Objectives

When you have completed this module, you will be able to do the following:

1. Identify electronic system components.
2. Describe the electrical characteristics of solid-state devices.
3. Describe the basic materials that make up solid-state devices.
4. Describe and identify the various types of transistors and explain how they operate.
5. Describe and connect diodes.
6. Describe and connect light-emitting diodes (LEDs).
7. Describe and connect silicon-controlled rectifiers (SCRs).
8. Identify the leads of various solid-state devices.
9. Describe integrated circuits.
10. Identify a microprocessor and applicable pin numbers.
11. Explain the purpose of logic gates.
12. Build a simple circuit using an LED and a diode; a simple bridge rectifier circuit; and a holding circuit using an SCR, LED, and a pushbutton.

Prerequisites

Before you begin this module, it is recommended that you successfully complete *Core Curriculum*; *Electronic Systems Technician Level One*; and *Electronic Systems Technician Level Two*, Modules 33201-05 and 33202-05.

Required Trainee Materials

1. Pencil and paper
2. Appropriate personal protective equipment
3. Scientific calculator

1.0.0 ◆ INTRODUCTION

Solid-state devices operate many controls and related components used in electronic systems. In fact, many types of motor controls that were once operated either magnetically or mechanically have been replaced with solid-state electronic devices. Modern fire alarm and intrusion detection systems also use solid-state electronic devices and circuits. Therefore, technicians in every branch of the industry should have a basic knowledge of electronics in order to install, maintain, and troubleshoot electronic control devices properly.

Electronics is a science that deals with the behavior and effect of electron flow in specific substances such as **semiconductors**. Electronics can be distinguished from electricity by thinking in terms of the voltages and currents used.

AC voltages are used in many electrical circuits. In electronic circuits, low-level DC voltages such as 5V, 10V, 15V, and 24V are used. Current is measured in milliamps (mA) and even microamps (μA).

2.0.0 ◆ SEMICONDUCTOR FUNDAMENTALS

Semiconductors are the basis for what is known as solid-state electronics. Solid-state electronics are, in turn, the basis for all modern microminiature electronics such as the tiny integrated circuit and microprocessor chips used in computers and controllers.

The ability to control the amount of conductivity in semiconductors makes them ideal for use in integrated circuits. In order to understand how semiconductors work, it is necessary to review the principles of conductors and insulators.

2.1.0 Conductors

Conductors are so called because they readily carry electrical current. Good electrical conductors are usually also good heat conductors.

In each atom, there is a specific number of electrons that can be contained in each orbit or shell. The outer shell of an atom is the valence shell and the electrons contained in the valence shell are known as valence electrons. If the valence shell is not full, these electrons can be easily knocked out of their orbits, becoming free electrons.

Conductors are materials that have only one or two valence electrons in their atoms, as shown in *Figure 1*. An atom that has only one valence electron makes the best conductor because the electron is loosely held in orbit and is easily released to create current flow.

Gold and silver are excellent conductors, but they are too expensive to use on a large scale. However, in special applications requiring high conductivity,

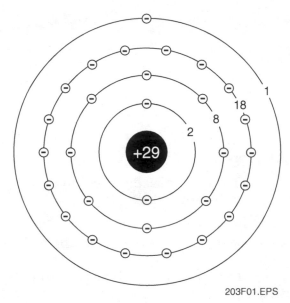

203F01.EPS

Figure 1 ◆ Atom of copper (a conductor).

Electron Orbits or Shells

The depiction of fixed orbits (shells) for electrons is a classical and convenient method used for notation only. Modern quantum mechanics indicates that electrons occupy specific areas of space around an atom, depending on their energy level, rather than fixed orbits. The specific areas are called orbitals, and the electrons in a particular orbital space may be passing around the nucleus in any number of random circular or elliptical paths within that space. Each electron at a particular energy level is defined in a wave function called an atomic orbital. The wave function is obtained by solving a quantum mechanics equation known as the Schrödinger equation. As shown here, when the areas of all the theoretical orbitals for all atoms are combined, they appear as a cloud around the nucleus of an atom with the cloud being denser toward the center.

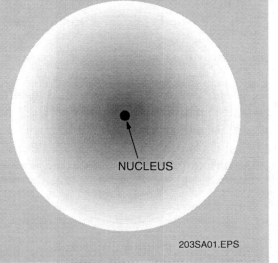

NUCLEUS

203SA01.EPS

The Number of Electrons in Each Classical Shell of an Atom

The first (innermost) shell of an atom has a maximum of two electrons, and the second shell has a maximum of eight electrons. Unless it is the outermost shell, the third shell can have up to 18 electrons and the fourth and fifth shells can have up to 32 electrons. The maximum number of electrons in the outermost shell cannot exceed eight.

contacts may be plated with gold or silver. You would be most likely to find such conductors in precision devices where small currents are common, and a high degree of accuracy is essential.

Copper is the most widely used conductor because it has excellent conductivity and is much less expensive than precious metals such as gold and silver. Copper is used as the conductor in most types of wire and provides the current path on printed circuit boards. Aluminum is also used as a conductor, but it is not as good as copper.

2.2.0 Insulators

As you already know, insulators are materials that resist (and sometimes totally prevent) the passage of electrical current. Rubber, glass, and some plastics are common insulators. The atoms of insulating materials are characterized by having more than four valence electrons in their atomic structures. *Figure 2* shows the structure of an insulator atom. Note that it has eight valence electrons; this is the maximum number of electrons for the outer shell of an atom. Therefore, this atom has no free electrons and will not easily pass electric current.

2.3.0 Semiconductors

Semiconductors are materials that are neither good conductors nor good insulators. The materials used as semiconductors, such as germanium and silicon, have more free electrons than an insulator, but less than a conductor. Silicon (*Figure 3*) is more commonly used because it withstands heat better.

The factor that makes semiconductors valuable in electronic circuits is that their conductivity can be readily controlled. Semiconductors can be made to have positive or negative characteristics by adding certain impurities in a process known as doping.

When a substance with five valence electrons, such as indium or gallium, is added to the semiconductor material, the semiconductor material will no longer be electrically neutral. Instead, it will take on a positive charge. This is known as a **P-type material**.

When substances like arsenic or antimony, which have three valence electrons, are added to the semiconductor material, the material takes on a negative charge and is known as an **N-type material**.

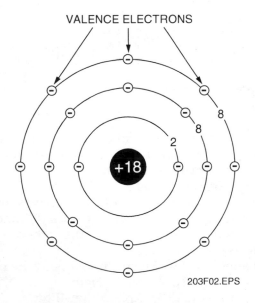

Figure 2 ◆ Atom of argon (an insulator).

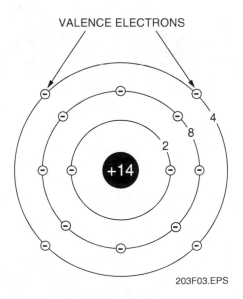

Figure 3 ◆ Atom of silicon (a semiconductor).

The Periodic Table

Some versions of the periodic table of elements, such as the portion of one table shown here, contain a vertical list of numbers beside each element. The list represents the numbers of electrons in each of the classical shells of the atom for the element. The total of the electrons for an element equals the atomic number above the element symbol. The period number for each row corresponds to the number of classical shells that exist for the elements in the row. The group numbers across the top with an A or B are an older method of notation for each column of elements. The group number A elements are the representative elements that have a corresponding number of electrons in the outer shell with a corresponding valence even though an inner shell may be incomplete (except He). These representative elements become more stable from left to right as the number of electrons in the outer shell increases. The group number B elements are the transition elements. These elements sometimes have incomplete inner shells as well as incomplete outer shells. For these elements, the common valence is generally represented by the group number. Be aware that in quantum mechanics theory, the classical main shells shown in the table are made up of various energy-level sub-shells and the electrons that are allocated to the main shell are assigned to the various sub-shells.

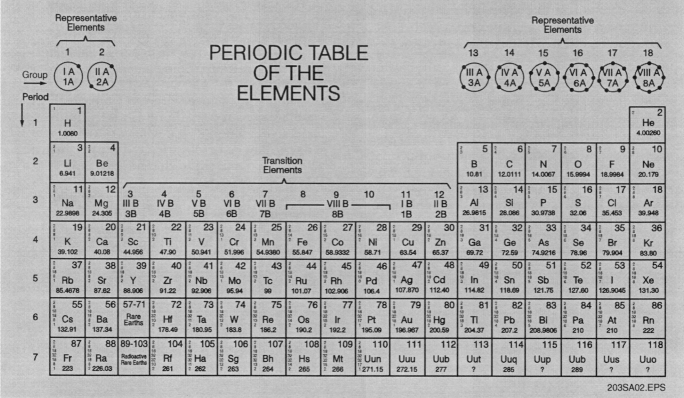

203SA02.EPS

3.0.0 ◆ DIODES

A **diode** is made by joining a piece of P-type material with a piece of N-type material, as shown in *Figure 4*. The contacting surface is called the PN junction.

Diodes allow current to flow in one direction, but not in the other. This unidirectional current capability is the distinguishing feature of the diode. The activity occurring at the PN junction of the materials is responsible for the unidirectional characteristic of the diode.

Current does not normally flow across a PN junction. However, if a voltage of the correct polarity is applied, current will flow. This is known as **forward bias** and is shown in *Figure 5*. Note that the positive side of the voltage source is connected to the P material (the anode) and the negative side to the N material (the cathode). This causes conventional current flow in the direction of the anode arrow. Electron flow is in the opposite direction. Also shown in *Figure 5* is a PN junction with **reverse bias**. Note that the polarity of the applied voltage is opposite that of the forward

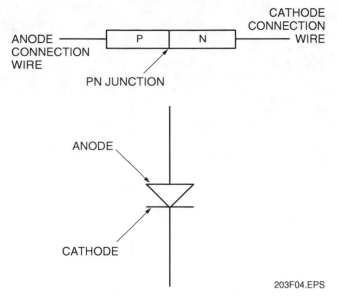

Figure 4 ◆ Material structure of a diode.

biased junction. No current or electron flow will occur in the reverse biased arrangement unless the reverse bias voltage exceeds the breakover voltage of the junction. This approach is used to create solid-state switching devices. Discrete component diodes are normally used in **rectifier** circuits and analog signal processing circuits.

As shown in *Figure 5*, a diode conducts current only when the voltage at its anode is positive with respect to the voltage at its cathode (forward bias). When the voltage at the anode is negative with respect to the cathode (reverse bias), current will not flow unless the voltage is so high that it overwhelms the diode. Most circuits using diodes are designed so that the diode will not conduct current unless the anode is positive with respect to the cathode.

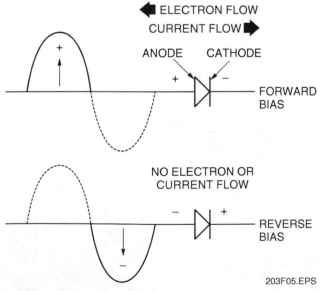

Figure 5 ◆ Forward and reverse bias.

3.1.0 Rectifiers

Modern systems rely heavily on electronic controls, which use low-level DC voltages. Some systems also use special controls powered by DC motors when very precise control is required. The electricity furnished by the power company is AC; it must be converted to DC to be suitable for most electronic circuits. The process of converting AC to DC is known as rectification and is accomplished using rectifier diodes.

In a half-wave rectifier (*Figure 6*), the single diode conducts current only when the AC applied to its anode is on its positive half-cycle. The result is a pulsating DC voltage.

Figure 7 shows a full-wave rectifier with a special, center-tapped transformer. In this circuit, one of the diodes conducts on each half-cycle of the AC input, producing a smoother pulsating DC voltage. Filter capacitors can be used to eliminate most of the ripple.

A bridge rectifier (*Figure 8*) contains four diodes, two of which conduct on each half-cycle. The bridge rectifier provides a smoother DC output and is the type most commonly used in electronic circuits. An advantage of the bridge rectifier is that it does not need a center-tapped transformer. A filter and voltage regulator added to the output of the rectifier provide the precise, stable DC voltage needed for electronic devices.

In a three-phase power system, a three-phase rectifier is used (*Figure 9*). A three-phase rectifier contains six diodes to produce high-efficiency DC power.

3.2.0 Diode Identification

Most manufacturers' catalogs featuring solid-state devices have hundreds of semiconductor diodes and their specifications listed. *Table 1* shows the specifications for two diodes. Note that the diodes are designated 1N34A and 1N58A. Manufacturers have agreed upon certain standard designations for diodes having the same characteristics—the same as for wire sizes and types of insulation.

In *Table 1*, the peak inverse voltage (PIV) is the reverse bias at which avalanche breakover occurs. The ambient temperature rating is the range of temperatures over which the diode will operate and still maintain its basic characteristics. Forward current values are given for both the average current (that current at which the diode is usually operated) and the peak current (that current which, if exceeded, will damage the diode). The only difference between these two diodes is in the peak inverse voltage. Therefore, the 1N34A could be substituted for the 1N58A in applications involving signals of less than 60V peak-to-peak.

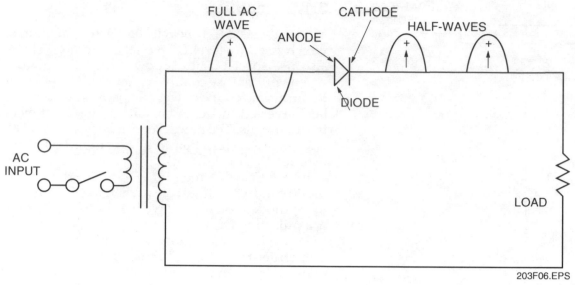

Figure 6 ◆ Half-wave rectifier.

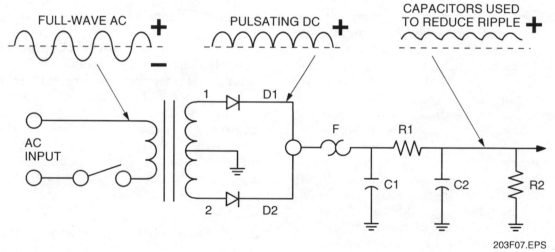

Figure 7 ◆ Full-wave rectifier.

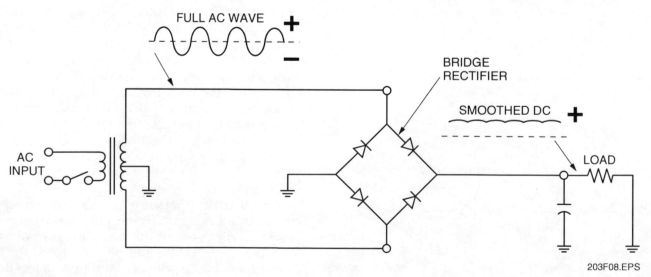

Figure 8 ◆ Bridge rectifier.

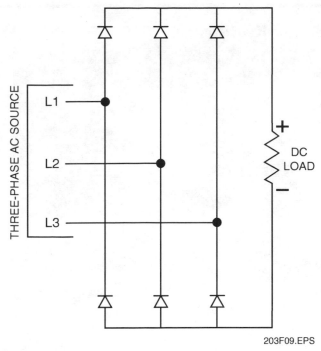

THREE-PHASE AC SOURCE

L1

L2

L3

+

DC LOAD

−

203F09.EPS

Figure 9 ◆ Three-phase rectifier.

In general, there are two basic types of diodes: the silicon diode and the germanium diode. In most cases, silicon diodes have higher PIV and current ratings and wider temperature ranges than germanium diodes. Consequently, the silicon diode will be the type most often encountered in motor and HVAC solid-state controls or power supplies.

PIV ratings for silicon can be in the neighborhood of 1,000V, whereas the maximum value for germanium is closer to 400V. Silicon can be used for applications where the temperature may rise to about 200°C (400°F), whereas germanium has a much lower maximum rating (100°C). The disadvantage of silicon, however, is the higher forward bias voltage required to reach operational conditions.

Diodes are used for many functions in solid-state devices. The most common use of silicon diodes is in constructing rectifiers. Germanium diodes are used mainly in signal and control circuits.

There are several ways in which diodes are marked to indicate the cathode and anode. See *Figure 10*. Note that in some cases diodes are marked with the schematic symbol, or there may be a band at one end to indicate the cathode. Other types of diodes use the shape of the diode housing to indicate the cathode end; for example, the cathode end is either beveled or enlarged to ensure proper identification. When in doubt, the polarity of a diode may be checked with an ohmmeter, as shown in *Figure 11*. A forward bias will show a low resistance; a reverse bias will show a high resistance.

CAUTION

The battery voltage of certain ohmmeters may exceed the PIV of the diode being tested. This may result in destruction of the diode.

Table 1 Diode Characteristics

Type	Peak Inverse Voltage (PIV)	Ambient Temperature Range (°C)	Forward Peak (mA)	Current Average (mA)	Capacitance (µF)
1N34A	60	−50 to +75	150	50	150
1N58A	100	−50 to +75	150	50	150

INSIDE TRACK

Silicon Power Rectifier Diodes

Silicon power rectifier diodes are used in high-current DC applications including un-interruptible power supplies (UPS), welders, and battery chargers such as those used for electric forklifts. The stud-mounted silicon power rectifier diode shown here is rated for 300A and a PIV of 600V. The cathode is the threaded section, and it is screwed into or secured with a nut to an insulated heat sink that is at the DC potential. The flexible anode lead is connected to the AC power source. These diodes are also made with the anode as the threaded portion and the cathode as the flexible lead.

203SA03.EPS

Typical Low-Voltage Bridge Rectifiers

The low-voltage bridge rectifiers shown here are rated for 4A and 25A with a PIV of 50V. The metallic heat sink mating surface of the 25A rectifier must be coated with a heat sink compound and then bolted to a grounded heat sink. The heat sink compound helps to transmit heat rapidly from the rectifier to the heat sink. In these rectifiers, the cases and heat sink are insulated from the AC and DC voltages.

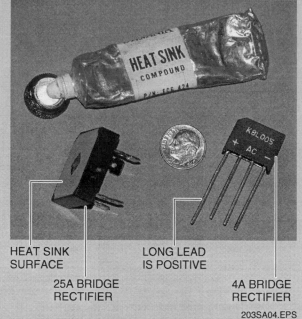

HEAT SINK SURFACE

25A BRIDGE RECTIFIER

LONG LEAD IS POSITIVE

4A BRIDGE RECTIFIER

203SA04.EPS

Axial-Lead Silicon Diodes

One of the most common diodes used is the tubular, axial-lead silicon diode. It is used in many low-voltage rectifiers for electronic circuits. The diode shown here is rated at 1A with a PIV of 1,400V.

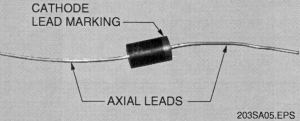

CATHODE LEAD MARKING

AXIAL LEADS

203SA05.EPS

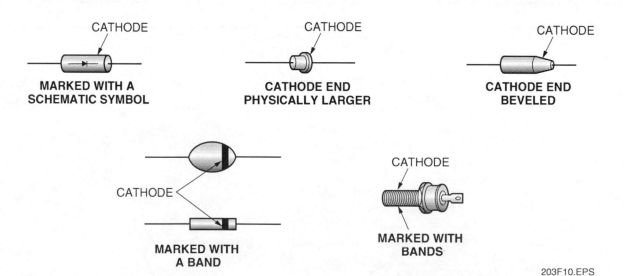

CATHODE

MARKED WITH A SCHEMATIC SYMBOL

CATHODE

CATHODE END PHYSICALLY LARGER

CATHODE

CATHODE END BEVELED

CATHODE

MARKED WITH A BAND

CATHODE

MARKED WITH BANDS

203F10.EPS

Figure 10 ◆ Methods used to identify diodes.

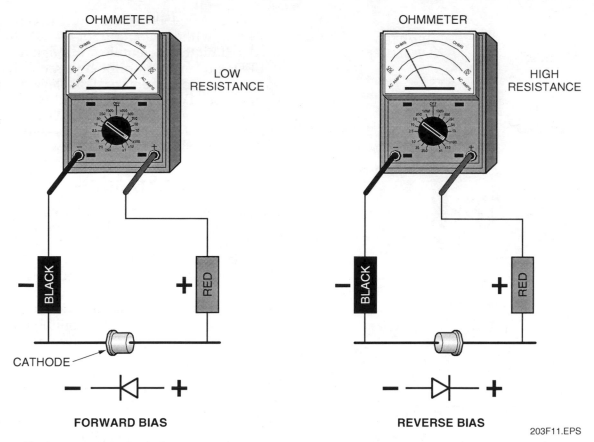

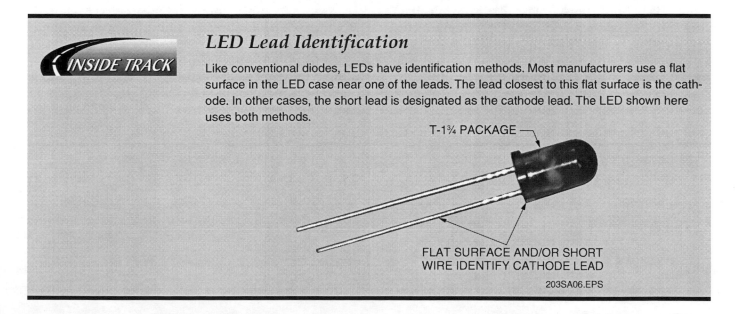

Figure 11 ◆ Testing a diode with an ohmmeter.

4.0.0 ◆ LIGHT-EMITTING DIODES

A light-emitting diode (LED) is, as the name implies, a diode that will give off visible light when it is energized. All energized, forward-biased LEDs give off some energy in the form of photons. In some types of diodes, the number of photons of light energy emitted is sufficient to create a visible light source.

The process of giving off light by applying an electrical source of energy is called electroluminescence. *Figure 12* shows that the conducting surface connected to the P-type material is much smaller to permit the emergence of the maximum number of photons of light energy in an LED. Note also in *Figure 13* that the symbol for an LED is similar to a conventional diode except that an arrow is pointing away from the diode.

LED Lead Identification

Like conventional diodes, LEDs have identification methods. Most manufacturers use a flat surface in the LED case near one of the leads. The lead closest to this flat surface is the cathode. In other cases, the short lead is designated as the cathode lead. The LED shown here uses both methods.

T-1¾ PACKAGE

FLAT SURFACE AND/OR SHORT WIRE IDENTIFY CATHODE LEAD

203SA06.EPS

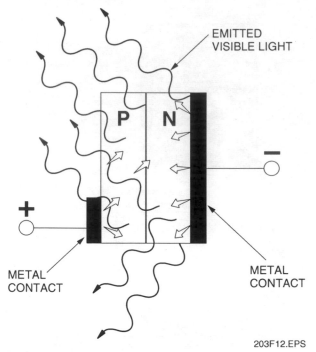

Figure 12 ◆ Process of electroluminescence in an LED.

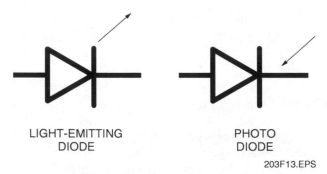

LIGHT-EMITTING DIODE

PHOTO DIODE

203F13.EPS

Figure 13 ◆ Schematic symbols for an LED and photo diode.

When used in a circuit, an LED is generally operated at about 20mA or less. For example, if an LED is to be connected to a 9V DC circuit, a current-limiting resistor must be connected in series with the LED. Ohm's law may be used to calculate the required resistance as follows:

$$R = \frac{E}{I}$$

Where:

R = resistance (ohms)

E = voltage or emf (volts)

I = current (amperes)

$$R = \frac{9VDC}{0.020A}$$

R = 450 ohms

Therefore, a 450Ω resistor or the closest standard size without going under 450Ω should be used to limit the current flow through the LED.

LEDs are used as pilot lights on electronic equipment and as numerical displays. Many programmable HVAC controls use LEDs to indicate when a process is in operation. LEDs are also used in the opto-isolation circuit of solid-state relays for both motor controls and HVAC control systems.

5.0.0 ◆ PHOTO DIODES

A solid-state device activated by light is a photo diode. The schematic symbol for the photo diode is exactly like a standard LED except that the arrow is reversed, as shown in *Figure 13*. A photo diode must have light in order to operate. It acts similarly to a conventional switch; that is, light turns the circuit on, and the absence of light opens the circuit. Photo diodes are commonly used to provide automatic control of outdoor lighting.

6.0.0 ◆ TRANSISTORS

A transistor is made by joining three layers of semiconductor material, as shown in *Figure 14*. Regardless of the type of solid-state device being used, it is made by the joining of P-type and N-type materials to form either a NPN or PNP transistor. Each type has two PN junctions.

Control voltages are applied to the center layer, which is known as the base. These voltages control when and how much the transistor conducts. You will not encounter many transistors as discrete components. In today's self-contained electronic integrated circuit devices and microprocessors, there may be thousands of transistors. They are so tiny that they can only be seen with a high-powered microscope.

Years ago, when computers were first invented, they required a roomful of electronic equipment. With the evolution of microminiature circuits, it became possible to perform the same work with a single, tiny microprocessor chip containing thousands of microscopic electronic devices such as amplifiers and switches.

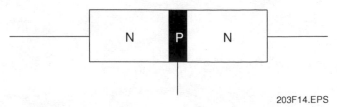

203F14.EPS

Figure 14 ◆ Material arrangement in a NPN transistor.

Transistor Electron Flow and Current Flow

In a transistor, electrons always flow against the direction of the arrow, while conventional current flow is in the direction of the arrow.

6.1.0 NPN Transistors

By sandwiching a very thin piece of P-type germanium between two slices of N-type germanium, a NPN transistor is formed. A transistor made in this way is called a junction transistor. The symbol for this type of transistor showing the three elements (emitter, base, and collector) is illustrated in *Figure 15*.

The current flow is in the direction of the arrow. Since the arrow points away from the base, current flows from the base to the emitter or from the P-type material to the N-type material.

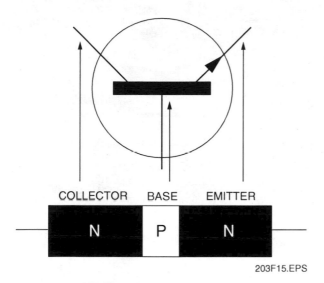

Figure 15 ◆ NPN transistor characteristics and schematic symbol.

6.2.0 PNP Transistors

A PNP transistor is formed by placing N-type germanium between two slices of P-type germanium.

The schematic symbol for the PNP transistor is almost identical to that of the NPN transistor. The only difference is the direction of the emitter arrow. In the NPN transistor, it points away from the base; in the PNP transistor, it points toward the base.

Current flow in a PNP transistor is from the P-type germanium to the N-type germanium, and since the arrow in *Figure 16* points toward the base, the current flow is in the direction of the arrow—from emitter to base. This is the reverse of the NPN transistor discussed previously.

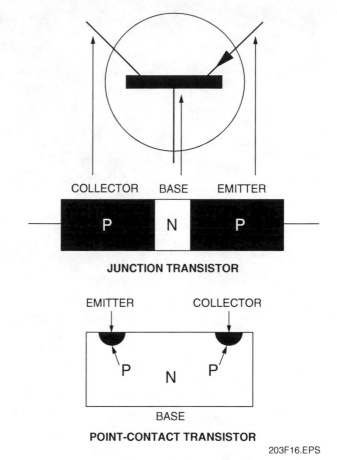

Figure 16 ◆ PNP transistor characteristics and schematic symbol.

Note also the point-contact transistor shown in *Figure 16*. Junction and point-contact transistors are almost identical in operation. The main difference is in the method of assembly.

6.3.0 Identifying Transistor Leads

Transistors are manufactured in a variety of configurations. Those with studs and heat sinks, as shown in *Figures 17A* and *17B*, are high-power devices. Those with a small can (top hat) or plastic body are low- to medium-power devices (*Figure 17C*).

Whenever possible, transistor casings will have some marking to indicate which leads are connected to the emitter, collector, or base of the transistor. A few of the methods commonly used are indicated in *Figure 18*.

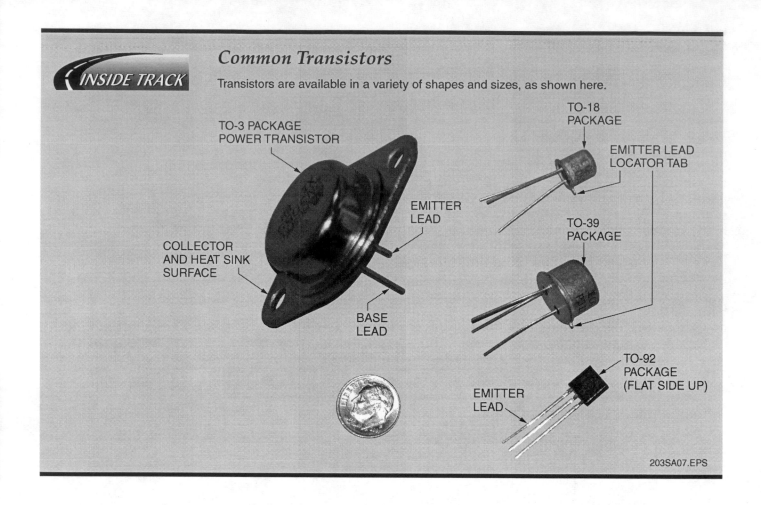

Common Transistors

Transistors are available in a variety of shapes and sizes, as shown here.

TO-3 PACKAGE POWER TRANSISTOR

EMITTER LEAD

COLLECTOR AND HEAT SINK SURFACE

BASE LEAD

TO-18 PACKAGE

EMITTER LEAD LOCATOR TAB

TO-39 PACKAGE

TO-92 PACKAGE (FLAT SIDE UP)

EMITTER LEAD

203SA07.EPS

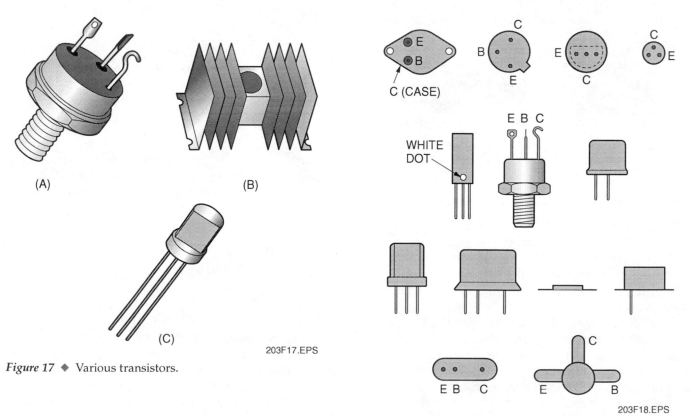

(A)

(B)

(C)

Figure 17 ◆ Various transistors.

203F17.EPS

C

E

B

C (CASE)

B

C

E

E

C

C

E

WHITE DOT

E B C

E B   C

E B   C

C

E   B

203F18.EPS

Figure 18 ◆ Lead identification of transistors.

6.4.0 Field-Effect Transistors

Field-effect transistors (FETs) control the flow of current with an electric field. There are two basic types of FETs: the **junction field-effect transistor (JFET)** and the metal-oxide semiconductor field-effect transistor (MOSFET). A MOSFET is also referred to as an insulated gate field-effect transistor (IGFET).

There are two types of JFETs: an N-channel and a P-channel. *Figure 19* shows the schematic symbols for these two types and also denotes the terms for each JFET lead.

The difference between these two devices is in the polarity of the voltage to which the transistor is connected. In an N-channel device, the drain is connected to the more positive voltage, and the source and gate are connected to a more negative voltage. A P-channel device has the drain connected to a more negative voltage, and the source and gate are connected to a more positive voltage.

FET devices are used in applications requiring a high-impedance (non-circuit-loading) input and/or lower power consumption. MOSFETs are primarily used in integrated circuit logic devices because of their low power consumption and reasonably good switching speeds.

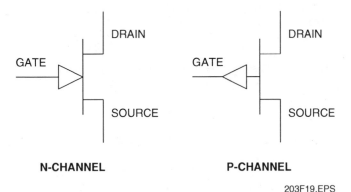

N-CHANNEL   **P-CHANNEL**

203F19.EPS

Figure 19 ◆ Junction field-effect transistor symbols.

7.0.0 ◆ SILICON-CONTROLLED RECTIFIERS

Silicon-controlled rectifiers (SCRs), along with **diacs** and **triacs**, belong to a class of semiconductors known as thyristors. One characteristic of SCRs is that they act as an open circuit until a triggering current is applied to their gate. Once that happens, the SCR acts as a low-resistance current path from anode to cathode. It will continue to conduct, even if the gate signal is removed, until either the current is reduced below a certain level or the SCR is turned off. These devices are used for a variety of purposes, including the following:

- AC power controllers
- Emergency lighting circuits
- Lamp dimmers
- Motor speed controls
- Ignition systems

The SCR is similar to a diode except that it has three terminals. Like a common diode, current will only flow through the SCR in one direction. However, in addition to needing the correct voltage polarity at the anode and cathode, the SCR also requires a gate voltage of the same polarity as the voltage applied to the anode. Once fired, the SCR will remain on until the cathode-to-anode current falls below a value known as the holding current. Once the SCR is off, another positive gate voltage must be applied before it will start conducting again.

The SCR is made from four adjoining layers of semiconductor material in a PNPN arrangement, as shown in *Figure 20*. The SCR symbol is also shown. Note that it is the same as the diode symbol except for the addition of a gating lead.

There is also a light-activated version of the SCR known as an LASCR. Its symbol is the same as that of the regular SCR, with the addition of two diagonal arrows representing light, similar to the photo diode previously covered.

The ability of an SCR to turn on at different points in the conducting cycle can be used to vary the amount of power delivered to a load. This type of variable control is called phase control. With such control, the speed of an electric motor, the brilliance of a lamp, or the output of an electric resistance heating unit can be controlled.

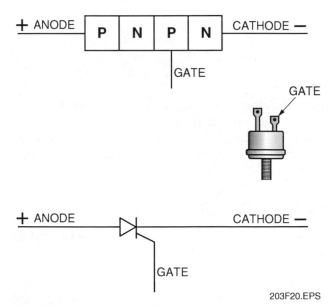

203F20.EPS

Figure 20 ◆ SCR characteristics and symbol.

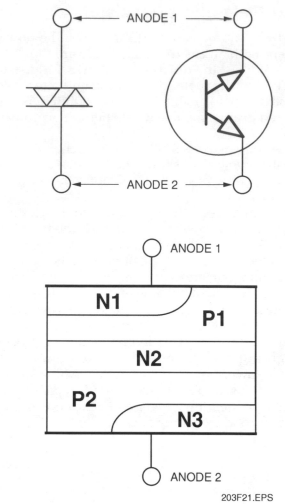

Typical SCR Package

SCRs are available in a wide variety of capacities and packages, as shown here.

INSIDE TRACK

203SA08.EPS

Figure 21 ◆ Basic diac construction and symbols.

203F21.EPS

8.0.0 ◆ DIACS

A diac can be thought of as an AC switch. Because it is bidirectional, current will flow through it on either half of the AC waveform. One major use of the diac is as a control for a triac. Although the diac is not gated, it will not conduct until the applied voltage exceeds its breakover voltage.

The basic construction and symbols for a diac are shown in *Figure 21*. Note that the diac has two symbols, either of which may be found on schematic diagrams.

9.0.0 ◆ TRIACS

The triac can be viewed as two SCRs turned in opposite directions with a common gate (*Figure 22*). It could also be viewed as a diac with a gate terminal added. One important distinction, however, is that the voltage applied across the triac does not have to exceed a breakover voltage in order for conduction to begin.

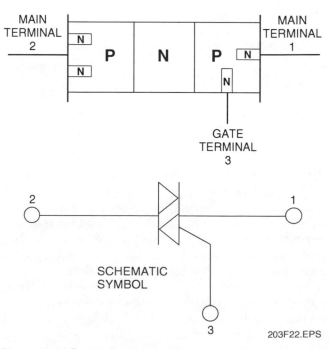

Figure 22 ◆ Basic triac construction and symbol.

203F22.EPS

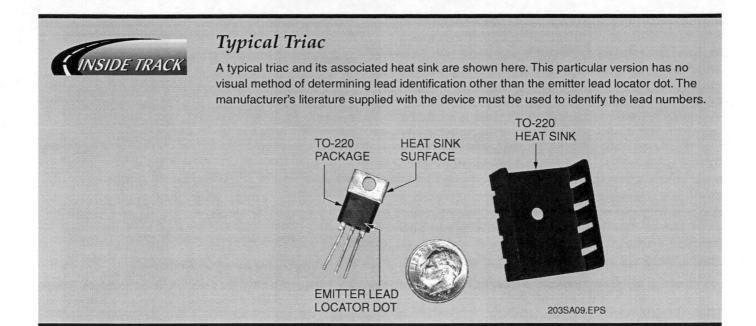
Like SCRs, triacs are used in phase control applications to control the average power applied to loads. Examples include light dimmers and photocell light switches.

10.0.0 ◆ PRINTED CIRCUIT BOARDS

Most of the electronic circuits you will encounter will be mounted on printed circuit (PC) boards. In many systems, all the control circuits—relays, capacitors, diodes, integrated circuits (ICs), etc.—are located on a single PC board (see *Figure 23*). The components are mounted on the top of the board; their electrical leads are soldered to terminal points on the board. There is very little, if any, wiring on the PC board. Instead, a copper foil is bonded to the bottom of the board. The desired circuit is imprinted on the foil by a machine, and the copper is then chemically etched away from the unprinted areas. The printed copper acts as the conductor between the components on the circuit. Instead of a wiring harness and plug to connect the circuit to the outside, an edge connector is often built into the board. The edge connector then is plugged into a connector mounted on the hardware.

Some electronic printed circuits are packaged in sealed modules. This method is common with electronic control devices that perform a specific function and can be used in a number of different systems.

A very important feature that distinguishes packaged electronic controls from circuits built of discrete (separate) components is that the electronic circuit is treated as a black box; that is, if there is a control circuit failure, the entire board or module is replaced. This is because the components of

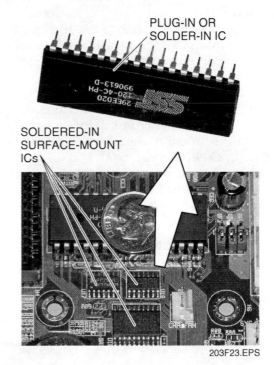

PLUG-IN OR SOLDER-IN IC

SOLDERED-IN SURFACE-MOUNT ICs

203F23.EPS

Figure 23 ◆ Printed circuit board.

modern PCs are very difficult to troubleshoot or replace. In conventional circuits, on the other hand, you have to analyze the circuit and isolate the fault to the failed component; a bad relay, for example.

When an electronically controlled system is not working, it is tempting to just replace the control module or PC board without checking anything. This will result in one of three outcomes, only one of which is desirable. The one good outcome is that it might fix the problem. A more likely outcome is that it won't. Electronic printed circuit boards are very reliable, and they have no moving parts, so it is not that common for them to fail. The worst possible outcome is that something external to the circuit board caused it to fail, and will also cause its replacement to fail. It can be very embarrassing to explain to a customer why you charged them for a repair that didn't work.

Before replacing an electronic printed circuit board, the troubleshooter must verify that the circuit has actually failed, and determine if an outside source caused the failure. To do this, you must first verify that the printed circuit board or module is receiving the necessary supply voltages and control signals, and that they are at the proper levels. Once that is done, the outputs need to be verified. If the device is receiving the required inputs, but fails to produce the expected outputs, it can be assumed that the device has failed.

10.1.0 Integrated Circuits

Most electronic printed circuit boards make use of integrated circuits that may be soldered to the board or plugged into sockets on the board. An integrated circuit (IC) (*Figure 24*) is a hermetically sealed case containing a tiny wafer of semiconductor material. The surface of the wafer contains microminiature electronic circuits designed to perform a specific function or functions. To get a perspective on what microminiature means, think about a multi-function digital wristwatch. All the complex timekeeping, calendar, and display functions are contained on a single integrated circuit chip that you might have trouble finding if you looked inside the watch.

10.2.0 Microprocessors

Microminiaturization enables tiny devices smaller than the tip of your little finger to perform work that, in the early days of computers, used to take a

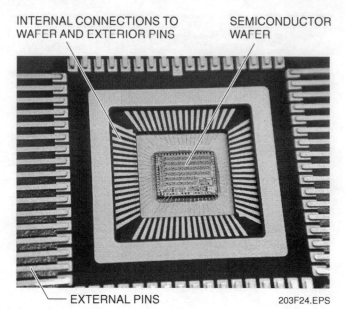

INTERNAL CONNECTIONS TO WAFER AND EXTERIOR PINS   SEMICONDUCTOR WAFER

EXTERNAL PINS   203F24.EPS

Figure 24 ◆ Interior of an integrated circuit (IC).

roomful of electronic equipment to do. The semiconductor makes microminiaturization possible.

Semiconductors are materials in which the capacity to conduct electricity can be controlled by varying the voltage applied. In this case, we are talking about low-level DC voltages in the range of 5V to 15V. Heat, light, and pressure are also used to control current flow in semiconductors. Silicon and germanium are the two most widely used semiconductor materials.

Some integrated circuits can be programmed to perform complex tasks such as decision-making and mathematical calculations. These are known as microprocessors (*Figure 25*). They are the brains of the personal computer and are used to perform logical and analytical functions in many computerized systems.

In microprocessor-controlled systems, integrated circuits, such as random-access memory (RAM),

Integrated Circuit Uses and Classifications

Integrated circuits are used in a variety of devices that include microprocessor controllers, audio amplifiers, and video equipment. Integrated circuits are sometimes classified by the number of transistors or other electronic components contained within them:

- *Small-scale integration (SSI)* – Up to 100 electronic components per chip
- *Medium-scale integration (MSI)* – From 100 to 3,000 electronic components per chip
- *Large-scale integration (LSI)* – From 3,000 to 100,000 electronic components per chip
- *Very large-scale integration (VLSI)* – From 100,000 to 1,000,000 electronic components per chip
- *Ultra large-scale integration (ULSI)* – More than 1 million electronic components per chip

Figure 25 ◆ A microprocessor IC.

203F25.EPS

read-only memory (ROM), and microprocessors are usually mounted with other components, including diodes, resistors, and capacitors, on PC boards or encapsulated modules. The major advantage of a microprocessor-controlled system is its ability to provide very precise control. The microprocessor collects inputs such as temperature or pressure from external sensors located at strategic points in the system. Information on the status of safety devices may also be supplied to the microprocessor. The microprocessor can evaluate the information and change system operation to meet changing conditions. Conventional relay-based controls are very limited in this sense; it would take many relays and hundreds of feet of wiring to accomplish even the most basic logic functions performed by a microprocessor.

RAM and ROM ICs

RAM ICs are referred to as volatile (temporary) memory. Computer programming software or data can be electrically stored in a powered RAM IC, but when power is removed, all the programming software or data is lost (erased). ROM ICs are referred to as non-volatile memory. They retain all data even when power is removed. Various types of ROM ICs are used to store data and/or computer-programming software that, in this case, is called firmware because it is retained when power is removed. Non-volatile ROM ICs can be any of the following types:

- *ROMs* – ROM ICs are inexpensive and manufactured with the desired data or firmware designed into the IC. However, they cannot be reprogrammed or changed in any way after the IC is made. The data or firmware is usually tested using one of the following programmable types of ROMs before permanent ROMs are manufactured.

- *PROMs* – PROMs are programmable ROMs. PROM ICs can be programmed only once after the IC is manufactured. Once the data or firmware is stored in the IC, it is permanent and cannot be erased or changed.

- *EPROMs* – EPROMs are erasable PROMs. The can be programmed and reprogrammed many times using special tools to erase and rewrite any data or firmware stored in the EPROM IC. To accomplish this the IC must be removed from the PC board.

- *EEPROMs and Flash Memory* – EEPROMs are electrically erasable PROMs. Like EPROM ICs, they can be programmed and reprogrammed many times. However, they can be erased and reprogrammed while they are in place on the PC board, but the process is accomplished slowly, one byte at a time. Flash memory, shown in the figure, is a much faster type of EEPROM. In this type of EEPROM, data or programming is erased in selected areas or throughout the entire IC at one time. Then, new data or programming is usually written in at 512 bytes at a time instead of one byte at a time.

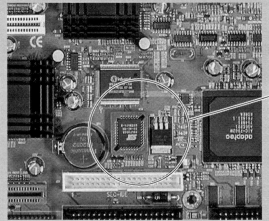

FLASH MEMORY CHIP

203SA10.EPS

10.3.0 Diagnostic Capability

Another important feature of microprocessor-controlled systems is their ability to recognize, isolate, and report faults. In small systems, the microprocessor receives sensor information such as temperatures and pressures and analyzes this information to locate a fault. The microprocessor is programmed to recognize patterns and relate those patterns to system components. Once the problem is isolated, the system may use a digital readout to identify where the problem is located. Larger, more complex systems may have programmed tests that the technician can select to help isolate a malfunction.

10.4.0 Electrostatic Discharge Sensitivity

We've all experienced discharges of static electricity after walking across a carpeted floor and touching a light switch or doorknob. While seemingly harmless, these discharges can seriously damage or even destroy delicate components on electronic devices. Before touching any electronic device, ground yourself by touching the equipment chassis or a metal electrical conduit. Never assume that you are grounded because you don't feel the static electricity like you would if you dragged your feet across a carpet and touched a metal doorknob. Most electrostatic discharge occurs at voltages below the range of human sensitivity. Make it a habit to ground yourself often when you are working with sensitive electronic equipment.

11.0.0 ◆ OPERATIONAL AMPLIFIERS

Operational amplifiers, called op amps, are one of the most useful analog devices that have been developed in integrated circuitry. They are usually packaged as one, two, or four individual amplifiers within an IC. An op amp is an almost perfect analog circuit building block. They have very high input impedance, a low output impedance, and very high gain. The schematic and schematic symbol of a single op amp is shown in *Figure 26*. Older op amps were made with bi-polar transistor technology and contained leads (pins) on the IC for the addition of external compensation capacitors to prevent the amplifier from going into oscillation at very high gains. Most new op amps are made with both JFET and bi-polar transistors and have internal compensation that eliminates the need for separate compensation pin-outs. Op amps can be configured externally as inverting or non-inverting amplifiers, buffer amplifiers, differential amplifiers, summing amplifiers, or voltage comparators.

Instrumentation Amplifier

A typical instrumentation amplifier using op amps is shown in the diagram. In this application, balanced and cascaded op amps perform differential amplification with very high input impedance for both inputs. The output is determined by the formula given on the diagram.

$$v_0 = \left(1 + \frac{2R_2}{R_1}\right) \frac{R_4}{R_3} (v_1 - v_2)$$

203SA11.EPS

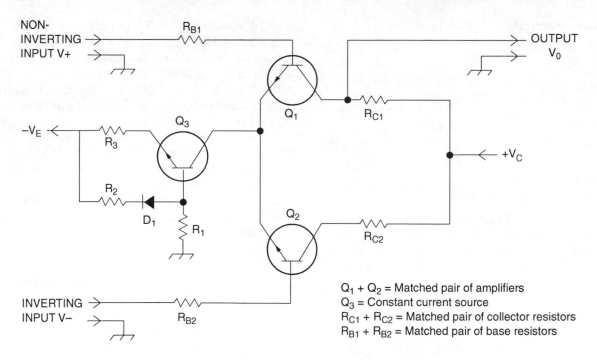

$Q_1 + Q_2$ = Matched pair of amplifiers
Q_3 = Constant current source
$R_{C1} + R_{C2}$ = Matched pair of collector resistors
$R_{B1} + R_{B2}$ = Matched pair of base resistors

SCHEMATIC OF AN IC OP AMP

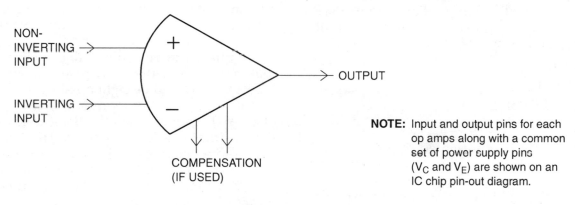

NOTE: Input and output pins for each
op amps along with a common
set of power supply pins
(V_C and V_E) are shown on an
IC chip pin-out diagram.

SCHEMATIC SYMBOL FOR AN IC OP AMP

203F26.EPS

Figure 26 ◆ Operational amplifier schematic and symbol.

12.0.0 ◆ BASIC DIGITAL GATES

Because most decision or control circuits use digital gates for logic functions, the gates are normally made in IC circuit form so that many consistent gates can be made and contained in a single IC. Like an op amp, an individual digital logic gate is made up of a number of semiconductors and other components configured to perform the logic function. To keep power consumption and heat generation low, most modern digital logic gates are made up of semiconductors in the form of transistor-transistor logic (TTL), MOSFET logic, or complementary metal-oxide semiconductor (CMOS) logic, depending on the speed and power required.

In digital logic there are three basic elements: the AND gate, the OR gate, and the inverter. What they do is very simple; however, it is essential that you understand them. By interconnecting a number of these gates into circuits, they can perform various increasingly complex functions such as addition of two numbers, counting, multiplication or division of any two numbers, keeping the time of day, and even running a whole computer.

The most common gates and circuits are as follows:

- AND gate
- OR gate
- Amplifier
- Inverter
- NAND gate
- NOR gate
- Exclusive OR gate
- Combination logic circuits

The following sections depict and describe the operation of these gates and circuits using their representative logic diagram symbols instead of the actual schematics of their components that are contained within an IC.

12.1.0 AND Gate

The AND gate is a device whose output is a logic 1 if both of its inputs are a logic 1. If only one input is a 1 with the other a logic 0, the output will be a 0. The numbers 1 and 0 are used to represent the two different levels of logic. Forms you may encounter that are used to describe logic levels of 1 and 0 are as follows:

- On and Off
- H (high) and L (low)
- T (true) and F (false)
- Yes and No

The AND gate is represented by the symbol shown in *Figure 27*, where the two inputs are on the left of the symbol, marked A and B, and the output is on the right of the symbol, marked C. These inputs and outputs may be marked E, F, and G; X, Y, and Z; or any other combination of letters.

To visualize the AND gate, use the light bulb circuit in *Figure 27* as an analogy. In this circuit, both switches A and B must be closed for the light bulb to be on. If only one of the switches is closed, the light bulb will be off. The two switches are therefore analogous to the AND gate inputs A and B, while the light bulb corresponds to output C.

The various combinations of input states of an AND gate and its response to these inputs can be expressed in a table. *Table 2* shows the input and output combinations and is called a **truth table**, which is an important tool in digital logic. You will frequently use one to represent the operation of many kinds of digital circuits. Referring to *Table 2* to determine the input-output relationships for the AND gate.

The two columns on the left show the states of the inputs to the AND gate and the column on the right shows the corresponding output. If you relate this table to the lamp circuit, a 0 represents the lamp off or open switch condition, and a 1 represents the lamp on or closed switch condition. If you read horizontally along the lines of this table, you will see the response of the output (or lamp) to all combinations of the inputs. The number of possible states for the AND gate truth table is determined by the number of inputs. The AND gate has two inputs so there is a possible combination of four states (2^2). If the AND gate contained three inputs there would be eight possible states (2^3).

Another method of representing the operation of an AND gate is called the Boolean logic equation or simply the logic equation. The AND statement represents the combination of variables by logic multiplication. The symbol used to represent the AND function is [•]. The AND function in *Figure 27* can be written as follows:

A and B = C

$A \times B = C$ (not frequently used)

A • B = C

(A)(B) = C

AB = C

Table 2 Truth Table – AND Gate

INPUTS		OUTPUT
A	B	C
0	0	0
0	1	0
1	0	0
1	1	1

203T02.EPS

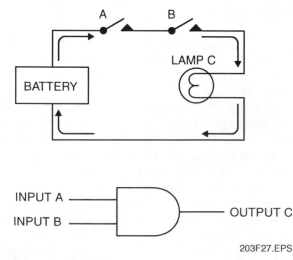

Figure 27 ◆ AND gate.

203F27.EPS

Referring again to the lamp circuit, the full logic equation that relates the output C to the inputs A and B is C = A • B. It reads C equals A and B. This equation means that both A and B must be logic 1 if C is to be logic 1.

12.2.0 OR Gate

The OR gate is a device whose output is a logic 1 if either or both of its inputs are a logic 1. The OR gate is shown by the symbol in *Figure 28* with the two inputs A and B again on the left and the output C on the right.

To visualize the OR gate, use the light bulb circuit with the switches connected in parallel rather than in series. The bulb can now be turned on by closing either switch A, switch B, or both.

The truth table for the OR gate is shown in *Table 3*, and the logic equation is C = A + B. This equation reads C = A or B. Note the differences between the truth tables and the logic equations of the OR and AND gates. The symbol for the OR function is [+], but this should not be confused with the plus sign of mathematics; the Boolean symbols have functions that are unique to Boolean algebra.

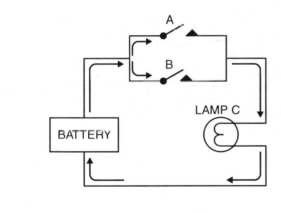

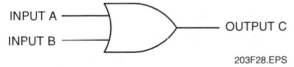

Figure 28 ◆ OR gate.

203F28.EPS

Table 3 Truth Table – OR Gate

| INPUTS | | OUTPUT |
A	B	C
0	0	0
0	1	1
1	0	1
1	1	1

203T03.EPS

12.3.0 Amplifier

The amplifier (*Figure 29*) is a device whose output assumes the high state if and only if the input assumes the high state. They are used to restore logic signal levels or to convert logic signals into usable power to drive solenoids, contactors, lamps, and other electrical devices. The Boolean equation for the amplifier is A = B.

12.4.0 Inverter

The simplest element of digital logic is the inverter (*Figure 30*), which is different from the AND and OR gates in that it has only a single input. As a result, it does not perform a decision-making function dependent on a combination of inputs. Instead, the inverter simply converts a logic 1 at its input to a logic 0 at its output and, conversely, a logic 0 to a logic 1. The inverter can be represented by either of the symbols shown in *Figure 30*. The logic equation for the inverter is C = $\overline{A}$; it reads C equals A not (or not A).

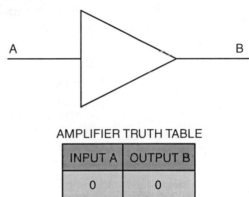

AMPLIFIER TRUTH TABLE

INPUT A	OUTPUT B
0	0
1	1

203F29.EPS

Figure 29 ◆ Amplifier symbol and truth table.

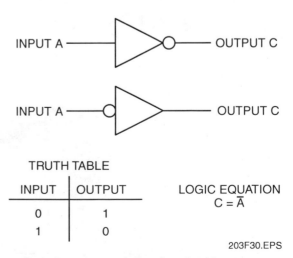

TRUTH TABLE

INPUT	OUTPUT
0	1
1	0

LOGIC EQUATION
C = $\overline{A}$

203F30.EPS

Figure 30 ◆ Inverter symbol and truth table.

The inverter, also known as the NOT function, produces an output that is always opposite of the input. Inverters may also be used to perform the same function as described for the amplifier.

12.5.0 NAND Gate

Another type of logic gate often found in digital circuits is the NAND gate (NOT AND) gate. As the name implies, a NAND gate is an AND gate with an inverter on its output. The truth table and symbol for a two-input NAND gate are shown in *Figure 31*.

The truth table shows that the output of a NAND gate assumes the 0 state if and only if all inputs assume the 1 state. In comparing the NAND gate truth table to that of the AND gate, you will see that in each case the outputs of the NAND are opposite those of the AND for the same input conditions. This is represented in the logic equation by placing a NOT symbol (a bar) over A • B, hence $\overline{A \bullet B}$. As shown, the inversion in the NAND gate is represented by a small circle at the output of the device.

If you look at the truth table for the NAND gate shown in *Figure 32*, it is apparent that by inverting (changing 1s to 0s and 0s to 1s) all the inputs to the NAND gate, you have exactly the truth table of the OR gate.

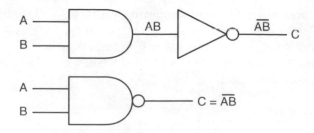

INPUTS		OUTPUT
A	B	C
0	0	1
0	1	1
1	0	1
1	1	0

203F31.EPS

Figure 31 ◆ NAND gate.

In other words, if the inputs (A, B) applied to the NAND are already individually inverted ($\overline{A}$, $\overline{B}$), it will perform the OR function. If these inputs are not inverted with the output inverted, it will perform as an AND. This dual function is represented by having two symbols for the NAND.

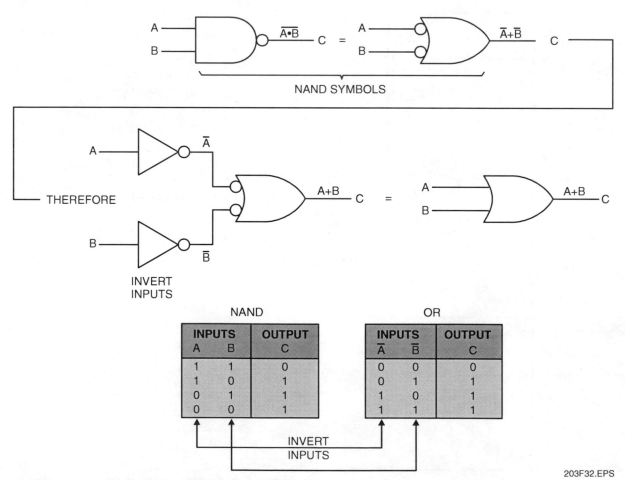

Figure 32 ◆ Two forms of NAND gate symbols.

12.6.0 NOR Gate

Another logic gate often found in digital circuits is the NOR gate. *Figure 33* shows the NOR gate composition along with its symbol and truth table.

As you can see, the NOR gate is an OR gate with an inverter on its output. When comparing the NOR gate truth table to that of the OR gate, you will see that, in each case, the outputs of the NOR gate are opposite those of the OR for the same input conditions. This is represented in the logic equation by placing a NOT symbol over A + B, hence $\overline{A + B}$. As shown, the inversion in the NOR

gate is represented by a small circle at the output of the device. Also from the truth table, you can see that the output of a NOR gate assumes the 0 state when any input assumes the 1 state.

By inspecting the NOR truth table in *Figure 34*, you can see that inverting all the inputs in the NOR truth table yields the AND truth table. NOR can, therefore, not only be used as an OR with an inverter on the output but also as an AND if the inputs (A, B) are already in the individual inverted form ($\overline{A}$, $\overline{B}$). The two symbols for the NOR gate are shown in *Figure 34*. Note again that the small circle indicates the inversion function.

(THE LOGIC EQUATION READS: C EQUALS A OR B NOT)

INPUTS		OUTPUT
A	B	C
0	0	1
0	1	0
1	0	0
1	1	0

203F33.EPS

Figure 33 ◆ NOR gate.

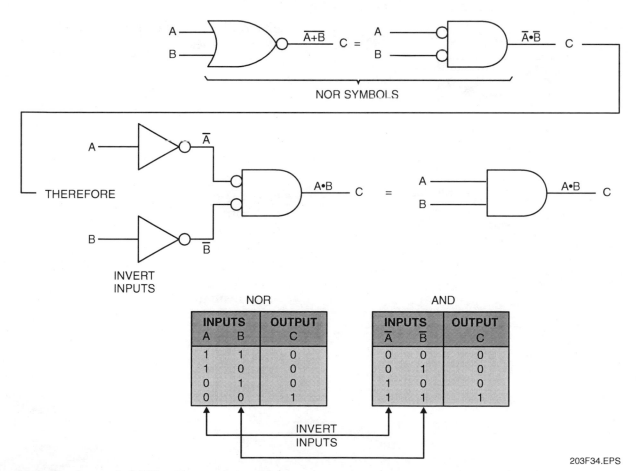

NOR				AND		
INPUTS		OUTPUT		INPUTS		OUTPUT
A	B	C		$\overline{A}$	$\overline{B}$	C
1	1	0		0	0	0
1	0	0		0	1	0
0	1	0		1	0	0
0	0	1		1	1	1

203F34.EPS

Figure 34 ◆ Two forms of a NOR gate.

12.7.0 Exclusive OR Gate

There is one more gate that needs to be considered: the exclusive OR gate (XOR), shown in *Figure 35*. The exclusive OR gate has a logic high output when either of its inputs is high but not when both are high.

The exclusive OR functions like the OR gate with one exception. When the OR receives two high inputs, the output is also high; however, when the exclusive OR receives two high inputs, the output is low. The exclusive OR gate is quite useful because its output is high only when the inputs are different. The logic equation for this gate introduces a new symbol [⊕] called exclusive OR. Therefore, the logic equation for the gate in *Figure 35* is $C = A \oplus B$.

The exclusive OR has been presented thus far as a single logic circuit with two inputs and one output. The exclusive OR gate is actually composed of many individual logic gates. *Figure 36* shows the exclusive OR logic circuit.

The output expression for this circuit is $C = A\overline{B} + \overline{A}B$. This logic circuit can be used to verify the truth table given previously for the exclusive OR gate.

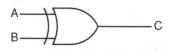

INPUTS		OUTPUT
A	B	C
0	0	0
0	1	1
1	0	1
1	1	0

203F35.EPS

Figure 35 ◆ Exclusive OR symbol and truth table.

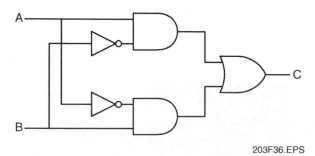

203F36.EPS

Figure 36 ◆ Exclusive OR logic circuit.

Summary

Solid-state control devices are rapidly replacing electromechanical controls in all areas of the electrical industry. Common solid-state component devices include the following:

- Diodes
- LEDs
- Transistors
- SCRs
- Diacs
- Triacs

Electronic circuits rely heavily on the ability of semiconductor materials to inhibit or promote the flow of electrons. Some semiconductor devices are used to convert AC to DC, while others are used as switching devices. Still others are used to amplify signals. In computer circuits, semiconductor devices are used to create gating circuits that are the fundamental components of digital logic. With the ever-increasing number of solid-state devices in use, all technicians should become familiar with the types of solid-state devices available and their applications.

Review Questions

1. Electronic circuits commonly use DC voltages of _____ or less.
   a. 24V
   b. 60V
   c. 120V
   d. 240V

2. The conductor commonly used to print circuits on a printed circuit board is _____.
   a. aluminum
   b. silver
   c. ceramic
   d. copper

3. An atom with five or more valence electrons is _____.
   a. an insulator
   b. easily combined with water
   c. a semiconductor
   d. a conductor

4. Which of the following can be added as an impurity to a semiconductor material to create an N-type semiconductor?

   a. Arsenic
   b. Germanium
   c. Gallium
   d. Indium

5. A PN junction solid-state device is more commonly known as a _____.

   a. transistor
   b. diode
   c. half-wave rectifier
   d. full-wave rectifier

6. The purpose of a rectifier is to _____.

   a. convert DC into AC
   b. step down voltage
   c. convert AC into DC
   d. step up voltage

7. A half-wave rectifier is constructed of _____ diode(s).

   a. one
   b. two
   c. three
   d. seven

8. What are the two basic types of transistors?

   a. AND and OR
   b. ASCII and SCICA
   c. Digital and analog
   d. NPN and PNP

9. What are the two materials normally used to construct transistors?

   a. L and O materials
   b. P and N materials
   c. A and D materials
   d. Silver and aluminum

10. The term thyristor includes which of the following groups of components?

   a. Transistors, FETs, and SCRs
   b. Diacs, triacs, and SCRs
   c. Diodes, diacs, and SCRs
   d. Diacs, triacs, and op-amps

11. Which of the following is a characteristic of a diac?

   a. It will conduct on either half of the AC waveform.
   b. The voltage applied to the anode and the gate junction must be of the same polarity.
   c. It will continue conducting after the gating voltage is removed.
   d. It has no breakover voltage rating.

12. A triac can be viewed as _____ turned in opposite directions with a common gate.

   a. three diacs
   b. two SCRs
   c. six SCRs
   d. three diodes

13. The two most widely used semiconductor materials are silicon and _____.

   a. selenium
   b. gallium
   c. copper
   d. germanium

14. An analog circuit building block with very high input impedance, low output impedance, and very high gain is a(n) _____.

   a. operational amplifier
   b. MOSFET
   c. microprocessor
   d. control module

15. A device whose output assumes the high state if and only if the input assumes the high state is a(n) _____.

   a. NAND gate
   b. Inverter
   c. NOR gate
   d. amplifier

Trade Terms Introduced in This Module

Diac: A three-layer diode designed for use as a trigger in AC power control circuits, such as those using triacs.

Diode: A semiconductor device that allows current to flow in only one direction.

Field-effect transistor (FET): A transistor that controls the flow of current through it with an electric field.

Forward bias: Forward bias exists when voltage is applied to a solid-state device in such a way as to allow the device to conduct easily.

Junction field-effect transistor (JFET): A field-effect transistor formed by combining layers of semiconductor material.

N-type material: A material created by doping a region of a crystal with atoms from an element that has more electrons in its outer shell than the crystal.

P-type material: A material created when a crystal is doped with atoms from an element that has fewer electrons in its outer shell than the natural crystal. This combination creates empty spaces in the crystalline structure. The missing electrons in the crystal structure are called holes and are represented as positive charges.

Rectifier: A device or circuit used to change AC voltage into DC voltage.

Reverse bias: A condition that exists when voltage is applied to a device in such a way that it causes the device to act as an insulator.

Semiconductor: A material that is neither a good insulator nor a good conductor. Such materials contain four valence electrons and are used in the production of solid-state devices.

Silicon-controlled rectifier (SCR): A device that is used mainly to convert AC voltage into DC voltage. To do so, however, the gate of the SCR must be triggered before the device will conduct current.

Solid-state device: An electronic component constructed from semiconductor material. Such devices have all but replaced the now-obsolete vacuum tube in electronic circuits.

Triac: A bidirectional triode thyristor that functions as an electrically controlled switch for AC loads.

Additional Resources

This module is intended to be a thorough resource for task training. The following reference works are suggested for further study. These are optional materials for continued education rather than for task training.

American Electricians' Handbook, 2002. Terrell Croft and Wilford I. Summers. New York, NY: McGraw-Hill.

National Electrical Code® Handbook, Latest Edition. Quincy, MA: National Fire Protection Association.

Solid-State Fundamentals for Electricians, 2001. Gary Rockis. Homewood, IL: American Technical Publishers.

Figure Credits

Biosman, Inc., www.biosman.com   203SA10

DPA Components International   203F24

International Rectifier Corporation   203SA03

Powerex, Inc.   203SA08

Topaz Publications, Inc.   203SA04–203SA07, 203SA09, 203F23, 203F25

NCCER makes every effort to keep these textbooks up-to-date and free of technical errors. We appreciate your help in this process. If you have an idea for improving this textbook, or if you find an error, a typographical mistake, or an inaccuracy in NCCER's Contren® textbooks, please write us, using this form or a photocopy. Be sure to include the exact module number, page number, a detailed description, and the correction, if applicable. Your input will be brought to the attention of the Technical Review Committee. Thank you for your assistance.

Instructors – If you found that additional materials were necessary in order to teach this module effectively, please let us know so that we may include them in the Equipment/Materials list in the Annotated Instructor's Guide.

Write: Product Development and Revision
National Center for Construction Education and Research
P.O. Box 141104, Gainesville, FL 32614-1104

Fax: 352-334-0932

E-mail: curriculum@nccer.org

Craft _____ Module Name _____

Copyright Date _____ Module Number _____ Page Number(s) _____

Description _____

_____

_____

_____

(Optional) Correction _____

_____

_____

(Optional) Your Name and Address _____

_____

_____

Basic Test Equipment

COURSE MAP

This course map shows all of the modules in the second level of the *Electronic Systems Technician* curriculum. The suggested training order begins at the bottom and proceeds up. Skill levels increase as you advance on the course map. The local Training Program Sponsor may adjust the training order.

ELECTRONIC SYSTEMS TECHNICIAN LEVEL TWO

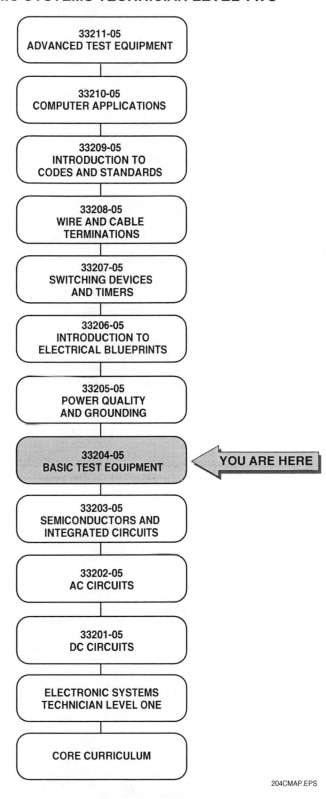

33211-05
ADVANCED TEST EQUIPMENT

33210-05
COMPUTER APPLICATIONS

33209-05
INTRODUCTION TO
CODES AND STANDARDS

33208-05
WIRE AND CABLE
TERMINATIONS

33207-05
SWITCHING DEVICES
AND TIMERS

33206-05
INTRODUCTION TO
ELECTRICAL BLUEPRINTS

33205-05
POWER QUALITY
AND GROUNDING

33204-05
BASIC TEST EQUIPMENT   ◁ YOU ARE HERE

33203-05
SEMICONDUCTORS AND
INTEGRATED CIRCUITS

33202-05
AC CIRCUITS

33201-05
DC CIRCUITS

ELECTRONIC SYSTEMS
TECHNICIAN LEVEL ONE

CORE CURRICULUM

204CMAP.EPS

Figures

Basic Test Equipment

Objectives

When you have completed this module, you will be able to do the following:

1. Explain the operation of and describe the following pieces of test equipment:
   - Ammeter
   - Voltmeter
   - Ohmmeter
   - Volt-ohm-milliammeter (VOM)
   - Continuity tester
   - Voltage tester
2. Explain how to read and convert from one scale to another using the above test equipment.
3. Explain the importance of proper meter polarity.
4. Define frequency and explain the use of a frequency meter.
5. Explain the difference between digital and analog meters.
6. Use an analog or digital multimeter to measure AC and DC voltage, AC and DC current, and resistance.
7. Use a clamp-on ammeter to measure AC current.
8. Perform a continuity test on a basic circuit to test for open shorts.

Prerequisites

Before you begin this module, it is recommended that you successfully complete *Core Curriculum; Electronic Systems Technician Level One;* and *Electronic Systems Technician Level Two*, Modules 33201-05 through 33203-05.

Required Trainee Materials

1. Pencil and paper
2. Appropriate personal protective equipment

1.0.0 ◆ INTRODUCTION

Electronic test instruments and meters are used in the following applications:

- Verifying proper operation of instruments and associated equipment
- Calibrating electronic instruments and associated equipment
- Troubleshooting electrical/electronic circuits and equipment

For these applications, specific test equipment is selected to analyze circuits and to determine specific characteristics of discrete components.

The test equipment a technician chooses for a specific task depends on the type of measurement and the level of accuracy required. Additional factors that may influence selection include the following:

- The portability of the test equipment
- The amount of information that the test equipment provides
- The likelihood that the test equipment may damage the circuit or component being tested because some test equipment can generate enough voltage or current to damage an instrument or electronic circuit

This module focuses on the basic test equipment that you will be required to use in your job as an electronic systems technician. The intent is to familiarize you with the use and operation of such equipment and to provide you with practical experience involving that equipment. Upon completion of this module, you should be able to select the appropriate test equipment and use that equipment effectively to perform an assigned task.

2.0.0 ◆ METERS

Meters are used to measure voltage, current, and resistance. The components of conventional meters usually work on direct current (DC). Mechanical frequency meters are an exception. A meter that measures alternating current (AC) has a built-in rectifier to change the AC to DC and resistors to correct for the various ranges.

Today, many meters are solid-state digital systems; they are superior because they are more precise and have no moving parts. These meters will work in any position, unlike mechanical meters, which must remain in one position to be read accurately.

In 1882, a Frenchman named Arsene d'Arsonval invented the galvanometer. This meter used a stationary permanent magnet and a moving **coil** (*Figure 1*) to indicate current flow on a calibrated scale. The early galvanometer was very accurate but could only measure very small currents. Over the following years, many improvements were made that extended the range of the meter and increased its ruggedness. The **d'Arsonval meter movement** is the most commonly used meter movement today.

A moving-coil meter movement operates on the electromagnetic principle. In its simplest form, the moving-coil meter uses a coil of very fine wire wound on a light aluminum frame. A permanent magnet surrounds the coil. The aluminum frame is mounted on pivots to allow it and the coil to rotate freely between the poles of the permanent magnet. When current flows through the coil, it becomes magnetized, and the polarity of the coil is such that it is repelled by the field of the permanent magnet. This will cause the coil frame to rotate on its pivots, and the distance it rotates is determined by the amount of current that flows through the coil. By attaching a pointer to the coil frame and adding a calibrated scale, the amount of current flowing through the meter can be measured.

The d'Arsonval meter movement uses this same principle of operation (*Figure 2*). As the current flow increases, the magnetic field around the coil increases, the amount of coil rotation increases, and the pointer swings farther across the meter scale.

3.0.0 ◆ AMMETER

The ammeter is used to measure current. Most models will measure only small amounts of current. The typical range is in microamperes, µA (0.000001A) or milliamperes, mA (0.001A). Very few ammeters can measure more than 10mA. To increase the range to the ampere level, a shunt is used. To measure above 10mA, a shunt with an extremely low resistance is placed in series with the load, and the meter is connected across the shunt to measure the resulting voltage drop proportional to current flow. A shunt has a very large wattage rating in order to carry a large current.

The meter is connected in parallel with the shunt (*Figure 3*). Shunts located inside the meter case (internal shunts) are generally used to measure values up to 30 amps; shunts located away

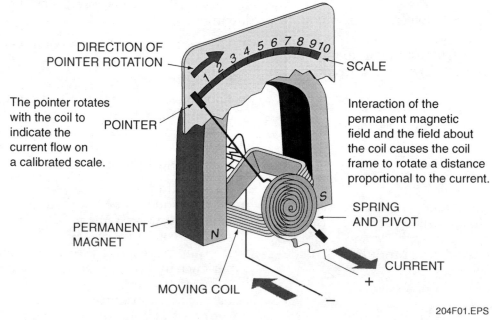

DIRECTION OF POINTER ROTATION

SCALE

The pointer rotates with the coil to indicate the current flow on a calibrated scale.

POINTER

Interaction of the permanent magnetic field and the field about the coil causes the coil frame to rotate a distance proportional to the current.

PERMANENT MAGNET

SPRING AND PIVOT

CURRENT

MOVING COIL

204F01.EPS

Figure 1 ◆ Moving-coil meter movement.

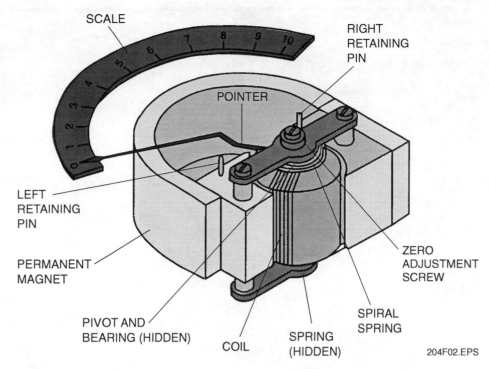

Figure 2 ◆ d'Arsonval meter movement.

from the meter case (external shunts) with leads going to the meter are generally used to measure values greater than 30 amps. Above 30 amps of current, the heat generated could damage the meter if an internal shunt were used. The use of a shunt allows the ammeter to derive current in amps by actually measuring the voltage drop across the shunt. Ammeter connections are shown in *Figure 4*.

Never connect an ammeter in parallel with a load. Because of the low resistance in the ammeter, this will cause a short circuit, probable damage to the meter and/or the circuit, and personal injury.

When connecting an ammeter in a DC circuit, you must observe proper polarity. In other words, you must connect the negative terminal of the meter to the negative or low-potential point in the circuit, and connect the positive terminal of the meter to the positive or high-potential point in the circuit (*Figure 5*). Current must flow through the meter from minus (–) to plus (+). If you connect the meter with the polarities reversed, the meter coil will move in the opposite direction, and the pointer might strike the left retaining pin. You will not obtain a current reading, and you might bend the pointer of the meter.

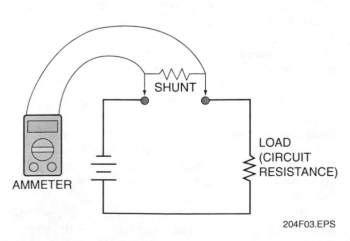

Figure 3 ◆ Ammeter shunt.

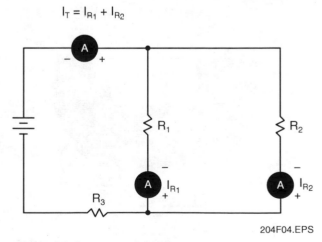

Figure 4 ◆ Ammeter connections.

Control Circuit Troubleshooting

Most control circuit troubleshooting can be done using a multimeter (VOM/DMM) and an AC clamp-on ammeter. Some types of clamp-on instruments like the one shown here incorporate the functions of a clamp-on ammeter and multimeter into one instrument.

204SA01.EPS

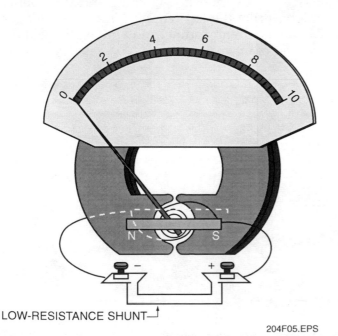

LOW-RESISTANCE SHUNT

204F05.EPS

Figure 5 ◆ DC ammeter.

It is not very practical to use an ammeter that has only one range. A multirange ammeter is one containing a basic meter movement and several shunts that can be connected across the meter movement. See *Figure 6*.

A range switch is normally used to select the particular shunt for the desired current range. Sometimes, however, separate terminals for each range are mounted on the meter case. Some multi range ammeters have only one set of values on the scale, even though they measure several different current ranges. For example, if the scale is calibrated in values from 0 to 1 milliamp (mA), and the range switch is in the 1mA position, read the current directly. However, if the range switch is in the 10mA position, multiply the scale reading by 10 to find the amount of current flowing through the circuit. See *Figure 7*.

Other current meters have a separate set of values on the calibrated scale that corresponds to the different positions of the range switch. In this case, be sure that you read the set of values that

The range of this 1mA meter movement has been extended to measure 0-10mA, 0-100mA, and 0-1A by using multiple shunts.

RANGE SWITCH

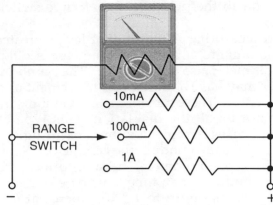

A range switch provides the simplest way of setting the meter to the desired range.

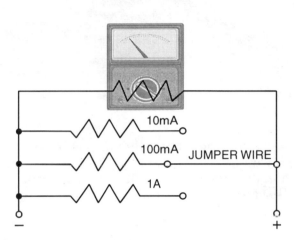

JUMPER WIRE

When separate terminals are used to select the desired range, a jumper must be connected from the positive terminal to connect the shunt across the meter movement.

204F06.EPS

Figure 6 ◆ Multirange ammeter.

Some multirange current meters have only one set of values marked on the scale.

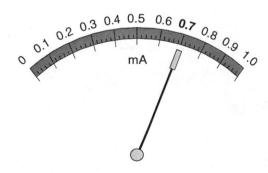

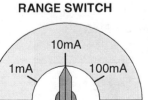

RANGE SWITCH

To find the current flowing in the circuit, multiply the scale reading by the range switch setting:
current = 0.7 × 10mA = 7mA.

204F07.EPS

Figure 7 ◆ Multirange, single-scale ammeter.

correspond to the position of the range switch (*Figure 8*).

When measuring AC current at levels greater than one ampere, a clamp-on ammeter is often used. This meter also measures DC. The clamp-on ammeter may have a mechanical movement or a digital readout. These meters clamp over the wire and do not break the insulation. This type of ammeter senses current flow by measuring the magnetic field surrounding the conductor. When using this type of ammeter, be sure to measure only one conductor at a time. This type of meter will often measure up to 1,000 amperes at 600 volts. *Figure 9* shows a digital clamp-on ammeter in use.

4.0.0 ◆ VOLTMETER

The basic current meter movement, whether AC or DC, can also be used to measure voltage (electromotive force or emf). The meter coil has a fixed resistance, and, therefore, when current flows through the coil, a voltage drop will develop across this resistance. According to Ohm's law, the voltage drop will be directly proportional to the amount of current flowing through the coil. Also, the amount of current flowing through the coil is directly proportional to the amount of voltage applied to it. Therefore, by calibrating the meter scale in units of voltage instead of current, the voltage in a circuit can be measured.

Since a basic current meter movement has a low coil resistance and low current-handling capabilities, its use as a voltmeter is very limited. In fact, the maximum voltage that could be measured with a one milliamp meter movement is one volt (*Figure 10*).

The voltage range of a meter movement can be extended by adding a resistor, called a multiplier

204F09.EPS

Figure 9 ◆ Clamp-on ammeter in use.

resistor, in series. The value of this resistor must be such that, when added to the meter coil resistance, the total resistance limits the current to the full-scale current rating of the meter for any applied voltage (*Figure 11*).

Voltmeters must be in parallel with the circuit component being measured. On the higher ranges, the amount of current flowing through the meter is much lower due to its very high total resistance. However, an inaccurate reading will

Some multirange current meters have a set of values for each range switch position.

RANGE SWITCH

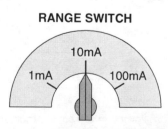

To find the current flowing in the circuit, read the meter scale that corresponds to the position of the range switch: current = 7mA.

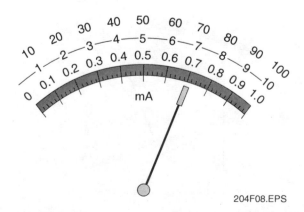

204F08.EPS

Figure 8 ◆ Multirange, multiscale ammeter.

Since the voltage across the meter coil resistance is proportional to the current flowing through the coil, the 1mA current meter movement can measure voltage directly by calibrating the meter scale in the units of voltage that produce the current through the coil.

$$E = I_R R_M = 0.001 \times 1000 = 1V$$

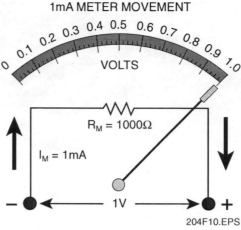

Figure 10 ◆ 1mA meter movement.

By connecting a multiplier resistor in series with the meter resistance, the range of a basic meter movement can be extended to measure voltages higher than the $I_M R_M$ voltage drop across the meter coil.

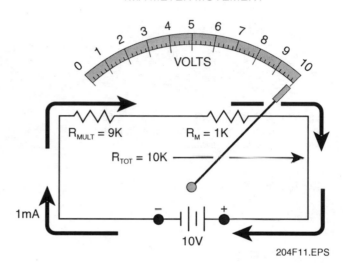

Figure 11 ◆ Adding a multiplier resistor.

result if the voltmeter is placed in series rather than in parallel with a circuit component. When connecting a DC voltmeter, always observe the proper polarity. The negative lead of the meter must be connected to the negative or low-potential end of the component, and the positive lead to the positive or high-potential end of the component. As is the case when using an ammeter, if you connect a voltmeter to the component with opposing polarities, the meter coil will move to the left, and the pointer could be bent.

In an AC circuit, the voltage constantly reverses polarity, so there is no need to observe polarity when connecting the voltmeter to a component in an AC circuit.

As with ammeters, it is impractical to have a voltmeter that will only measure one range of voltages; therefore, multirange voltmeters are also used. To make a voltmeter capable of measuring multiple ranges, you need several multiplier resistors that are switch-selectable for the different

ranges desired. Reading the scale of a voltmeter is as simple as reading the scale of an ammeter. Some multirange voltmeters have only one range of values marked on the scale, and the scale reading must be multiplied by the range switch setting to obtain the correct voltage (*Figure 12*).

Other voltmeters have different ranges on the scale for each setting of the range switch. In this case, be sure that you read the set of values that correspond to the position of the range switch (*Figure 13*).

Some digital voltmeters are autoranging. These types of voltmeters do not have a range switch. The internal construction of the meter itself will select the proper resistance for the current being detected. However, when using a voltmeter that is not autoranging, always start with the highest voltage setting, and work down until the indication reads somewhere between half-scale and three-quarter scale. This will give a more accurate reading and prevent damage to the voltmeter.

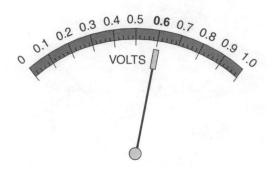

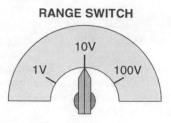

To find the voltage across a component, multiply the scale reading by the setting of the range switch: voltage = 0.6 × 10 = 6.0V.

204F12.EPS

Figure 12 ◆ Multirange, single-scale voltmeter.

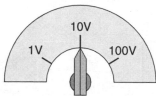

RANGE SWITCH

To find the voltage across a component, simply read the meter scale that corresponds to the position of the range switch: voltage = 6V.

Figure 13 ◆ Multirange, multiscale voltmeter.

204F13.EPS

5.0.0 ◆ OHMMETER

An ohmmeter is a device that measures the resistance of a circuit or component. It can also be used to locate open circuits or shorted circuits. Basically, an ohmmeter consists of a DC current meter movement, a low-voltage DC power source (usually a battery), and current-limiting resistors, all of which are connected in series (*Figure 14*).

This combination of devices allows the meter to calculate resistance by deriving it using Ohm's law. Before measuring the resistance of an unknown resistor or electrical circuit, the test leads are shorted together. This causes current to flow through the meter movement, and the pointer deflects to the right. This means that there is zero resistance present across the input terminals of the ohmmeter, and when zero resistance is present, the pointer will deflect full scale. Therefore, full-scale deflection of an ohmmeter indicates zero resistance. Most ohmmeters have a zero adjustment knob. This is used to correct for the fact that as the batteries of the meter age, their output voltage decreases. As the voltage drops, the

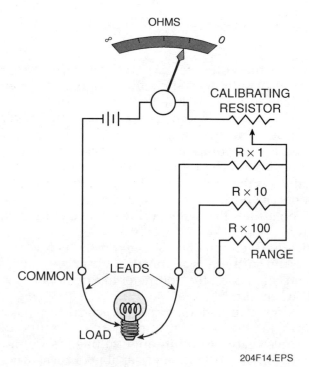

204F14.EPS

Figure 14 ◆ Ohmmeter schematic.

current through the circuit decreases, and the meter will no longer deflect full scale. By correcting for this change before each use, the internal resistance of the meter, along with the resistance of the leads, is lowered and nulled to deflect the pointer full scale.

After the ohmmeter is adjusted to zero, it is ready to be connected in a circuit to measure resistance. The circuit must be verified as de-energized by using a voltmeter prior to taking a reading with an ohmmeter. If the circuit were energized, its voltage could cause a damaging current to flow through the meter. This can cause damage to the meter and/or circuit as well as personal injury.

When making resistance measurements in circuits, each component in the circuit can be tested individually by removing the component from the circuit and connecting the ohmmeter leads across it. Actually, the component does not have to be removed from the circuit completely. Usually, the part can be isolated effectively by disconnecting one of its leads from the circuit. However, this

Using an Ohmmeter

Ohmmeters are commonly used to check for the mechanical integrity of circuits. For example, suppose that you wanted to determine if a wire in a house circuit had been broken by a sheetrock screw. How would you connect an ohmmeter to test this house circuit, and what reading would you get?

Know What You're Testing

Voltage detectors should only be used to tell whether or not voltage is present. If you need to know the value of the voltage, use a voltmeter.

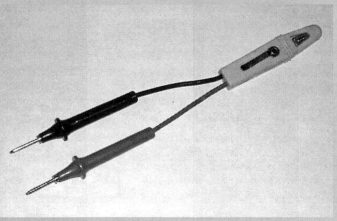

204SA02.EPS

method can still be somewhat time consuming. Some manufacturers provide charts that list the resistance readings that should be obtained from various test points to a reference point in the equipment. There are usually many parts of the circuit between the test point and the reference point, so if you get an abnormal reading, you must begin checking smaller groups of components, or individual components, to isolate the defective one. If resistance charts are not available, be very careful to ensure that other components are not in parallel with the component being tested.

6.0.0 ◆ VOLT-OHM-MILLIAMMETER

The volt-ohm-milliammeter (VOM), also known as the multimeter, is a multipurpose instrument. It is a combination of the three meters discussed: the milliammeter, the voltmeter, and the ohmmeter. One common analog multimeter is shown in *Figure 15*. There are many different models of this basic multimeter. To prevent having to discuss each and every meter, one version will be explained here. Any controls or functions on your meter that are not covered here should be reviewed in the applicable owner's manual.

The typical volt-ohm-milliammeter is a rugged, accurate, compact, and easy-to-use instrument. The instrument can be used to make accurate measurements of DC and AC voltages, current, resistance, **decibels**, and output voltage. The output voltage function is used for measuring the AC component of a mixture of AC and DC voltages. This occurs primarily in amplifier circuits.

This meter has the following features: a 0–1 volt DC range, 0–500 volt DC and AC ranges, a transit position on the range switch, rubber plug bumpers on the bottom of the case to reduce sliding, and an externally accessible battery and fuse compartment.

6.1.0 Specifications

The specifications of the sample multimeter shown in *Figure 15* are as follows:

- DC voltage:
  Sensitivity: 20KV per volt
  Accuracy: 1¾ percent of full scale
- AC voltage:
  Sensitivity: 5KV per volt
  Accuracy: 3 percent of full scale

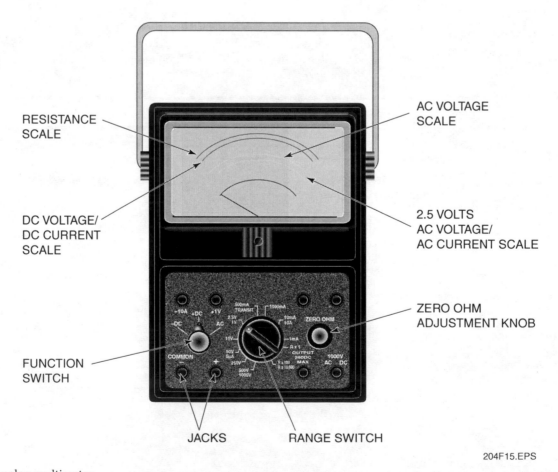

Figure 15 ◆ Analog multimeter.

204F15.EPS

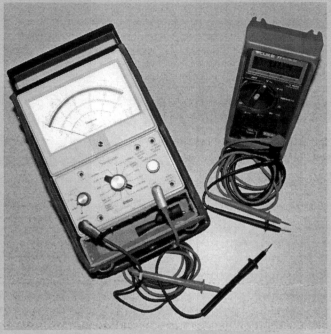

204SA03.EPS

- DC current:
    250mV to 400mV drop
    Accuracy: 1¾ percent of full scale
- Resistance:
    Accuracy: 1.75 degree of arc
    Nominal open circuit voltage 1.5V
    (9V on the 10KV ohm range)
- Nominal short circuit current:
    1V range: 1.25mA
    100V range: 1.25mA
    10KV range: 75mA
- Meter frequency response:
    Up to 100kHz

6.2.0 Overload Protection

In the sample multimeter, a 1A, 250V fuse is provided to protect the circuits on the ohmmeter ranges. It also protects the milliampere ranges from excessive overloads. If the instrument fails to indicate, the fuse may be burned out. The fuse is mounted in a holder in the battery and fuse compartment. A spare fuse is located in a well between the 1 terminal of the D cell and the side of the case.

Access to the compartment is obtained by loosening the single captivating screw on the compartment cover. To replace a burned-out fuse, remove it from the holder, and replace it with a fuse of the exact same type. When removing the fuse from its holder, first remove the battery.

In addition to the fuse, a varistor protects the indicating instrument circuit. The varistor limits the current through the moving coil in case of overload.

The fuse and varistor will prevent serious damage to the meter in most cases of accidental overload. However, no overload protection system is

completely foolproof, and misapplication on high-voltage circuits can damage the instrument. Care and caution should always be exercised to protect both you and the VOM.

6.2.1 Care of a Meter

In addition to the actual steps used in making a measurement, observe these precautions while using a multimeter:

- Keep the instrument in a horizontal position when storing, and keep it away from the edge of a workbench, shelf, or other area where it may be knocked off and damaged.
- Avoid rapid or extreme temperature changes. For example, do not leave the meter in your truck during hot or cold weather. Such temperature changes will advance the aging of the meter components and adversely affect meter life and accuracy.
- Avoid overloading the measuring circuits of the instrument. Develop a habit of checking the range position before connecting the test leads to a circuit. Even slight overloads can damage the meter, though it may not be noticeable in blown fuses or a bent needle. Slight overloads will advance the aging of components, again causing changes in meter life and accuracy.
- Place the range switch in the TRANSIT position when the instrument is not in use or when it is being moved. This reduces the swinging of the pointer when the meter is carried. Not every meter has a TRANSIT position, but if the meter does, it should be used. Random, uncontrolled swings of the meter movement may damage the movement, bend the needle, or reduce its accuracy.
- If the meter has not been used for a long period of time, rotate the function and range switches in both directions to wipe the switch contacts. Most switch contacts are plated with copper or silver. Over a period of time, these materials will oxidize (tarnish). This will create a high resistance through the switch, causing a large inaccuracy. Rotating through the switch positions will clean the tarnish off and provide good electrical contact.

6.3.0 Making Measurements

The procedures used for making DC and AC voltage measurements vary depending on the expected value of the measurement. The following sections describe the procedures for making various measurements.

CAUTION

When using a VOM, make sure your leads match the function. The meter can be in the voltage function and even display volts, but if the leads are in the amp jack, the meter will short circuit if connected to a voltage source.

6.3.1 Measuring DC Voltage, 0–250 Millivolts

Step 1 Set the function switch to +DC.

Step 2 Plug the black test lead into the – (COMMON) jack and the red test lead into the 50µA/250mV jack.

Step 3 Set the range switch to 50V.

Step 4 Connect the black test lead to the negative side of the circuit being measured and the red test lead to the positive side of the circuit.

Step 5 Read the voltage on the black scale marked DC, and use the figures marked 0–250. Read directly in millivolts.

6.3.2 Measuring DC Voltage, 0–1 Volt

Step 1 Set the function switch to –DC. Plug the black test lead into the – (COMMON) jack and the red test lead into the +1V jack.

Step 2 Set the range switch to 1V (dual position with 2.5V).

Step 3 Connect the black test lead to the negative side of the circuit being measured and the red test lead to the positive side of the circuit.

Step 4 Read the voltage on the black scale marked DC. Use the figures marked 0–10 and divide the reading by 10.

6.3.3 Measuring DC Voltage, 0–2.5 Through 0–500 Volts

WARNING!

Be extremely careful when working with higher voltages. Do not touch the instrument test leads while power is on in the circuit being measured.

Step 1 Set the function switch to +DC.

Step 2 Set the range switch to one of the five voltage range positions marked 2.5V, 10V, 50V,

250V, or 500V. When in doubt as to the voltage present, always use the highest voltage range as a protection for the instrument. If the voltage is within a lower range, the switch may be set for the lower range to obtain a more accurate reading.

Step 3 Plug the black test lead into the – (COMMON) jack and the red test lead into the + jack.

Step 4 Connect the black test lead to the negative side of the circuit being measured and the red test lead to the positive side of the circuit.

Step 5 Read the voltage on the black scale marked DC. For the 2.5V range, use the 0–250 figures and divide by 100. For the 10V, 50V, and 250V ranges, read the figures directly. For the 500V range, use the 0–50 figures and multiply by 10.

6.3.4 Measuring AC Voltage, 0–2.5 Through 0–500 Volts

CAUTION
When measuring line voltage such as from a 120V, 240V, or 480V source, be sure that the range switch is set to the proper voltage position.

Step 1 Set the function switch to AC.

Step 2 Set the range switch to one of the five voltage range positions marked 2.5V, 10V, 50V, 250V, or 500V. When in doubt as to the actual voltage present, always use the highest voltage range as a protection to the instrument. If the voltage is within a lower range, the switch may be set for the lower range to obtain a more accurate reading.

Step 3 Plug the black test lead into the – (COMMON) jack and the red test lead into the + jack.

Step 4 Connect the test leads across the voltage source (in parallel with the circuit).

Step 5 Turn on the power in the circuit being measured.

Step 6 For the 2.5V range, read the value directly on the AC scale marked 2.5V. For the 10V, 50V, and 250V ranges, read the red scale marked AC and use the black figures immediately above the scale. For the 500V range, read the red scale marked AC and use the 0–50 figures. Multiply that reading by 10.

6.3.5 Measuring Output Voltage

It is often desired to measure the AC component of an output voltage where both AC and DC voltage levels exist. This occurs primarily in amplifier circuits. The meter has a 0.1µF, 400V capacitor in series with the OUTPUT jack. The capacitor blocks the DC component of the current in the test circuit but allows the AC or desired component to pass on to the indicating instrument circuit. The blocking capacitor may alter the AC response at low frequencies but is usually ignored at audio frequencies.

CAUTION
When using OUTPUT, do not apply it to a circuit where the DC voltage component exceeds the 400V rating of the blocking capacitor.

Step 1 Set the function switch to AC.

Step 2 Plug the black test lead into the – (COMMON) jack and the red test lead into the OUTPUT jack.

Step 3 Set the range switch to one of the range positions marked 2.5V, 10V, 50V, or 250V.

Step 4 Connect the test leads across the circuit being measured, with the black test lead attached to the ground side.

Step 5 Turn on the power in the test circuit. Read the output voltage on the appropriate AC voltage scale. For the 2.5V range, read the value directly on the AC scale marked 2.5V. For the 10V, 50V, or 250V ranges, use the red scale marked AC, and read the black figures immediately above the scale.

6.3.6 Measuring Decibels

For some applications, mockup audio frequency voltages are measured in terms of decibels. The decibel scale (dB) at the bottom of the dial is marked from –20 to +10.

NOTE
The decibel is a ratio of two values. It is commonly used to state the gain or loss of audio signals. Technically, a decibel is one-tenth of a Bel, a value that was named after Alexander Graham Bell, the inventor of the telephone. The concept and use of decibels is explained in detail in the Level Four module, *Audio Systems*.

Step 1 To measure decibels, read the dB scale in accordance with the instructions for measuring AC. For example, when the range switch is set to the 2.5V position, read the dB scale directly.

Step 2 The dB readings on the scale are referenced to a 0dB power level of 0.001W across 600Ω, or 0.775VAC across 600Ω.

Step 3 For the 10V range, read the dB scale and add +12dB to the reading. For the 50V range, read the dB scale and add +26dB to the reading. For the 250V range, read the dB scale and add +40dB to the reading.

6.4.0 Direct Current Measurements

This section describes how to use the multimeter to make direct current measurements. These measurements include voltage drop, different levels of direct current, and resistance.

6.4.1 Voltage Drop

The voltage drop across the meter on all milliampere current ranges is approximately 250mV measured at the jacks. An exception is the 0–500mA range with a drop of approximately 400mV. This voltage drop will not affect current measurements. In some transistor circuits, however, it may be necessary to compensate for the added voltage drop when making measurements.

6.4.2 Measuring Direct Current, 0–50 Microamperes

CAUTION

Never connect the test leads directly across voltage when the meter is used as a current-indicating instrument. Always connect the instrument in series with the load across the voltage source.

Step 1 Set the function switch to +DC.

Step 2 Plug the black test lead into the – (COMMON) jack and the red test lead into the +50μA/250mV jack.

Step 3 Set the range switch to 50μA (dual position with 50V).

Step 4 Open the circuit in which the current is being measured. Connect the instrument in series with the circuit. Connect the red test lead to the positive side and the black test lead to the negative side.

Step 5 Read the current on the black DC scale. Use the 0–50 figures to read directly in microamperes.

NOTE

In all direct current measurements, be certain the power to the circuit being tested has been turned off before disconnecting test leads and restoring circuit continuity.

6.4.3 Measuring Direct Current, 0–1 Through 0–500 Milliamperes

Step 1 Set the function switch to +DC.

Step 2 Plug the black test lead into the – (COMMON) jack and the red test lead into the + jack.

Step 3 Set the range switch to one of the four range positions (1mA, 10mA, 100mA, or 500mA).

Step 4 Open the circuit in which the current is being measured. Connect the VOM in series with the circuit. Connect the red test lead to the positive side and the black test lead to the negative side of the part of the circuit you are measuring.

Step 5 Read the current in milliamperes on the black DC scale. For the 1mA range, use the 0–10 figures and divide by 10. For the 10mA range, use the 0–10 figures and multiply by 10. For the 500mA range, use the 0–50 figures and multiply by 10.

6.4.4 Measuring Direct Current, 0–10 Amperes

Step 1 Plug the black test lead into the –10A jack and the red test lead into the +10A jack.

Step 2 Set the range switch to 10A (dual position with 10mA).

Step 3 Open the circuit in which the current is being measured. Connect the instrument in series with the circuit. Connect the red test lead to the positive side and the black test lead to the negative side.

NOTE

The function switch has no effect on polarity for the 10A range.

Step 4 Read the current on the black DC scale. Use the 0–10 figures to read directly in amperes.

CAUTION
When using the 10A range, never remove a test lead from its panel jack while current is flowing through the circuit. Otherwise, damage may occur to the plug and jack.

6.4.5 Zero Ohm Adjustment

When resistance is measured, the VOM batteries furnish power for the circuit. Since batteries are subject to variation in voltage and internal resistance, the instrument must be adjusted to zero prior to measuring a resistance.

Step 1 Set the range switch to the desired ohms range.

Step 2 Plug the black test lead into the – (COMMON) jack and the red test lead into the + jack.

Step 3 Connect the ends of the test leads to short the VOM resistance circuit.

Step 4 Rotate the ZERO OHM control until the pointer indicates zero ohms. If the pointer cannot be adjusted to zero, one or both of the batteries must be replaced.

Step 5 Disconnect the ends of the test leads, and connect them to the component being measured.

6.4.6 Measuring Resistance

CAUTION
Before measuring resistance, be sure power is off to the circuit being tested. Disconnect the component from the circuit before measuring its resistance.

Step 1 Set the range switch to one of the resistance range positions:
- Use R × 1 for resistance readings from 0 to 200 ohms.
- Use R × 100 for resistance readings from 200 to 20,000 ohms.
- Use R × 10,000 for resistance readings above 20,000 ohms.

Step 2 Set the function switch to either the –DC or +DC position. The operation is the same in either position.

Current Testing
There are specialized clamp-on ammeters available that can be used to measure direct current in the milliamp range. These meters are used for checking electronic signals in the 4–20m range.

Step 3 Adjust the ZERO OHM control for each resistance range.
- Observe the reading on the OHMS scale at the top of the dial. Note that the OHMS scale reads from right to left for increasing values of resistance.
- To determine the actual resistance value, multiply the reading by the factor at the switch position (K on the OHMS scale equals one thousand).

Step 4 If there is a forward and backward resistance such as in diodes, the resistance should be relatively low in one direction (for forward polarity) and higher in the opposite direction.

CAUTION
Check that the OHMS range being used will not damage any of the semiconductors.

Step 5 If the purpose of the resistance measurement is to check a semiconductor in or out of a circuit (forward and reverse bias resistance measurements), check the following prior to making the measurement:
- The polarity of the voltage at the input jacks is identical to the input jack markings. Therefore, be certain that the polarity of the test leads is correct for the application.
- Ensure that the range selected will not damage the semiconductor (use R × 100 or below).
- Refer to the meter specifications, and review the limits of the semiconductor according to the manufacturer's ratings.

Step 6 Rotate the function switch between the two DC positions to reverse polarity. This will determine if there is a difference between the resistance in the two directions.

Step 7   The resistance of such diodes will measure differently from one resistance range to another on the same VOM with the function switch in a given position. For example, a diode that measures 80Ω on the R × 1 range may measure 300Ω on the R × 100 range. The difference in values is a result of the diode characteristic and does not indicate any fault in the VOM.

7.0.0 ◆ DIGITAL METERS

Digital meters have revolutionized the test equipment world. Improved accuracy is very easily attainable, more functions can be incorporated into one meter, and both autoranging and automatic polarity indication can be used. Technically, digital multimeters are classified as electronic multimeters; however, digital multimeters do not use a meter movement. Instead, a digital meter's input circuit converts a current into a digital signal, which is then processed by electronic circuits and displayed numerically on the meter face.

A major limitation with many meters that use meter movements is that the scale reading must be estimated or interpolated if the meter pointer falls between scale divisions. Digital multimeters eliminate the need to estimate these readings by displaying the reading as a numerical display.

With digital meters, technicians must revise the way the indications are viewed. For example, if a technician were reading the AC voltage on a normal wall outlet with an analog voltmeter, any indication within the range of 120VAC would be considered acceptable. But, when reading with a digital meter, the technician might think something was wrong if the meter showed a reading of 114.53VAC. Bear in mind that the digital meter is very precise in its reading, sometimes more precise than is called for, or usable. Also, be aware that the indicated parameter may change with the range used. This is primarily due to the change in accuracy and where the meter is rounding off.

There are many types of digital multimeters. Some are bench-type multimeters, while others are designed to be handheld. Most types of digital multimeters have an input impedance of 10 megohms and above. They are very sensitive to small changes in current and are therefore very accurate.

An example of a digital meter is shown in *Figure 16*. The internal operation of this meter is basically the same as that of other digital meters. The following sub-sections discuss the operation and use of this particular meter. For specific instructions, always refer to the owner's manual supplied with your meter.

7.1.0 Features

The sample meter shown in *Figure 16* offers the following features:

- *Autorange/manual range modes* – The meter features autoranging for all measurement ranges. Press the RANGE button to enter manual range mode. A flashing symbol may be used to show that you are in the manual range mode. Press the RANGE button as required to select the desired range. To switch back to auto range, press the ON/CLEAR button once (clear mode) or select another function.
- *Automatic off* – The meter turns itself off after one hour of non-use. The current draw while the meter is turned off does not affect battery life. If the meter turns itself off while a parame-

204F16.EPS

Figure 16 ◆ Digital multimeter.

ter is being monitored, press the ON/CLEAR button to turn it on again. To protect against electrical damage, the meter also turns itself off if a test lead is inserted into the 10A jack while the meter is in any mode other than A ⎓ (DC) or A~ (AC).

- *Dangerous voltage indication* – The meter shows the symbol for any range over 20V. In the autoranging mode, the meter also beeps when it changes to any range over 20V.
- *Out of Limits (OL)* – The meter displays OL and a rapidly flashing decimal point (position determined by range) when the measured value is greater than the limit of the instrument or selected range.
- *Audible acknowledgment* – The meter acknowledges each press of a button or actuation of the selector switch with a beep.

7.2.0 Operation

This section will discuss the use of various controls and explain how measurements should be taken.

7.2.1 Dual Function ON/CLEAR Button

Press the ON/CLEAR button to turn the meter on. Operation begins in the autorange mode, and the

Check Your Meter Before Using It

Never assume that a blank meter represents a reading of zero volts. Some digital meters automatically cut off the power to preserve the battery.

Similarly, if you press the HOLD button on some meters when you are reading 0V, it will lock on that reading and will continue to read 0V, regardless of the actual voltage present. Always check the meter for proper operation before using it.

range for maximum resolution is selected automatically. Press the ON/CLEAR button again to turn the meter off.

7.2.2 Measuring Voltage

Step 1 Select V—DC or V—AC.

Step 2 Connect the test leads as shown in *Figure 17.*

Step 3 Observe the voltage reading on the display. Depending on the range, the meter displays units in mV or V.

Digital Multimeter Classifications

Newer digital multimeters are rated for safety according to voltage and current limitations and fault interrupting capacity. Never use a multimeter outside of the limits specified by the manufacturer.

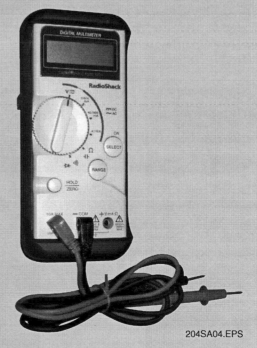

204SA04.EPS

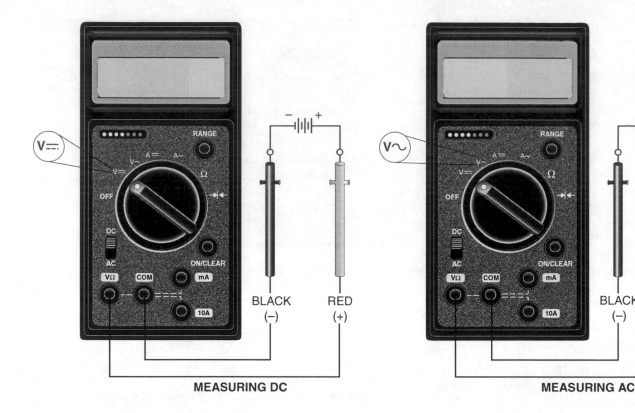

MEASURING DC **MEASURING AC**

204F17.EPS

Figure 17 ◆ Measuring voltage.

To avoid shock hazard or meter damage, do not apply more than 1,500VDC or 1,000VAC to the meter input or between any input jack and earth ground.

7.2.3 Measuring Current

Step 1 Select A $\overline{}$ or A~.

Step 2 Insert the meter in series with the circuit with the red lead connected to one of these jacks:

- The mA jack for input up to 200 milliamps
- The 10A jack for input up to 10 amps

Step 3 Make hookups as shown in *Figures 18* and *19*.

Step 4 Observe the current reading.

NOTE

The meter shuts itself off if a test lead is inserted into the 10A jack when the meter is in any function other than A $\overline{}$ or A~.

7.2.4 Measuring Resistance

When measuring resistance, any voltage present will cause an incorrect reading. For this reason,

you should discharge the capacitors in a circuit before you measure the resistance of that circuit.

Step 1 Select Ω (ohms).

Step 2 Connect the test leads as shown in *Figure 20*.

Step 3 Observe the resistance reading.

7.2.5 Diode/Continuity Test

Step 1 Select →⊩⧗ (diode/continuity).

Step 2 Choose one of the following:

- *Forward bias* – Connect to the diode, as shown in *Figure 21A*. The meter will

Measuring Resistance

The actual resistance value measured for resistive and inductive loads can vary widely depending on the type of device. Ideally, the exact resistance value for the device can be found in the manufacturer's service literature. Another way to judge the resistance reading is by comparing the resistance of the device being tested with that of a similar device that is known to be good.

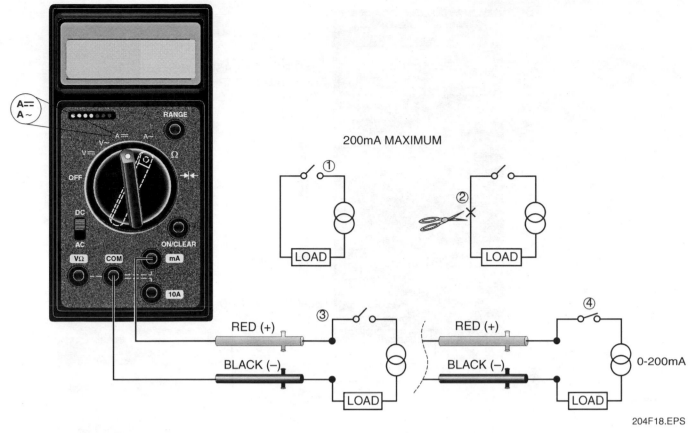

Figure 18 ◆ Measuring current (mA).

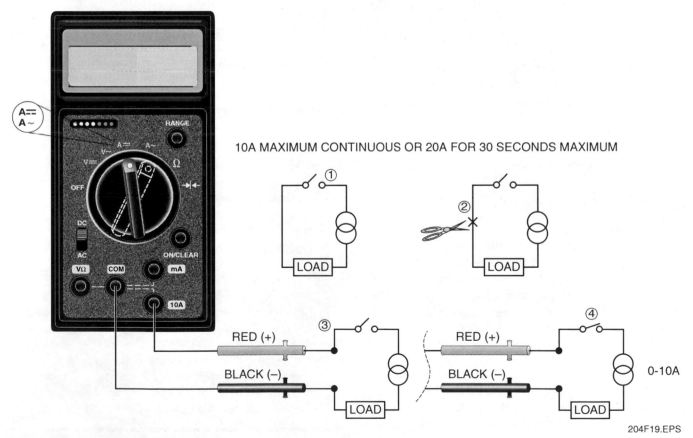

Figure 19 ◆ Measuring current (10A).

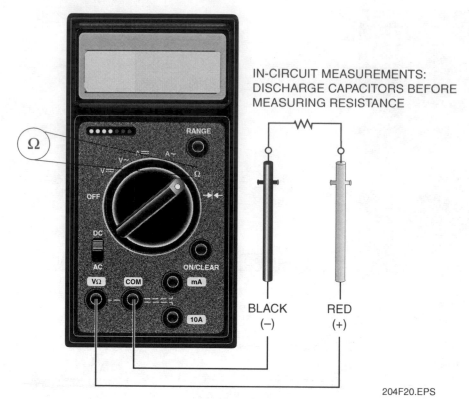

IN-CIRCUIT MEASUREMENTS:
DISCHARGE CAPACITORS BEFORE
MEASURING RESISTANCE

BLACK
(–)

RED
(+)

204F20.EPS

Figure 20 ◆ Measuring resistance.

display one of the following: the forward voltage drop (VF) of a good diode (<0.7V), a very low reading for a shorted diode (<0.3V), or OL for an open diode.

- *Reverse bias or open circuit* – Reverse the leads to the diode. The meter displays OL, as shown in *Figure 21B*. It does not beep.
- *Continuity* – The meter beeps once if the circuit resistance is less than 150 ohms, as shown in *Figure 21C*.

7.2.6 Transistor Junction Test

Test transistors in the same manner as diodes by checking the two diode junctions formed between the base and emitter, and the base and collector of the transistor. *Figure 22* shows the orientation of these effective diode junctions for PNP and NPN transistors. Also check between the collector and emitter to determine if a short is present.

7.2.7 Display Test

To test the LCD display, hold the ON/CLEAR button down when turning on the meter. Verify that the display shows all segments, as shown in *Figure 23*.

7.3.0 Maintenance

The following sections discuss the necessary maintenance for a typical multimeter (*Figure 24*). Maintenance must be performed in accordance with the manufacturer's requirements.

Understand Diodes

A diode is a two-terminal semiconductor device that behaves like a check valve. Forward bias means that you are testing that the diode conducts properly when it is closed. Reverse bias means that current does not flow through the diode.

Digital Meters

Precision measurements are required when adjusting and calibrating equipment such as power supplies, or when determining the percentage of voltage or current imbalance that exists in a three-phase system. Precision is not important if making measurements just to determine the absence or presence of voltage or current, as is commonly done when troubleshooting.

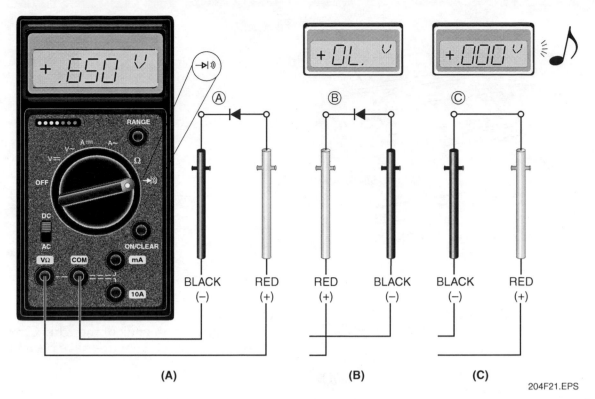

(A)   **(B)**   **(C)**

204F21.EPS

Figure 21 ◆ Diode/continuity test.

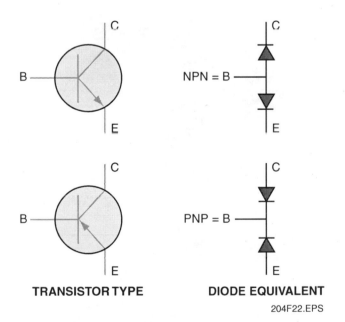

TRANSISTOR TYPE   **DIODE EQUIVALENT**

204F22.EPS

Figure 22 ◆ Transistor junction test.

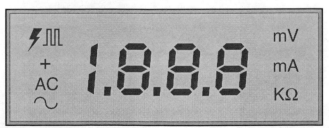

204F23.EPS

Figure 23 ◆ Display test.

7.3.1 Battery Replacement

Replace the battery as soon as the meter's decimal point starts blinking during normal use; this indicates that fewer than 100 hours of battery life remain. Remove the case back, and replace the battery with the same or equivalent 9V alkaline battery.

7.3.2 Fuse Replacement

Some meters use two input protection fuses for the mA and 10A inputs. Remove the case back to gain access to the fuses. Replace with the same type only. The large fuse should be readily available. A spare for the smaller fuse is included in the case. If necessary, this fuse must be reordered from the factory.

7.3.3 Calibration

Have a qualified technician calibrate the meter once a year. The procedure for calibrating a meter is as follows:

Step 1   Remove the case back (Figure 24). Turn the meter ON and select V ⎓.

Step 2   Apply +1.900VDC +/−0.0001V to the V-n input (negative to COM).

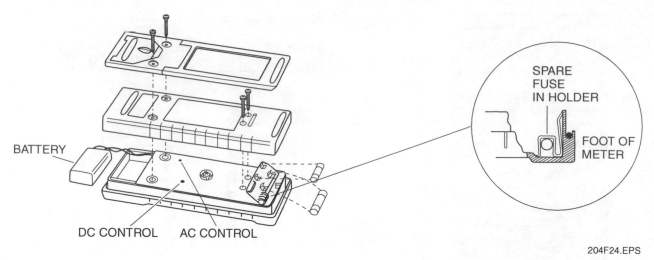

SPARE
FUSE
IN HOLDER

FOOT OF
METER

204F24.EPS

Figure 24 ◆ Meter maintenance.

Step 3   Adjust the DC control through the hole in the circuit board for a display of +1.900V.

Step 4   Select V.

Step 5   Apply +1.900VAC +/–0.002VAC @ 60Hz to the V-n input.

Step 6   Adjust the AC control through the hole in the circuit board for a display of +1.900V.

Step 7   Reassemble the meter.

8.0.0 ◆ CONTINUITY TESTER

A continuity tester (*Figure 25*) is a simple device consisting mainly of a battery and either an audible or visual indicator. It can be used in place of an ohmmeter to test the continuity of a wire and to identify individual wires contained in a conduit or other raceway. To test the continuity of a wire, strip the insulation off of the end of the wire to be tested at one end of the conduit run, then connect (short) the wire to the metal conduit. At the other end of the conduit run, clip the alligator clip lead of the tester to the conduit and touch the probe to the end of the wire under test. If the tester audible alarm sounds or the indicator light comes on, there is continuity. Note that this only indicates that there is continuity between the two points being tested; it does not indicate the actual value of the resistance. If there is no indication, the wire is open.

To identify individual wires in a conduit run, touch the tester probe to the wires in the conduit one at a time until the tester audible alarm sounds or the indicator lights. Then, put matching identification tags on both ends of the wire. Continue this procedure until all the wires have been identified.

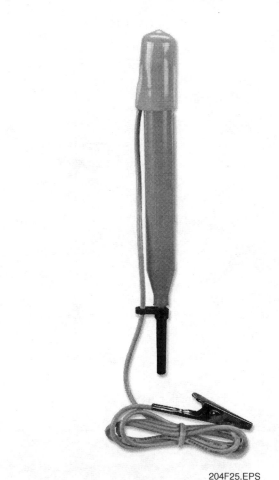

204F25.EPS

Figure 25 ◆ Continuity tester.

Summary

As an electronic systems technician, you will use electrical test equipment nearly every day as part of your installation, checkout, and troubleshooting activities. Basic electrical test equipment is used primarily to measure voltage, current, and resistance, and to check continuity of wires and cables.

It is important that you know which equipment item to use in a given situation and how to use it properly. Although analog meters are still around, you will most likely use meters with a digital readout. Digital meters are both more accurate and easier to use than analog meters.

The multimeter is a combination testing device that can be used to measure voltage, current, resistance, and decibels. It will most likely be your primary electrical tester. Other basic test devices used in low-voltage work include go/no-go voltage testers and continuity testers.

ESTs use other, most sophisticated test equipment to check cables, devices, and networks. This advanced test equipment will be covered in a later module within this level.

Review Questions

1. An ammeter is used to measure _____.
   a. current
   b. voltage
   c. resistance
   d. insulation value

2. Ammeters use a _____ to measure values higher than 10mA.
   a. multiplier resistor
   b. rectifier bridge
   c. transformer coil
   d. shunt resistor

3. Voltmeters use a _____ to measure values higher than the maximum range of the meter movement.
   a. multiplier resistor
   b. rectifier bridge
   c. transformer coil
   d. shunt resistor

4. Electromotive force (emf) is measured using a(n) _____.
   a. ammeter
   b. wattmeter
   c. voltmeter
   d. ohmmeter

5. A voltmeter is connected _____ with the circuit being tested.
   a. in parallel
   b. in series

6. An ohmmeter must be used in an energized circuit.
   a. True
   b. False

7. A VOM can be used to measure decibels, voltage, resistance, and current.
   a. True
   b. False

8. When in doubt as to the voltage present, which range should be used to protect the VOM?
   a. Lowest
   b. Highest
   c. Middle
   d. It does not matter with today's precision instruments.

9. Although digital meters are easy to read, they are less accurate than analog meters.
   a. True
   b. False

10. A continuity tester can be used in place of a(n) _____ to identify individual wires contained in a conduit or other raceway.
    a. voltmeter
    b. ohmmeter
    c. ammeter
    d. power factor meter

Trade Terms Introduced in This Module

Coil: A number of turns of wire, especially in spiral form, used for electromagnetic effects or for providing electrical resistance.

Continuity: An uninterrupted electrical path for current flow.

d'Arsonval meter movement: A meter movement that uses a permanent magnet and moving coil arrangement to move a pointer across a scale.

Decibel: One-tenth of a Bel. A decibel is a logarithmic measurement of electrical, acoustic, or power ratios.

Additional Resources

This module is intended to present thorough resources for task training. The following reference works are suggested for both instructors and motivated participants interested in further study. These are optional materials for continued education rather than for task training.

Electronics Fundamentals: Circuits, Devices, and Applications, 2000. Thomas L. Floyd. New York, NY: Prentice Hall.

Principles of Electric Circuits: Conventional Current Version, 2002. Thomas L. Floyd. New York, NY: Prentice Hall.

Figure Credits

Amprobe Instruments   204F25

Extech Instruments Corporation   204SA01

Topaz Publications, Inc.   204F09, 204SA02–204SA04

NCCER makes every effort to keep these textbooks up-to-date and free of technical errors. We appreciate your help in this process. If you have an idea for improving this textbook, or if you find an error, a typographical mistake, or an inaccuracy in NCCER's Contren® textbooks, please write us, using this form or a photocopy. Be sure to include the exact module number, page number, a detailed description, and the correction, if applicable. Your input will be brought to the attention of the Technical Review Committee. Thank you for your assistance.

Instructors – If you found that additional materials were necessary in order to teach this module effectively, please let us know so that we may include them in the Equipment/Materials list in the Annotated Instructor's Guide.

Write: Product Development and Revision
National Center for Construction Education and Research
P.O. Box 141104, Gainesville, FL 32614-1104

Fax: 352-334-0932

E-mail: curriculum@nccer.org

Craft   Module Name
_____

Copyright Date   Module Number   Page Number(s)
_____

Description
_____

_____

_____

_____

(Optional) Correction
_____

_____

_____

(Optional) Your Name and Address
_____

_____

_____

Power Quality and Grounding

COURSE MAP

This course map shows all of the modules in the second level of the *Electronic Systems Technician* curriculum. The suggested training order begins at the bottom and proceeds up. Skill levels increase as you advance on the course map. The local Training Program Sponsor may adjust the training order.

ELECTRONIC SYSTEMS TECHNICIAN LEVEL TWO

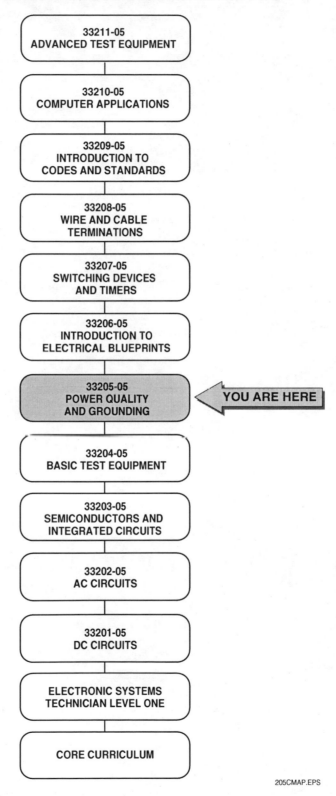

33211-05
ADVANCED TEST EQUIPMENT

33210-05
COMPUTER APPLICATIONS

33209-05
INTRODUCTION TO
CODES AND STANDARDS

33208-05
WIRE AND CABLE
TERMINATIONS

33207-05
SWITCHING DEVICES
AND TIMERS

33206-05
INTRODUCTION TO
ELECTRICAL BLUEPRINTS

33205-05
POWER QUALITY
AND GROUNDING

YOU ARE HERE

33204-05
BASIC TEST EQUIPMENT

33203-05
SEMICONDUCTORS AND
INTEGRATED CIRCUITS

33202-05
AC CIRCUITS

33201-05
DC CIRCUITS

ELECTRONIC SYSTEMS
TECHNICIAN LEVEL ONE

CORE CURRICULUM

205CMAP.EPS

Figures

Power Quality and Grounding

Objectives

When you have completed this module, you will be able to do the following:

1. Explain the purpose of grounding.
2. Determine the *National Electrical Code®* (NEC®) requirements for electrical system and telecommunications equipment grounding.
3. Locate the cause of a ground fault.
4. Recognize and describe the purpose of the components used for grounding and bonding a telecommunications system in a typical commercial multi-story building.
5. Recognize and describe the purpose of the components used in a lightning protection system for a typical high-rise building.
6. Define clean, pure power relating to AC power quality and describe the types of abnormalities and their sources that can cause poor AC power quality.
7. Recognize and describe the purpose for using various AC power system protection and conditioning devices, including:
   - Isolation transformers
   - Surge protecting devices
   - Power line conditioners
   - Harmonic and noise suppression filters
   - Motor and engine-generator sets
   - Uninterruptible power supplies
8. Select and test DC power supplies used in electronic equipment.
9. Recognize different types of storage batteries and describe the advantages and disadvantages of each type.
10. Describe cable shielding and grounding techniques used to minimize EMI.
11. Describe the types of problems that can be caused by static electricity and how to prevent them.
12. Use a ground resistance tester to test for a low-voltage ground.

Prerequisites

Before you begin this module, it is recommended that you successfully complete *Core Curriculum; Electronic Systems Technician Level One;* and *Electronic Systems Technician Level Two,* Modules 33201-05 through 33204-05.

Required Trainee Materials

1. Pencil and paper
2. Appropriate personal protective equipment
3. Scientific calculator
4. Copy of the latest edition of the *National Electrical Code®*

1.0.0 ◆ INTRODUCTION

Power quality involves powering and grounding sensitive electronic equipment used in telecommunications, security, CATV, and other low-voltage equipment in a manner that promotes optimum operation of the equipment. The quality of power needed to supply this complex equipment is much more critical today than in the past.

Note: The designations "National Electrical Code®" and "NEC®," where used in this document, refer to the *National Electrical Code®,* which is a registered trademark of the National Fire Protection Association, Quincy, MA. *All National Electrical Code® (NEC) references in this module refer to the 2002 edition of the NEC®.*

Ideally, for alternating current (AC)-powered equipment, the input power should be of constant voltage and **frequency** with perfect sinusoidal waveshapes, and be free of **harmonics**, noise, and **transients**. For direct current (DC)-powered equipment, the DC power should be a ripple-free voltage free of noise and transients.

The first part of this module provides a brief introduction to power systems and their related grounding schemes. This information is presented as background information for a better understanding of power quality. The second part of this module focuses on the conditions that can cause poor power quality and the various equipment and devices that can be used to correct these conditions in order to maintain good power quality.

2.0.0 ◆ TYPICAL ELECTRICAL GENERATION AND DISTRIBUTION SYSTEMS

Normally, the electronic systems technician is not involved in the installation or maintenance of a building's utility power and branch distribution circuits. However, in order to understand the subject of power quality and grounding as it applies to the installation and maintenance of low-voltage equipment, it is important to be familiar with the basic characteristics of electrical power generation and distribution circuits and their related grounding schemes.

2.1.0 Utility Power Generation, Transmission, and Distribution

The essential elements of an AC electrical generation and distribution system include generating stations, **transformers**, substations, transmission lines, and distribution lines. *Figure 1* shows these elements and their relationship. The path begins at the electric utility company's generating station where AC power is produced at voltages typically ranging from 2,400V to 13,200V by large generators powered by water, coal, oil, or nuclear fuel. Then, the power plant uses transformers to further increase (step up) the voltage for transmission on a network of high-voltage wires (transmission lines) to substations located along the transmission line. This voltage can be anywhere between 115,000V and 1,000,000V, depending on a combination of factors, including the transmission distance.

At each substation, the voltage is reduced (stepped down) to voltages below 35,000V for distribution to users in the local area or region. Substations can be small buildings or fenced-in areas containing switches, transformers and other elec-

trical equipment and structures. They are convenient places used to monitor the system and adjust circuits.

Electrical devices called **voltage regulators** and **capacitor banks** are also installed in substations. Voltage regulators maintain the system voltage at a constant level as the demand for electricity changes. A **capacitor** or capacitor bank momentarily stores energy in order to help reduce energy losses and improve voltage regulation. Within the substation, rigid tubular or rectangular bars, called busbars or buses, are used as conductors.

A network of power lines called the distribution system delivers the electrical energy from the substation to the individual customers via conductors called feeders. At key locations within the distribution system, the voltage is stepped down by pole- or pad-mounted transformers to a level needed by the customer. Distribution lines carry either three-phase or single-phase power. Single-phase power is normally used for residential and small commercial occupancies, while three-phase power serves commercial and industrial users with large electrical loads such as air conditioning systems and manufacturing machines. The voltages available at any location depend on the type of transformer connection that is made between the utility and the user. In North America, the most common nominal voltages used are 120V, 208V, 240V, and 480V. Note that a nominal voltage value is assigned to a circuit or system for the purpose of conveniently designating its voltage class, such as 120/240V, 277/480V, and 600V. The actual voltage at which a circuit operates can vary from the nominal value within a range that permits satisfactory

Volt-Amperes and Watts

In the study of DC current, we learned that VA is equal to volts times amperes. There is only pure resistance to oppose current flow in DC circuits. However, in AC circuits, additional opposition to current flow can be present in the form of inductive reactance, caused by magnetic fields such as motors and coils, or capacitive reactance caused by the use of capacitors in the circuit. When these two components are added to the resistance of the circuit, not all of the applied power results in work, because some is lost due to the additional opposition to current flow. The ratio of usable power to applied power is referred to as the power factor. Watts is equal to VA times the power factor.

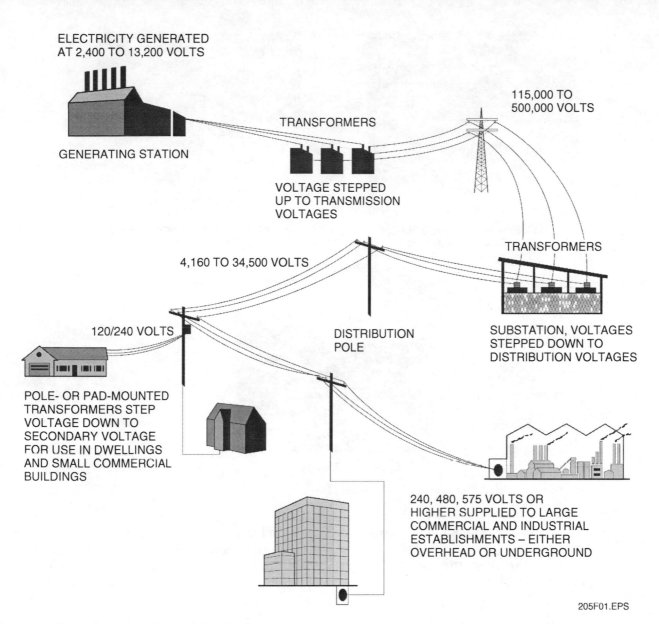

ELECTRICITY GENERATED
AT 2,400 TO 13,200 VOLTS

GENERATING STATION

TRANSFORMERS

VOLTAGE STEPPED
UP TO TRANSMISSION
VOLTAGES

115,000 TO
500,000 VOLTS

TRANSFORMERS

4,160 TO 34,500 VOLTS

120/240 VOLTS

DISTRIBUTION
POLE

SUBSTATION, VOLTAGES
STEPPED DOWN TO
DISTRIBUTION VOLTAGES

POLE- OR PAD-MOUNTED
TRANSFORMERS STEP
VOLTAGE DOWN TO
SECONDARY VOLTAGE
FOR USE IN DWELLINGS
AND SMALL COMMERCIAL
BUILDINGS

240, 480, 575 VOLTS OR
HIGHER SUPPLIED TO LARGE
COMMERCIAL AND INDUSTRIAL
ESTABLISHMENTS – EITHER
OVERHEAD OR UNDERGROUND

205F01.EPS

Figure 1 ◆ Parts of a typical electrical distribution system.

operation of the equipment. The capacity of a given system is the amperage that is available for use at the nominal voltage. This can be stated in volt-amperes (VA) or watts (W).

Most power companies use transmission systems that include both overhead and underground installations. In general, the terms and devices are the same for both. In the case of the underground system, distribution transformers are installed at or below ground level. Those mounted on concrete pads are called padmounts, while those installed in underground vaults are called submersibles. Buried conductors (cables) are insulated to protect them from soil chemicals and moisture. Many overhead conductors do not require such protective insulation.

2.2.0 Premises Wiring

The wiring between the connection at the distribution system at the customer's site and the equipment to be powered is called the premises wiring. *Figure 2* shows the basic components of a typical three-phase overhead service used to connect a building or premises to the utility's distribution system. In this example, note that the high-voltage lines terminate on a power pole near the building being served (*Figure 2A*). A bank of transformers is mounted on the pole to reduce the transmission voltage to the level required for building use. From the secondary of the transformers, the service feeders enter parallel service masts (*Figure 2B*) above the building, and down

(A) 3-PHASE POWER POLE WITH TRANSFORMERS

(B) PARALLEL SERVICE MASTS ABOVE BUILDING

CONDUITS FROM SERVICE MASTS

SERVICE EQUIPMENT

CONDUIT HOLDING GROUNDING CONDUCTOR

(C) SERVICE EQUIPMENT

205F02.EPS

Figure 2 ◆ Three-phase overhead service.

into the service equipment (*Figure 2C*) mounted on the wall of the building.

The components forming the connection between the pole and the building include the following:

- *Service drop* – This includes overhead conductors through which the electrical service is supplied between the last power company pole and the point of their connection to the service-entrance conductors located at the building. Note that when the service conductors to the building are routed underground, these conductors are known as the service lateral.
- *Service entrance* – This includes all the components between the point of termination of the overhead service drop or underground service lateral and the building's main disconnecting device, except the metering equipment.
- *Service conductors* – These are conductors between the point of termination of the overhead service drop or underground service lateral and the main disconnecting device in the building or premises.
- *Service equipment* – This is necessary equipment, usually consisting of a circuit breaker or

switch and fuses and their accessories, located near the point where supply conductors enter a building. It provides the main control and cutoff means for the electric supply to the building.

2.3.0 Characteristics of Alternating Current (AC) Power

Commercial utility power sources provide three-phase AC power. With AC power, the electrical current periodically reverses (alternates) its direction of flow because the polarity of the voltage source constantly changes. When an AC circuit is connected to a load (any device that consumes power), the alternating current that flows through the load follows the same alternating pattern as the voltage source. Commercial AC power provided to buildings is in the form of a continuous sine wave (*Figure 3*).

As shown, a sine wave cycle consists of one variation of the sine wave from zero to maximum (positive), back through zero, to minimum (negative), and back to zero. The number of times this cycle occurs in one second is called the frequency and is expressed in hertz (Hz). In North America the frequency of utility power is 60Hz. Other countries may have different frequencies, with 50Hz being the most common. The most common method of expressing the voltage or current amplitude of a sine wave is by its root-mean-square (rms) value. The rms value is 0.7071 times the zero-to-peak (maximum) value of the sine wave. For example, if the maximum value of a sine wave is 170V, then the rms value is 120V (170V × 0.7071). Note that rms value is also commonly called the effective value.

When three-phase electrical power is generated, three sine waves are produced; each sine wave is 120 degrees out of phase with the others, as shown in *Figure 4*. Each of the waveforms generated is called a phase, hence the terms single-phase and three-phase. Phase is the relationship in time between two waveforms of the same frequency. In order to measure phase difference, the time required to complete one cycle is divided

into 360 electrical degrees. In *Figure 4*, waveform A (Phase A) starts 120 degrees before waveform B (Phase B), and 240 degrees before waveform C (Phase C). With this relationship, Phase A is said to be 120 degrees out of phase with Phase B. Also, Phase A is leading Phase B by 120 degrees. Conversely, Phase B is lagging Phase A by 120 degrees. If two waveforms match each other in electrical degrees (not shown), they are said to be in phase.

3.0.0 ◆ OVERVIEW OF PREMISES ELECTRICAL SYSTEM GROUNDING

The grounding system is a major part of any premises electrical system. Grounding of an electrical system is required to limit the voltage entering the system from surges in power from the

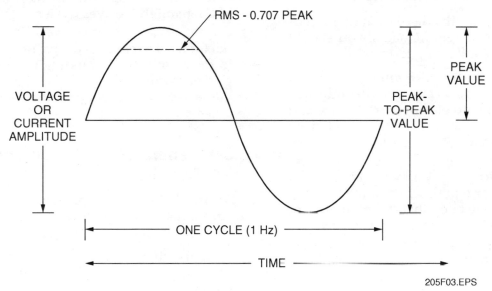

Figure 3 ◆ AC sine wave.

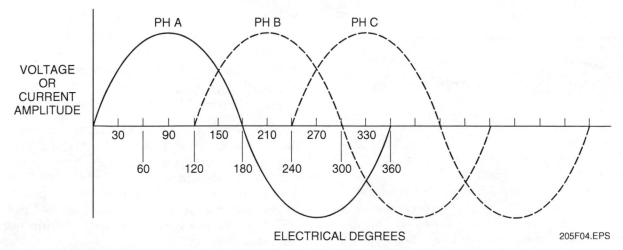

Figure 4 ◆ Relationship of three-phase power sine waves.

---

supply transformer or from lighting strikes. It also serves to hold all parts of the electrical system, including non-current-carrying enclosures, at zero potential to ground. High voltages can accidentally enter an electrical system as a result of lighting strikes, the breakdown of insulation in the supply transformer, chance contact between high-voltage supply lines and the incoming service wires, etc. Should any of these conditions occur, several thousand volts can be present in the electrical system. Without proper system grounding, the high current related to the high voltage can create so much heat that the insulation on conductors would melt and electrical equipment would be destroyed. In a system that is properly **grounded**, the excess voltage and current are rapidly directed to earth, thereby preventing these problems.

3.1.0 Grounding System Terminology

In order to understand the discussion of grounding systems and their requirements, you should be familiar with the following terms:

- *Bonding* – This is the permanent joining of metallic parts to form an electrically conductive path that will ensure electrical continuity and the capacity to conduct safely any current likely to be imposed on it.
- *Ground* – This is a conducting connection (whether intentional or accidental) between an electrical circuit or equipment and the earth, or to some conducting body that serves in place of earth. When the term ground is used, we are talking about ground potential or earth ground.
- *Grounded* – This means connected to earth or to some conducting body that serves in place of earth.
- *Grounded conductor* – Describes a system or circuit conductor that is intentionally grounded. If

a conductor is connected to the earth or some conducting body that serves in place of the earth, such as a driven ground rod (electrode) or cold water pipe, the conductor is considered a grounded conductor.

- *Grounding electrode* – This is a conductor placed in the earth, providing a connection to a circuit.
- *Effectively grounded* – This means intentionally connected to the earth through a ground connection or connections of sufficiently low **impedance** and having sufficient current-carrying capacity to prevent the buildup of voltages that may result in undue hazards to connected equipment or to persons.
- *Equipment grounding conductor* – This is the conductor used to connect the non-current-carrying metal parts of equipment, raceways, and other enclosures to the system grounded conductor, the grounding electrode conductor, or both, at the service equipment or at the source of a **separately derived system**.
- *Grounding electrode conductor* – This is the conductor used to connect the grounding electrode to the equipment grounding conductor, to the grounded conductor, or to both, of the circuit at the service equipment or at the source of a separately derived system.

3.2.0 General NEC® Grounding Requirements

NEC Article 250 covers the general requirements for grounding and bonding of electrical installations. It covers the following topics:

- Systems, circuits, and equipment that are required, permitted, or not permitted to be grounded
- Circuit conductors to be grounded on grounded systems
- Location of grounding connections
- Types and sizes of grounding and bonding conductors and electrodes
- Methods of grounding and bonding
- Conditions under which guards, isolation, or insulation may be substituted for grounding

NEC Section 250.4(A)(1) defines the general requirements for grounding electrical systems. It requires grounded electrical systems to be connected to earth in a manner that will limit the voltage imposed by lightning, line surges, or unintentional contact with higher voltage lines and will stabilize the voltage to earth during normal operation.

NEC Section 250.4(A)(2) defines grounding of electrical equipment. It requires that conductive

Potential to Earth Ground

Some people think electricity is naturally attracted to ground. Due to its ionic composition, lightning is attracted to ground; however, manufactured electricity initially has no difference of potential (voltage) or shock hazard to ground until these systems are physically grounded at the generating station or other points in the system. Without this ground potential, it would be difficult to troubleshoot and measure voltage values, and circuit breakers and fuses would not operate properly.

materials (metal boxes, metal conduit, etc.) enclosing electrical conductors or equipment, or forming part of such equipment, shall be connected to earth so as to limit the voltage to ground on these materials.

Where the electrical system is required to be grounded, these conductive materials shall be connected together and to the supply system grounded conductor in the manner specified by *NEC Article 250*. Where the electrical system is not solidly grounded, conductive materials shall be connected together in a manner that establishes an effective path for fault current [*NEC Section 250.4(A)(3)*].

NEC Section 250.4(A)(4) defines bonding of electrically conductive materials and other equipment. It requires that electrical conductive materials, such as metal water piping, metal gas piping, and structural steel members, that are likely to become energized shall be connected together in a manner that establishes an effective path for fault current.

NEC Section 250.4(A)(5) requires that the fault current path to ground from circuits, equipment, and metal enclosures shall meet the following conditions:

- Be permanent and electrically continuous
- Have the capacity to conduct safely any fault current likely to be imposed on it
- Have sufficiently low impedance to limit the voltage to ground and to facilitate the operation of the circuit protective devices

3.3.0 System and Equipment Grounding

As implied by the general NEC® requirements outlined above, there are basically two kinds of grounding for electrical wiring: system grounding and equipment grounding. These grounding schemes have different purposes and involve different electrical paths. System grounding relates to service-entrance equipment and its interrelated and bonded components; that is, one system current-carrying conductor is grounded to limit voltages due to lightning, line surges, or unintentional contact with higher voltage lines, and to stabilize the voltage to ground during normal operation.

Equipment grounding conductors are used to connect non-current-carrying metal parts of equipment, conduit, outlet boxes, and other enclosures to the system ground conductor, the grounding electrode conductor, or both, at the service entrance or at the source of a separately derived system. The purpose of the grounding conductor is to carry any fault current back to the source through earth ground.

3.3.1 Basic Grounding System

To better understand a complete grounding system, we will examine a conventional single-phase residential/light commercial system beginning at the power company's high-voltage lines and transformer, as shown in *Figure 5*. The pole-mounted transformer is fed with one leg of a three-phase sys-

Figure 5 ◆ Pole-mounted distribution transformer.

tem, and a solidly grounded neutral from the source. It is transformed and stepped down to a three-wire, 120/240V, single-phase electric service suitable for residential/commercial use. A wiring diagram of the transformer connections is shown in *Figure 6*. Note that the voltage between Leg A and Leg B is 240V.

By connecting a third wire (neutral) on the secondary winding of the transformer between the other two, the 240V splits in half, giving 120V between either Leg A or Leg B and the neutral conductor. Consequently, 240V is available to power 240V loads such as motors, electric water heaters, etc., while 120V is available for lights and other 120V loads. The conductors for Legs A and B are ungrounded conductors, while the neutral is a grounded conductor. If only 240V were connected, the neutral or grounded conductor would carry no current. However, since 120V loads are present, the neutral will carry the unbalanced load and become a current-carrying conductor.

For example, if Leg A carries 60A and Leg B carries 50A, the neutral would carry only 10A (60A – 50A = 10A). This is why the NEC® sometimes allows the neutral conductor in an electric service to be smaller than the ungrounded conductors.

The grounding of the neutral is done at both the power supply pole and the building's service panel. The typical pole-mounted service entrance is normally routed by a grounded (neutral) messenger cable from a point on the pole to a point on the building being served, terminating at the service drop. Service-entrance conductors are routed between the meter housing and the main service switch or panelboard. The neutral bus in the main panelboard is the point in the premises where most systems are grounded. See *Figure 7*.

3.3.2 Grounding Electrode System

The grounding electrode system, also known as the earthing system, consists of a grounding field (earth), a grounding electrode, and a grounding electrode conductor. This section covers the grounding field and grounding electrode. The methods of grounding an electric service are covered in **NEC Section 250.50**. *Figure 8* shows examples of NEC®-approved grounding electrodes.

In general, all of the following grounding electrodes (if available) and any made (rod, pipe, plate) electrodes must be bonded together to form the grounding electrode system:

- An underground water pipe in direct contact with the earth for no less than 10'
- The metal frame of a building where effectively grounded

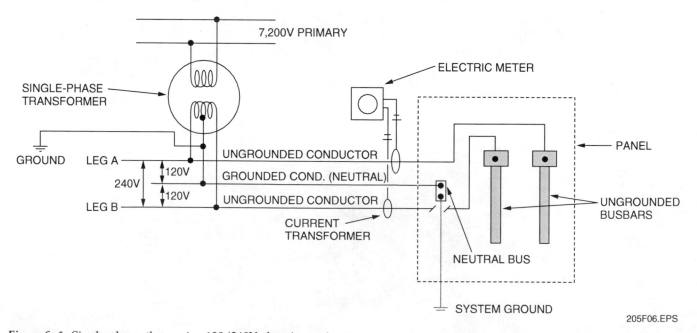

Figure 6 ◆ Single-phase, three-wire, 120/240V electric service.

205F06.EPS

NEUTRAL BUSES

GROUNDING CONDUCTOR FROM OUTSIDE SERVICE ENTRANCE

205F07.EPS

Figure 7 ◆ Neutral bus in interior residential panel (rough-in stage).

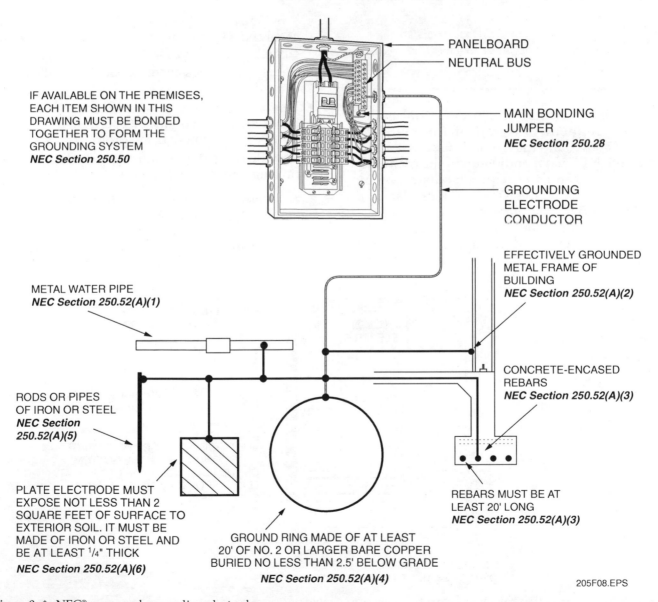

PANELBOARD

NEUTRAL BUS

IF AVAILABLE ON THE PREMISES, EACH ITEM SHOWN IN THIS DRAWING MUST BE BONDED TOGETHER TO FORM THE GROUNDING SYSTEM
NEC Section 250.50

MAIN BONDING JUMPER
NEC Section 250.28

GROUNDING ELECTRODE CONDUCTOR

METAL WATER PIPE
NEC Section 250.52(A)(1)

EFFECTIVELY GROUNDED METAL FRAME OF BUILDING
NEC Section 250.52(A)(2)

RODS OR PIPES OF IRON OR STEEL
NEC Section 250.52(A)(5)

CONCRETE-ENCASED REBARS
NEC Section 250.52(A)(3)

PLATE ELECTRODE MUST EXPOSE NOT LESS THAN 2 SQUARE FEET OF SURFACE TO EXTERIOR SOIL. IT MUST BE MADE OF IRON OR STEEL AND BE AT LEAST 1/4" THICK
NEC Section 250.52(A)(6)

GROUND RING MADE OF AT LEAST 20' OF NO. 2 OR LARGER BARE COPPER BURIED NO LESS THAN 2.5' BELOW GRADE
NEC Section 250.52(A)(4)

REBARS MUST BE AT LEAST 20' LONG
NEC Section 250.52(A)(3)

205F08.EPS

Figure 8 ◆ NEC®-approved grounding electrodes.

- An electrode encased by at least 2" of concrete, located within and near the bottom of a concrete foundation or footing that is in direct contact with the earth

> **NOTE**
> This electrode must be at least 20' long and must be made of electrically conductive coated steel reinforcing bars or rods of not less than ½" in diameter, or consisting of at least 20' of bare copper conductor not smaller than No. 4 AWG wire size.

- A ground ring encircling the building or structure, in direct contact with the earth at a depth not less than 2.5' below grade

> **NOTE**
> This ring must consist of at least 20' of bare copper conductor not smaller than No. 2 AWG wire size.

Grounding systems used in industrial buildings will frequently use all of the methods described above, and the methods used will often surpass the NEC®, depending upon the manufacturing process and the requirements made by plant engineers. *Figure 9* shows a floor plan of a typical industrial grounding system.

In some structures, only the water pipe must be supplemented by an additional electrode. This is specified in *NEC Section 250.52*. *Figure 10* shows a typical electric service and the available grounding electrodes. The building in *Figure 10* has a metal underground water pipe that is in direct contact with the earth for more than 10', so this is one approved grounding source. The building also has a metal underground gas piping system, but this may not be used as a grounding electrode [*NEC Section 250.52(B)*].

NEC Section 250.53(D)(2) further states that the underground water pipe must be supplemented by an additional electrode of a type specified in *NEC Section 250.52*. Since a grounded metal building frame, concrete-encased electrode, or ground ring is not available in this application, *NEC Section 250.52* must be used in determining the supplemental electrode. In most cases, this supplemental electrode will consist of either a driven ground rod (*Figure 11*) or pipe electrode, the specifications for which are as follows:

- Provide a low-impedance path to ground for personal and equipment protection and effective circuit relaying
- Withstand and dissipate repeated fault and surge currents
- Provide corrosion resistance to various soil chemistries to ensure continuous performance for the life of the equipment being protected
- Provide rugged mechanical properties for easy driving with minimum effort and rod damage

An alternate method to the pipe or rod method is a plate electrode. Each plate electrode must expose not less than two square feet of surface to the surrounding earth. Plates made of iron or steel must be

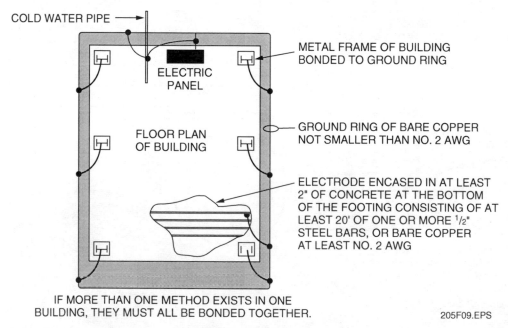

Figure 9 ◆ Floor plan of the grounding system for an industrial building.

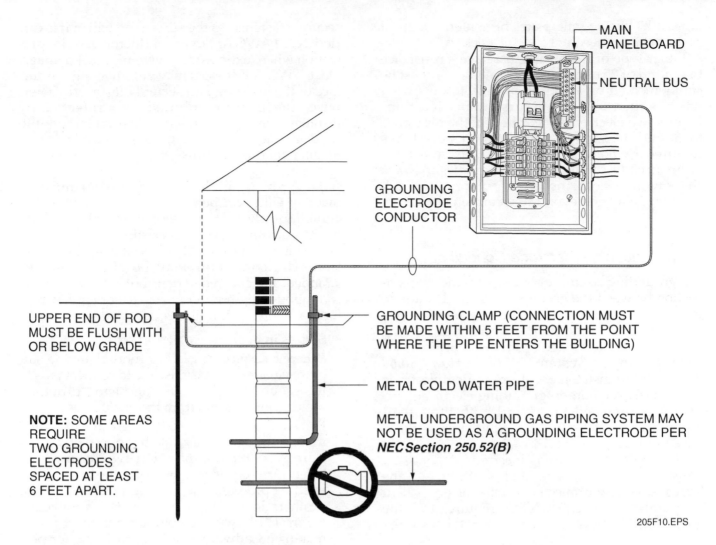

MAIN PANELBOARD

NEUTRAL BUS

GROUNDING ELECTRODE CONDUCTOR

UPPER END OF ROD MUST BE FLUSH WITH OR BELOW GRADE

GROUNDING CLAMP (CONNECTION MUST BE MADE WITHIN 5 FEET FROM THE POINT WHERE THE PIPE ENTERS THE BUILDING)

METAL COLD WATER PIPE

NOTE: SOME AREAS REQUIRE TWO GROUNDING ELECTRODES SPACED AT LEAST 6 FEET APART.

METAL UNDERGROUND GAS PIPING SYSTEM MAY NOT BE USED AS A GROUNDING ELECTRODE PER *NEC Section 250.52(B)*

205F10.EPS

Figure 10 ◆ Grounding requirements for non-industrial buildings.

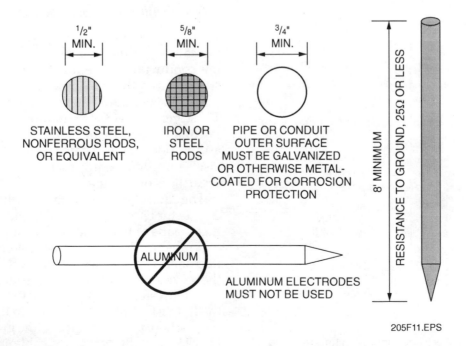

1/2" MIN.

STAINLESS STEEL, NONFERROUS RODS, OR EQUIVALENT

5/8" MIN.

IRON OR STEEL RODS

3/4" MIN.

PIPE OR CONDUIT OUTER SURFACE MUST BE GALVANIZED OR OTHERWISE METAL-COATED FOR CORROSION PROTECTION

8' MINIMUM

RESISTANCE TO GROUND, 25Ω OR LESS

ALUMINUM

ALUMINUM ELECTRODES MUST NOT BE USED

205F11.EPS

Figure 11 ◆ Requirements for ground rods.

at least ¼" thick, while plates of nonferrous metal such as copper need only be .06" thick.

Either type of electrode must have a resistance to ground of 25 ohms (Ω) or less. If not, it must be supplemented by an additional electrode spaced not less than 6' from the first. Many locations require two electrodes regardless of the resistance to ground. This is not a NEC® requirement, but is required by some power companies and local ordinances in some cities and counties. Always check with the local inspection authority for such rules that may go beyond the requirements of the NEC®.

3.3.3 Grounding Electrode Conductor

The grounding electrode conductor is the sole connection between the grounding electrode and the grounded system conductor for a grounded system, or the sole connection between the grounding electrode and the service equipment enclosure for an ungrounded system. *NEC Section 250.64* describes the installation of grounding electrode conductors. A common grounding electrode conductor is required to ground both the circuit grounded conductor and the equipment grounding conductor (*Figure 12*).

The size of a grounding electrode depends on the service-entrance size; that is, the size of the largest service-entrance conductor or equivalent for parallel connectors. NEC® requires that the grounding electrode conductor or its enclosure be securely fastened to the surface on which it is carried. No. 4 AWG or larger conductors require protection where exposed to severe physical damage. No. 6 AWG conductors may be run along the surface of the building and securely fastened. Otherwise, the conductor must be protected by installation in rigid or intermediate metal conduit, rigid nonmetallic conduit, EMT, or cable armor. Smaller conductors must be protected in conduit or armor.

The grounding electrode conductor must be made of either copper, aluminum, or copper-clad aluminum. The material selected must be resistant to any corrosive condition existing at the installation, or it must be suitably protected against corrosion. The conductor may be either solid or stranded, and covered or bare, but it must be in one continuous length without a splice or joint, except for the following conditions:

- Splices in busbars are permitted.
- Where a service consists of more than one single enclosure, it is permissible to connect taps to the grounding electrode conductor, provided the taps are made within the enclosures.
- Per *NEC Section 250.64(C)*, grounding electrode conductors may only be spliced by means of irreversible compression-type connectors listed for the purpose or using the exothermic welding process. One popular manufacturer of exothermic welding components is Cadweld®. *Figure 13* shows the components used when making a Cadweld® joint. A procedure for performing the weld is provided in *Appendix A*.

3.3.4 Equipment Grounding System

The equipment grounding system, also known as the safety ground, consists of equipment grounding conductors and equipment bonding. *Figure 14* summarizes the equipment grounding rules for most types of equipment that must be grounded. These general NEC® regulations apply to all installations except for specific equipment (special applications) as indicated in the code. The NEC® also lists specific equipment that is to be grounded regardless of voltage.

The equipment grounding conductor (EGC) is the conductor used to connect the non-current-carrying metal parts of fixed/portable equipment, raceways, and other enclosures to the system grounded conductor and/or the grounding electrode conductor at the service equipment or at the source of a separately derived system. The EGC or path must extend from the farthest point on the circuit to the service equipment where it is connected to the grounded conductor. Most metallic raceways, cable sheaths, and cable armor that are continuous and use proper fittings

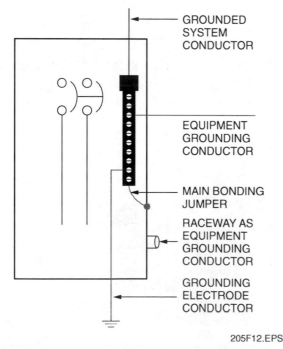

GROUNDED SYSTEM CONDUCTOR

EQUIPMENT GROUNDING CONDUCTOR

MAIN BONDING JUMPER

RACEWAY AS EQUIPMENT GROUNDING CONDUCTOR

GROUNDING ELECTRODE CONDUCTOR

205F12.EPS

Figure 12 ◆ Grounding electrode connection for grounded system.

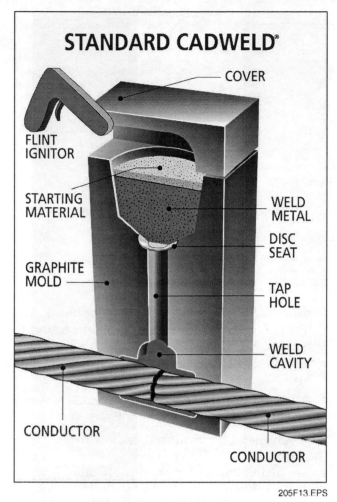

STANDARD CADWELD®

COVER

FLINT
IGNITOR

STARTING
MATERIAL

WELD
METAL

DISC
SEAT

GRAPHITE
MOLD

TAP
HOLE

WELD
CAVITY

CONDUCTOR

CONDUCTOR

205F13.EPS

Figure 13 ◆ Exothermic method of splicing grounding conductors (Cadweld®).

may serve as the EGC. A separate grounding conductor is needed when plastic conduit, nonmetallic-sheathed cable, or other wiring methods are used that are not approved grounding methods. ECGs carry fault current from the load to the grounded terminal bar of the service equipment. The size of the EGC is determined from the size of the overcurrent device protecting that particular system. *Figure 15* shows an EGC and its path to ground when a ground fault occurs in a portable power tool.

The difference between a short circuit and ground fault failure is often misunderstood. Short circuits are defined as conducting connections, accidental or intentional, between any of the current-carrying conductors of an electrical system and not through the intended load. A failure may occur because of a connection from one phase conductor to another phase conductor (line-to-line) or from one phase conductor to the grounded conductor or neutral (*Figure 16*).

A ground fault is defined as a conducting connection, accidental or intentional, between any conductors of an electrical system and the normally non-current-carrying conducting material enclosing the conductors, as well as any conductors that are grounded or may become grounded. In a ground fault, there is a connection from a current-carrying phase conductor to the conductor enclosure such as a metal conduit, a metal box, or motor frames (*Figure 17*). A ground fault typically occurs when the conductor insulation fails or when a wire comes lose from its terminal point.

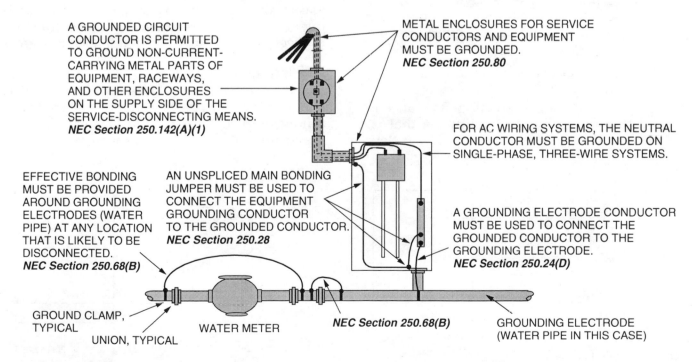

A GROUNDED CIRCUIT CONDUCTOR IS PERMITTED TO GROUND NON-CURRENT-CARRYING METAL PARTS OF EQUIPMENT, RACEWAYS, AND OTHER ENCLOSURES ON THE SUPPLY SIDE OF THE SERVICE-DISCONNECTING MEANS.
NEC Section 250.142(A)(1)

METAL ENCLOSURES FOR SERVICE CONDUCTORS AND EQUIPMENT MUST BE GROUNDED.
NEC Section 250.80

FOR AC WIRING SYSTEMS, THE NEUTRAL CONDUCTOR MUST BE GROUNDED ON SINGLE-PHASE, THREE-WIRE SYSTEMS.

EFFECTIVE BONDING MUST BE PROVIDED AROUND GROUNDING ELECTRODES (WATER PIPE) AT ANY LOCATION THAT IS LIKELY TO BE DISCONNECTED.
NEC Section 250.68(B)

AN UNSPLICED MAIN BONDING JUMPER MUST BE USED TO CONNECT THE EQUIPMENT GROUNDING CONDUCTOR TO THE GROUNDED CONDUCTOR.
NEC Section 250.28

A GROUNDING ELECTRODE CONDUCTOR MUST BE USED TO CONNECT THE GROUNDED CONDUCTOR TO THE GROUNDING ELECTRODE.
NEC Section 250.24(D)

GROUND CLAMP, TYPICAL

UNION, TYPICAL

WATER METER

NEC Section 250.68(B)

GROUNDING ELECTRODE (WATER PIPE IN THIS CASE)

205F14.EPS

Figure 14 ◆ Bonding jumpers are required in some cases to ensure continuity.

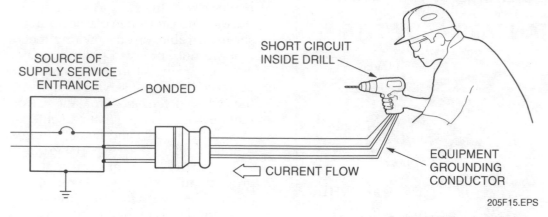

Figure 15 ◆ In the event of a ground fault, current flows through the equipment conductor back to ground.

3.3.5 Bonding

Equipment bonding covers the interconnection of metallic non-current-carrying parts of an electrical system. Such parts include metallic conduit, outlet boxes, enclosures and frames on motors, and other electrically operated equipment. These items are bonded together for the following purposes:

- To limit the voltage to ground on metallic enclosures and conduit
- To ensure operation of overcurrent devices in case of ground faults

Bonding means conductively joining all metal parts of the wiring system: boxes, cabinets, enclosures, and conduit. It ensures having good, continuous metallic connections throughout the grounding system. Bonding is required for the following:

- All conduit connections at the electrical service equipment
- Points where a nonconducting substance is used that might impair continuity; that is, where fault current may not be able to flow

past the nonconductor. Such bonding is done by connecting a conductive jumper around the nonconducting substance
- All service equipment enclosures whether inside or outside the building
- All metallic components of the electrical system that are normally non-current-carrying

When metal conduit and enclosures are used, the conduit itself may serve as the grounding means if the conduit provides an uninterrupted path back to the service entrance. However, for additional protection, a grounding conductor should be installed with the circuit. Circuits having conductors enclosed in nonmetallic sheaths or plastic conduit must have an additional grounding wire. This wire must be interconnected between outlets. When a grounding conductor is properly bonded to the metal service equipment, a fault from line-to-ground on the load side of a circuit protection device (fuse or circuit breaker) will follow a low-resistance path to a grounded conductor at the transformer. This would allow the circuit protection device to open rapidly.

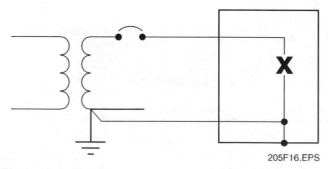

Figure 16 ◆ Short circuit.

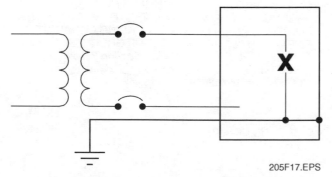

Figure 17 ◆ Ground fault.

3.3.6 Bonding and Grounding for Specific Equipment

The NEC® also contains several articles pertaining to the bonding and grounding for specific types of equipment. Like *NEC Article 250*, these articles are concerned mainly with grounding and bonding issues relating to the protection of people and property from electrical hazards. Some of the more common of these articles are referenced as follows:

- *NEC Article 640* – *Audio Signal Processing, Amplification, and Reproduction Equipment*
- *NEC Article 700* – *Emergency Systems*
- *NEC Article 701* – *Legally Required Standby Systems*
- *NEC Article 720* – *Circuits and Equipment Operating at Less Than 50 Volts*
- *NEC Article 725* – *Class 1, Class 2, and Class 3 Remote-Control, Signaling, and Power-Limited Circuits*
- *NEC Article 727* – *Instrumentation Tray Cable: Type ITC*
- *NEC Article 760* – *Fire Alarm Systems*
- *NEC Article 770* – *Optical Fiber Cables and Raceways*
- *NEC Article 800* – *Communications Circuits*
- *NEC Article 810* – *Radio and Television Equipment*
- *NEC Article 820* – *Community Antenna Television and Radio Distribution Systems*
- *NEC Article 830* – *Network-Powered Broadband Communications Systems*

Note that the installation of many types of specialized equipment often involves the installation of additional grounding and bonding other than that specifically required by NEC®. These additional requirements, normally specified by the equipment manufacturer and/or covered in the American National Standards Institute/Telecommunications Industry Association/Electronic Industries Alliance (ANSI/TIA/EIA) standards, do not replace the requirements for electrical power bonding and grounding, but supplement them with additional bonding and grounding designed to enhance equipment performance and/or reliability. Some examples of bonding and grounding techniques commonly used to enhance the performance and/or reliability of equipment are described in various sections later in this module.

3.3.7 Typical Telecommunications System Grounding and Bonding

An example of how the grounding and bonding of a specialized equipment system is interfaced with the building's electrical grounding system is discussed here. For our example, we will describe the building grounding scheme used with the typical telecommunications system shown in *Figure 18*. The bonding and grounding of other types of systems would be accomplished in somewhat the same way.

NEC Sections 800.10 and *800.11* cover the requirements for communication cabling that enters buildings. *NEC Section 800.10* covers conductors entering a building from aboveground locations and *NEC Section 800.11* covers underground circuits entering a building. Any underground communications conductors run in the same raceway, handhole enclosure, or manhole containing electric light, power, Class 1, or non-power-limited fire alarm conductors must be suitably separated from these conductors per *NEC Section 800.11(A)*. A separate grounding electrode must be installed in accordance with *NEC Section 800.40*, where required by *NEC Section 800.33*. When making such connections, the service provider or Authority Having Jurisdiction (AHJ) should be consulted as to the specific grounding requirements and size of wires to be used. A description of an AHJ is covered later in the *Introduction to Codes and Standards* module.

The interface between the building's electrical grounding system and the telecommunications grounding system begins at the telecommunications main grounding busbar (TMGB) shown directly bonded to the building's electrical service panel and exterior wall. A telecommunications bonding backbone (TBB) conductor connected to the TMGB is used to interconnect the telecommunications grounding busbars (TGBs) located in the telecommunications closets within the building. The TBB is designed to interconnect busbars and is not intended to have equipment bonding conductors spliced on to it. Each TBB should be a continuous conductor from the TMGB out to the farthest TGB. Intermediate TGBs should be spliced on to the TBB with a short bonding conductor. The TBB is connected to the busbars with a two-bolt attachment to provide a secure connection. Larger

INSIDE TRACK

Securing Bonding and Grounding Strips

Always remove and thoroughly clean any paint or other contamination from the mounting surface before securing any bonding or grounding strip to equipment housings. Failure to prepare the surface may impede fault current flow to ground.

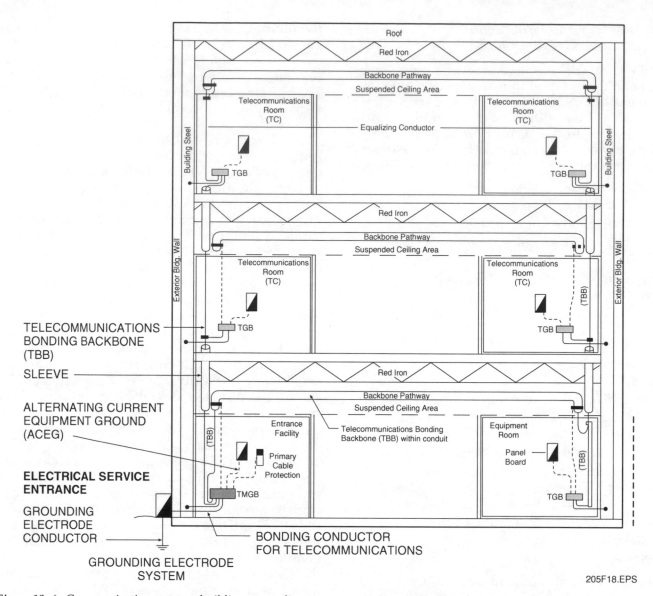

Figure 18 ◆ Communications system building grounding.

telecommunications closets may require more than one TGB. Multiple TGBs within a closet are allowed provided they are all bonded together.

In larger multi-story buildings that have more than one telecommunications closet on each floor, the potentials between the closets need to be equalized. When applicable, a telecommunications equalizing conductor is installed between TGBs to interconnect all the closets on the same

Equalizing Conductor

According to *ANSI/ TIA/EIA-607*, the equalizing conductor must be no less than No. 6 AWG in size and may be as large as AWG 3/0. It must be green insulated stranded copper cable.

floor. The equalizing conductor is not required for every floor, but is installed on every third floor and the top floor.

Another bonding conductor, called a coupled bonding conductor (CBC), provides equalization like a TBB and also provides a different form of protection through electromagnetic coupling (close proximity) with the telecommunications cable (*Figure 19*). The CBC is generally considered part of an installed telecommunications cable and not part of a grounding and bonding infrastructure. Some **private branch exchange (PBX)** equipment manufacturers specify a CBC between their equipment and exposed circuit protectors. The CBC can be a cable shield or a separate copper conductor tie wrapped at regular intervals to an unshielded cable. To work properly, the CBC must be connected directly to the protector ground and the PBX ground.

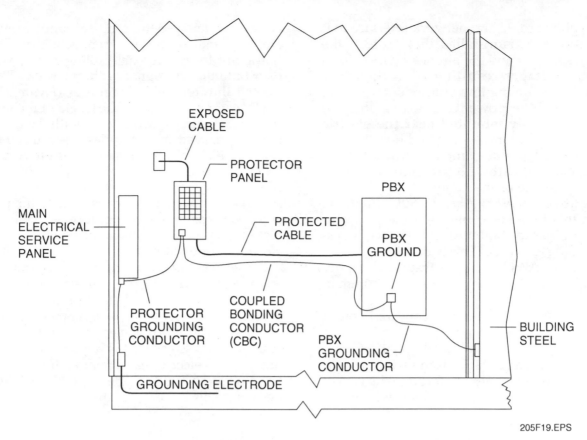

Figure 19 ◆ Communications small system grounding.

Some PBX manufacturers and other large equipment manufacturers require single-point equipment grounding. When equipment is grounded through only one point, surges that are conducted through the building ground will not pass through the equipment. A lot of smaller equipment relies solely on a receptacle (safety) equipment grounding conductor, power cord, and plug for adequate grounding.

Telephone and other telecommunications companies normally provide primary protectors where incoming wires and cables are exposed to lightning. These protectors serve to protect the incoming wires and cables from hazardous voltages resulting from lightning strikes on the incoming lines or accidental contact with electric light or power conductors. Telecommunications networks that have premise system wires and/or cables that run between buildings should also have primary protectors installed if these system wires or cables can be exposed to lightning strikes or accidental contact with electric light or power conductors. A primary protector is required at each end of such an interbuilding communications circuit. These primary protectors must be located in, on, or immediately next to the building being served and as close as practical to the point at which the exposed conductors enter and attach. The primary

protectors typically consist of an arrester connected between each line conductor and ground in an appropriate mounting. In some installations, secondary protectors are also installed in series with the indoor communications wiring and cables between the primary protector and the equipment. This secondary protector provides additional protection by safely limiting currents to less than the current-carrying capacity of the listed indoor communications wires, cables, and terminal equipment. Protective devices and their applications are discussed in detail later in this module.

3.3.8 Typical Residential Telecommunications Grounding and Bonding

Grounding and bonding requirements for telecommunication systems installed in houses and other residential applications are fairly straightforward. The grounding and bonding requirements for telephone and community antenna television (CATV) systems are typical examples and are briefly described here.

For residential telephone systems, the telephone company installs the service line (cable pair) to a residence and terminates it at a standard network interface unit (*Figure 20*). Between the incoming phone cable and network interface unit a protector device is installed that protects the incoming cable pair and the premises wiring from hazardous voltages resulting from lightning strikes on the incoming line or accidental contact with electric light or power conductors. The protector is commonly mounted near the electric meter, where proper grounding and bonding can be done. The telephone company grounds their incoming cable sheath and protector in accordance with the one of the many approved methods described in *NEC Article 800, Part IV*. Typically, this involves connecting the protector ground terminal to the building's electrical system grounding electrode. This connection is made using a No. 14 AWG or larger insulated copper grounding electrode conductor. Another common method is to fasten the protector device to the grounded metal service raceway conduit in order to establish the proper ground.

When installing a CATV system, the CATV company runs a coaxial cable from their overhead or underground cable network through the wall of a residence at some convenient point where it is typically connected to a multiset coupler unit or an amplifier that serves as a connection or distribution point for the coaxial cabling runs needed to distribute the TV signal to the modular TV jacks located throughout the premises (*Figure 21*). The CATV company grounds the shield of their incoming coaxial cable in accordance with the one of the many approved methods described in *NEC Section 800.40*. Typically, this involves grounding the outer conductive shield of the incoming coaxial cable via a grounding block to the main electrical system electrode at a point as close to the point of entry as possible. This connection is made using a No. 14 AWG or larger insulated copper grounding electrode conductor.

Sometimes the method used for grounding a telephone, CATV, or other telecommunications system can result in a dedicated grounding electrode for the telephone, CATV, or other system and another grounding electrode for the electrical system. In these cases, it is required by the NEC® that the two electrodes be bonded together with a bonding jumper not smaller than No. 6 AWG copper wire. This bonding of the telephone or CATV

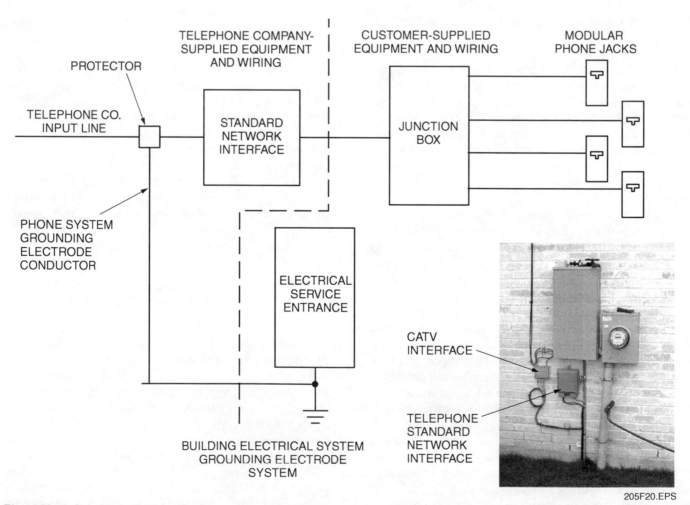

Figure 20 ◆ Standard network interface.

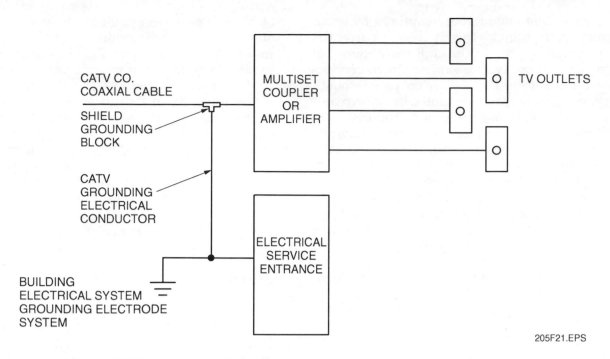

Figure 21 ◆ Basic residential CATV system grounding and bonding.

system electrode to the main electrical service grounding electrode is necessary in order to minimize the possibility that a difference of potential might exist between the two grounding systems.

4.0.0 ◆ LIGHTNING PROTECTION

A building's lightning protection equipment and devices are designed for the exclusive purposes of intercepting, directing, and dissipating direct lightning strikes into the ground. The conventional lightning protection system consists of multiple rooftop air (lightning) terminals placed on the highest points of the structure where lightning is most likely to strike, down conductors, equalizing conductors, and solidly and permanently grounded terminals that surround the building (Figure 22). Some systems are designed into the building structure so that the structural steel serves as the down conductors and related equalizing grid. Lightning is substantially prevented from directly striking within and under the tips of the lightning terminals, thus forming a zone of protection around the building.

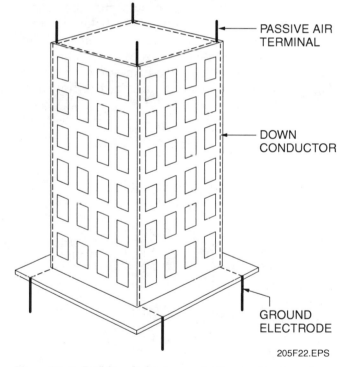

Figure 22 ◆ Building lightning protection system (passive system).

Commercial and Industrial Lightning Protection Standards

Air terminals should be spaced approximately 20' apart around the perimeter of the building's rooftop. Ground rods should be 10' in length and driven fully into the ground.

Once the lightning has been captured by an air terminal or terminals, the discharge current is safely routed to ground through the network of down conductors. Typically, these down conductors are made from flat copper strips or smooth copper or aluminum woven cables. The down conductors for the lightning protection ground system are connected to ground electrodes separate from, but similar in construction to, the ground electrodes used for the building electrical system power ground. The lightning system ground electrodes are bonded to the power distribution system ground electrode(s), and the ground electrodes of other building systems, such as the telecommunication system, in order to provide for ground potential equalization. Creating an **equipotential ground plane** under transient conditions is essential for the safety of equipment and personnel. Proper grounding also contributes to the reduction of electrical noise and provides a reference for circuit conductors to stabilize their voltage to ground during normal operation.

The conventional lightning protection system described above is considered a passive system. This is because the air terminals must be placed where lightning is most likely to strike, but they are attractive to lightning only by their positioning, hence they are passive. Active attraction protection systems are also used for some structures (*Figure 23*). These systems attract the lightning

strike to a preferred point by the use of a special spherical-shaped air terminal that generates a pre-ionized path (streamer). This streamer, which is emitted upward from the terminal, acts to intercept the lightning downstroke and route it to the air terminal. From the air terminal, the energy is routed safely to ground in the same manner as the conventional system. Some advantages of the active-type system are that fewer air terminals are required and only one down conductor is needed per terminal. Typically, one active air terminal can provide up to 300' of protection.

5.0.0 ◆ CAUSES OF POOR AC POWER QUALITY

As explained previously, the AC power provided by the power utility has certain characteristics such as voltage and frequency. Under certain conditions, such as severe weather or changes in utility or building load conditions, these characteristics and others can become distorted, thus resulting in poor power quality. This section describes the many conditions that can contribute to poor power quality. The methods and devices used in a building and related equipment to help protect against and/or eliminate the effects of poor power quality are described in following sections. Some of the common types of power abnormalities that can occur include the following:

- Voltage transients/surges
- Voltage swell/sag
- Overvoltage/undervoltage
- Voltage interruptions
- Frequency variations
- Harmonics
- Noise/**electromagnetic interference (EMI)**

5.1.0 Voltage Transients and Surges

Voltage transients (*Figure 24*), also called spikes, are instantaneous, dramatic increases in voltage, typically caused by nearby lightning strikes. Lightning strikes can induce high voltages in outdoor electric lines during lightning storms. These induced voltages produce traveling surges of high voltage that travel along the lines and into electrical equipment located both indoors and outdoors. The high induced voltages can puncture the insulation of equipment and be dangerous to life. Transients can also occur when utility power comes back on line after having been knocked out in a storm, or as a result of a car accident.

Surges are short-duration variations in power (less than one half-cycle of the normal voltage waveform) with a rapid increase in voltage or cur-

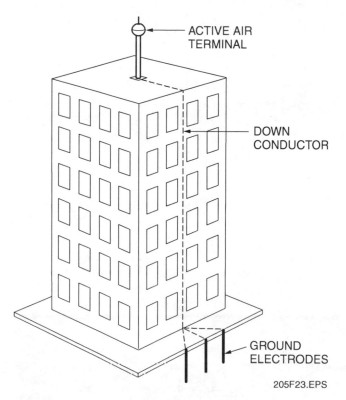

ACTIVE AIR TERMINAL

DOWN CONDUCTOR

GROUND ELECTRODES

205F23.EPS

Figure 23 ◆ Building lightning protection system (active system).

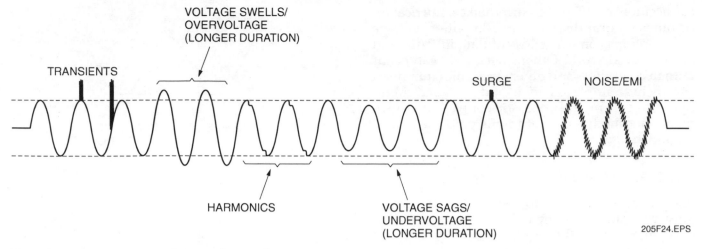

Figure 24 ◆ Common types of power abnormalities.

rent that can often be up to thousands of volts in amplitude. Surges typically result because of the interaction between electrical energy stored in power utility and/or user circuit load inductances and capacitances when electrical loads are switched out of a circuit. For example, when current flow is interrupted at peak flow in **reactive loads** such as transformers, electric motors, etc., these devices are left with considerable stored energy, which can cause surges to occur. These surges can have a peak voltage value as high as ten times the normal peak. Like transients, the high voltages caused by switching surges can puncture the insulation of equipment and be dangerous to life if the voltage generated becomes high enough.

5.2.0 Voltage Swells and Sags

A voltage swell is a relatively short-term (up to about one minute) increase in the nominal rms voltage and related current. Swells can upset electric controls such as variable-speed motor drives, which can trip because of their built-in protective circuitry.

Voltage sags, also called brownouts, are relatively short-term decreases in the nominal rms

voltage and related current lasting from .05 cycles to one minute. According to a Bell Labs study, sags are the most common power problem, accounting for 87 percent of all power disturbances. Sags are typically caused by start-up power demands of many electrical devices, including motors, compressors, elevators, and shop tools.

5.3.0 Overvoltage and Undervoltage

Overvoltage is an increase in the nominal rms voltage that can last anywhere from one minute to several hours. Similarly, undervoltage is a decrease in the nominal rms voltage that can last anywhere from one minute to several hours. With an overvoltage or undervoltage condition, the voltage seldom rises or falls as dramatically as it does with swells and sags. Because these events can last for hours, they cause stress on computers and controllers, as well as motors and other conventional loads. Undervoltages are frequently caused by deliberate utility voltage reductions. In a procedure known as rolling brownouts, the utility systematically lowers voltage levels in certain areas for hours or days at a time. Hot summer days, when air conditioning requirements are at their peak, will often prompt the use of rolling brownouts. Undervoltages can affect the output of capacitor banks installed to help minimize power losses. Overvoltage has fewer immediate effects, but over time will tend to shorten the life of the power distribution and other equipment.

5.4.0 Voltage Interruptions

Voltage interruptions, commonly called blackouts, are a total absence of voltage on one or more phase conductors for a period of time. Blackouts are typically caused by excessive demand on the power grid, lightning storms, ice on power lines,

Voltage Sag in Your Home

Whenever your central air conditioning unit automatically turns on, the lights in your house may temporarily dim. This is a typical example of voltage sag caused by the high power start-up demand of the AC compressor, which is driven by an AC motor. The startup of any large AC motor will generally cause such a sag.

car accidents, backhoes, earthquakes, hurricanes, or other natural disasters. In computers, voltage interruptions can cause loss of data in RAM and any unsaved work. Other effects that can occur both in computers and other electronic equipment can include improper logic and memory circuit operation in microprocessor-controlled equipment, unscheduled shutdowns, and/or equipment damage.

5.5.0 Frequency Variations

Frequency variations do not commonly occur; however, the frequency can vary from the standard due to a badly regulated source or utility company correction/switching.

5.6.0 Harmonics

A harmonic is a sinusoidal wave having a frequency that is a multiple of the fundamental system frequency. For example, harmonics of a 60Hz sine wave would be 120Hz (second harmonic), 180Hz (third harmonic), 300Hz (fifth harmonic) and so on. When one or more harmonic components are added to the fundamental frequency, a distorted (non-sinusoidal) waveform is produced (*Figure 25*). Since these distorted waveforms are no longer simply related, they are said to be non-linear waveforms. Loads that produce harmonics are called non-linear loads.

Harmonics are especially prevalent whenever there is a large amount of equipment such as switching power supplies, personal computers, adjustable speed drives, and medical test equipment that draw current in short pulses. Electronic ballasts used in fluorescent and high-intensity discharge (HID) lighting fixtures are also common sources of harmonic generation. All these devices are designed to draw current only during a controlled portion of the incoming voltage waveform, which causes harmonics to be generated in the load currents. Any loads sharing a transformer or a branch circuit with a heavy harmonic load can be affected by the voltage and current harmonics generated. This can result in overheated transformers and neutrals, as well as tripped circuit breakers or blown fuses.

Currents resulting from harmonics cause overheating in wires, cables, motors, and transformers as a result of skin effect. Skin effect is the increase in the AC resistance of a conductor as frequency increases. In a conductor, the current density is greatest near the surface or skin. The higher the frequency of the harmonics, the less skin depth available in the conductor, resulting in excess heat.

A serious problem can arise from harmonic distortion in three-phase, four-wire power distribu-

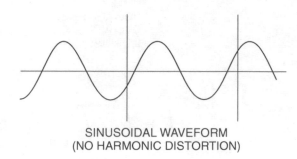

SINUSOIDAL WAVEFORM
(NO HARMONIC DISTORTION)

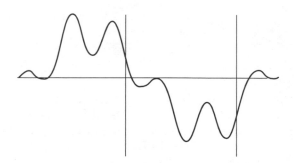

DISTORTED CURRENT WAVEFORM

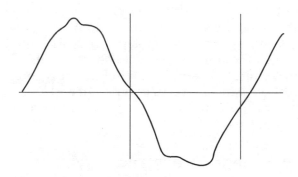

DISTORTED VOLTAGE WAVEFORM

205F25.EPS

Figure 25 ◆ Non-sinusoidal waveform resulting from harmonic distortion.

tion systems. In these systems, neutral conductors can be severely affected by non-linear loads connected to the 120V branch circuits. Under normal conditions for a balanced linear load, the fundamental 60Hz portion of the phase currents will cancel in the neutral conductor. In a four-wire system with single-phase, non-linear loads, certain odd-numbered harmonics called triplens that are odd multiples of the third harmonic (3rd, 9th, 15th, and so on), do not cancel, but rather add together in the neutral conductor.

In systems with a large number of single-phase, non-linear loads, triplen harmonics can cause the neutral current to actually exceed the phase current. This can cause excessive overheating because there is no circuit breaker in the neutral conductor to limit the current as there are in the phase conductors. When this high neutral current reaches the power system transformer, it is reflected into the primary winding where it circulates and causes overheating and resultant transformer failures. Excessive current in the neutral conductor can also cause higher-than-normal voltage drops between the neutral conductor and ground at 120V outlets.

Remember that harmonics are not transients. The distorted waveform caused by transients may contain high-frequency components, but these occur only briefly after there has been an abrupt change in the power system. These frequencies have no relation to the fundamental system frequency; therefore, they are not harmonics. They are the natural frequencies of the system as determined by the circuit inductance and capacitance involved at the time of the switching operation.

5.7.0 Noise/Electromagnetic Interference

Noise is defined as any unwanted electrical signals that are induced onto or superimposed on power or signal lines. Note that voltage and current transients/surges are commonly referred to by some as noise signals. Noise interference appears between the terminals of a circuit. Noise is classified in two ways: common-mode noise and normal-mode noise. Common-mode noise occurs between the line and ground or neutral and ground (*Figure 26*), but not between each line. The noise signals on each of the current-carrying conductors are in phase and equal in magnitude; thus, no voltage signal is generated between the conductors by the noise. Normal-mode noise, also called traverse-mode noise, occurs between the current-carrying conductors (line-to-line) or line-to-neutral. A voltage is generated between the ground and neutral lines because noise is only present on two of the conductors.

The terms electromagnetic interference (EMI) and **radio-frequency interference (RFI)** are used interchangeably and are sometimes used in the same context as the term noise. EMI is a more general term than RFI. EMI is stray electrical energy radiated from electrical and electronic systems, including the related cables. The presence of EMI can cause undesirable distortion or interference to signals in other nearby cables or electronic equipment. Digital computing devices are known producers of high EMI emissions. Electrical cables can both be a transmitter and receiver of EMI. EMI generators emit or radiate an electromagnetic noise field which can be picked up by the following:

- Telecommunications and power lines
- Power supplies
- Interconnections
- Radio, TV, and closed-circuit camera receivers
- Computers
- Telecommunications and data systems
- Antennas

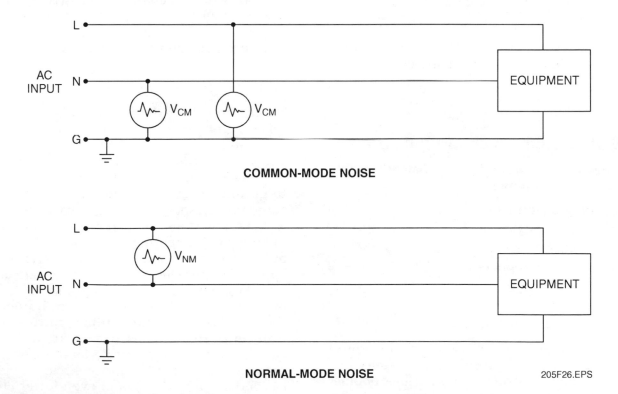

Figure 26 ◆ Common-mode and normal-mode noise.

Electrical cables are vulnerable to receiving EMI noise from nearby sources. The transfer of noise can occur over one or more paths by radiation, conduction, and/or inductive and **capacitive coupling**. Note that optical fibers neither emit nor receive EMI. Improper bonding, shielding, and grounding of cable shields and equipment can increase the susceptibility of a cable or device to EMI.

In the United States, the Federal Communications Commission (FCC) is responsible for specifying EMI limits to prevent unacceptable levels of electromagnetic pollution (interference) being released into the environment. FCC regulations establish the maximum permissible emissions of electronic devices. For computing devices, the FCC separates its regulations for digital EMI into two categories: Class A computing devices (industrial and commercial) and Class B computing devices (residential). The FCC rules set limits on two kinds of emissions: conductive emissions and radio frequency (RF) emissions. Conductive emissions travel through the wires in the power cord. RF emissions radiate from devices into space.

6.0.0 ◆ POWER SYSTEM PROTECTION AND CONDITIONING EQUIPMENT

Power system protection and conditioning equipment is used to correct or prevent one or more power abnormalities and can be divided into the following three broad categories:

- Power filters and regulators
- Motor and engine generators
- Static uninterrupted power supplies

These categories are discussed in the following sections.

6.1.0 Power Filters and Regulators

Equipment that filters and/or regulates the utility line includes the following:

- Isolation transformers
- Surge protecting devices
- Voltage regulators
- Power line conditioners
- Harmonic filters

6.1.1 Isolation Transformers

An isolation transformer is used to avoid a direct electric connection between a piece of electric/electronic equipment and the power lines or other source of power. Isolation transformers reject com-

mon-mode (line-to-ground) electrical noise disturbances, but have limited ability to prevent normal-mode (line-to-line) noise. In an isolation transformer, there is no electrical connection between the primary and secondary windings. The ratio of turns in the primary winding as compared to the secondary winding is usually 1:1. They are designed to provide magnetic coupling (flux coupling) between one or more pairs of isolated circuits, without introducing significant coupling of any other signals. Both shielded or unshielded static and portable models are available.

Shielded isolation transformers (*Figure 27*) are electrically isolated and have a grounded electrostatic shield between the primary and secondary windings that directs unwanted signals to ground, thus preventing the electrical disturbances from being transmitted to the load circuits. Shielded isolation transformers can reduce common-mode noise (line-to-ground or neutral-to-ground) and fast transients.

In addition to preventing the transmission of unwanted noise signals and transients, isolation transformers are used for grounding. Since one side of the AC power line is grounded, connecting a piece of grounded equipment to the lines could result in a short circuit unless precautions are taken to ensure that the grounded side of the equipment is connected to the grounded side of the power lines. There are also certain types of equipment, such as those with built-in power supplies, from which a lethal shock could be received when the chassis is exposed. By using an isolation transformer, the chassis becomes a **floating ground**, making it nearly impossible to receive a shock under such conditions.

6.1.2 Surge Protecting Devices

Surge protecting devices are also commonly called lightning arresters, transient suppressors, surge protectors, surge arresters, or transient voltage surge suppressors. They limit peak surge voltages and divert power surges and transients to ground. There are many different kinds of surge suppressors. The designs vary widely due to the wide range of currents and voltage rise times typical of lightning and transients, the level of protection needed, and the different circuit applications in which the suppressors are used.

A stage concept of surge suppressor installation is normally used throughout a building. This involves the installation of high current surge/transient protectors at the electrical service-entrance panel (primary protectors), lower current protectors installed at branch-circuit subpanels (secondary protectors), and even lower current

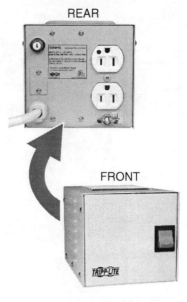

REAR

FRONT

TYPICAL STATIC ISOLATION TRANSFORMER
(15 TO 300kVA)

TYPICAL PORTABLE ISOLATION TRANSFORMER
(HOSPITAL GRADE, 250 TO 1800 WATTS)

205F27.EPS

Figure 27 ◆ Shielded isolation transformers.

protectors (tertiary protectors) installed at the various point-of-use branch circuit receptacles, or in the equipment connected to the branch circuit receptacles. A staged scheme of surge protection is used because the surge suppressor device installed at the service-entrance panel guards against lightning or switching transients imposed on the utility system, but it does not protect the system branch circuits against internally generated transients. Surge suppression devices installed at the individual branch circuit level provide additional protection against currents caused by power faults that are too low in voltage to operate the primary protectors. They also protect against surges generated on other branch circuits, but they do not protect against surges originating on the same circuit. This problem is solved by providing surge protection at the point of use on the branch circuit.

Primary and secondary surge protection devices are made in the form of carbon blocks, gas tubes, or **solid-state devices**. Some contain two or more types of protective devices to meet the needs of specific applications. **Tertiary surge protection devices** are normally solid-state devices. These devices are described in the following text.

Lightning arresters – Lightning/surge arresters (*Figure 28*) are used as primary protectors to limit voltages caused by lightning or switching transients to a safe value and provide a path to ground to dissipate the surge. Two examples of such applications are shown in *Figures 29* and *30*. To provide this protection satisfactorily, arresters must perform the following functions:

- They must not allow the passage of current to ground as long as the voltage is normal.
- When the voltage rises to a predetermined threshold, they must provide a path to ground to dissipate the surge without a further rise in voltage of the circuit.
- As soon as the voltage has been reduced below the threshold of the arrester, the arrester will stop the flow of current to ground and reseal itself so as to insulate (isolate) the conductor from ground.
- Arresters must not be damaged by the discharge and must be capable of automatically repeating their action as frequently as required.

205F28.EPS

Figure 28 ◆ Modular DIN-RAIL transient voltage surge suppressor.

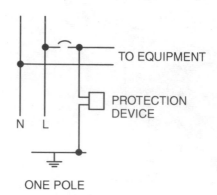

TO EQUIPMENT

PROTECTION DEVICE

N L

NOTE:
The best use of these devices is to place them at the power distribution panel. They may be placed at the equipment or load but they provide less protection. Install as far from the load to be protected as possible.

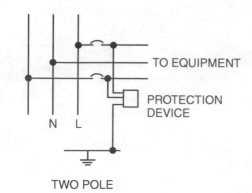

TO EQUIPMENT

PROTECTION DEVICE

N L

ONE POLE

TWO POLE

205F29.EPS

Figure 29 ◆ Example of a lightning/surge suppression device used to protect AC power lines.

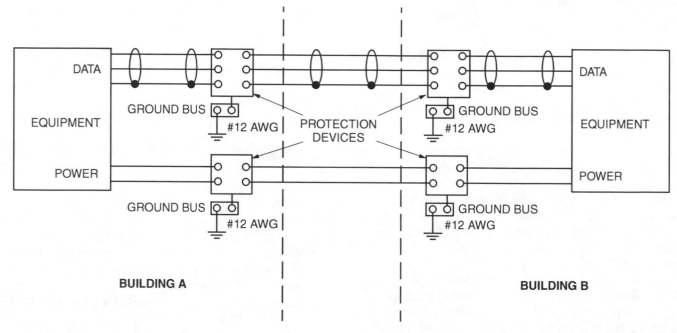

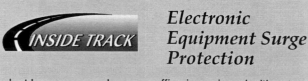

NOTE:

Suppression should be placed as far from the equipment to be protected as possible. It is preferred that they be located just inside the building at the the penetration of the wall. These devices typically do not have their own enclosure and therefore an appropriate electrical enclosure is required. Multiple suppression devices may be placed within the same enclosure.

205F30.EPS

Figure 30 ◆ Example of lighting/surge suppression devices used to protect data and DC power lines.

The carbon-block lightning arrester is the oldest and most basic kind of arrester. It consists of carbon blocks separated by an air gap. It is connected across a current-carrying ungrounded conductor and ground. During normal operation, the air gap isolates the ungrounded conductor from ground. The width of the air gap is such that if a surge or transient appears on the ungrounded conductor, it will arc across the air gap at about 300V to 1,000V, conducting the surge current to ground. When the surge current drops low enough, the arc stops and the arrester once again isolates the ungrounded conductor from ground. A fail-safe function will cause the carbon blocks to short permanently to ground should an extended hot surge or permanent fault current overheat them. Carbon blocks tend to wear out quickly under extreme conditions. Today, the use of carbon-block arresters has diminished in favor of gas tube and solid-state surge protector and arrester devices.

Gas tube arresters are improved arresters used in primary and secondary protector applications that basically operate the same way as carbon-block arresters, arcing over a gas-filled gap to a grounding conductor. They have a higher reliability and tighter tolerances on arc breakdown voltage and are typically designed to arc at a lower voltage, providing better protection than carbon-block arresters. Another type, the dual-gap, provides a common arc chamber that grounds both wires of a pair together and minimizes metallic surges that would otherwise occur from individual arrester operation. Gas tube arresters are often used in two-stage lightning arrester units to provide high-energy protection for a **varistor** or other device used as a protector in the second stage.

Varistor-type lightning arresters – Varistors (voltage-dependent resistors) are solid-state devices whose electrical resistance changes as the applied voltage changes. They can be used to provide surge protection at all levels. With normal voltage applied, the resistance of the varistor is high,

allowing the signal to pass on to the circuit being protected. When a high-voltage transient occurs, the varistor resistance falls sharply, thus providing a low-resistance path to ground for surge current. Varistors clamp the surge voltage at a particular level as the current through them increases by several orders of magnitude. They do not create a short across a circuit, and current from the circuit does not flow through the device after the surge terminates. Because varistors are available for a wide range of system voltages, and some can handle surge currents as high as 70,000A, they are often used in power circuit applications. There are several types of varistors; however, the metal-oxide varistor (MOV) is the most commonly used type of varistor for surge protection (*Figure 31*) applications.

Combined varistor/spark gap suppressors – A single varistor cannot be used to provide voltage limiting during very high surge conditions because the heat would damage it. This can be corrected by placing a spark gap in series with the varistor, as shown in *Figure 32*. The spark gap prohibits current flow through the varistor when no transient is present. When a transient appears, it arcs across the gap, which in turn activates the varistor. There is no excessive system current due to the spark gap because of the varistor. The varistor also provides an automatic reset to extinguish the arc in the spark gap.

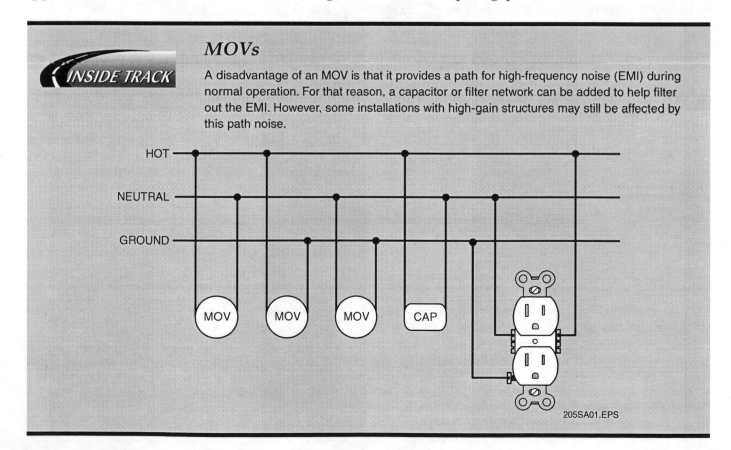

MOVs

INSIDE TRACK

A disadvantage of an MOV is that it provides a path for high-frequency noise (EMI) during normal operation. For that reason, a capacitor or filter network can be added to help filter out the EMI. However, some installations with high-gain structures may still be affected by this path noise.

205SA01.EPS

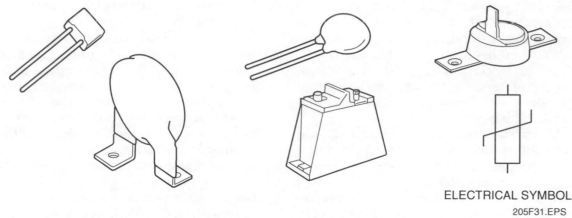

ELECTRICAL SYMBOL

205F31.EPS

Figure 31 ◆ Metal-oxide varistors (MOVs).

The valve-block type arrester is an example of a typical varistor/spark gap hybrid. It consists of a spark gap and varistor (valve block). The spark overvoltage is the level at which the gas in the spark gap is ionized to an active or conductive state. The varistor prevents the spark gap from short circuiting the AC line once it is ionized.

Semiconductor surge suppression devices – Many specialized surge protectors are used to provide point-of-use protection for radio, telecommunications, and signal lines. Individual surge protection devices, commonly called surge strips, incorporate duplex receptacles along with jacks for modem and network connections (*Figure 33*). These surge strips provide protection for the devices plugged into them from transients, power surges, and spikes. Most also have noise filtering circuitry to reduce or eliminate EMI/RFI noise interference both in the power and telephone lines.

In addition to the MOV, many other semiconductor devices are used in surge strips and other surge protection devices to provide a low-impedance path to ground for transients generated by lightning or other sources. Two widely used devices are diodes and breakdown diodes. Diodes are devices that conduct electricity in a single direction. This property is used to protect a device from surges and voltage spikes caused by inductive kickbacks. Breakdown diodes operate with a reverse current. There are two kinds of breakdown diodes in common use: zener diodes and avalanche diodes. They have a high ratio of reverse-to-forward resistance until their breakdown occurs as a result of an applied surge or transient. When the reverse voltage exceeds a certain level, the diode electrically breaks down. After breakdown, the voltage drop across the zener or avalanche diode remains essentially constant and is independent of the current.

6.1.3 Voltage Regulators

A voltage regulator is a device that adjusts the line voltage to a load up or down, as necessary, providing that the input voltage remains within the range of regulator control. A voltage regulator is typically used where the utility line voltage is reliable but of poor quality. There are several types of voltage regulators in use today with tap-changing, buck-boost, and constant-voltage regulators being used almost exclusively, rather than older and slower-acting electromechanical types.

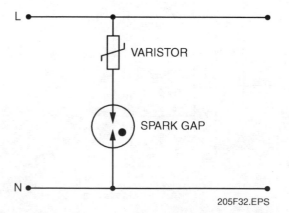

205F32.EPS

Figure 32 ◆ Varistor/spark gap hybrid protective device.

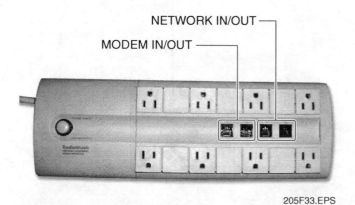

205F33.EPS

Figure 33 ◆ Combination surge strip.

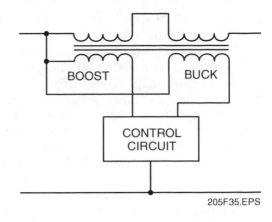

205F35.EPS

Figure 35 ◆ Simplified diagram of a buck-boost regulator.

Number of Taps

Most tap-changing voltage regulators are available in three- to six-tap arrangements, with the greater number of taps having the tighter regulation. For instance, a three-tap regulator is generally used when extreme highs or lows in line conditions occur. Six-tap arrangements are found mostly in applications involving electronic equipment.

Tap-changing regulators – Tap-changing regulators (*Figure 34*) adjust for varying input voltages by automatically transferring taps on an isolating or **autotransformer** type of transformer. Depending on the design, tap switching occurs at the zero current or voltage point of the output wave. The number of taps determines the magnitude of the steps and the range of regulation possible. A good regulator typically has at least four taps below normal and two taps above normal. The taps are usually around 4 to 10 percent steps, depending on the specific design. Response time is typically one-tenth of a second.

A major advantage of the tap-changing regulator is that its only impedance is the transformer or autotransformer and the semiconductor switches. It introduces little harmonic distortion under steady-state operation and minimizes load-induced disturbances as compared to regulators with higher series impedance. In its usual configuration, with an isolating transformer and wide undervoltage capability, it provides for both common-mode noise isolation and regulation.

Buck-boost transformer regulators – A buck-boost regulator (*Figure 35*) uses semiconductor control of a buck-boost transformer in combination with special filters to provide a regulated sinusoidal output even when supplying nonlinear loads such as computer systems. This is done in a

smooth, continuous manner, eliminating the steps inherent in the tap-changing regulator. Units can be equipped with an input isolating transformer containing an electrostatic shield to provide for voltage step down and common-mode noise attenuation when needed. Power is fed to the regulator, which either adds to (boosts) or subtracts from (bucks) the incoming voltage so that the output is maintained at a constant level for 15 to 20

Buck-Boost Transformers

A small buck-boost transformer is shown here. It is a single-phase compound-filled unit rated at 0.05kVA. However, depending on the percentage of voltage buck or boost required, it can be used in circuits with much higher loads.

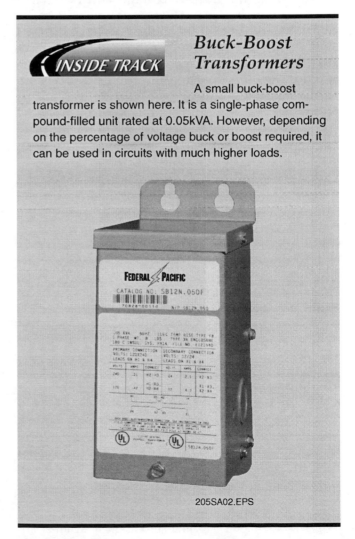

205SA02.EPS

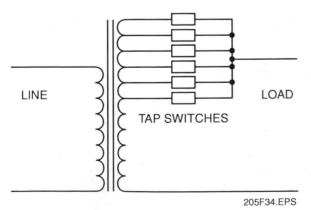

205F34.EPS

Figure 34 ◆ Simplified diagram of a tap-changing regulator.

POWER QUALITY AND GROUNDING

5.29

percent variations of input voltage. This is done by comparing the output boost or buck so that the desired level is maintained. A filter provides a path for nonlinear currents generated by the load and by the regulator itself and produces a sine wave output with low total harmonic distortion.

Constant-voltage transformer regulators – The constant-voltage transformer (CVT) regulator uses a saturating transformer with a **resonant circuit** made up of the transformer's inductance and a capacitor. Such an arrangement is referred to as a ferroresonant regulator (*Figure 36*). The regulator maintains a nearly constant voltage on the output for input voltage swings of 20 to 40 percent. These units are reliable because they contain no moving parts or active electronic parts. If these units are built with isolation (and shielding), they can provide common-mode noise reduction and a separately derived source for local power grounding. They also attenuate normal-mode noise and surges.

Note that the load current can cause the unit to go out of resonance if it gets too high. Often, these units can supply 125 to 200 percent of their full load rating. If inrush or starting currents exceed these limits, the output voltage will be greatly reduced, which may not be compatible with many loads. The other devices on the output of the CVT will see this sag in the voltage and may shut down due to an undervoltage. CVTs should be oversized if they are expected to provide heavy starting or inrush currents.

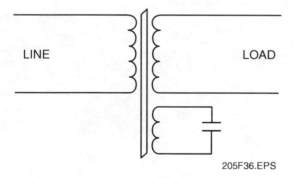

Figure 36 ◆ Simplified diagram of a ferroresonant regulator.

CVTs draw current all of the time. This current is due to the resonant circuit and causes these units to be less efficient at light loads, as compared to other types of regulators. Some of the units are quite noisy and require special enclosures before they can be installed in office-type environments.

6.1.4 Power Line Conditioners

Power line conditioners are units that incorporate two or more condition and protection methods in a single unit. Sometimes they have a power distribution means complete with main and branch circuit breakers. Normally, power conditioners do not provide protection against power abnormalities or against frequency variations or power blackouts. There are a wide variety of power line conditioners of various designs and made for specific applications.

One type of power conditioning device is a magnetic synthesizer (*Figure 37*). This is a ferroresonant device that consists of nonlinear inductors and capacitors in a parallel resonant circuit with six saturating pulse transformers. The units draw power from the source and generate their output voltage waveform by combining the pulses of the saturating transformers in a stepped-wave manner. These units generally use shielding to attenuate common-mode disturbances from the pulse transformers. Additional filtering is included to eliminate self-induced harmonics. This filtering can handle reasonable levels of harmonic distortion at the input or output.

The regulator has an inherent current-limiting characteristic that limits maximum current at full voltage to the range of 150 to 200 percent of the rated current. Beyond that load, the voltage drops off rapidly, producing a very high current at short circuit. This is a limitation with large inrush and starting currents. Sudden large load changes, even within the unit rating, can cause significant voltage and frequency transients in the output of this type of line conditioner. These units are best applied when the load does not make large step changes.

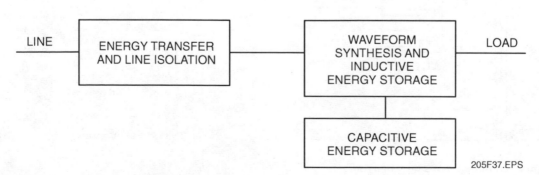

Figure 37 ◆ Simplified diagram of a magnetic synthesizer.

The resonant circuit has stored energy, which allows it to ride through outages of one half cycle, or slightly more if the outage is not a fault close to the input that drains the stored energy. Magnetic synthesizers are usually large and heavy and can be acoustically noisy without special packaging. Some of the larger units display good efficiencies as long as they are operated at close to full load. Depending upon the design, the synthesizer may introduce some current distortion on its input due to its nonlinear elements.

6.1.5 Harmonic and Noise Suppression Filters

Harmonic filters typically formed by series inductors and parallel-connected capacitors are used to reduce voltage waveform distortion that affects sensitive electronic equipment. These filters screen out harmonics and high-frequency noise. They also serve to reduce harmful harmonics that can cause heating of power conductors, transformers, and motors. Most of the better surge protecting devices, voltage regulators, and power line conditioners include harmonic filters.

Noise suppression (EMI/RFI) filters, formed by various circuit networks consisting of inductors and capacitors, clean up input signals, and waveforms distorted by noise. Noise suppression filter designs can be grouped into four types based on their application:

- *Low-pass filters* – These pass all frequencies below a specified cutoff frequency with little or no loss, but discriminates strongly against higher frequencies (*Figure 38*).
- *High-pass filters* – These pass all frequencies above a specified frequency with little or no loss but discriminates strongly against lower frequencies.
- *Bandpass filters* – These reject or greatly attenuate frequencies below and above a selected band (pass band) of frequencies while allowing signals within the pass band to easily pass through.
- *Band rejection filters* – These reject or greatly attenuate signals within a given band (stop band) of frequencies, while passing frequencies below and above this range.

6.2.0 Motor and Engine Generator Sets

AC motor-generator sets are devices used to regenerate power or generate power, respectively. AC motor-generator sets (M-Gs) provide electrical isolation from the utility line by using the utility power to drive a motor, which in turn drives a generator that produces an AC waveform at the required voltage. The AC motor-generator set normally consists of either an induction or synchronous AC motor mechanically coupled to and driving one or more generators (*Figure 39*). The motor is mounted on a common platform and on the same shaft with the generator or generators which it drives. This mechanical coupling electrically isolates the motor from the generator, and consequently, the utility line from the load.

In addition to providing isolation, the AC motor-generator set can convert electrical energy from one voltage or frequency to another voltage or frequency. Note that motor-generator sets are also made that can convert AC to DC or DC to AC. AC motor-generator sets are commonly used to convert 60Hz power to 400Hz power for computer applications. Since the motor and generator are mounted on the same shaft and rotate at the same speed, the desired frequency change is obtained by having a proper ratio exist between the number of poles in the motor and the number of poles in the generator. The ratio between the frequency applied to the motor and the frequency which the generator produces is exactly equal to the ratio between the number of poles on the motor and the number of poles on the generator.

Motor-generator sets protect the load from voltage sags, swells, and surges. They are able to bridge short-term sags or outages due to the rotational momentum of the rotating elements. When equipped with large flywheels to add inertia, some motor-generators can ride through short outages of up to several seconds.

6.3.0 Static Uninterruptible Power Supply

Most of the power system regulation and protection devices described up to this point are suitable only for use to protect equipment such as lighting systems, motors, air conditioning units, and similar equipment that is reasonably robust with

INSIDE TRACK

M-G Replacement

DC motors that are powered by motor-generator sets (M-Gs) are generally used in applications where motor speed and torque must be regulated. However, as budgets permit, these types of systems are gradually being replaced by AC motors powered and controlled by solid-state AC drive systems to obtain improved efficiency.

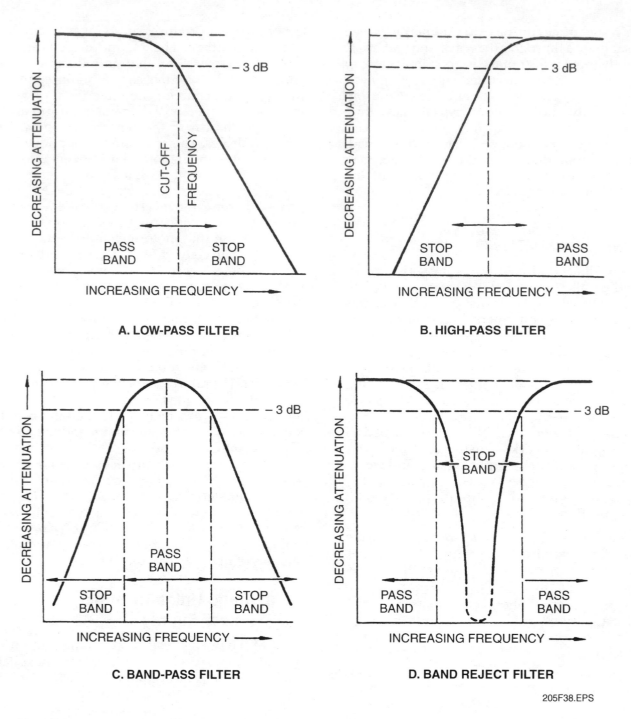

Figure 38 ◆ Frequency responses of four common types of filter circuits.

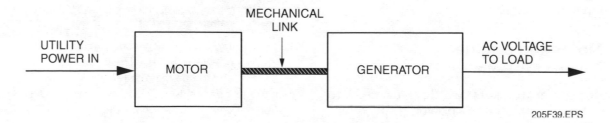

Figure 39 ◆ Block diagram of motor-generator set.

regard to power problems. These types of equipment can normally withstand momentary power surges or interruptions and/or brownouts without being damaged. On the other hand, computers and sensitive telecommunications electronic equipment are much more susceptible to damage caused by these hazards. For example, consider what happens in a computer/data center and network environment when a power surge caused by a lighting strike to a nearby transformer occurs. If the surge is powerful enough, it can travel instantaneously through wiring, network data lines, phone lines, etc. into the data center computers and related peripheral equipment. This can result in damaged modems or motherboards and/or chips causing a loss of data. In the event that the voltage drops low enough or blacks out, hard disk drives can crash, destroying the data stored on them. In all cases, works-in-progress stored in computer RAM and cache memory are lost instantly. For this reason, the power protection/conditioning equipment used with computer and other electronic systems needs to be capable of compensating for such power problems as sags, blackouts, spikes, surges, and noise if it is to protect these systems and their associated peripherals. Static uninterruptible power supplies (UPS) are designed and used to provide such protection.

A UPS is an electronically controlled, solid-state power control system. It provides regulated AC power to critical loads in the event of a partial or total failure of the normal source of power. Today, the term UPS commonly refers to a system consisting of a dedicated stationary-type battery, rectifier/battery charger, static inverter, and accessories such as a transfer switch. The period of time that a UPS can carry the load is a function of its design and the capacity of the battery. Note that in the past, the term UPS was commonly used to describe a static inverter often used in large DC systems.

There is a wide diversity in UPS configurations that provide for voltage regulation, line conditioning, lightning protection, redundancy, EMI management, extended run time, load transfer to other units, and other such features required to protect the critical loads against failure of the normal AC source, or against other power system disturbances. UPS systems are available in a number of designs and sizes, ranging from less than 100W to several megawatts. They may provide single-phase or three-phase power at frequencies of 50Hz, 60Hz, or 400Hz. Depending on their cost and sophistication, static UPS systems provide power that may vary from a near-perfect sine wave with less than 1 percent total harmonic distortion to a power wave that is essentially a square wave. There are two basic types of UPS systems used in a variety of configurations: double-conversion UPS systems and single-conversion UPS systems.

6.3.1 Double-Conversion UPS Systems

A double-conversion UPS system is also known as an online UPS. This type of system is the static electrical equivalent to the motor-generator set. In double-conversion systems, the incoming AC power is first rectified and converted to DC (*Figure 40*). The DC is then supplied as input power to a DC regulator and then to an AC converter (inverter), as well as charger power to a bank of batteries. The inverter output is AC, which is used to power the critical loads. The battery bank is connected to the DC regulator input to the inverter, and provides continuous power to it any time the incoming line fails or is outside of its specification. Because the batteries are always in the circuit, there is no switching and no breaks in either the input to the inverter or the output from it. These systems are typically used for mission critical applications, such as large computer systems that require high productivity and high systems availability.

The double-conversion system has several advantages:

- It has excellent frequency stability.
- It has a high degree of protection and isolation from variations in incoming line voltage and frequency, including power surges, spikes, transients, line noise, power sags, brownouts, and blackouts.
- It has zero transfer time.
- It provides quiet operation.
- Sophisticated systems can provide a sinusoidal output waveform with low distortion and EMI.

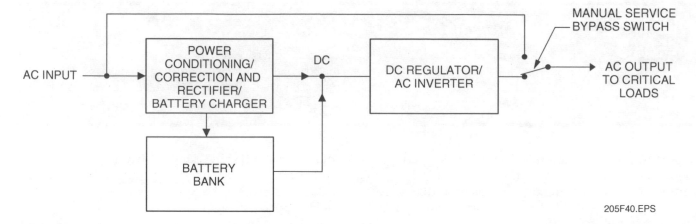

205F40.EPS

Figure 40 ◆ Simplified block diagram of a double-conversion UPS system.

In lower-power UPS applications (0.1 to 20 kW), the double-conversion UPS can have some of the following disadvantages. Note that some of the units on the market are specifically designed to overcome these disadvantages:

- Lower efficiency
- Large DC power supply is required (typically 1.5 times the full rated load rating of the UPS)
- Output noise isolation line-to-load can be poor
- Greater heat dissipation that can shorten the service life of the unit
- Excessive battery ripple current resulting in reduced battery life
- Poor power factor reflected to the AC input

6.3.2 Single-Conversion UPS Systems

There are a number of versions of single-conversion UPS systems. Systems are considered single-conversion systems when, at any point in time, power is only being converted once from AC to DC or DC to AC. The simplest versions of single-conversion systems are standby systems, also known as offline UPS systems (*Figure 41*). In this type of system the AC input power is fed to a battery charger and to the AC loads through an automatic transfer switch. The automatic switch detects AC line voltage and frequency variations. The loads are subjected to these variations as well as line noise until certain thresholds are exceeded. Once the power variation thresholds are reached, the switch transfers the loads to a stable inverter output obtained from the batteries. The switching time can range from 4 to 25 milliseconds and the output can vary considerably from a sinewave output. Offline UPS systems are usually single-phase systems because their generator incompatibility makes them unsuitable for three-phase

applications. Some of these systems are available with optional power conditioners to increase the level of power protection. These systems are cost-effective for non-critical equipment that requires basic power protection and backup.

Somewhat better versions of the system are known as line-interactive UPS systems (*Figure 41*). These systems are hybrids of the offline UPS system. One type uses a tap-changing voltage regulator that momentarily switches to the battery backup inverter when the regulator changes taps due to power fluctuations caused by voltage changes. When the power fluctuations reach a certain threshold, the unit goes to battery backup full time. Because the batteries are used during switching, they must be constantly recharged via a trickle charger or by a three-stage charging technique that extends battery life that can be shortened by trickle charging. These units pass perturbations in frequency and power quality directly to the load until they switch to battery power only. These types of line-interactive systems, like offline systems, are normally used for single-phase applications with non-computer linear loads such as monitors, heaters, and lights.

A more advanced but less efficient type of line-interactive UPS uses an inductor or ferroresonant transformer and a four-quadrant converter that charges the batteries as well as converts DC from the batteries back to AC power. The transformer provides voltage regulation and power conditioning for disturbances such as line noise. It also maintains a small energy reserve to power through very short power sags and interruptions. Typically, these units are constant-voltage devices that adjust to shifting loads or varying input voltages by changing the output phase angle. Because dynamic load changes or voltage variations cause power to be extracted from the batteries as needed until

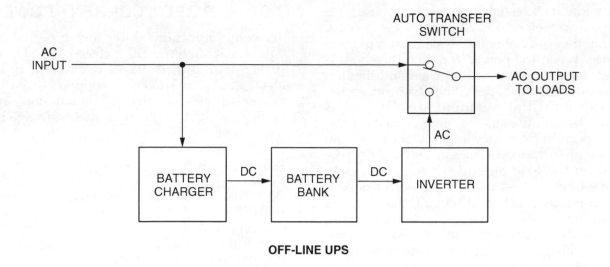

OFF-LINE UPS

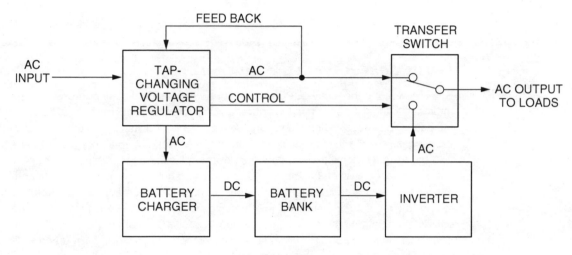

TAP-CHANGING LINE-INTERACTIVE UPS

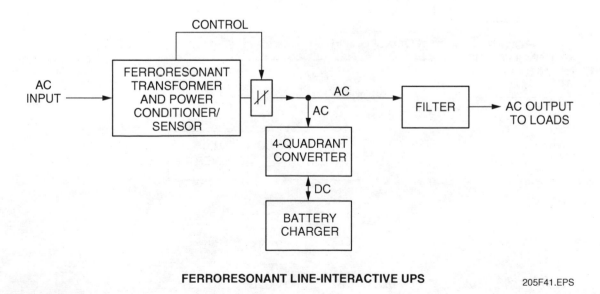

FERRORESONANT LINE-INTERACTIVE UPS

205F41.EPS

Figure 41 ◆ Typical single-conversion UPS systems.

power variation thresholds are reached, these units are sometimes referred to as online UPS systems. Note that they are not online systems in the classical sense. When the power variation threshold is reached, the AC contactor is opened and the unit operates only from battery backup through the converter until the AC input stabilizes or is restored. As with the tap-changing regulator type, perturbations in frequency and power quality can be passed on to the loads unless additional filtering is used on the load side of the transformer to improve quality. Because the converter is constantly connected to the AC circuit, the resulting frequent hits on the batteries can shorten battery life unless three-stage charging is used to extend their life. With load-side filtering, these systems can be used with computer networking devices such as hubs, routers, and servers.

7.0.0 ◆ DIRECT CURRENT POWER

Telecommunications and other types of control equipment often use DC instead of AC. DC is unidirectional and, unlike AC current, does not change polarity. The DC is supplied by either DC power supplies, which can be stand-alone units or an integral part of the equipment, storage batteries, or both.

7.1.0 DC Power Supplies

Both linear and nonlinear power supplies are used in or with telecommunications and other equipment. Basically, a linear power supply is one where the magnitude of the voltage output is directly proportional to the input. Conversely, the output of a nonlinear power supply has an output

Delta Conversion UPS Systems

Some newer line-interactive UPS systems use what is called delta conversion to modify the input voltage. As shown in the figure, a delta four-quadrant converter is connected to the battery bank bus. When the voltage varies between certain limits, the delta converter exchanges power with the main four-quadrant converter. The main converter is synchronized with the AC power or with an internal reference and is used for both power correction and outage protection. When the input voltage is present but not nominal, the delta converter injects a phased voltage into the buck-boost transformer to add or subtract from the input to create a regulated output voltage. When the input power exceeds thresholds, the input contactor opens and the main converter supplies full AC output power from the battery bank until the input power is within specifications. Either converter is used under different conditions to charge the battery bank.

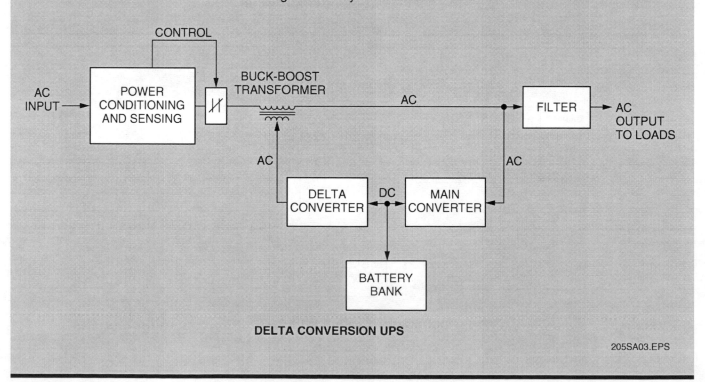

DELTA CONVERSION UPS

205SA03.EPS

that does not rise or fall in direct proportion to its input. DC power supplies can be stand-alone units or they can be an integral part of an item of equipment.

7.1.1 Linear Power Supplies

There are many types and designs of linear DC power supplies ranging from simple half-wave, rectifier-type units to complex amplifier-regulated units consisting of several solid-state devices. The more complex the unit, generally the better the regulation. Linear DC power supplies used in telecommunications, computers, and other equipment usually are of the full-wave rectifier or full-wave bridge rectifier type (*Figure 42*). In these power supplies, conversion of the AC input voltage to DC voltage is done by rectifiers or diodes, as shown. Full-wave or bridge rectifiers operate in a manner that passes the positive half cycle of an AC

input, then inverts the negative half cycle of the AC input and passes it so that it flows in the same direction as the positive half cycle to produce an unregulated rippled (pulsating) DC voltage. Following this, the unregulated DC voltage is passed through a filter network that smooths out the rippled DC in order to provide a stable and constant (regulated) DC voltage.

7.1.2 Nonlinear Power Supplies

Switching power supplies, also called switching regulators, are the main type of nonlinear power supply used in or with communication, computer, and other equipment. Compared with linear supplies, switching power supplies have significant advantages in efficiency, size, and weight. There are many types and designs of switching power supplies available. *Figure 43* shows a basic circuit for a typical switching power supply.

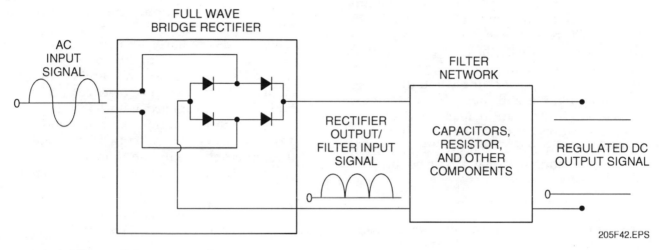

205F42.EPS

Figure 42 ◆ A full-wave DC power supply.

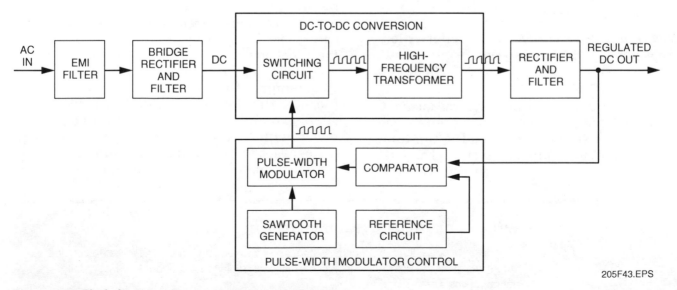

205F43.EPS

Figure 43 ◆ Block diagram of a typical switching power supply.

As shown, the input AC is applied through an EMI filter, then is rectified by a bridge rectifier and filtered to produce an unregulated DC voltage. The resulting DC voltage is then applied to a DC-to-DC conversion circuit. There, the DC input voltage is chopped by high-frequency square wave control pulses applied to a switching (chopper) circuit connected in series with the primary winding of a high-frequency transformer. The resulting square wave output produced by the switching circuit is coupled through the transformer to the secondary winding where it is stepped up or stepped down in amplitude. Following this, the square waves are passed through a second full-wave rectification stage and another filtering stage to produce the desired regulated DC output voltage.

The DC output voltage is regulated by varying the width of switching pulses applied to the switching circuit, typically a transistor or silicon-controlled rectifier (SCR). This is done through a pulsewidth-modulator circuit consisting of a triangular-wave (sawtooth) generator, a pulsewidth modulator, a DC reference source, and a comparator. The comparator senses the DC output voltage of the supply and compares it to a reference source. Any deviation of the power supply output voltage from its design value will cause the comparator control voltage output to increase or decrease, which in turn causes the pulsewidth modulator to produce wider or narrower pulsewidth switching pulses for application to the switching circuit.

7.1.3 Selecting a Power Supply

Power supplies for a specific application are normally selected by the system design engineer. However, it is useful for the technician to understand some of the factors that must be considered when choosing a power supply:

- *Code requirements* – NEC® and Underwriters Laboratories (UL) have specific requirements for power supplies. Be sure to take them into account.
- *Load requirements* – The most critical factor in selecting a power supply is determining the load it must support. The load is the total power requirements for the system. The load requirement includes the power you will need

to operate the system in normal circumstances. It may be desirable to oversize the unit to account for a power margin for protection and/or future expansion. For example, if your actual load requirements are 400W and you want to oversize the unit by 50 percent, then you would select a power supply capable of supplying 600W (400W + 200W).

- *Frequency* – 60Hz power is the most commonly used power in the U.S.; 50Hz power may be used elsewhere.
- *Voltage* – Equipment containing microprocessors and similar components is designed to operate at specific voltages. Refer to the manufacturer's literature for the equipment involved to determine the voltage tolerances. The 120V system is typical of North America and includes 208V, 208V three-phase, 240V, and 490V three-phase. The 230V system is typical of most of the world and includes 400V and 400V three-phase.
- *Desired run time (UPS)* – This is the run time a UPS-type power supply can supply the load during a power outage. Power outages lasting a few seconds are fairly common. Accidents and storms can disrupt utility power for many hours. Usually, outages are limited to four to eight hours, but longer outages can occur. Some typical run times used with UPSs that supply smaller equipment are: computer workstation, 15 minutes; department network server, 20 minutes; internetworking equipment, 30 minutes; and telephone system, one hour to two hours.

7.1.4 Power Supply Testing

Occasionally it is necessary to test a power supply to see if it is operating correctly and providing the required noise-free DC output. A complete checkout of a power supply normally involves performing voltage, current, regulation, and ripple checks in a bench-testing environment. However, this is not always practical.

An online check of a power supply can be done to determine if the power supply is operating properly. This test should be done with the power supply connected in the system normally and under normal load conditions. Basically, the test involves locating a suitable measurement point (terminal board, connector pin, etc.) where the

Other Considerations in Selecting a UPS

In addition to code, load, frequency, voltage, and desired run time, don't forget to consider the efficiency of the unit, which includes initial cost, replacement of batteries, and daily operating costs.

power supply output voltage can be accessed for testing. Then, the output voltage should be measured with a VOM and the ripple checked with an oscilloscope.

Remember that all switching power supplies will have ripple no matter how good the regulation and filtering. To measure the ripple voltage (sometimes called output noise) with an oscilloscope, the scope AC/DC mode controls should be set to the AC mode, because this blocks the DC output of the supply. Adjust the scope controls to produce two or three stationary cycles on the screen, then measure the peak amplitude of the ripple on the scope's voltage-calibrated vertical scale. Once the value for the ripple is known, the percentage of ripple relative to the full output voltage can be calculated. For example, if 0.03V of ripple is measured, with a 5V output, the ratio is 0.03/5 or 0.006, which can be converted to a percentage ($0.006 \times 100 = 0.6$ percent). The percent of ripple voltage should be small compared to the supply voltage. The specifications for the power supply being tested should be consulted to determine exactly what the maximum ripple should be.

If desired, the frequency of the ripple can be measured using the scope's horizontal scale and the formula: frequency = 1/period of a complete cycle. For example, if the total period is 16.666 milliseconds, then the frequency equals 60Hz.

7.2.0 Storage Batteries

Storage batteries (*Figure 44*) are widely used in starting systems for generator sets, emergency lighting, uninterruptible power supplies, and many other applications. There are two main classes of batteries: lead acid and alkaline. A storage battery stores energy in chemical form and converts it to electrical form during the discharge chemical reaction. Storage batteries consist of a number of cells. The voltage depends on the plate material and type of **electrolyte**. For a lead-acid cell, the voltage is a nominal 2V and for an alkaline cell, 1.2V. Cells are connected in series to increase battery capacity.

There are many different types of batteries within the two categories. It is important to understand the characteristics of each type to apply, operate, and maintain them properly. A summary of the characteristics of each type is given in *Table 1*.

The service life stated for the different types of batteries listed in *Table 1* and in battery manufacturers catalog/data sheets are for operation under ideal conditions and may not be achieved under less than ideal conditions. Ideal conditions include operation in the proper environment and with proper maintenance.

Battery voltage is computed on the basis of 2.0V per cell for lead acid types and 1.2V per cell for the alkali type.

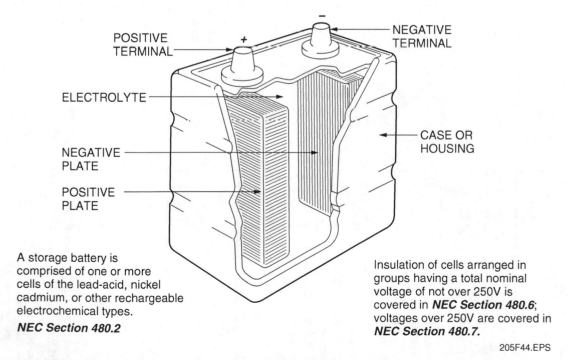

POSITIVE TERMINAL

NEGATIVE TERMINAL

ELECTROLYTE

CASE OR HOUSING

NEGATIVE PLATE

POSITIVE PLATE

A storage battery is comprised of one or more cells of the lead-acid, nickel cadmium, or other rechargeable electrochemical types. **NEC Section 480.2**

Insulation of cells arranged in groups having a total nominal voltage of not over 250V is covered in **NEC Section 480.6**; voltages over 250V are covered in **NEC Section 480.7**.

205F44.EPS

Figure 44 ◆ Storage battery.

Table 1 Characteristics of Standby Power Batteries

	Planté Lead Acid	Tubular Lead Acid (Lead Selenium)	Pasted Plate Lead Acid			Valve-Regulated Lead Acid	Vented Pocket Plate Nickel Cadmium	Valve-Regulated Nickel Cadmium
			Antimony	Selenium	Calcium			
Expected Reliable Service Life (Years)	25 Plus	15	10-15	12-15	10-15	10-15	25 Plus	25 Plus
Cycle Life to 80% D.O.D. at 77°F (25°C)	1000	1500	800	1000	50-100	100	1000 / 2500 Double Pocket Plate	1000
Approximate Watering Required on Float (Months)	18	18	12 New 1 Old	18	18	Not Required	18	Not Required
Capacity at End of Life (%)	100	80	80	80	80	80	80	80
Recommended Temperature Range	50°-90°F (10°-32°C)	50°-90°F (10°-32°C)	50°-80°F (10°-27°C)	50°-90°F (10°-32°C)	50°-80°F (10°-27°C)	50°-80°F (10°-27°C)	-40°-140°F (-40°-60°C)	-40°-140°F (-40°-60°C)
Recommended Charge Voltage	2.23	2.23	2.20	2.23	2.25	2.25-2.28	1.45-1.47	1.42-1.47
Recharge Time from 100% Discharge at Recommended Float Voltage	3 Days	3 Days	3 Days	3 Days	6-7 Days	3 Days	12-14 Days	12-14 Days
Equalize Charge Required When Floated at Recommended Float Voltage	Never	Never	6 Months at End of Life	Never	Never	Never	Never	Never
Storage Time at 77°F (Filled) Before Freshening Charge	3-6 Months	3-6 Months	3-6 Months	3-6 Months	3-6 Months	3-6 Months	2 Years	2 Years
Availability of Flame Retardant (UL94) Cases	No	No	Yes	No	Yes	Yes	Yes (R.R. Only)	No
Typical Case Material	Transparent S.A.N.	Transparent S.A.N.	Transparent S.A.N.	Transparent S.A.N.	Transparent S.A.N.	Opaque ABS or Polypropylene	Translucent Polypropylene	Opaque Polypropylene
Electrolyte	Sulfuric Acid (H_2SO_4)	Sulfuric Acid (H_2SO_4)	Sulfuric Acid (H_2SO_4)	Sulfuric Acid (H_2SO_4)	Sulfuric Acid (H_2SO_4)	Sulfuric Acid (H_2SO_4)	Potassium Hydroxide (KOH)	Potassium Hydroxide (KOH)
Vibration Resistance	Low	Good	Medium	Medium	Medium	Good	Excellent	Excellent
Vented Gas Composition	Hydrogen, Oxygen, Acid Vapor	Hydrogen, Oxygen, Acid Vapor	Hydrogen, Oxygen, Acid Vapor	Hydrogen, Oxygen, Acid Vapor	Hydrogen, Oxygen, Acid Vapor	Little Gas Under Normal Operation	Hydrogen, Oxygen	Little Gas Under Normal Operation

Note: Individual products vary with manufacturer. Check with the manufacturer for precise data on specific products.

205T01.EPS

7.2.1 Lead Acid Batteries

Lead acid batteries derive their name from the lead plate material and acid electrolyte used. Because lead is soft, a hardening agent, usually calcium or antimony, is normally added to the lead. Lead acid batteries fall into four categories:

* *Planté designs (Manchex or rosette antimony)* – This is the oldest design of any rechargeable battery. This type of battery uses a positive plate made of pure lead casting. The plate itself appears like a car radiator with fine platelike layers giving the plate an extremely large surface area. This gives the plate its high performance and long life. The plate sheds active material to the bottom of the cell jar and new lead is exposed to the charging reaction to become new active material. One variation of this design is the Manchex or rosette plate, which uses pure lead buttons pushed into a heavy honeycomb-type lead antimony grid. The lead buttons are formed to become active material.

    Planté cells have typical life spans of over 25 years with proper charging and maintenance and in temperature-controlled environments. Given proper maintenance, they have the longest life of all lead acid batteries. The major disadvantages to their use are high initial costs and the use of a larger floor area than other designs.

* *Pasted-plate designs (calcium, antimony, selenium)* – The pasted-plate battery uses a lead alloy grid covered with a mixture of lead active material paste. Development of this type of battery allowed lead acid batteries to be made lighter, more compact, and more importantly, more economical. There are several lead alloys used to make these types of batteries. They can be grouped into three categories: lead antimony, lead calcium, and lead selenium. It is important to keep in mind that each alloy changes the battery's performance and operational characteristics.

    – *Lead antimony pasted-plate lead acid batteries.* The lead antimony battery is the oldest pasted-plate type of battery. The antimony battery has two major advantages, one of which is the ability to withstand repeated deep discharge cycles. The battery will give the user about 800 to 1,000 cycles during its life, if required. Another advantage is its predictable life. One can expect 15 reliable years with proper maintenance and a temperature-controlled environment.

    The main disadvantage of the lead antimony battery is the condition of antimonial poisoning of the negative plate. Oxidized antimony gradually adheres to the negative plate over time. This lowers the resistance to charging and the current required to fully charge the battery increases over the life of the battery. This causes the battery to consume more water, and causes a gradual rise in the watering requirements to maintain the battery. The rate of antimony poisoning is also greatly affected by high temperature. This is a particularly important consideration for non-temperature-controlled environments.

    – *Lead calcium pasted-plate lead acid batteries.* Lead calcium batteries were developed because they displayed remarkably stable float charge characteristics over their life, unlike the antimony alloy. Unfortunately, the calcium battery does not cycle well at all. The battery will likely exhibit a marked reduction in capacity after less than 100 cycles. These batteries are the least appropriate for applications such as photovoltaics where the battery may be exposed to deep discharge.

    – *Lead selenium pasted-plate lead acid batteries.* By reducing the amount of antimony in lead antimony batteries, the antimonial poisoning effect is essentially eliminated. However, grids cast with antimony in low concentrations lose some of the required rigidity. Work on optimizing the low antimony alloy grid has resulted in the development of the lead selenium battery. The addition of small amounts of selenium into the low antimony grids restored the necessary hardness of the grid. The lead selenium battery has the advantage of delivering stable float charge characteristics over its life like the lead calcium battery, and repeated deep discharges, 1,000 cycles, over its life like the lead antimony battery. High-temperature environments will reduce the life of this battery as with all lead acid-type batteries.

- *Tubular-plate design (antimony, selenium)* – The tubular plate battery uses an alloy grid that resembles a long-toothed comb placed inside a polyester pan pipe. The space between the lead splines and the polyester pipe is filled with active materials. Tubular plate lead acid batteries are most tolerant to cycle service, delivering as many as 1,500 cycles. Their cost is marginally more than pasted-plate batteries. They are best suited to duties such as photovoltaics because of the high number of cycles required. They can be found in many telecommunication backup systems as well.

- *Valve regulated or recombination (sealed or gel cell, absorbed electrolyte)* – There has been a great deal of interest and marketing in battery types that are sealed and called maintenance free. These batteries are neither. The proper term for these type of batteries is the valve-regulated or recombination battery. This is because under normal conditions, the gas generated on charge is recombined to form water inside the cell.

It is true that this type of battery does not have to replenish the electrolyte reserve, but it is necessary to carry out all of the other inspections and periodic tests required of vented cell batteries, such as checking voltages, cleaning and tightening connections, and performing an annual load capacity test. They can only be considered maintenance free if they are completely replaced on a periodic interval, typically one year, without testing or inspection. This can be costly. It is also true that these batteries are sealed, but only under normal operation. In an overcharge or high-temperature situation, these batteries will dispel charging gases (hydrogen and oxygen) to the atmosphere via the valve. Hence the name valve-regulated batteries. Under normal operating conditions, the valve is closed, and the gas generated on charge is recombined to form water inside the cell.

Regardless of the design, with all lead batteries the chemical reaction that takes place to store and produce electricity is the same. The lead acid battery is a secondary, or storage type battery. In secondary batteries, electricity or power must be put into the battery to start the chemical reaction. When the battery is being charged, power is stored chemically in the electrodes and electrolyte of the battery. When the battery is called upon to provide power to its load, the resulting chemical reaction is reversed, discharging the battery.

7.2.2 Alkaline Batteries

The most common type of alkaline battery is the nickel cadmium battery. Nickel cadmium batteries have been widely used for years. The nickel cadmium battery is the most rugged of the battery technologies, mainly because of its durability in extreme temperature environments. Its reliability has been demonstrated both in arctic and tropical regions. However, the initial cost of the nickel cadmium battery may be difficult to justify for a large number of applications. It is in the extremes of cold or heat, and where maintenance practices are not all that they might be, that the nickel cadmium battery is well worth the investment.

Nickel cadmium batteries fall into three main categories: pocket plate, fiber plate, and sintered plate.

- *Pocket-plate nickel cadmium batteries* – Pocket-plate nickel cadmium batteries represent over 90 percent of the nickel cadmium batteries used for stationary applications. The active materials in the plates are held in pockets. The pocket-plate nickel cadmium battery is only available as a vented type of battery.

- *Fiber-plate nickel cadmium batteries* – The fiber-plate nickel cadmium battery is the newest design of nickel cadmium battery for stationary applications. This type of cell has a high power-to-weight ratio. The fiber-plate nickel cadmium battery is only available as a vented type battery.

- *Sintered-plate nickel cadmium batteries* – The sintered-plate battery is used in some emergency lighting units, particularly fluorescent or decorator types. It is available in vented or valve-regulated models.

7.3.0 Battery and Battery Charger Operation

Standby batteries used in most applications are normally operated in the float mode of operation where the battery, battery charger, and load are connected in parallel (*Figure 45*). The charging equipment should be sized to provide all the power normally required by the loads plus enough additional power to keep the battery at full charge. Relatively large intermittent loads will draw power from the battery. This power is restored to the battery by the charger when the intermittent load ceases. If AC input power to the system is lost, the battery instantly carries the full load. If the battery and charger are properly matched to the load and to each other, there is no noticeable voltage dip when the system goes to full battery operation.

When AC charging power is restored, a constant-potential type of charger will deliver more current than is necessary if the battery were fully charged. Some constant-potential chargers will automatically increase their output voltage to the equalized setting after power to the charger is restored. Similarly, a constant-current charger may automatically increase its output to the high rate. The battery charger must be sized so that it can supply the load and restore the battery to full charge within an acceptable time. The increased current delivered by a constant-potential charger during this restoration period will decrease as the battery approaches full charge.

The battery charger is an important part of the DC power system, and consideration must be given to redundant chargers on critical systems. The charger should be operated only with the battery connected to the DC bus. Without this connection, excessive ripple on the system could occur, which could affect the connected equipment causing improper operation or failure of the equipment. If a system must be able to operate without the battery connected, then a device called

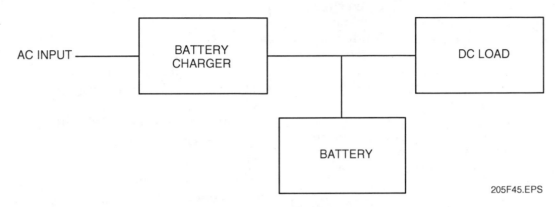

205F45.EPS

Figure 45 ◆ Simplified block diagram of a stationary battery in float operation.

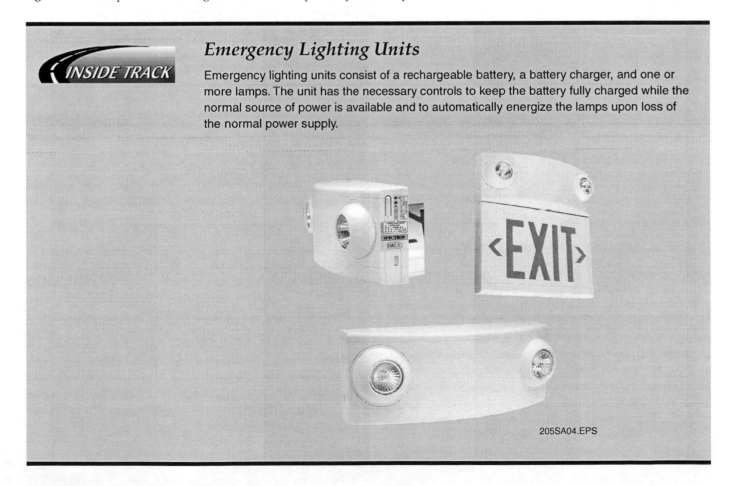

Emergency Lighting Units

Emergency lighting units consist of a rechargeable battery, a battery charger, and one or more lamps. The unit has the necessary controls to keep the battery fully charged while the normal source of power is available and to automatically energize the lamps upon loss of the normal power supply.

205SA04.EPS

a battery eliminator should be used. Also note that a battery charger's output must be derated for altitude when it is installed at locations above 3,300', and for temperature, when the ambient temperature exceeds 122°F. The battery charger manufacturer can provide these derating factors.

8.0.0 ◆ CABLE SHIELDING AND GROUNDING TECHNIQUES USED TO MINIMIZE EMI

One widely used method for preventing the penetration of electromagnetic interference (EMI) into sensitive equipment is to enclose the equipment and associated cabling entirely in a shielded environment. A proper shield is metallic, or other conductive material, and is well sealed. Conduction EMI can occur when a signal induces current into an exposed power, ground, control or signal cable entering or leaving a shielded electronic assembly. In this situation, the EMI signal reradiates inside the equipment housing, and is picked up by the electronics within the housing. One common method used to combat EMI is achieved by shielding the power, control and signal cables entering and leaving an assembly.

8.1.0 Cable Shields

A cable shield is a metallic covering or envelope enclosing an insulated conductor, cable core, or individual group of conductors within a core. Shields are made of foil, wire strands, or braided metal (*Figure 46*). They are usually tinned copper, bare copper, aluminum, or other electrically conductive material. Cable shields have the following characteristics:

- They reduce the radiated signal from the cable to an acceptable level.
- They reduce the effects of electrical hazards when properly grounded and bonded.
- They minimize the effect of external electromagnetic signals on the circuits within the shielded cable.

Metal conduit is the best possible shield. It displays very good shielding properties at all frequencies. It should be pointed out that, while conduits are used as shields in some very specialized applications, their rigid nature makes them inappropriate for most normal cable applications. The selection of a cable shield depends on the following considerations:

- The nature of the signal to be transmitted (frequency range affects the performance of most fields)

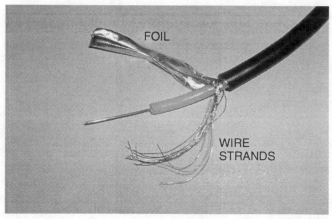

205F46.EPS

Figure 46 ◆ Coaxial cable shielding.

- Electromagnetic interference fields through which the cable will run
- FCC regulations (any cable operating within a given system must be designed to meet the EMI radiation limits of that system)
- Physical environment and specific mechanical requirements

Table 2 provides a comparison of some types of cable shields.

8.2.0 Preventing Ground Loops

Impulse noise due to electrical arcs, lightning strikes, electrical motors, and other devices can interfere with the operation of communications, computer, and other sensitive electronic equipment. Shielding of signal and data lines can help somewhat, but it is not the whole answer. Improper cable-shielding practices can also cause induced noise problems. If too many grounds are used, a problem referred to as a ground loop can result. *Figure 47A* shows an example where standard grounding causes ground loops. In this example, the shielded source, shielded input lines, the electronic circuit, and the DC power supply are all grounded to different points on the ground plane. Power supply DC current flows from the power supply at point A to the electronic circuit at point E. Since the ground plane exhibits a very low resistance path, current flow in the ground plane will cause voltage drops (E1 through E4) to occur between each of the ground points much like in a resistive voltage divider network. These voltages are seen by the electronics circuit as valid signals, and can become especially troublesome. The solution to this problem is to use single-point grounding as shown in *Figure 47B*.

Remember that grounding cable shields is a controversial subject. Always follow the equip-

Table 2 Comparison of Cable Shields

Characteristic	Single-Layer Braid	Multiple-Layer Braid[2]	Foil	Solid-Wall Conduit[3]	Flexible Conduit
Shield effectiveness[1]					
Audio frequency	Good	Good	Good	Excellent	Good
Radio frequency	Good	Excellent	Excellent	Excellent	Poor
Normal coverage	60–95 percent	95–97 percent	100 percent	100 percent	95–97 percent
Fatigue life	Good	Good	Fair	Poor	Fair
Tensile strength	Excellent	Excellent	Poor	Excellent	Fair

Notes:

[1] In the shield effectiveness ratings:
    Poor = less than 20 dB
    Fair = 20–40 dB
    Good = 40–60 dB
    Excellent = over 60 dB

[2] The effectiveness of single-layer and multiple-layer braids against magnetic fields is poor.

[3] For foil and conduit to effectively shield against magnetic fields, a high permeability material must be used.

ment manufacturer's directions when grounding cable shields. Some manufacturers require grounding inside the equipment enclosure, others outside the equipment, and still others require grounding at the device, not at its controller. Generally, instrumentation cable shields should only be grounded at one end. For longer runs and interfacility runs, cable shields are generally grounded at both ends. Cable shields should be electrically continuous.

9.0.0 ◆ STATIC ELECTRICITY PROBLEMS AND THEIR PREVENTION

Static electricity is the charge produced by the transfer of electrons from one object to another. Static electricity is generated as the result of electron movement between two different materials that come in contact with each other and are then separated. When the two materials are good con-

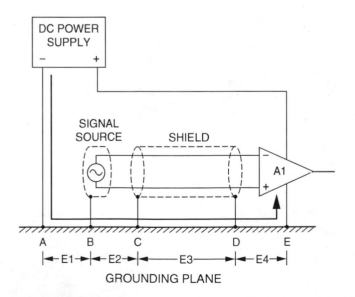

(A) STANDARD GROUNDING CAUSES GROUND LOOPS DUE TO VOLTAGE (IR) DROPS

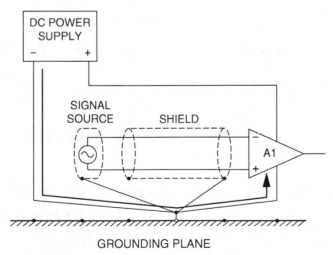

(B) SINGLE-POINT GROUNDING ELIMINATES GROUND LOOPS

205F47.EPS

Figure 47 ◆ Standard grounding versus single-point grounding.

ductors, the excess electrons in one will return to the other before the separation is complete. However, if one of them is an insulator, both will become charged by the loss or gain of electrons, unless grounded. An example of this is when you get an electric shock from touching a doorknob after walking across a carpet. Static electricity changes can have potentials that exceed several thousand volts. Some examples are as follows:

- Walking across a carpet – 35,000V
- Walking over a vinyl floor – 2,000V
- Vinyl envelopes for work instructions – 7,000V
- Picking up a common poly bag – 20,000V
- Work chair padded with polyurethane foam – 18,000V

When static electric discharges occur physically close to sensitive electronic equipment, they can create malfunctions and at times even damage semiconductor circuits and devices. Most electronic equipment is designed to be insensitive to static discharge. After installation, equipment can be protected with a conductive material shielding by limiting cutouts in the equipment enclosure to 4"; fitting butt joints and housing separation points with high-frequency shields or separating them by at least 4"; protecting unshielded inputs and outputs with high-frequency filters; separately grounding conducting operating controls; and electrically connecting metal frames and parts.

Some precautions that will minimize the problems associated with static electricity include the following:

- Maintain the room humidity at a minimum of 40 percent.
- Ensure that all carpets, as well as the clothing of operating personnel, are made of natural materials.
- If carpets are made of synthetic materials, they should have a conductive backing.
- Ionize the air to increase conductivity and ensure that equipment operators wear conductive footwear.
- Use anti-static materials whenever possible to increase conductivity.

- Ship and handle sensitive components in containers made of conductive plastic.

Transients can also be created by static electricity from your body. It is important when handling printed circuit (PC) boards to avoid touching components, the printed circuit, and the connector pins. You should ground yourself before touching a PC board. Disposable grounding wrist straps are sometimes supplied with boards containing electrostatic-sensitive components. If a wrist strap is not supplied with the equipment, a wrist strap grounding system similar to the one shown in *Figure 48* should be used. It allows you to handle static-sensitive components without fear of electrostatic discharge (ESD). Boards not in use should be kept in their special conductive plastic storage bags.

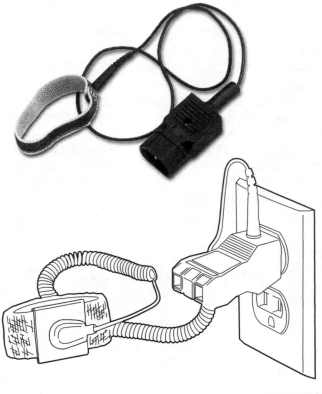

205F48.EPS

Figure 48 ◆ ESD wrist strap.

Why High Voltage Levels of Static Electricity Don't Cause Death

Static electricity in itself is generally not harmful to humans because there is no current flow, only a discharge or balancing of electrons and protons. Static electricity is harmful to humans when injuries result due to the sudden reactions when the shock occurs, or if the static electricity presents an ignition source for flammable materials or vapors.

10.0.0 ◆ TESTING FOR EFFECTIVE GROUNDS

An earth ground resistance tester (*Figure 49*) may be used to make soil resistivity measurements or to measure the resistance to earth of the installed grounding electrode system.

NOTE

An ordinary ohmmeter cannot be used to measure the resistance of a grounding electrode to earth because it is not adequate for the levels of current and voltage.

WARNING!

Ground testers can be hazardous to both personnel and equipment if improperly used. Always check with your supervisor before using a ground tester.

205F49.EPS

Figure 49 ◆ Earth ground resistance tester.

One use of the ground tester is for testing electrical systems after they are installed and before normal voltage is applied. This test is made after all the conductors, fuses or circuit breakers, panel-

Fall-of-Potential Testers

Newer models of ground testers, such as the one shown here, typically incorporate a digital readout, but the circuitry and principles of operation are very similar to those of older types. Another difference between the two types is that newer models often refer to the electrodes and terminals as X for the grounding electrode and Y and Z for the auxiliary electrodes. Some older models used the letter X as the grounding electrode, and the numbers 1 and 2 for the auxiliary electrodes.

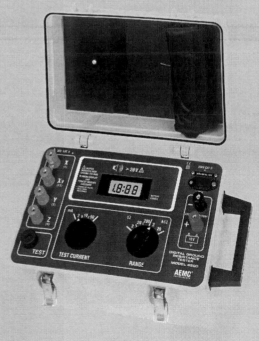

205SA05.EPS

boards, and outlets are in place and connected. The current used for testing is produced by a small generator within the ground tester that generates DC power, either by turning a crank handle (also a part of the ground tester), or by using a small electric DC motor within the ground tester.

The test is made by connecting the terminals to the points between which the test is to be made and then rapidly turning the handle on the ground tester. The resistance in ohms can then be read from the meter dial. Satisfactory insulation resistance values will vary under different conditions, and the charts supplied with the ground tester should be consulted for the proper value for a particular installation.

10.1.0 Measuring Earth Resistance

An earth ground is commonly used as an electrical conductor for system returns. Although the resistivity of the earth is high compared to a metal conductor, its overall resistance can be quite low because of the large cross-sectional area of the electrical path.

Connections to the earth are made with grounding electrodes, ground grids, and ground mats. The resistance of these devices varies proportion-ately with the earth's resistivity, which in turn depends on the composition, compactness, temperature, and moisture content of the soil.

A good grounding system limits system-to-ground resistance to an acceptably low value. This protects personnel from a dangerously high voltage during a fault in the equipment. Furthermore, equipment damage can be limited by using this ground current to operate protective devices.

Ground testers measure the ground resistance of a grounding electrode or ground grid system. Some of the major purposes of ground testing are to verify the adequacy of a new grounding system, detect changes in an existing system, and determine the presence of hazardous step voltage and touch voltage.

In addition to personnel safety considerations, ground testing also provides information for equipment insulation ratings. Equipment can be damaged by an overvoltage that exceeds the rating of the insulation system. *Figure 50* depicts a poorly grounded system where the ground resistance (R_1) is 10Ω.

Assuming a power source resistance (R_2) of 40Ω, a short circuit between the 5,000V power line and the steel tower would produce 100A of short circuit current (I).

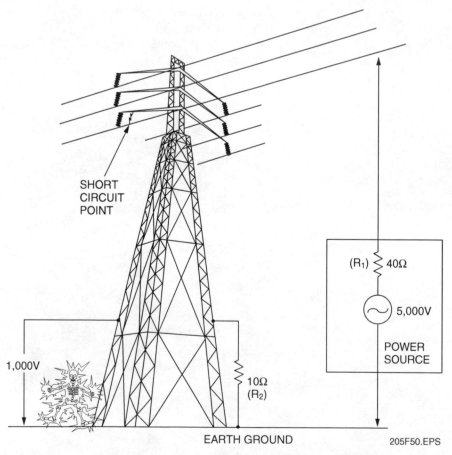

Figure 50 ◆ Poorly grounded system.

$$I = \frac{E}{R_1 + R_2} = \frac{5,000}{40 + 10} = 100A \text{ of short circuit current}$$

A person touching the tower would be subjected to the voltage (E') developed across the ground resistance:

E' = IR = 100 × 10 = 1,000 volts
between power and ground

Statistics vary widely concerning what may be considered a dangerous voltage. This depends largely on body resistance and other conditions. However, to limit the touch voltage for this situation to 100V, the ground resistance for the tower would have to be less than 1Ω.

10.2.0 How the Ground Tester Works

Several methods are used for measuring the resistance to earth of a grounding electrode; one of the most common is the fall-of-potential method (*Figure 51*).

In this method, auxiliary electrodes 1 and 2 are placed at sufficient distances from grounding electrode X. A current (I) is passed through the earth between the grounding electrode X and auxiliary current electrode 2 and is measured by the ammeter. The voltage drop (E) between the grounding electrode X and the auxiliary potential electrode 1 is indicated on the voltmeter. Resistance (R) can therefore be calculated as follows:

$$R = \frac{E}{I}$$

Certain problems may arise in measuring with the simple system shown in *Figure 51*:

- Natural currents in the soil caused by electrolytic action can cause the voltmeter to read either high or low, depending on polarity.
- Induced currents in the soil, instrument, or electrical leads can cause vibration of the meter pointer, interfering with readability.
- Resistance in the auxiliary electrode and electrical leads can introduce error into the voltmeter reading.

Most ground testers use a null balance metering system. Unlike the separate voltmeter and ammeter method, this instrument provides a readout directly in ohms, thus eliminating calculation. Although the integrated systems of the ground testers are sophisticated, they still perform the basic functions for fall-of-potential testing.

10.3.0 Current Supply

As in the simple circuit, the ground tester also has a current supply circuit (*Figure 52*). This may be traced from grounding electrode X through terminal X, potentiometer R1, the secondary of power transformer T1, terminal 2, and auxiliary current electrode 2. This produces a current in the earth between electrodes X and 2.

When switch S1 is closed, battery B energizes the coil of vibrator V. Vibrator reed V1 begins oscillating, thereby producing an alternating current in the primary and secondary windings of T1.

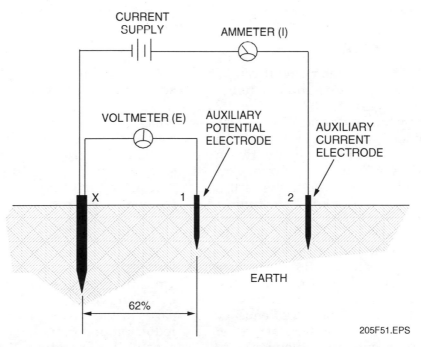

Figure 51 ◆ Fall-of-potential method of testing.

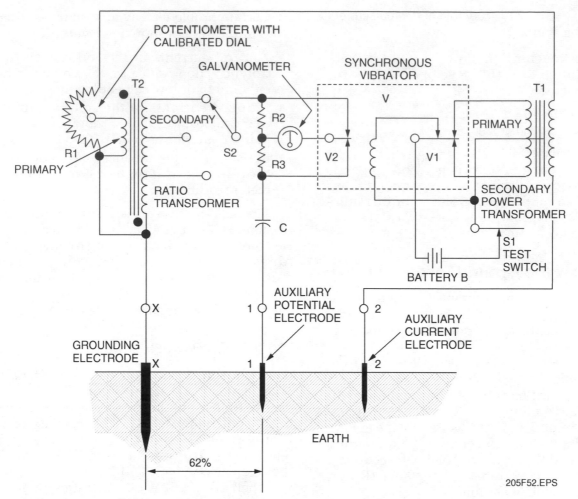

Figure 52 ◆ Three-point testing using a ground tester.

10.4.0 Voltmeter Circuit

Refer to *Figure 52*. The voltmeter circuit can be traced from grounding electrode X through terminal X, the T2 secondary, switch S2, resistors R2 and R3 (paralleled by the meter and V2 contacts), capacitor C, terminal 1, and auxiliary potential electrode 1. The current in the earth between grounding electrode X and auxiliary current electrode 2 creates a voltage drop due to the earth's resistance. With auxiliary potential electrode 1 placed at any distance between grounding electrode X and auxiliary current electrode 2, the voltage drop causes a current in the voltmeter circuit through balanced resistors R2 and R3. The voltage drop across these resistors causes galvanometer M to deflect from zero center scale.

Vibrator reed V2 operates at the same frequency as V1, thereby functioning as a mechanical rectifier for galvanometer M. The vibrator is tuned to operate at 97.5 hertz (Hz), a frequency unrelated to commercial power line frequencies and their harmonics. Thus, currents induced in the earth by power lines are rejected by most ground testers and have virtually no effect on their accuracy. Stray direct current in the earth is blocked out of the voltmeter circuit by capacitor C.

The current in the primary of T2 can be adjusted with potentiometer R1. Primary current in T2 induces a voltage in the secondary of T2, which is opposite in polarity to the voltage drop caused by current in the voltmeter circuit.

With R1 adjusted so the primary and opposing secondary voltages of T2 are equal, current in the voltmeter circuit is zero, and the galvanometer reads zero. The resistance of grounding electrode X can then be read on the calibrated dial of the potentiometer.

With no current in the voltmeter circuit, the lead resistance of the auxiliary potential electrode 1 has no bearing on accuracy. (With no current, there is no voltage drop in the leads.) Resistance to earth of the current electrode results only in a reduction of current, and consequently, a loss of sensitivity. Therefore, the auxiliary electrodes need only be inserted into the earth 6" to 8" to

The negative battery terminal is connected alternately across first one and then the other half of the primary winding.

make sufficient contact. In some locations, where the soil is very dry, it may be necessary to pour water around the current electrode to lower the resistance to a practical value.

10.5.0 Nature of Earth Electrode Resistance

Current in a grounding system is primarily determined by the voltage and impedance of the electrical equipment. However, the resistance of the grounding system is very important in determining the voltage rise between the ground electrode and the earth as well as the voltage gradients that will occur in the vicinity when current is present.

Three components constitute the resistance of a grounding system:

• Resistance of the conductor connecting the ground electrode

• Contact resistance between the ground electrode and the soil

• Resistance of the body of earth immediately surrounding the electrode

Resistance of the connecting conductor can be dealt with separately since this is a function of the conductor cross-sectional area and length. Contact resistance is usually negligible if the electrode is free from paint or grease. Therefore, the main resistance is that of the body of earth immediately surrounding the electrode. Current from a grounding electrode flows in all directions via the surrounding earth. It is as though the current flows through a series of concentric spherical shells, all of equal thickness.

The shell immediately surrounding the electrode has the smallest cross-sectional area, and therefore its resistance is highest. As the distance from the electrode is increased, each shell becomes correspondingly larger; thus, the resistance becomes smaller. Finally, a distance from the electrode is reached where additional shells do not add significantly to the total resistance. From a theoretical viewpoint, total resistance is included only when the distance is infinite. For practical purposes, only that volume which contributes the major part of the resistance need be considered. This is known as the effective resistance area and depends on electrode diameter and driven depth.

10.6.0 Three-Point Testing Procedure

To measure resistance of a grounding electrode with most ground testers, an auxiliary current electrode and an auxiliary potential electrode are required (see *Figures 53* and *54*). The current electrode is placed a suitable distance from the grounding electrode under test, and the potential electrode is then placed at 62 percent of the current electrode distance.

> **NOTE**
> The 62-percent figure has been arrived at by empirical data gathered by many authorities on ground resistance measurement, and in some cases, it has been computed based on analysis of an equivalent hemisphere.

If the current electrode is too near the grounding electrode, their effective resistance areas will overlap, as shown in *Figure 53*. If a series of measurements are made with the potential electrode driven at various distances in a straight line between the current electrode and grounding electrode, the readings will yield a curve as illustrated in *Figure 53*.

In *Figure 54*, a curve is plotted with the current electrode at a sufficient distance from the grounding electrode. Note that the curve is relatively flat between points B and C, which are usually considered to be at ±10' with respect to the 62-percent point. Usually, a tolerance is established for the maximum allowable deviation for the second and third reading with relation to the initial reading at 62 percent. This tolerance is a certain ± percent of the initial reading, such as ±1 percent, ±2 percent, and so on.

No definite distance from the current electrode can be forecast since the optimum distance is based on the homogeneity of the earth, depth of the grounding electrode, diameter, etc. However, for a starting point for a single driven grounding electrode, the effective radius of the equivalent hemisphere can be computed. The curve of *Figure 55* can then be used for initial placement of the auxiliary electrodes. Using the practical method of moving the auxiliary potential electrode 10' to either side of the 62-percent point, a curve can be plotted to determine if the current electrode spacing is adequate, as depicted by the curve flattening out between points B and C in *Figure 54*.

For example, assume that a grounding electrode 1" (2.54cm) in diameter is buried 10' (3.0m) deep. The equivalent hemisphere radius is 1.7' (51.8cm). Using *Figure 55*, the current electrode would be established at approximately 90' (27m), and the potential electrode at approximately 55'(17m) for the initial reading. The results obtained when the potential electrode is moved (to 45' [14m] and then to 65' [20m]) will determine if the current electrode, and consequently the potential electrode, must be spaced at a greater distance.

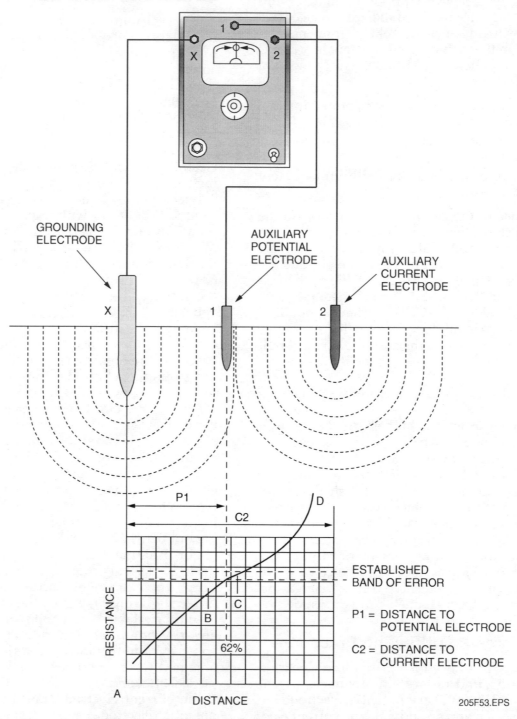

Figure 53 ◆ Plotted curve showing insufficient electrode spacing.

205F53.EPS

It is usually advisable to plot a complete curve for each season of the year. See *Figure 56*. These curves should be retained for comparison purposes. Measurements at established intervals in the future need only be made at the 62-percent point and, if desired, 10' on each side, providing there is no erratic deviation from the original curve. Serious deviation, other than seasonal, could mean corrosion has eaten away some of the electrode.

10.6.1 Operating Procedure

Establish the proper location for the auxiliary current electrode and auxiliary potential electrode as follows:

Step 1   Determine the effective radius of the equivalent hemisphere in feet for the grounding electrode depth in feet, as well as the grounding electrode diameter in inches, using the curves in *Figure 57*.

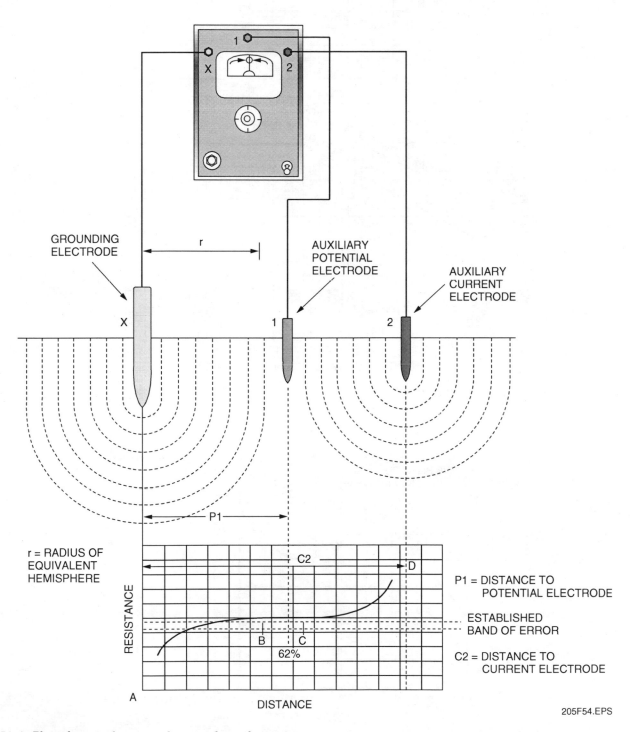

Figure 54 ◆ Plotted curve showing adequate electrode spacing.

Step 2   Using this effective radius and the data in *Figure 55*, determine the distances for the auxiliary current electrode No. 2 and auxiliary potential electrode No. 1 from the ground electrode.

Step 3   Position the ground tester in a suitable location near the grounding system to be tested.

Step 4   Using a test lead of less than 0.05V resistance, connect the tester terminal X to the grounding electrode or grounding grid to be measured, terminal 1 to the auxiliary potential electrode, and terminal 2 to the auxiliary current electrode (less than 1Ω).

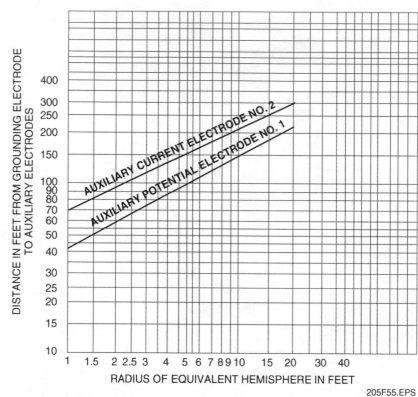

Figure 55 ◆ Auxiliary electrode distance/radii chart.

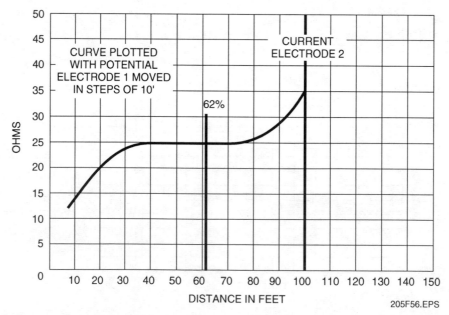

Figure 56 ◆ Typical grounding resistance curve to be recorded and retained.

Step 5 If the general range of resistance to be measured is anticipated, then set the MULTIPLY BY switch to a multiplying factor within this range. If unknown, select the highest multiplier range first.

Step 6 Move the TEST switch to the ADJ position. Rotate the OHMS control knob until the galvanometer indicates balance at center scale.

> **NOTE**
> Meter instructions will vary depending on the instrument used. Consult the manufacturer's operating manual.

Step 7 Should the galvanometer indication remain to the right of center scale with the OHMS control at zero, set the MULTIPLY BY switch to the adjacent lower range.

Step 8 Should the galvanometer indication remain to the left of center scale with the OHMS control fully clockwise, set the MULTIPLY BY switch to the adjacent higher range.

Step 9 After balance is achieved with the TEST switch in the ADJ position, move this switch to the READ position and rebalance the galvanometer using the OHMS control.

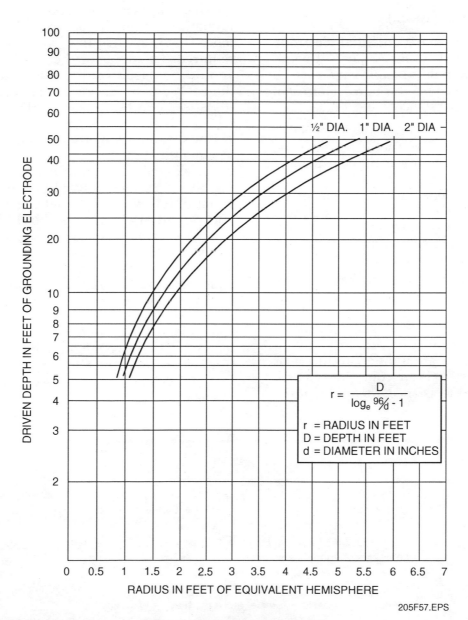

$$r = \frac{D}{\log_e \frac{96D}{d} - 1}$$

r = RADIUS IN FEET
D = DEPTH IN FEET
d = DIAMETER IN INCHES

205F57.EPS

Figure 57 ◆ Ground rod depth versus equivalent hemisphere radii.

Step 10 Read the resistance on the calibrated OHMS scale. If this indication is less than one-tenth of full range, set the MULTIPLY BY switch to the adjacent lower range and rebalance the galvanometer. This provides better readability and resolution.

Step 11 Multiply the OHMS scale indication by the factor indicated by the MULTIPLY BY switch. This product is the resistance of the grounding system connected to terminal X. Record this resistance value. (If balance cannot be achieved under any conditions, check the test leads. Ensure that the leads are properly connected, as described in Step 4. If the trouble persists, the grounding system may be inadequate due to a high resistance beyond the range of the tester.)

Step 12 Move the auxiliary potential electrode 10' on each side of the 62-percent point, as shown by points B and C in *Figure 54*, and repeat Steps 5 through 11 for each. Record this resistance value.

Step 13 Compare the resistance values obtained in Steps 11 and 12. They should be within the established tolerance band as explained previously. If they are not within the established tolerance, move the auxiliary current and potential electrodes to a point farther away from the ground under test and repeat Steps 4 through 12. Record the resistance values obtained.

A single driven rod is an economical and simple means of making a grounding electrode. In general, however, a single driven rod does not provide a sufficiently low resistance, and consequently, several rods must be driven and connected in parallel by a grounding conductor or cable. Although these rods are connected as they would be in parallel resistance, their total resistance does not follow the usual law for computing resistances in parallel. To attain the full effect of resistances in parallel, these rods would have to be spaced at such distances that the effective resistance areas immediately surrounding them would not overlap. The extent to which the areas overlap determines how much of their effectiveness is lost. The curves of *Figure 58* show a change in resistance for two, four, and eight parallel rods with various spacing.

When two, three, or four rods are used in parallel, they are usually driven in a straight line and connected together. In cases where more than four are used, they are usually driven in a hollow-square formation and connected in parallel. Distances between the rods are usually made equal, as in the hollow-square formation of eight rods shown in *Figure 59*.

When measurements must be made on large formations or on substations contained within a fence that is part of the grounding system, the required distances for the auxiliary electrodes should be based on the diagonal (or longest dimension) of the entire grounding system. These distances can be determined from *Figure 60*, which is based on the computed hemisphere for various numbers of rods arranged in hollow-square formation.

For example, a hollow-square formation having 40' (12.2m) sides would have a diagonal of about 57' (17.4m). *Figure 60* shows that current and potential electrodes should be spaced at about 370' (113m) and 230' (70m), respectively.

Major difficulties are encountered when measuring large grid systems due to the fact that theory and computations are usually based on homogeneous soil. Differences in the soil will affect penetration depth and consequently distort the curve being plotted. In extremely large ground grid systems using several configurations of electrodes, the true center from which to properly space the current and potential electrodes is difficult to determine.

Excessive Electrode Resistance During Testing

Excessive electrode resistance or transient noise during testing may indicate an incorrect measurement. If this problem occurs, try the following remedies:

- Check the integrity of all test connections between the leads and the electrodes.
- Make sure that the ground electrodes Y and Z are properly inserted in a quality ground and completely buried, if possible.
- Select a lower test current setting.
- If stray currents are suspected, try moving both electrodes Y and Z in an arc relative to the ground electrode; for example, move them 90° and test again.

When measuring the resistance of any grounding system, single-driven electrodes, or grid systems, it is important to realize that other common grounding connections will influence the total measured resistance. For instance, if a ground grid system consisting of multiple-driven rods connected in parallel is in turn connected to a fence surrounding a substation, the fence can be considered connected in parallel with the grid system. The total resistance measured will be that of the entire grounding system. Whether these grounds can be isolated and measured separately is primarily a matter of regulations, the danger involved while they are isolated, and practicality.

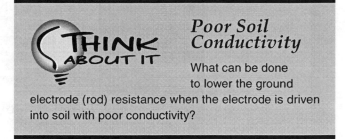

THINK ABOUT IT

Poor Soil Conductivity

What can be done to lower the ground electrode (rod) resistance when the electrode is driven into soil with poor conductivity?

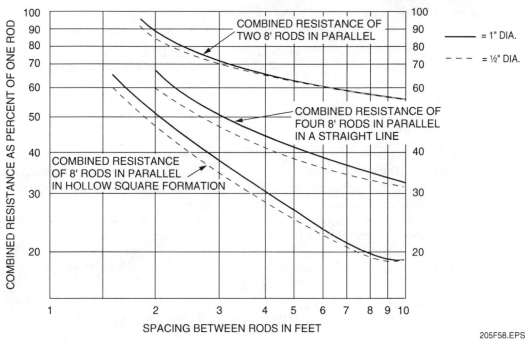

Figure 58 ◆ Combined resistances of rods in parallel.

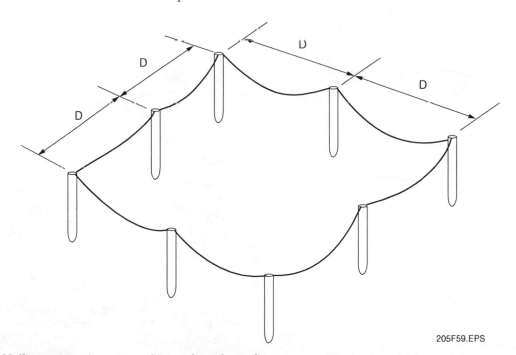

Figure 59 ◆ Hollow-square formation of grounding electrodes.

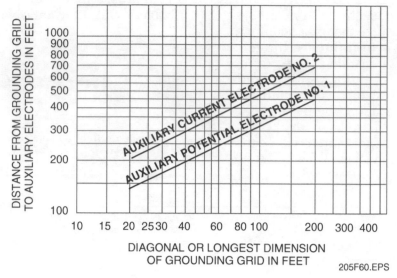

Figure 60 ◆ Auxiliary electrode distances versus longest grid dimension.

10.6.2 Electrode Arrangements

The angle at which the electrodes are placed with respect to a ground grid system is primarily a matter of choice or requirement due to surrounding obstructions. However, there are pros and cons as to whether the electrodes should be placed at 90 degrees with respect to the sides of the grid or diagonally from a corner. Many times, one placement as opposed to the other has yielded different test results. This may very well be due to the homogeneity of the soil of one area versus the other. Curves have also been plotted with the current electrode on one side of the grounding electrode or ground grid system, and the potential electrode on the opposite side (180 degrees apart). This arrangement yields a curve similar to that shown in *Figure 61*. This arrangement may give erroneous results, because there is no way of checking for the presence of errors during the test.

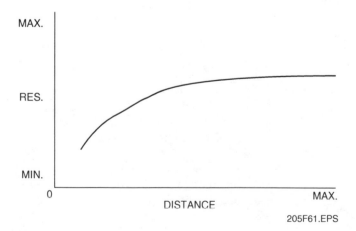

Figure 61 ◆ Plotted curve with current and potential electrodes spaced 180 degrees apart.

10.6.3 Equipotential Grounding

Safety grounds should be applied in such a way that a zone of equal potential is formed in the work area. This equipotential zone is formed

Grounding

A low-impedance ground is essential to the performance of any electrical protection system. The ground must dissipate electrical transients and surges in order to minimize the chance of damage or injury. Proper grounding, which includes bonding and connections, protects personnel from the danger of shock and protects equipment and buildings from hazardous voltages. Proper grounding also contributes to the reduction of electrical noise and provides a reference for circuit conductors to stabilize their voltage to ground during normal operation. Examine the grounding system of the electrical service at your home or workplace. Is it adequate? If not, how can it be corrected to meet current NEC® requirements?

Clamp-On Resistance Testings

Clamp-on resistance testers offer the ability to measure the resistance without disconnecting the ground. This type of measurement also has the advantage of including the bonding to ground and the overall grounding connection resistance.

when fault current is bypassed around the work area by metallic conductors. In this situation, the worker is bypassed by the low-resistance metallic conductors of the safety ground.

11.0.0 ◆ LOCATION OF GROUND FAULTS

As explained earlier, a ground fault is defined as a conducting connection, accidental or intentional, between any conductors of an electrical system and the normally non-current-carrying conducting material enclosing the conductors, that is, a short circuit. In a ground fault, there is a connection from a current-carrying phase conductor to the conductor enclosure such as a metal conduit, a metal box, or a motor frame. A ground fault typically occurs when the conductor insulation fails or when a wire comes loose from its terminal point.

Ground fault protection of equipment and personnel involves the use of ground fault circuit interrupter (GFCI) devices in the electrical system. These devices de-energize a circuit or portion of a circuit within an established time period when a current to ground exceeds some predetermined value (typically 4 to 6 mA) that is less than that required to operate the overcurrent protective device of the supply circuit. GFCI devices are available both in circuit breakers and receptacles (*Figure 62*), as well as in portable units.

GFCIs monitor the current imbalance between the ungrounded hot conductor and the grounded neutral conductor. Under normal conditions (no fault), the current flowing to an equipment, tool, or other load via the hot wire, and the current returning to the source along the grounded neutral are equal. When a ground fault is present, the current flowing in the hot conductor will exceed the current flowing in the return grounded neutral conductor by 4 to 6 mA or more. The GFCI circuit senses this imbalance and trips (opens) the circuit. The imbalance indicates that some of the current flowing in the circuit is being diverted to some path other than the normal return path along the grounded neutral conductor. If the other path is through a human body, the GFCI becomes a lifesaving device.

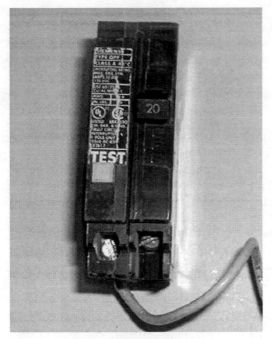

GFCI CIRCUIT BREAKER

GFCI RECEPTACLE

205F62.EPS

Figure 62 ◆ GFCI circuit breaker and receptacle.

Why Not Use One GFCI Breaker Instead of Individual GFCI Receptacles?

The cost of several GFCI receptacles that are required on one circuit is usually much greater than one GFCI circuit breaker that could be installed in the panel. However, GFCI circuit breakers are not generally installed to protect multiple devices because all receptacles would become de-energized by a single ground fault incident anywhere within the circuit.

GFCI circuit breakers and receptacles have a reset button to reset a circuit that has been tripped, and a test button to test the tripping circuitry to make sure the GFCI is working properly. Some GFCI receptacles have an indicator light to show if the circuit is energized. Pushing the test button places a small ground fault on the circuit. If operating properly, the GFCI should trip to the off position. In addition to the built-in test functions in GFCIs themselves, there are various independent test devices also available that can be used to test GFCI receptacles.

In the event that a GFCI circuit breaker or receptacle has tripped, it is necessary to locate the reason why. Note that nuisance tripping of GFCIs is known to occur. Sometimes this can be attributed to extremely long cable runs for the protected circuit. For example, a circuit supplied by a GFCI circuit breaker in the main panel could cause nuisance tripping if the branch circuit is 50' long or more. To locate a ground fault other than one caused by nuisance tripping, disconnect all electrical loads connected to the circuit, then reset the GFCI. If the GFCI does not trip, this is an indication that the fault exists in one of the loads that has been disconnected. The quickest thing to do to isolate the faulty load is to reconnect the various loads to the circuit one at a time. Continue to do this until the GFCI trips again. When this occurs, the last load connected to the circuit should be the faulty load.

If the GFCI trips with all the loads disconnected, this means the ground fault exists within the branch circuit wiring itself. To isolate the fault, a time domain reflectometer (TDR) or megohmmeter can be used per the manufacturer's instructions to measure the individual hot wires in the branch circuit. The TDR transmits a test pulse down the wire that is reflected back by the short

(ground fault) to ground. The time between the transmission and reflection is proportional to the distance to the fault.

If you are using the megohmmeter to measure the individual hot wires to ground, it is necessary to do a divide-by-two approach to aid in fault location. This means that the branch circuit wiring should be disconnected at about the halfway distance point in the circuit so that the fault can be isolated to either the first half or the second half of the circuit wiring, as indicated by a low resistance reading from the conductor under test to ground. Following this, the faulty half of the circuit is again divided by two to further isolate its faulty half. This process is continued as necessary until the specific location of the ground fault has been located and the faulty wiring replaced. Note that detailed information about time domain reflectometers and megohmmeters is given in the module entitled *Advanced Test Equipment*.

Summary

This module provides a general overview of power distribution and transmission systems. The problems of power quality were covered, with the prime consideration being on its effects on electric/electronic equipment. The different types of disturbances that can have varying effects on equipment, including harmonics, surges, high and low voltages, and noise were described. Also provided was information on some of the devices available to protect equipment from damage that can be caused by these disturbances. The subject of grounding was also covered. Grounding is the chief means of protecting life and property from electrical hazards and also ensures proper operation of the system as well as helping other protective devices to function properly.

1. The root-mean-square (rms) value of an AC voltage or current sinusoidal waveform is equal to _____ times the zero-to-peak (maximum) value of the sine wave.
   a. 0.623
   b. 1.000
   c. 0.7071
   d. 1.414

2. Which of the following statements defines the term bonding?
   a. The permanent joining of metallic parts to form an electrically conductive path that will ensure electrical continuity and the capacity to safely conduct any current likely to be imposed on it.
   b. A system or circuit conductor that is intentionally grounded.
   c. A conductor placed in the earth, providing a connection to a circuit.
   d. Intentionally connected to earth through a ground connection or connections of sufficiently low impedance and having sufficient current-carrying capacity to prevent the buildup of voltages that may result in undue hazards to connected equipment or to persons.

3. In a single-phase, three-wire 120/240V system, the _____ conductor(s) are grounded.
   a. hot
   b. current-carrying
   c. black
   d. neutral

4. All of the following are considered part of the system grounding scheme except _____.
   a. grounding electrode field
   b. grounding electrode
   c. equipment grounding conductor
   d. grounding electrode conductor

5. To be considered an NEC®-approved grounding conductor, an underground water pipe must be in direct contact with the earth for no less than _____ feet.
   a. 5
   b. 10
   c. 15
   d. 20

6. To determine the requirements for grounding fire alarm systems, you would look in _____.
   a. *NEC Article 640*
   b. *NEC Article 700*
   c. *NEC Article 760*
   d. *NEC Article 800*

7. A down connector connected to a lightning protection system air terminal or grid at one end, is connected directly to _____ at the other end.
   a. the power distribution system grounding electrode conductor
   b. the power distribution system ground electrode(s)
   c. a lightning protection system ground electrode
   d. an equipotential ground plane

8. An increase in the nominal rms voltage and related current for less than one minute is called a(n) _____.
   a. frequency variation
   b. voltage swell
   c. surge
   d. overvoltage

9. According to a Bell Labs study, the most common power abnormality experienced is _____.
   a. voltages interruptions
   b. voltage sags
   c. voltage transients
   d. frequency variations

10. A power abnormality that causes overheating in transformers, motors, or wires, because of skin effect is _____.
    a. electromagnetic interference (EMI)
    b. frequency variations
    c. harmonics
    d. overvoltage

11. A type of varistor widely used in lightning arresters/surge suppressors at all stages of surge protection is the _____.
    a. metal-oxide varistor (MOV)
    b. zener diode
    c. avalanche diode
    d. gas tube arrester

12. A harmonic/noise suppression filter that passes all frequencies below a specified frequency with little or no loss but discriminates strongly against higher frequencies is a _____ filter.
    a. low-pass
    b. high-pass
    c. bandpass
    d. band rejection

13. Surge protecting devices work to _____.
    a. provide a grounded electrostatic shield
    b. open the related circuit breaker when peak surge voltages occur
    c. reduce noise
    d. limit peak surge voltages and divert transients to ground

14. A(n) _____ provides electrical isolation from the utility line and operates to regenerate power.
    a. supplemental DC generator set
    b. power line conditioner
    c. AC motor-generator set
    d. uninterruptible power supply

15. The simplest version of a single-conversion UPS system is a _____ system.
    a. standby
    b. line-interactive
    c. ferroresonant
    d. delta-conversion

16. A full-wave rectifier and a full-wave bridge rectifier are types of _____ DC power supplies.
    a. switching
    b. linear
    c. nonlinear
    d. uninterruptible

17. Given proper maintenance, _____ lead acid batteries have the longest expected life.
    a. tubular
    b. Planté
    c. pasted-plate
    d. valve-regulated

18. The valve-regulated battery, often referred to as a sealed battery, will vent and release gas under _____ conditions.
    a. undercharge
    b. overcharge
    c. low-temperature
    d. electrolyte overfilling

19. A type of shielding that has poor effectiveness against radio frequency noise is _____.
    a. single-layer braid
    b. multiple-layered braid
    c. solid-wall conduit
    d. flexible conduit

20. To minimize the problems associated with static electricity, you should maintain room humidity at a minimum of _____ percent.
    a. 20
    b. 30
    c. 40
    d. 50

Trade Terms Introduced in This Module

Autotransformer: A type of transformer in which parts of one winding are common to both the primary and secondary circuits. For example, to operate a step-down transformer, the whole winding acts as the primary winding, and only part of the winding acts as the secondary. To operate a step-up transformer, part of the winding acts as the primary and the whole winding acts as the secondary.

Capacitor: An electrical device consisting of two conducting surfaces separated by an insulating material or dielectric such as air, paper, mica, glass, plastic film, or oil. It stores electrical energy, blocks the flow of direct current, and permits the flow of alternating current to a degree dependent on the capacitance of the device and the frequency of the AC voltage applied across it.

Capacitor bank: A number of capacitors connected together in series, parallel, or series-parallel.

Capacitive coupling: Energy transfer from one circuit or conductor to another when mutual capacitance is created between the two.

Electrolyte: A substance in which the conduction of electricity is accompanied by chemical action. The paste which forms the conducting medium between electrodes of a dry cell or storage cell.

Electromagnetic interference (EMI): Any electrical or electromagnetic interference that causes undesirable signals in electrical or electronic equipment.

Equipotential ground plane: A bonded and grounded system that is in direct contact with earth and provides negligible impedance (resistance) to ground.

Floating ground: A referenced ground that is not connected to the earth.

Frequency: The rate at which a phenomenon is repeated. It refers to the number of cycles per second of an alternating current (AC) or radio frequency (RF) signal. The basic unit of frequency is the hertz (Hz), which is one cycle per second.

Grounded: Connected to earth or to some other conducting body that serves in place of the earth.

Harmonics: Sinusoidal waves having a frequency that is an integral multiple of the fundamental frequency. For example, a wave with twice the frequency of the fundamental is called the second harmonic.

Impedance: The total resistance to current flow as a result of resistance and reactance.

Photovoltaics: Capable of generating a voltage when exposed to visible or other light radiation.

Power quality: The concept of powering and grounding sensitive electronic equipment in a manner that is suitable to the optimum operation of that equipment.

Private branch exchange (PBX): A private telephone switchboard.

Radio-frequency interference (RFI): Any electrical signal capable of interfering with the proper operation of electrical or electronic equipment. The frequency range of such interference may include the entire electromagnetic spectrum.

Reactive load: A circuit that contains both inductance and capacitance and is therefore tuned to resonance at a certain frequency. The resonant frequency can be raised or lowered by changing the inductance and/or capacitance values.

Caldwell Procedure

IP-X | CONDENSED INSTRUCTIONS

CADWELD® ELECTRICAL CONNECTIONS

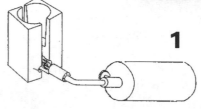

1

**BEFORE FIRST CONNECTION OF THE DAY, DRY THE MOLD BY HEATING WITH A TORCH

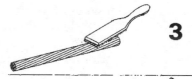

2

**DRY CONDUCTORS TO BE WELDED WITH A TORCH

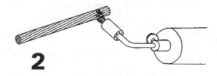

3

**CLEAN DRIED CABLE ENDS WITH A BRUSH TO REMOVE DIRT AND OXIDES.

4

**WHEN WELDING TO A STEEL SURFACE, USE A RASP OR ERICO APPROVED GRINDING WHEEL SBS2333 TO REMOVE PAINT, RUST AND MILL SCALE FROM AREA TO BE WELDED. (BRIGHT METAL SHOWING)

5

**POSITION MOLD OVER CONDUCTOR WITH CONDUCTOR ENDS UNDER CENTER OF TAP HOLE. GAP DISTANCE, IF REQUIRED, IS NOTED ON MOLD TAG.

**LOCK MOLD HANDLES.

6

**INSERT ROUND METAL DISK (PACKAGED WITH WELD METAL) IN BOTTOM OF CRUCIBLE. MAKE SURE IT COVERS TAP HOLE.

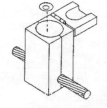

QUESTIONS? CALL CADWELD AT 1–800–248–WELD

ERICO®️ products inc. 34600 Solon Rd. • Cleveland (Solon). Ohio 44139

205A01.EPS

Additional Resources

This module is intended to present thorough resources for task training. The following reference works are suggested for both instructors and motivated participants interested in further study. These are optional materials for continued education rather than for task training.

American Electricians' Handbook, 2002. Terrell Croft and Wilford I. Summers. New York, NY; McGraw-Hill.

National Electrical Code® Handbook, Latest Edition. Quincy, MA: National Fire Protection Association.

Telecommunications Cabling Installation Manual, Latest Edition. Tampa, FL: BICSI. www.bicsi.org

Telecommunications Distribution Methods Manual, Latest Edition. Tampa, FL: BICSI. www.bicsi.org

Figure Credits

AEMC Instruments	205SA05
Copyright © 2004 by Building Industry Consulting Services International, Inc. (BICSI)	205F18, 205F19
Dual-Lite	205SA04
Erico, Inc.	205F13, Appendix A
Federal Pacific	205SA02
Liebert Corporation	205F27A
Middle Atlantic Products, Inc.	205SA01
Phoenix Contact, Inc.	205F28
Stored Energy Systems, LLC	Table 1
Topaz Publications, Inc.	205F02, 205F05, 205F07, 205F20 (photo), 205F33, 205F46, 205F48 (top), 205F49, 205F62
Tripp Lite	205F27B

NCCER makes every effort to keep these textbooks up-to-date and free of technical errors. We appreciate your help in this process. If you have an idea for improving this textbook, or if you find an error, a typographical mistake, or an inaccuracy in NCCER's Contren® textbooks, please write us, using this form or a photocopy. Be sure to include the exact module number, page number, a detailed description, and the correction, if applicable. Your input will be brought to the attention of the Technical Review Committee. Thank you for your assistance.

Instructors – If you found that additional materials were necessary in order to teach this module effectively, please let us know so that we may include them in the Equipment/Materials list in the Annotated Instructor's Guide.

Write: Product Development and Revision
National Center for Construction Education and Research
P.O. Box 141104, Gainesville, FL 32614-1104

Fax: 352-334-0932

E-mail: curriculum@nccer.org

Craft _____ Module Name _____

Copyright Date _____ Module Number _____ Page Number(s) _____

Description _____

_____

_____

_____

_____

(Optional) Correction _____

_____

_____

(Optional) Your Name and Address _____

_____

_____

Introduction to Electrical Blueprints

COURSE MAP

This course map shows all of the modules in the second level of the *Electronic Systems Technician* curriculum. The suggested training order begins at the bottom and proceeds up. Skill levels increase as you advance on the course map. The local Training Program Sponsor may adjust the training order.

ELECTRONIC SYSTEMS TECHNICIAN LEVEL TWO

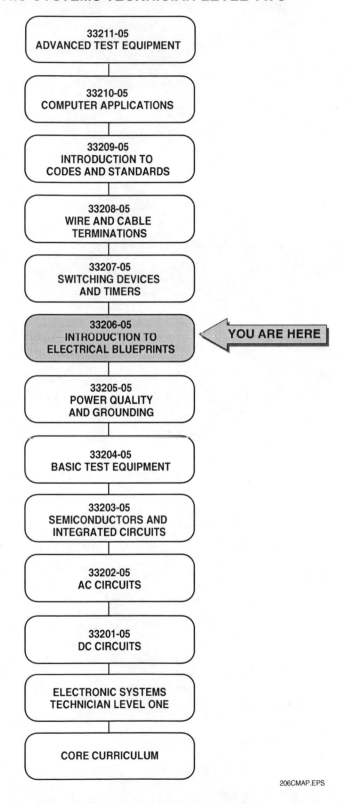

33211-05
ADVANCED TEST EQUIPMENT

33210-05
COMPUTER APPLICATIONS

33209-05
INTRODUCTION TO
CODES AND STANDARDS

33208-05
WIRE AND CABLE
TERMINATIONS

33207-05
SWITCHING DEVICES
AND TIMERS

33206-05
INTRODUCTION TO
ELECTRICAL BLUEPRINTS

YOU ARE HERE

33205-05
POWER QUALITY
AND GROUNDING

33204-05
BASIC TEST EQUIPMENT

33203-05
SEMICONDUCTORS AND
INTEGRATED CIRCUITS

33202-05
AC CIRCUITS

33201-05
DC CIRCUITS

ELECTRONIC SYSTEMS
TECHNICIAN LEVEL ONE

CORE CURRICULUM

206CMAP.EPS

Figures

Tables

Introduction to Electrical Blueprints

Objectives

When you have completed this module, you will be able to do the following:

1. Explain the basic layout of a blueprint.
2. Describe the information included in the title block of a blueprint.
3. Identify the types of lines used on blueprints.
4. Identify common symbols used on blueprints.
5. Understand the use of architect's and engineer's scales.
6. Demonstrate the use of an architect's scale.
7. Interpret electrical drawings, including site plans, floor plans, and detail drawings.
8. Read equipment schedules found on electrical blueprints.
9. Describe the type of information included in electrical specifications.
10. Look for devices on blueprints and perform a take-off.
11. Identify the type of infrastructure specification.

Prerequisites

Before you begin this module, it is recommended that you successfully complete *Core Curriculum*; *Electronic Systems Technician Level One*; and *Electronic Systems Technician Level Two*, Modules 33201-05 through 33205-05.

Required Trainee Materials

1. Pencil and paper
2. Appropriate personal protective equipment
3. Scientific calculator
4. Sample electrical blueprints

1.0.0 ◆ INTRODUCTION TO BLUEPRINT READING

In all large construction projects and in many of the smaller ones, a drawing set that includes site, architectural, mechanical, electrical, and structural drawings is normally prepared as required. An architect and, if necessary, special engineers, sub-contractors, or consultants are employed to prepare the drawing set and specification along with any additionally required detail drawings. The drawing set, sometimes called architectural, mechanical, electrical (AME) or architectural, mechanical, electrical, structural (AMES) drawings, is used for bidding and during construction. When construction is complete and all in-process changes have been incorporated, the drawing set is referred to as the as-built drawings for the project. **Architectural drawings**, identified as A drawings in the sheet number of the title blocks for the drawing set, include the following elements:

- A **site plan** indicating the location of the building on the property
- **Floor plans** showing the walls and partitions for each floor or level
- Elevations of all exterior faces of the building
- Several vertical cross sections to indicate clearly the various floor levels and details of the footings, foundation, walls, floors, ceilings, and roof construction
- Large-scale **detail drawings** showing such construction details as required

If not covered in the A drawings, structural drawings, identified as S drawings, may be included. These drawings provide the necessary detail for structural support components such as footings, foundations, floors, ceilings, and other

loadbearing elements, including concrete or structural steel supports. Electrical drawings, identified as E drawings, show the requirements for power, lighting, and other special electrical systems. While traditionally included in the E drawings, drawings for special electrical systems, such as alarm, sound, video, or security systems, also may be prepared as separate drawings. This is because of the technical complexity of the circuits, especially for large projects. Mechanical drawings, identified as M or ME drawings, include the heating, ventilating, air conditioning (HVAC), sprinkler, instrumentation, and plumbing system drawings.

1.1.0 Site Plans

Site plans provide a bird's-eye view of the building site. Site plans feature the property boundaries, the existing contour lines, the new contour lines (after grading), the location of the building on the property, new and existing roadways, all utility lines, and other pertinent details. The drawing **scale** is also shown. Descriptive notes may also be found on the site (plot) plan listing names of adjacent property owners, the land surveyor, and the date of the survey. A legend or symbol list is also included so that anyone who must work with the site plan can readily read the information. See *Figure 1*.

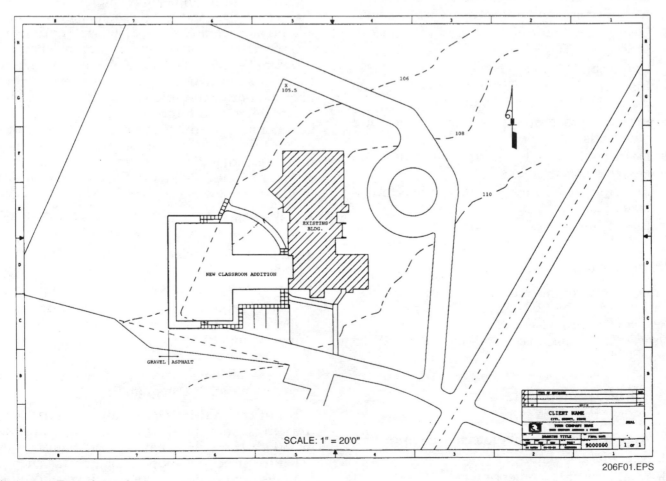

SCALE: 1" = 20'0"

206F01.EPS

Figure 1 ◆ Typical site plan.

1.2.0 Floor Plans

The **plan view** of any object is a drawing showing the outline and all details as seen when looking directly down on the object. It shows only two **dimensions**, length and width. The floor plan of a building is drawn as if a horizontal cut were made through the building—at about window height—and then the top portion removed to reveal the bottom part. See *Figure 2*.

If a plan view of a home's basement is needed, the part of the house above the middle of the basement windows is imagined to be cut away. By looking down on the uncovered portion, every detail and partition can be seen. Likewise, imagine the part above the middle of the first floor windows being cut away. A drawing that looks straight down at the remaining part would be called the first floor plan or lower level. A cut through the second floor windows would be called the second floor plan or upper level. See *Figure 3*.

1.3.0 Elevations

The elevation is an outline of an object that shows heights and may show the length or width of a particular side, but not depth. *Figures 4* and *5* show **elevation drawings** for a building.

> **NOTE**
>
> These elevation drawings show the heights of windows, doors, and porches, the pitch of roofs, and so on because all of these measurements cannot be shown conveniently on floor plans.

1.4.0 Sections

A section or **sectional view** (*Figure 6*) is a cutaway view that allows the viewer to see the inside of a structure. The point on the plan or elevation

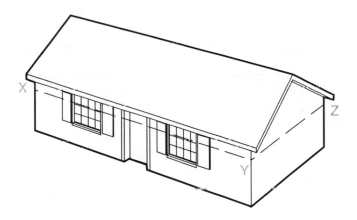

PERSPECTIVE VIEW SHOWING SECTION CUTS

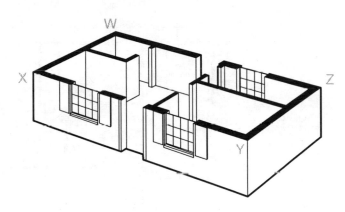

TOP HALF OF SECTION REMOVED

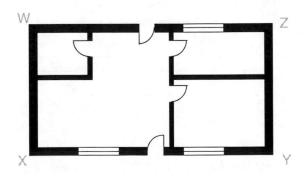

RESULTING FLOOR PLAN IS WHAT THE REMAINING STRUCTURE LOOKS LIKE WHEN VIEWED FROM ABOVE

206F02.EPS

Figure 2 ◆ Principles of floor plan layout.

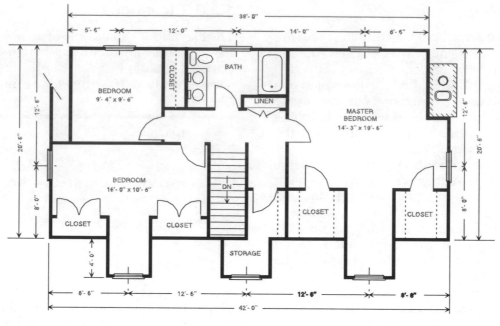

UPPER LEVEL

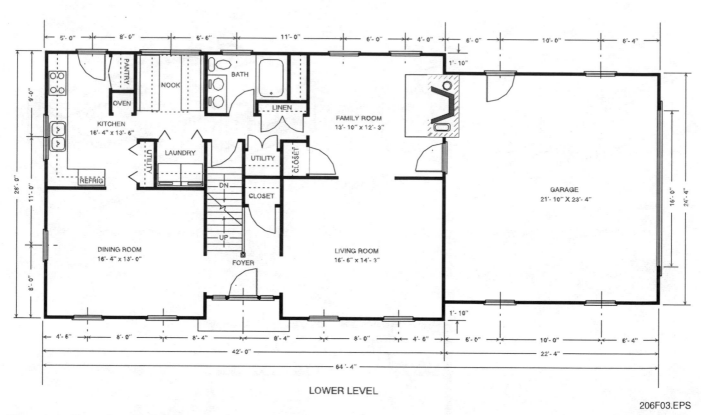

LOWER LEVEL

206F03.EPS

Figure 3 ◆ Floor plans of a building.

showing where the imaginary cut has been made is indicated by the section line, which is usually a dashed line. The section line shows the location of the section on the plan or elevation. It is necessary to know which of the cutaway parts is represented in the sectional drawing. To show this, arrow points are placed at the ends of the section lines.

In architectural drawings, it is often necessary to show more than one section on the same drawing. The different section lines must be distinguished by letters, numbers, or other designations placed at the ends of the lines. These section letters are generally large so as to stand out on the drawings. To further avoid confusion, the same letter is usually

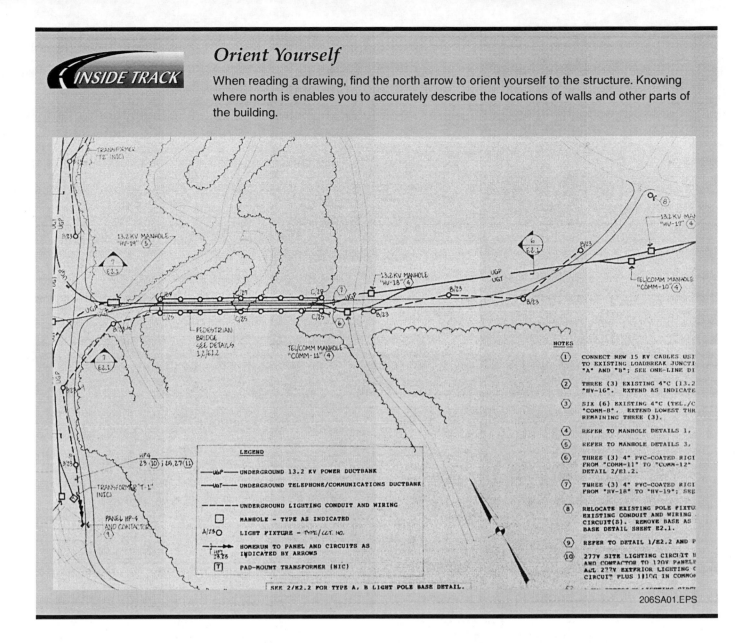

INSIDE TRACK

Orient Yourself

When reading a drawing, find the north arrow to orient yourself to the structure. Knowing where north is enables you to accurately describe the locations of walls and other parts of the building.

206SA01.EPS

---

placed at each end of the section line. The section is named according to these letters; for example, Section A-A, Section B-B, and so forth.

A longitudinal section is taken lengthwise while a cross section is usually taken straight across the width of an object. Sometimes, however, a section is not taken along one straight line. It is often taken along a zigzag line to show important parts of the object.

A sectional view, as applied to architectural drawings, is a drawing showing the building, or portion of a building, as though it were cut through on some imaginary line. This line may be either vertical (straight up and down) or horizontal. Wall sections are nearly always made vertically so that the cut edge is exposed from top to bottom. In some ways, the wall section is one of

the most important of all the drawings to construction workers, because it answers the questions as to how a structure should be built. The floor plans of a building show how each floor is arranged, but the wall sections tell how each part is constructed. Also, wall sections usually specify the building material. Electronic systems technicians need to know this information when determining wiring methods that comply with the *National Electrical Code® (NEC®)*.

1.5.0 Electrical Drawings

Electrical drawings show in a clear, concise manner exactly what is required of the electronic systems technician. The amount of data shown on such drawings should be sufficient, but not overdone.

FRONT ELEVATION

REAR ELEVATION

206F04.EPS

Figure 4 ◆ Front and rear elevations.

This means that a complete set of electrical drawings could consist of only one 8½" × 11" sheet, or it could consist of several dozen 24" × 36" (or larger) sheets, depending on the size and complexity of a given project. A **shop drawing**, for example, may contain details of only one piece of equipment, while a set of working drawings for an industrial installation may contain dozens of drawing sheets detailing the electrical system for lighting and power, along with equipment, motor controls, wiring diagrams, **schematic diagrams**, equipment **schedules**, and other pertinent data.

In general, the working drawings for a given project serve three distinct functions:

• They provide contractors with an exact description of the project so that materials and labor may be estimated to project a total cost of the project for bidding purposes.

• They give workers installation instructions for the electrical and electronic systems.

• They provide a map of the electrical and electronic systems once the job is completed to aid in maintenance and troubleshooting for years.

Electrical drawings from consulting engineering firms will vary in quality from sketchy, incomplete drawings to neat, precise drawings that are easy to understand. Few, however, will cover every detail of the electrical and electronic systems. Therefore, a good knowledge of installation practices must go hand-in-hand with interpreting electrical working drawings.

Some system contractors have electrical drafters prepare special supplemental drawings for use by the contractors' employees. On certain projects, these supplemental drawings can save supervision time in the field once the project has begun.

LEFT ELEVATION

RIGHT ELEVATION

206F05.EPS

Figure 5 ◆ Left and right elevations.

Section View Drawings

Section views provide important information not shown on other types of drawings. For example, section views show the structural members and materials used inside the walls and on the outside wall surfaces. The height, thickness, and shape of the walls are shown, all of which are important in the selection and installation of outlets and fixtures.

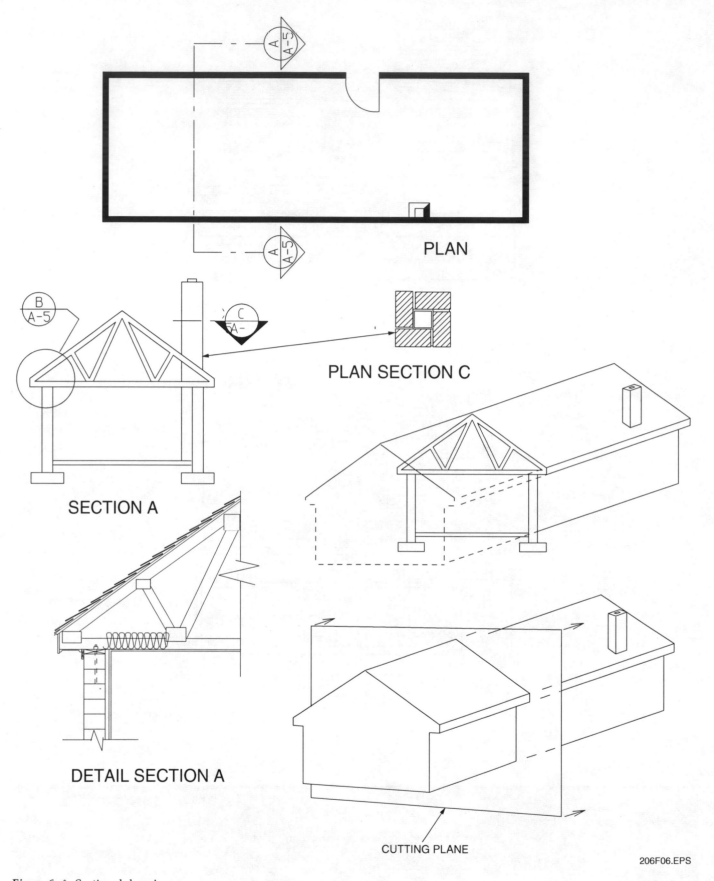

PLAN

PLAN SECTION C

SECTION A

DETAIL SECTION A

CUTTING PLANE

206F06.EPS

Figure 6 ◆ Sectional drawing.

2.0.0 ◆ BLUEPRINT LAYOUT

Although a strong effort has been made to standardize drawing practices in the building construction industry, **blueprints** prepared by different architectural or engineering firms will rarely be identical. Similarities, however, exist between most sets of blueprints, and with a little experience, you should have no trouble interpreting any set of drawings that might be encountered.

Most drawings used for building construction projects are drawn on sheets ranging from 11" × 17" to 24" × 36" in size. Each drawing sheet has border lines framing the overall drawing and one or more title blocks, as shown in *Figure 7*. The type and size of title blocks varies with each firm preparing the drawings. In addition, some drawing sheets will also contain a revision block near the title block, and perhaps an approval block. This information is normally found on each drawing sheet, regardless of the type of project or the information contained on the sheet.

You may hear drawings referred to as A-size, B-size, and so forth. These sizes refer to standard sheet dimensions used for construction drawings. *Table 1* shows the dimensions of the various sheets. When drawings are prepared on CAD systems, they are usually printed by plotters that are fed by long rolls of special paper. The standard widths for plotters are 24, 36, and 42 inches.

Table 1 Standard Drawing Dimensions

Size	Dimensions (inches)
A	8½ × 11
B	11 × 17
C	17 × 22
D	24 × 36
E	30 × 42
F	34 × 44

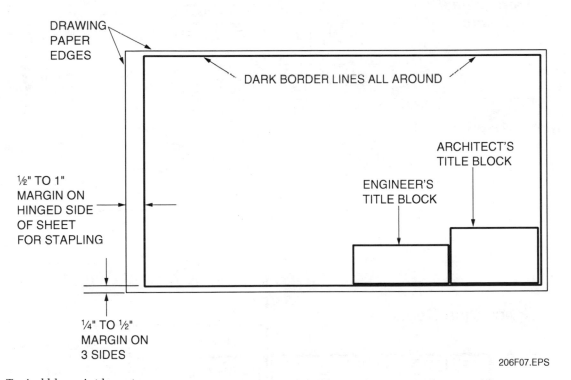

Figure 7 ◆ Typical blueprint layout.

Using a Drawing Set

Always treat a drawing set with care. It is best to keep two sets, one for the office and one for field use. After you use a sheet from a set of drawings, be sure to refold the sheet with the title block facing up.

2.1.0 Title Block

The architect's title block for a blueprint is usually boxed in the lower right-hand corner of the drawing sheet. The size of the block varies with the size of the drawing and with the information required. See *Figure 8*.

In general, the title block of an electrical drawing should contain the following information:

- Name of the project
- Address of the project
- Name of the owner or client
- Name of the architectural firm
- Date of completion
- Scale(s)

- Initials of the drafter, checker, and designer, with dates under each
- Job number
- Sheet number
- General description of the drawing

Every architectural firm has its own standard for drawing titles, and they are often preprinted directly on the tracing paper or else printed on a sticker, which is placed on the drawing.

Often, the consulting engineering firm will also be listed, which means that an additional title block will be applied to the drawing, usually next to the architect's title block. *Figure 9* shows completed architectural and engineering title blocks as they appear on an actual drawing.

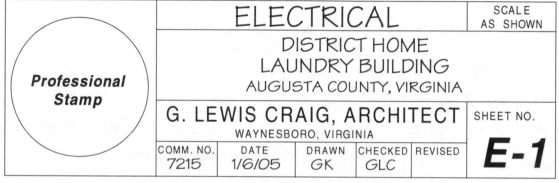

206F08.EPS

Figure 8 ◆ Typical architect's title block.

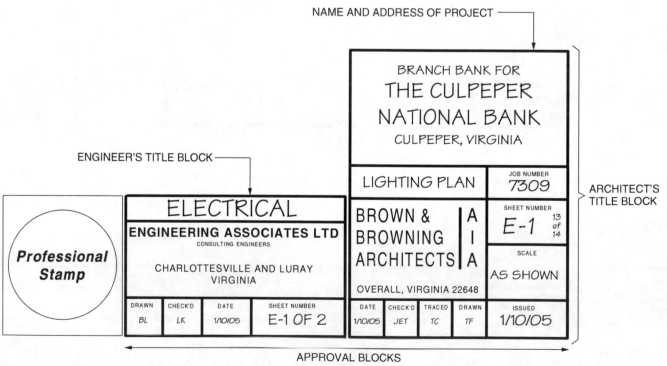

206F09.EPS

Figure 9 ◆ Title blocks.

ELECTRONIC SYSTEMS TECHNICIAN LEVEL TWO — TRAINEE MODULE 33206-05

2.2.0 Approval Block

The approval block, in most cases, will appear on the drawing sheet as shown in *Figure 10*. The various types of approval blocks (drawn, checked, etc.) will be initialed by the appropriate personnel. This type of approval block is usually part of the title block and appears on each drawing sheet.

On some projects, authorized signatures are required before certain systems may be installed, or even before the project begins. An approval block such as the one shown in *Figure 11* indicates that all required personnel have checked the drawings for accuracy and that the set meets with everyone's approval. Such an approval block usually appears on the front sheet of the blueprint set and may include the following information:

- *Professional stamp* – Registered seal of approval by the architect or consulting engineer
- *Design supervisor* – Signature of the person who is overseeing the design
- *Drawn (by)* – Signature or initials of the person who drafted the drawing and the date it was completed
- *Checked (by)* – Signature or initials of the person who reviewed the drawing and the date of approval
- *Approved* – Signature or initials of the architect/ engineer and the date of the approval
- *Owner's approval* – Signature of the project owner or the owner's representative along with the date signed

COMM. NO.	DATE	DRAWN	CHECKED	REVISED
7215	1/6/05	GK	GLC	

206F10.EPS

Figure 10 ◆ Typical approval block.

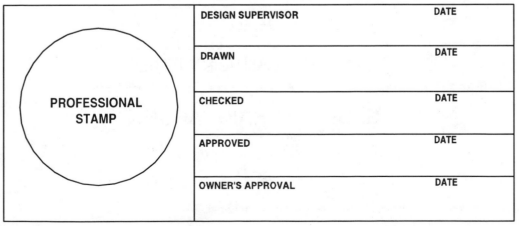

206F11.EPS

Figure 11 ◆ Alternate approval block.

2.3.0 Revision Block

Sometimes electrical drawings have to be partially redrawn or modified during the construction of a project. It is extremely important that such modifications are noted and dated on the drawings to ensure that the workers have an up-to-date set of drawings from which to work. In some situations, sufficient space is left near the title block for dates and descriptions of revisions, as shown in *Figure 12*.

In other cases, a revision block is provided near the title block, as shown in *Figure 13*.

NOTE

Architects, engineers, designers, and drafters have their own methods of showing revisions, so expect to find deviations from those shown here.

REVISIONS
1/8/05 - REVISED LIGHTING FIXTURE
NO. 3 IN. LIGHTING FIXTURE SCHEDULE

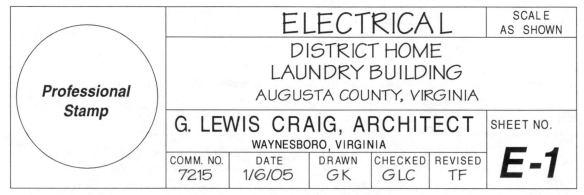

206F12.EPS

Figure 12 ◆ One method of showing revisions on working drawings.

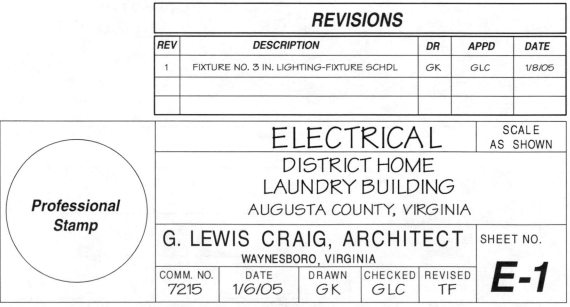

206F13.EPS

Figure 13 ◆ Alternate method of showing revisions on working drawings.

3.0.0 ◆ DRAFTING LINES

Drawings contain many types of drafting lines. To specify the meaning of each type of line, contrasting lines can be made by varying the width of the lines or breaking the lines in a uniform way.

Figure 14 shows common lines used on architectural drawings. However, these lines can vary. Architects and engineers have strived for a common standard for the past century, but unfortunately, their goal has yet to be reached. Therefore, you will find variations in lines and symbols from drawing to drawing, so always consult the legend or symbol list when referring to any drawing. Also, carefully inspect each drawing to ensure that line types are used consistently.

The drafting lines shown in *Figure 14* are used as follows:

- *Light full line* – This line is used for section lines, building background (outlines), and similar uses where the object to be drawn is secondary to the system being shown, such as HVAC or electrical.

- *Medium full line* – This type of line is frequently used for hand lettering on drawings. It is also used for some drawing symbols, circuit lines, etc.

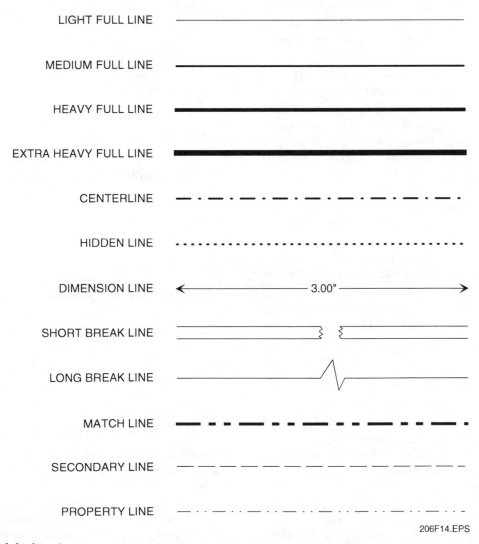

206F14.EPS

Figure 14 ◆ Typical drafting lines.

- *Heavy full line* – This line is used for borders around title blocks, schedules, and for hand lettering drawing titles. Some types of symbols are frequently drawn with a heavy full line.
- *Extra heavy full line* – This line is used for border lines on architectural/engineering drawings.
- *Centerline* – A centerline is a broken line made up of alternately spaced long and short dashes. It indicates the centers of objects such as holes, pillars, or fixtures. Sometimes, the centerline indicates the dimensions of a finished floor.
- *Hidden line* – A hidden line consists of a series of short dashes that are closely and evenly spaced. It shows the edges of objects that are not visible in a particular view. The object outlined by hidden lines in one drawing is often fully pictured in another drawing.
- *Dimension line* – These are thin lines used to show the extent and direction of dimensions. The dimension is usually placed in a break inside the dimension lines. Normal practice is to place the dimension lines outside the object's outline. However, it may sometimes be necessary to draw the dimensions inside the outline.
- *Short-break line* – This line is usually drawn freehand and is used for short breaks.
- *Long-break line* – This line, which is drawn partly with a straightedge and partly with freehand zigzags, is used for long breaks.
- *Match line* – This line is used to show the position of the cutting plane. Therefore, it is also called the cutting-plane line. A match or cutting-plane line is a heavy line with long dashes alternating with two short dashes. It is used on drawings of large structures to show where one drawing stops and the next drawing starts.
- *Secondary line* – This line is frequently used to outline pieces of equipment or to indicate reference points of a drawing that are secondary to the drawing's purpose.
- *Property line* – This is a light line made up of one long and two short dashes that are alternately spaced. It indicates land boundaries on the site plan.

Other uses of the lines just mentioned include the following:

- *Extension lines* – Extension lines are lightweight lines that start about 1/16 inch away from the edge of an object and extend out. A common use of extension lines is to create a boundary for dimension lines. Dimension lines meet extension lines with arrowheads, slashes, or dots. Extension lines that point from a note or other reference to a particular feature on a drawing are called leaders. They usually end in either an

arrowhead or a dot and may include an explanatory note at the end.
- *Section lines* – These are often referred to as cross-hatch lines. Drawn at a 45-degree angle, these lines show where an object has been cut away to reveal the inside.
- *Phantom lines* – Phantom lines are solid, light lines that show where an object will be installed. A future door opening or a future piece of equipment can be shown with phantom lines.

3.1.0 Electrical Drafting Lines

Besides the architectural lines shown in *Figure 14*, consulting electrical engineers, designers, and drafters use additional lines to represent circuits and their related components. Again, these lines may vary from drawing to drawing, so check the symbol list or legend for the exact meaning of lines on the drawing with which you are working. *Figure 15* shows lines used on some electrical drawings.

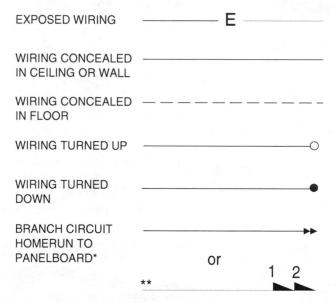

* Number of arrowheads indicates number of circuits. A number at each arrowhead may be used to identify circuit numbers.

** Half arrowheads are sometimes used for homeruns to avoid confusing them with drawing callouts.

206F15.EPS

Figure 15 ◆ Electrical drafting lines.

4.0.0 ◆ ELECTRICAL SYMBOLS

The electronic systems technician must be able to read and understand electrical working drawings. Consequently, they must have a knowledge of electrical symbols and their applications.

An electrical symbol is a figure or mark that stands for a component used in the electrical system. As an example, *Figure 16* shows a list of electrical

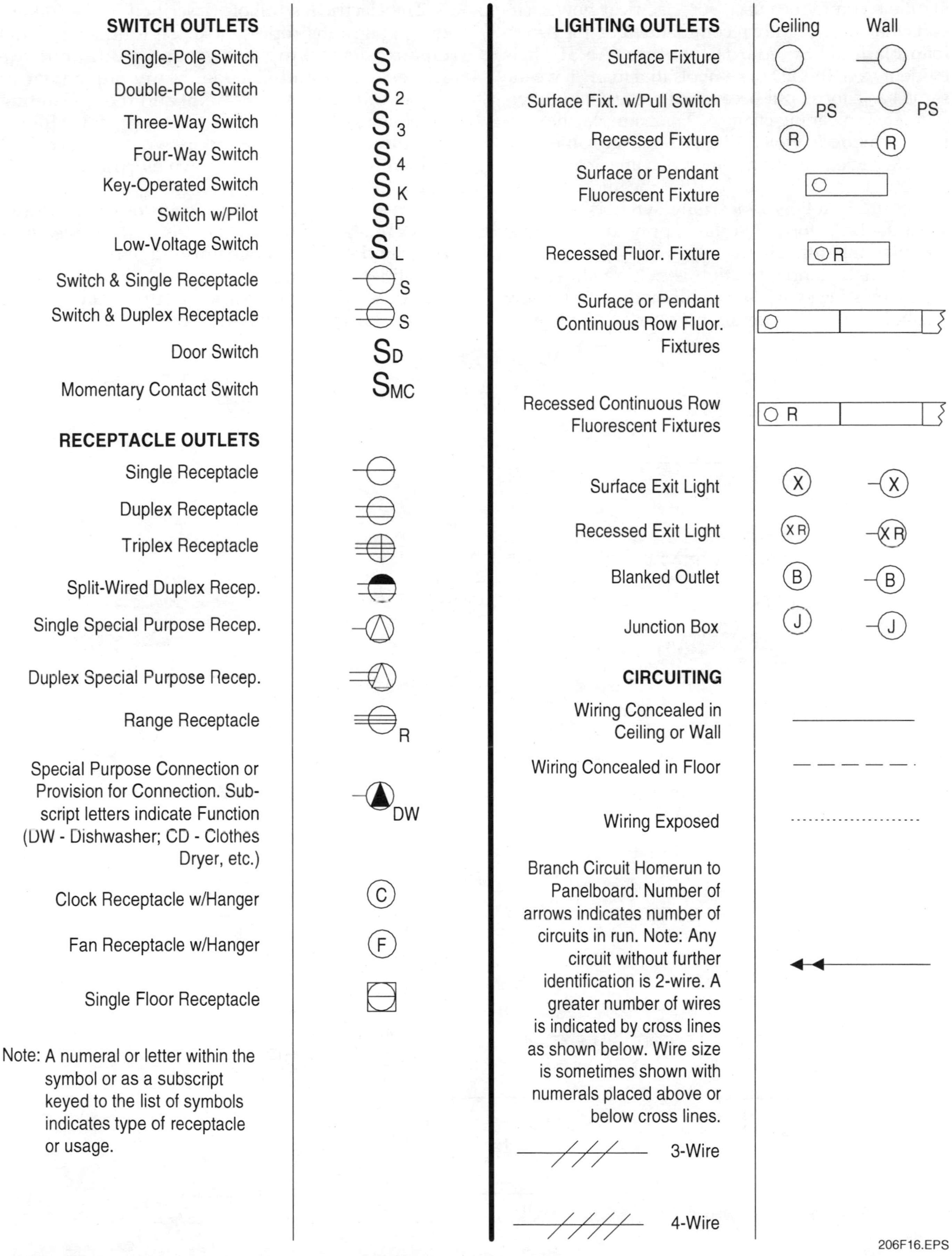

SWITCH OUTLETS

Single-Pole Switch	S
Double-Pole Switch	S₂
Three-Way Switch	S₃
Four-Way Switch	S₄
Key-Operated Switch	Sₖ
Switch w/Pilot	Sₚ
Low-Voltage Switch	Sₗ
Switch & Single Receptacle	⊖ₛ
Switch & Duplex Receptacle	⊖ₛ
Door Switch	S_D
Momentary Contact Switch	S_MC

RECEPTACLE OUTLETS

Single Receptacle

Duplex Receptacle

Triplex Receptacle

Split-Wired Duplex Recep.

Single Special Purpose Recep.

Duplex Special Purpose Recep.

Range Receptacle

Special Purpose Connection or Provision for Connection. Subscript letters indicate Function (DW - Dishwasher; CD - Clothes Dryer, etc.)

Clock Receptacle w/Hanger

Fan Receptacle w/Hanger

Single Floor Receptacle

Note: A numeral or letter within the symbol or as a subscript keyed to the list of symbols indicates type of receptacle or usage.

LIGHTING OUTLETS

Ceiling   Wall

Surface Fixture

Surface Fixt. w/Pull Switch

Recessed Fixture

Surface or Pendant Fluorescent Fixture

Recessed Fluor. Fixture

Surface or Pendant Continuous Row Fluor. Fixtures

Recessed Continuous Row Fluorescent Fixtures

Surface Exit Light

Recessed Exit Light

Blanked Outlet

Junction Box

CIRCUITING

Wiring Concealed in Ceiling or Wall

Wiring Concealed in Floor

Wiring Exposed

Branch Circuit Homerun to Panelboard. Number of arrows indicates number of circuits in run. Note: Any circuit without further identification is 2-wire. A greater number of wires is indicated by cross lines as shown below. Wire size is sometimes shown with numerals placed above or below cross lines.

3-Wire

4-Wire

Figure 16 ◆ ANSI electrical symbols.

206F16.EPS

symbols commonly used for electrical power circuits that are currently recommended by the American National Standards Institute (ANSI). It is evident from this list of symbols that many have the same basic form, but because of some slight difference, their meaning changes. For example, the electrical symbols in *Figure 17* each have the same basic form (a circle), but the addition of a line or an abbreviation gives each an individual meaning. A good procedure to follow in learning symbols is to first learn the basic form and then apply the variations for obtaining different meanings.

Although standardization is getting closer to a reality, existing symbols are still modified, and new symbols are created for almost every new project.

The electrical symbols described in the following paragraphs represent those found on actual electrical working drawings throughout the United States and Canada. Many are similar to those recommended by ANSI and the Consulting Engineers Council/US; others are not. Understanding how these symbols were devised will help you to interpret unknown electrical symbols in the future.

Some of the symbols used on electrical drawings are abbreviations, such as WP for weatherproof and AFF for above finished floor. Others are simplified pictographs, such as (A) in *Figure 18* for a double floodlight fixture or (B) for an infrared electric heater with two quartz lamps.

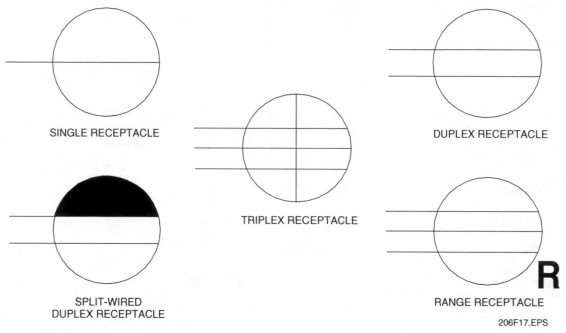

Figure 17 ◆ Example of symbol variations.

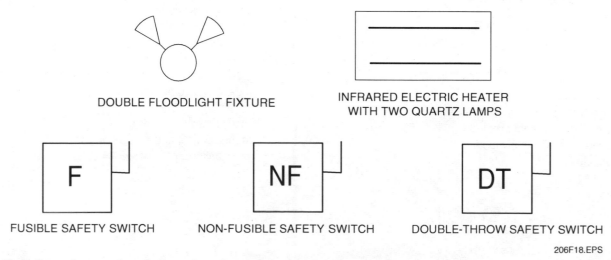

Figure 18 ◆ General types of symbols used on electrical drawings.

In some cases, the symbols are combinations of abbreviations and pictographs, such as (C) in *Figure 18* for a fusible safety switch, (D) for a nonfusible safety switch, and (E) for a double-throw safety switch. In each example, a pictograph of a switch enclosure has been combined with an abbreviation: F (fusible), DT (double-throw), and NF (nonfusible), respectively.

Lighting outlet symbols have been devised that represent incandescent, fluorescent, and high-intensity discharge lighting; a circle usually represents an incandescent fixture, and a rectangle is used to represent a fluorescent fixture. These symbols are designed to indicate the physical shape of a particular fixture, and while the circles representing incandescent lamps are frequently enlarged somewhat, symbols for fluorescent fixtures are usually drawn as close to scale as possible. The type of mounting used for all lighting fixtures is usually indicated in a lighting fixture schedule, which is shown on the drawings or in the **written specifications**.

The type of lighting fixture is identified by a numeral placed inside a triangle or other symbol, and placed near the fixture to be identified. A complete description of the fixtures identified by the symbols must be given in the lighting fixture schedule and should include the manufacturer, catalog number, number and type of lamps, voltage, finish, mounting, and any other information needed for proper installation of the fixture.

Switches used to control lighting fixtures are also indicated by symbols (usually the letter S followed by numerals or letters to define the exact type of switch). For example, S_3 indicates a three-way switch; S_4 identifies a four-way switch; and S_P indicates a single-pole switch with a pilot light.

Main distribution centers, panelboards, transformers, safety switches, and other similar electrical components are indicated by electrical symbols on floor plans and by a combination of symbols and semipictorial drawings in riser diagrams.

A detailed description of the service equipment is usually given in the panelboard schedule or in the written specifications. However, on small projects, the service equipment is sometimes indicated only by notes on the drawings.

Circuit and feeder wiring symbols are getting closer to being standardized. Most circuits concealed in the ceiling or wall are indicated by a solid line; a broken line is used for circuits concealed in the floor or the ceiling below a floor (e.g. a basement ceiling); exposed raceways are indicated by short dashes or the letter *E* placed in the same plane with the circuit line at various intervals. The number of conductors in a conduit or raceway system may be indicated in the panelboard schedule under the appropriate column, or the information may be shown on the floor plan.

Symbols for communication and signal systems, as well as symbols for light and power, are drawn to an appropriate scale and accurately located with respect to the building. This reduces the number of references made to the architectural drawings. Where extreme accuracy is required in locating outlets and equipment, exact dimensions are given on larger-scale drawings and shown on the plans.

Each different category in an electrical system is usually represented by a basic distinguishing symbol. To further identify items of equipment or outlets in the category, a numeral or other identifying mark is placed within the open basic symbol. In addition, all such individual symbols used on the drawings should be included in the symbol list or legend. The electrical symbols shown in *Figure 19* were modified by a consulting engineering firm for use on a small industrial electrical installation.

Drawing and Symbol Standardization

In an effort to standardize drawings and symbols used in design and construction projects, the Construction Specifications Institute (CSI) released a new Computer Aided Design (CAD) software program in 2005. The software program is entitled *U.S. National CAD Standard (NCS) Version 3.1 – Uniform Drawing System™ (UDS)*. This CAD standard allows project drawing sets and drawings prepared by subcontractors to be produced in a uniform format. The software also includes an extensive uniform symbol library. The CAD standard was developed based on understandings between CSI and the following organizations:

- Construction Specifications Canada (CSC)
- American Institute of Architects (AIA)
- National Institute of Building Sciences (NIBS)
- Sheet Metal and Air Conditioning Contractors National Association (SMACNA)
- U.S. General Services Administration (GSA)

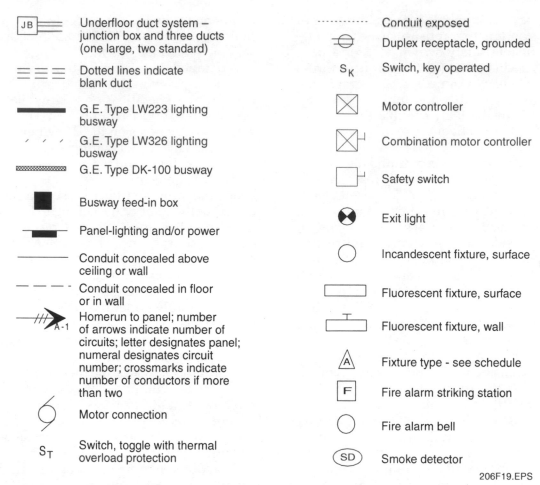

JB ☰	Underfloor duct system – junction box and three ducts (one large, two standard)	
☰ ☰ ☰	Dotted lines indicate blank duct	
▬▬▬	G.E. Type LW223 lighting busway	
⁄ ⁄ ⁄ ⁄	G.E. Type LW326 lighting busway	
▨▨▨▨	G.E. Type DK-100 busway	
■	Busway feed-in box	
▬	Panel-lighting and/or power	
———	Conduit concealed above ceiling or wall	
– – –	Conduit concealed in floor or in wall	
A-1	Homerun to panel; number of arrows indicate number of circuits; letter designates panel; numeral designates circuit number; crossmarks indicate number of conductors if more than two	
⟨	Motor connection	
S_T	Switch, toggle with thermal overload protection	
- - - - -	Conduit exposed	
⊖	Duplex receptacle, grounded	
S_K	Switch, key operated	
⊠	Motor controller	
⊠⌐	Combination motor controller	
☐⌐	Safety switch	
⊗	Exit light	
○	Incandescent fixture, surface	
▭	Fluorescent fixture, surface	
⊤	Fluorescent fixture, wall	
Ⓐ	Fixture type - see schedule	
F	Fire alarm striking station	
○	Fire alarm bell	
SD	Smoke detector	

206F19.EPS

Figure 19 ◆ Electrical symbols used by one consulting engineering firm.

5.0.0 ◆ SCALE DRAWINGS

In most electrical drawings, the components are so large that it would be impossible to draw them actual size. Consequently, drawings are made to some reduced scale; that is, all the distances are drawn smaller than the actual dimensions of the object itself, with all dimensions being reduced in the same proportion. For example, if a floor plan of a building is to be drawn to a scale of ¼" = 1'–0", each ¼" on the drawing would equal 1 foot on the building itself; if the scale is ⅛" = 1'–0", each ⅛" on the drawing equals 1 foot on the building, and so forth.

When architectural and engineering drawings are produced, the selected scale is very important. Where dimensions must be held to extreme accuracy, the scale drawings should be made as large as practical with dimension lines added. Where dimensions require only reasonable accuracy, the object may be drawn to a smaller scale (with dimension lines possibly omitted).

In dimensioning drawings, the dimensions written on the drawing are the actual dimensions of the building, not the distances that are measured on the drawing. To further illustrate this point, look at the floor plan in *Figure 20*; it is drawn to a scale of ½" = 1'–0". One of the walls is drawn to an actual length of 3½" on the drawing paper, but since the scale is ½" = 1'–0" and since 3½" contains 7 halves of an inch (7 × ½ = 3½"), the dimension shown on the drawing will therefore be 7'–0" on the actual building.

As shown in the previous example, the most common method of reducing all the dimensions (in feet and inches) in the same proportion is to choose a certain distance and let that distance represent one foot. This distance can then be divided into twelve parts, each of which represents an inch. If half inches are required, these twelfths are further subdivided into halves, etc. Now the scale represents the common foot rule with its subdivisions into inches and fractions, except that the scaled foot is smaller than the distance known as a foot and, likewise, its subdivisions are proportionately smaller.

When a measurement is made on the drawing, it is made with the reduced foot rule or scale; when a measurement is made on the building, it is made with the standard foot rule. The most common reduced foot rules or scales used in electrical

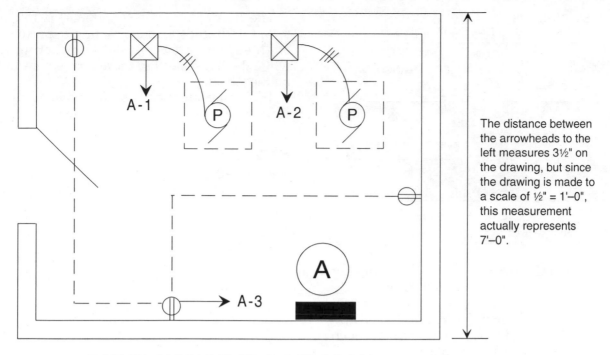

The distance between the arrowheads to the left measures 3½" on the drawing, but since the drawing is made to a scale of ½" = 1'–0", this measurement actually represents 7'–0".

PUMP HOUSE FLOOR PLAN
½" = 1'–0"

206F20.EPS

Figure 20 ◆ Typical floor plan showing drawing scale.

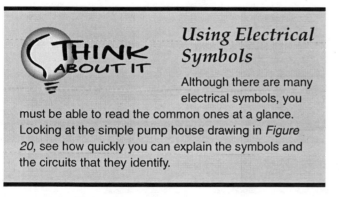

Using Electrical Symbols

Although there are many electrical symbols, you must be able to read the common ones at a glance. Looking at the simple pump house drawing in *Figure 20*, see how quickly you can explain the symbols and the circuits that they identify.

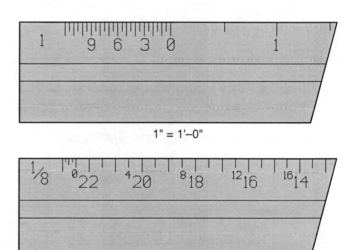

1" = 1'–0"

⅛" = 1'–0"

206F21.EPS

Figure 21 ◆ Two different configurations of architect's scales.

drawings are the architect's scale and the engineer's scale. Drawings may sometimes be encountered that use a metric scale, but using this scale is similar to using the architect's or engineer's scales. A metric conversion chart is included in *Appendix A*.

5.1.0 Architect's Scale

Figure 21 shows two configurations of architect's scales. The one on the top is designed so that 1" = 1'–0", and the one on the bottom has graduations spaced to represent ⅛" = 1'–0".

Note that on the one-inch scale in *Figure 22*, the longer marks to the right of the zero (with a numeral beneath) represent feet. Therefore, the distance between the zero and the numeral 1

equals one foot. The shorter mark between the zero and 1 represents ½ of a foot, or six inches.

Referring again to *Figure 22*, look at the marks to the left of the zero. The numbered marks are spaced three scaled inches apart and have the numerals 0, 3, 6, and 9 for use as reference points. The other lines of the same length also represent scaled inches, but are not marked with numerals. In use, you can count the number of long marks to the left

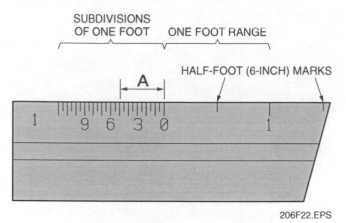

Figure 22 ◆ One-inch architect's scale.

of the zero to find the number of inches, but after some practice, you will be able to tell the exact measurement at a glance. For example, the measurement A represents five inches because it is the fifth inch mark to the left of the zero; it is also one inch mark short of the six-inch line on the scale.

The lines that are shorter than the inch line are the half-inch lines. On smaller scales, the basic unit is not divided into as many divisions. For example, the smallest subdivision on some scales represents two inches.

5.1.1 Types of Architect's Scales

Architect's scales are available in several types, but the most common include the triangular scale (*Figure 23*) and the flat scale. The quality of architect's scales also varies from cheap plastic scales (costing a dollar or two) to high-quality, wooden-laminated tools that are calibrated to precise standards.

The triangular scale (*Figure 24*) is frequently found in drafting and estimating departments or engineering and electrical contracting firms, while the flat scales are more convenient to carry on the job site.

Triangular architect's scales have 12 different scales (two on each edge) as follows:

- Common foot rule (12 inches)
- $\frac{1}{16}$" = 1'–0"
- $\frac{3}{32}$" = 1'–0"
- $\frac{3}{16}$" = 1'–0"
- $\frac{1}{8}$" = 1'–0"
- $\frac{1}{4}$" = 1'–0"
- $\frac{3}{8}$" = 1'–0"
- $\frac{3}{4}$" = 1'–0"
- 1" = 1'–0"
- $\frac{1}{2}$" = 1'–0"
- $1\frac{1}{2}$" = 1'–0"
- 3" = 1'–0"

Two separate scales on one face may seem confusing at first, but after some experience, reading these scales becomes second nature.

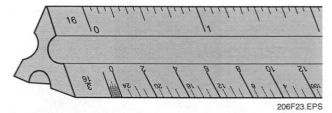

Figure 23 ◆ Typical triangular architect's scale.

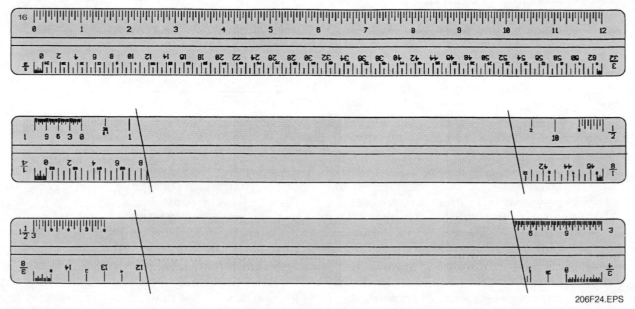

Figure 24 ◆ Various scales on a triangular architect's scale.

In all but one of the scales on the triangular architect's scale, each face has one of the scales placed opposite to the other. For example, on the one-inch face, the one-inch scale is read from left to right, starting from the zero mark. The half-inch scale is read from right to left, again starting from the zero mark.

On the remaining foot-rule scale ($\frac{1}{16}$" = 1'–0"), each $\frac{1}{16}$" mark on the scale represents one foot.

Figure 24 shows all the scales found on the triangular architect's scale.

The flat architect's scale shown in *Figure 25* is ideal for workers on most projects. It is easily and conveniently carried in the shirt pocket, and the four scales ($\frac{1}{8}$", $\frac{1}{4}$", $\frac{1}{2}$", and 1") are adequate for the majority of projects that will be encountered.

The partial floor plan shown in *Figure 25* is drawn to a scale of $\frac{1}{8}$" = 1'–0". The dimension in question is found by placing the $\frac{1}{8}$" architect's scale on the drawing and reading the figures. It can be seen that the dimension reads 24'–6".

Every drawing should have the scale to which it is drawn plainly marked on it as part of the drawing title. However, it is not uncommon to have several different drawings on one blueprint sheet where each drawing has a different scale. Therefore, always check the scale of each different view found on a drawing sheet.

5.2.0 Engineer's Scale

The civil engineer's scale is used in basically the same manner as the architect's scale, with the principal difference being that the graduations on the engineer's scale are decimal units rather than feet, as on the architect's scale.

The engineer's scale is used by placing it on the drawing with the working edge away from the user. The scale is then aligned in the direction of the required measurement. Then, by looking down at the scale, the dimension is read.

Civil engineer's scales commonly show the following graduations:

- 1" = 10 units
- 1" = 20 units
- 1" = 30 units
- 1" = 40 units
- 1" = 60 units
- 1" = 80 units
- 1" = 100 units

The purpose of this scale is to transfer the relative dimensions of an object to the drawing or vice versa. It is used mainly on site plans to determine distances between property lines, manholes, duct runs, direct-burial cable runs, and the like.

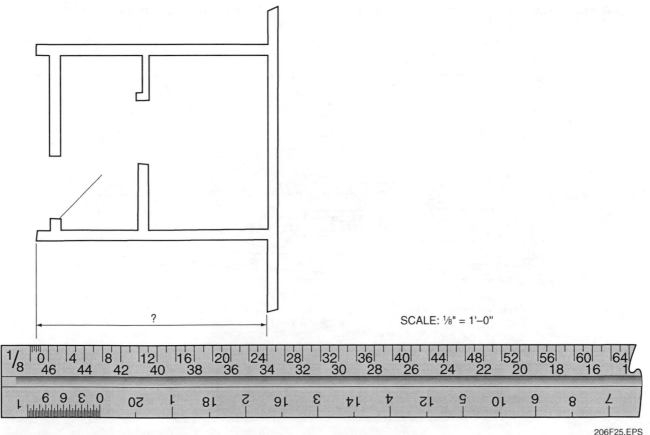

SCALE: $\frac{1}{8}$" = 1'–0"

206F25.EPS

Figure 25 ◆ Using the $\frac{1}{8}$" architect's scale to determine the dimensions on a drawing.

Site plans are drawn to scale using the engineer's scale rather than the architect's scale. On small lots, a scale of 1 inch = 10 feet or 1 inch = 20 feet is used. For a 1:10 scale, this means that one inch (the actual measurement on the drawing) is equal to 10 feet on the land itself.

On larger drawings, where a large area must be covered, the scale could be 1 inch = 100 feet or 1 inch = 1,000 feet, or any other integral power of 10. On drawings with the scale in multiples of 10, the engineering scale marked 10 is used. If the scale is 1 inch = 200 feet, the engineer's scale marked 20 is used, and so on.

Although site plans appear reduced in scale, depending on the size of the object and the size of the drawing sheet to be used, the actual dimensions must be shown on the drawings at all times. When you are reading the drawing plans to scale, think of each dimension in its full size and not in the reduced scale it happens to be on the drawing (*Figure 26*).

5.3.0 Metric Scale

The metric scale (*Figure 27*) is divided into centimeters (cm), with the centimeters divided into 10 divisions for millimeters (mm) or in 20 divisions for half millimeters. Scales are available with metric divisions on one edge while inch divisions are inscribed on the opposite edge. Many contracting firms that deal in international trade have adopted a dual-dimensioning system expressed in both metric and English symbols. Drawings prepared for government projects may also require metric dimensions.

6.0.0 ◆ ANALYZING ELECTRICAL DRAWINGS

The most practical way to learn how to read electrical construction documents is to analyze an existing set of drawings prepared by consulting or industrial engineers.

Engineers or electrical designers are responsible for the complete layout of electrical systems for most projects. Electrical drafters then transform the engineer's designs into working drawings, using either manual drafting instruments or computer-aided design (CAD) systems. The following is a brief outline of what usually takes place in the preparation of electrical design and working drawings:

- The engineer meets with the architect and owner to discuss the electrical needs of the building or project and to discuss various recommendations made by all parties.
- After that, an outline of the architect's floor plan is laid out.
- The engineer then calculates the required power and lighting outlets for the project; these are later transferred to the working drawings.

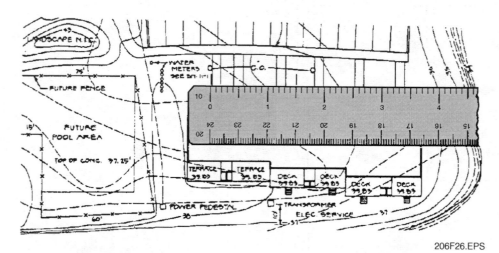

206F26.EPS

Figure 26 ◆ Practical use of the engineer's scale.

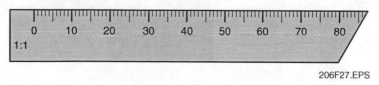

206F27.EPS

Figure 27 ◆ Typical metric scale.

- All communications and alarm systems are located on the plan, along with lighting and power panelboards.
- Circuit calculations are made to determine wire size and overcurrent protection.
- The main electric service and related components are determined and shown on the drawings.
- Schedules are then placed on the drawings to identify various pieces of equipment.
- Wiring diagrams are made to show the workers how various electrical components are to be connected.
- An electrical symbol list or legend is drafted and shown on the drawings to identify all symbols used to indicate electrical outlets or equipment.
- Various large-scale electrical details are included, if necessary, to show exactly what is required to complete the installation.
- Written specifications are then made to give a description of the materials and installation methods.

6.1.0 Development of Site Plans

In general practice, it is usually the owner's responsibility to furnish the architect/engineer with property and topographic surveys, which are made by a certified land surveyor or civil engineer. These surveys show the following:

- All property lines
- Existing public utilities and their location on or near the property, such as electrical lines, sanitary sewer lines, gas lines, water-supply lines, storm sewers, manholes, or telephone lines

A land surveyor does the property survey from information obtained from a deed description of the property. A property survey shows only the property lines and their lengths, as if the property were perfectly flat.

The topographic survey shows both the property lines and the physical characteristics of the land by using contour lines, notes, and symbols. The physical characteristics may include:

- The direction of the land slope
- Whether the land is flat, hilly, wooded, swampy, high, or low, and other features of its physical nature

All of this information is necessary so that the architect can properly design a building to fit the property. The electrical engineer also needs this information to locate existing electrical utilities and to route the new service to the building, provide outdoor lighting and circuits, etc.

7.0.0 ◆ TYPICAL SITE ELECTRICAL PLAN

A site's electrical work is sometimes shown on the architect's plot plan. However, when site work involves many trades and several utilities, such as gas, telephone, electric, television, water, and sewage, it can become confusing if all details are shown on one drawing sheet. In cases like these, it is best to have separate drawings devoted entirely to the electrical work. The first sheet of a typical multisheet site electrical plan is shown in *Figure 28*. This project is an office/warehouse building for Virginia Electric, Inc. The complete site electrical plan drawings consist of four 24" × 36" drawing sheets, along with a set of written specifications. The first two sheets will be discussed in this and the following sections. The specifications will be discussed later in the module.

The first sheet of the site electrical plan shown in *Figure 28* is the utilities plan and shows the location of the outside utilities for the project. It has the conventional architect's and engineer's title blocks in the lower right-hand corner of the drawing. These blocks identify the project and project owners, the architect, the engineer, and the title of the drawing sheet. They also show how this drawing sheet relates to the site electrical plan (E sheets), as well as the entire set of site drawings. Note the engineer's professional stamp of approval to the left of the engineer's title block. Similar blocks appear on all four of the electrical drawing E sheets.

When examining a set of electrical drawings for the first time, always look at the area around the title block. This is where most revision blocks or revision notes are placed. If revisions have been made to the drawings, make certain that you have a clear understanding of what has taken place before proceeding with the work.

Refer again to the drawing in *Figure 28* and note the North Arrow in the upper left corner. A North Arrow shows the direction of true north to help you orient the drawing to the site. Look directly down from the North Arrow to the bottom of the page and notice the drawing sheet function is identified in large type as the plot for utilities. Directly beneath, you can see that the drawing scale of 1" = 30' is shown. This means that each inch on the drawing represents 30 feet on the actual job site. This scale holds true for all drawings on the page unless otherwise noted.

An outline of the proposed building is indicated on the drawing by cross-hatched rectangles along with a callout stating, *Proposed Bldg. Fin. Flr. Elev. 590.0'*. This means that the finished floor level of the building is to be 590 feet above sea level, which in this part of the country will be about two feet above finished grade around the

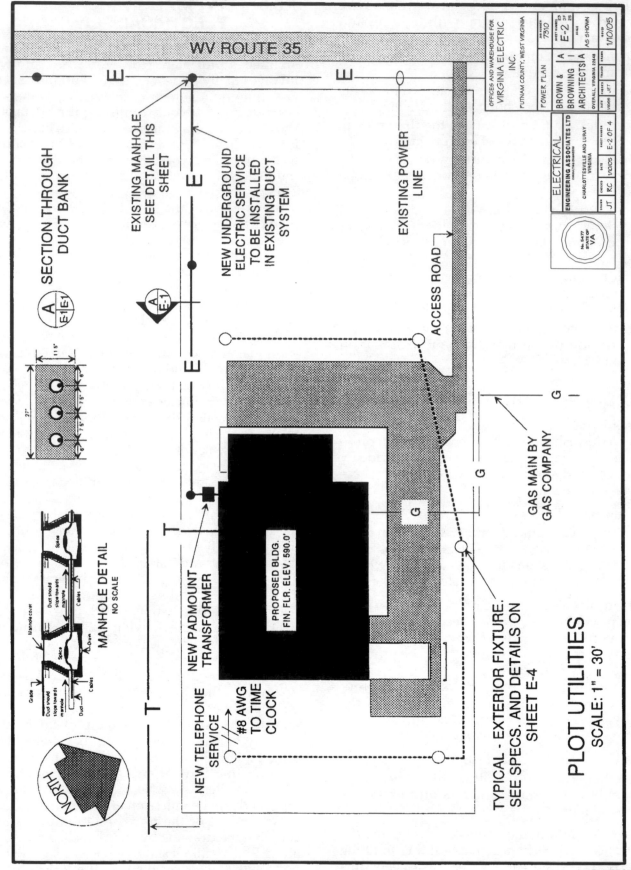

Figure 28 ◆ First sheet of a typical site electrical plan.

206F28.EPS

Interpreting Site Plans

Study *Figure 28* and explain as many of its features as you can. How much can be understood using common sense? What features require special information?

building. This information helps to position conduit sleeves and stub-ups for underground utilities at the correct height before the finished concrete floor is poured.

The shaded area represents asphalt paving for the access road, drives, and parking lot. Note that the access road leads into a highway, which is designated Route 35. This information further helps workers to orient the drawing to the building site.

Existing manholes are indicated by a solid circle, while an open circle is used to show the position of the five new pole-mounted lighting fixtures that are to be installed around the new building. Existing power lines are shown with a light solid line with the letter *E* placed at intervals along the line. The new underground electric service is shown in the same way, except the lines are somewhat wider and darker on the drawing. Note that this new high-voltage cable terminates into a padmount transformer near the proposed building. New telephone lines are similar except the letter *T* is used to identify the telephone lines.

The direct-burial underground cable supplying the exterior lighting fixtures is indicated with dashed lines on the drawing—shown connecting the open circles. A homerun for this circuit is also shown to a time clock.

The manhole detail shown to the right of the North Arrow may seem to serve very little purpose on this drawing since the manholes have already been installed. However, the dimensions and details of their construction will help the electrical contractor or supervisor to better plan the pulling of the high-voltage cable. The same is true of the cross section shown of the duct bank. The electrical contractor knows that three empty ducts are available if it is discovered that one of them is damaged when the work begins.

Although the electrical work will not involve working with gas, the main gas line is shown on the electrical drawing to let the electrical workers know its approximate location while they are installing the direct-burial conductors for the exterior lighting fixtures.

8.0.0 ◆ POWER PLANS

The second sheet (*Figure 29*) of the electrical plan is the power plan and shows the complete floor plan of the office/warehouse building with all interior partitions drawn to scale. Sometimes, the physical locations of all wiring and outlets are shown on one drawing; that is, outlets for lighting, power, signal and communications, special electrical systems, and related equipment are shown on the same plan. However, on complex installations, the drawing would become cluttered if everything were shown on a one-sheet floor plan. Therefore, most projects will have separate sheets for power, lighting, signal and communications, fire, video, and other circuits. Riser diagrams and details may be shown on other drawing sheets, or if room permits, on the applicable circuit sheets. Note that this sheet contains the key plan and symbol list for all the sheets of the electrical plan.

A closer look at this drawing reveals the title blocks in the lower right corner of the drawing sheet. These blocks list both the architectural and engineering firms, along with information to identify the project and drawing sheet. Also note that the floor plan is identified as Floor Plan "B"—Power and is drawn to a scale of ⅛" = 1'–0". There are no revisions shown on this drawing sheet.

8.1.0 Key Plan

A key plan (*Figure 30*) appears on the drawing sheet immediately above the engineer's title block on *Figure 29*. The purpose of this key plan is to identify that part of the project to which this sheet applies. In this case, the project involves two buildings: Building A and Building B. Since the outline of Building B is cross-hatched in the key plan, this is the building to which this drawing applies. Note that this key plan is not drawn to scale, but only to its approximate shape.

Although Building A is also shown on this key plan, a note below the key plan title states that there is no electrical work required in Building A.

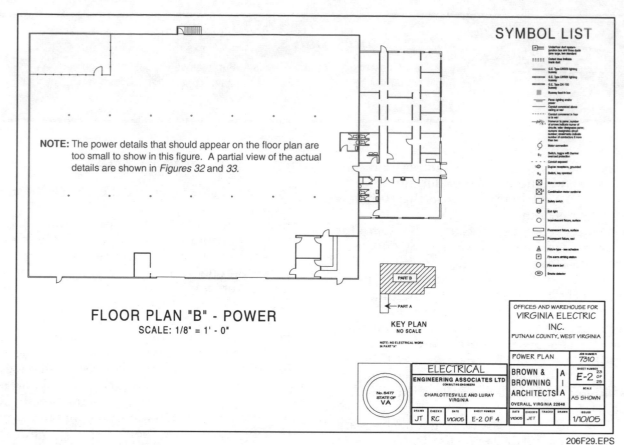

NOTE: The power details that should appear on the floor plan are too small to show in this figure. A partial view of the actual details are shown in *Figures 32* and *33*.

FLOOR PLAN "B" - POWER
SCALE: 1/8" = 1' - 0"

Figure 29 ◆ Second sheet of an electrical plan.

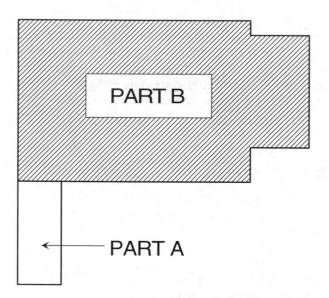

PART B

← PART A

KEY PLAN
NO SCALE

NOTE: NO ELECTRICAL WORK
IN PART "A"

206F30.EPS

Figure 30 ◆ Key plan appearing on electrical power plan.

On some larger installations, the overall project may involve several buildings requiring appropriate key plans on each drawing to help the workers orient the drawings to the appropriate building. In some cases, separate drawing sheets may be used for each room or area in an industrial project, again requiring key plans on each drawing sheet to identify applicable drawings for each room.

8.2.0 Symbol List

A symbol list (legend) appears on the electrical power plan (*Figure 29*), immediately above the architect's title block, to identify the various symbols used for power, lighting, and other circuits on this project. In most cases, the only symbols listed are those that apply to the particular project. In other cases, however, a standard list of symbols is used for all projects with the following note:

These are standard symbols and may not all appear on the project drawings; however, wherever the symbol on the project drawings occurs, the item shall be provided and installed.

Only electrical symbols that are actually used for the office/warehouse drawings are shown in

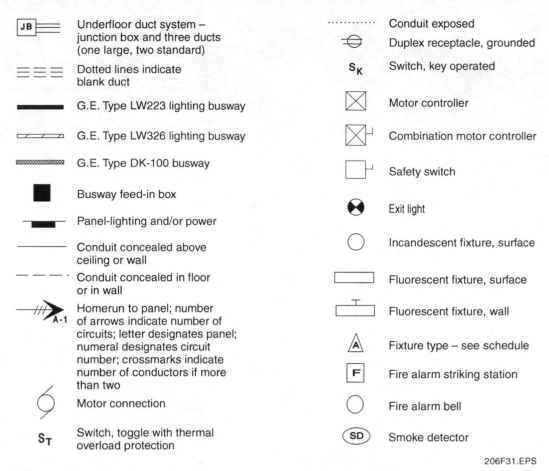

Symbol	Description
JB	Underfloor duct system – junction box and three ducts (one large, two standard)
	Dotted lines indicate blank duct
	G.E. Type LW223 lighting busway
	G.E. Type LW326 lighting busway
	G.E. Type DK-100 busway
	Busway feed-in box
	Panel-lighting and/or power
	Conduit concealed above ceiling or wall
	Conduit concealed in floor or in wall
A-1	Homerun to panel; number of arrows indicate number of circuits; letter designates panel; numeral designates circuit number; crossmarks indicate number of conductors if more than two
	Motor connection
S_T	Switch, toggle with thermal overload protection
	Conduit exposed
	Duplex receptacle, grounded
S_K	Switch, key operated
	Motor controller
	Combination motor controller
	Safety switch
	Exit light
	Incandescent fixture, surface
	Fluorescent fixture, surface
	Fluorescent fixture, wall
A	Fixture type – see schedule
F	Fire alarm striking station
	Fire alarm bell
SD	Smoke detector

206F31.EPS

Figure 31 ◆ Power plan electrical symbols list.

the list on the example electrical power plan. A close-up look at these symbols appears in *Figure 31*. Note that the symbol list (legend) contains a definition for each symbol.

8.3.0 Floor Plan

A somewhat enlarged, partial view of Floor Plan "B"—Power from *Figure 29* is shown in *Figure 32*. However, due to the size of the drawing in comparison with the size of the pages in this module, it is still difficult to see very much detail. This illustration is meant to show the partial layout of the power plan and how symbols and notes are arranged.

In general, this plan shows the service equipment (in plan view), receptacles, underfloor duct system, motor connections, motor controllers, electric heat, busways, and similar details. The electric panels and other service equipment are drawn close to scale. The locations of electrical outlets and other components are only approximated on the drawings because they have to be exaggerated to show up on the prints. For example, a common duplex receptacle is only about three inches wide. If such a receptacle were to be located on the floor plan of this

building (drawn to a scale of ⅛" = 1'–0"), even a small dot on the drawing would be too large to draw the receptacle exactly to scale. Therefore, the receptacle symbol is exaggerated. When such receptacles are scaled on the drawings to determine the proper location, a measurement is usually taken to the center of the symbol to determine the distance between outlets. Junction boxes, switches, and other electrical or electronic devices shown on the plan will be exaggerated in a similar manner. A partially enlarged view of *Figure 32* that is shown in *Figure 33* allows a better view of the drawing details.

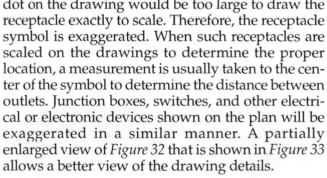

Watch Specified Dimensions

When devices are to be located at heights specified above the finished floor (AFF), be sure to find out the actual height of the flooring to be installed. Some materials, such as ceramic tile, can add significantly to the height of the finished floor.

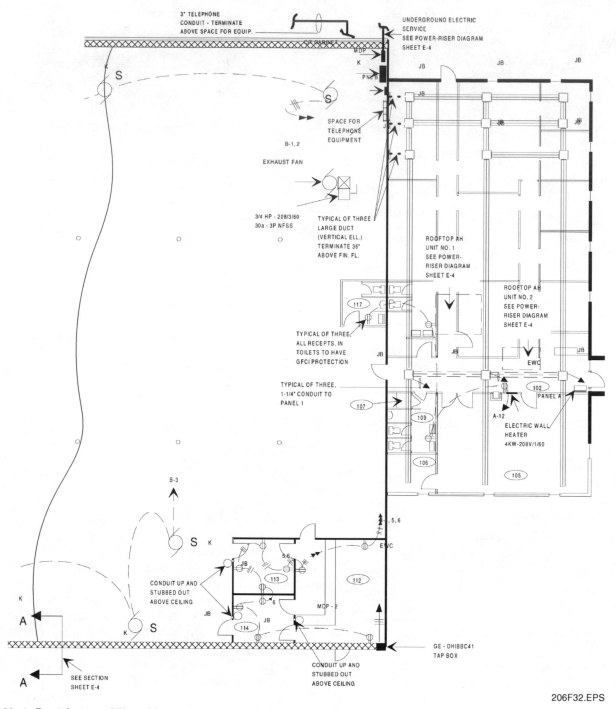

3" TELEPHONE
CONDUIT - TERMINATE
ABOVE SPACE FOR EQUIP.

UNDERGROUND ELECTRIC
SERVICE
SEE POWER-RISER DIAGRAM
SHEET E-4

MDP

K

PNL-B

SPACE FOR
TELEPHONE
EQUIPMENT

B-1, 2

EXHAUST FAN

3/4 HP - 208/3/60
30a - 3P NFSS

TYPICAL OF THREE,
LARGE DUCT
(VERTICAL ELL.)
TERMINATE 36"
ABOVE FIN. FL.

ROOFTOP AH
UNIT NO. 1
SEE POWER-
RISER DIAGRAM
SHEET E-4

ROOFTOP AH
UNIT NO. 2
SEE POWER-
RISER DIAGRAM
SHEET E-4

JB   JB   JB

JB

JB   JB

JB

TYPICAL OF THREE,
ALL RECEPTS. IN
TOILETS TO HAVE
GFCI PROTECTION

117

JB   JB   EWC   JB

102

PANEL A

TYPICAL OF THREE,
1-1/4" CONDUIT TO
PANEL 1

107   109   A-12

ELECTRIC WALL
HEATER
4KW-208V/1/60

106

105

B-3

5, 6

EWC

S   K

5,6

JB   113   112

CONDUIT UP AND
STUBBED OUT
ABOVE CEILING

JB

6

JB

MDP - 2

A

K

S

K   114   JB

A   SEE SECTION
SHEET E-4

GE - DH1BBC41
TAP BOX

CONDUIT UP AND
STUBBED OUT
ABOVE CEILING

206F32.EPS

Figure 32 ◆ Partial view of Floor Plan "B"—Power detail.

8.3.1 Notes and Building Symbols

Referring again to *Figure 32*, you will notice numbers placed inside an oval symbol in each room. These numbered ovals represent the room name or type and correspond to a room schedule in the architectural drawings. For example, room number 112 is designated as the lobby in the room schedule (not shown), room number 113 is designated as office No. 1, etc. On some drawings, these room symbols are omitted and the room names are written out on the drawings.

There are also several notes appearing at various places on the floor plan. These notes offer additional information to clarify certain aspects of the drawing. For example, only one electric heater is to be installed by the electrical contractor; this heater is located in the building's vestibule. Rather than have a symbol in the symbol list for this one heater,

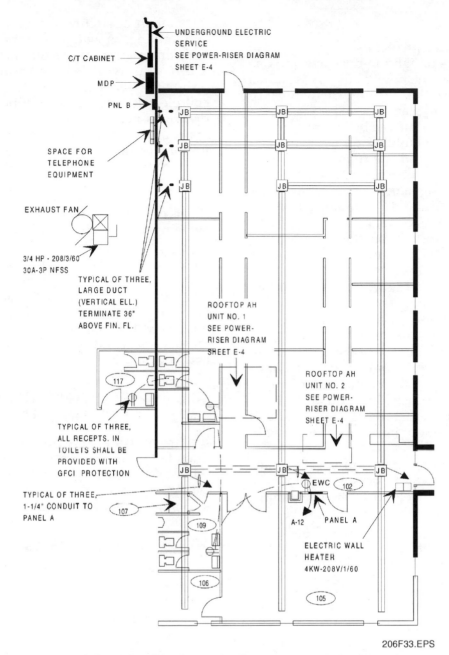

UNDERGROUND ELECTRIC
SERVICE
SEE POWER-RISER DIAGRAM
SHEET E-4

C/T CABINET

MDP

PNL B

SPACE FOR
TELEPHONE
EQUIPMENT

EXHAUST FAN

3/4 HP - 208/3/60
30A-3P NFSS

TYPICAL OF THREE,
LARGE DUCT
(VERTICAL ELL.)
TERMINATE 36"
ABOVE FIN. FL.

ROOFTOP AH
UNIT NO. 1
SEE POWER-
RISER DIAGRAM
SHEET E-4

ROOFTOP AH
UNIT NO. 2
SEE POWER-
RISER DIAGRAM
SHEET E-4

TYPICAL OF THREE,
ALL RECEPTS. IN
TOILETS SHALL BE
PROVIDED WITH
GFCI PROTECTION

TYPICAL OF THREE,
1-1/4" CONDUIT TO
PANEL A

EWC   102

A-12   PANEL A

ELECTRIC WALL
HEATER
4KW-208V/1/60

JB   JB   JB
JB   JB   JB
JB   JB   JB
JB   JB   JB

117
107
109
106
105

206F33.EPS

Figure 33 ◆ Enlarged partial view of Floor Plan "B"—Power detail.

a note is used to identify it on the drawing. Other notes on this drawing describe how certain parts of the system are to be installed. For example, in the office area (rooms 112, 113, and 114), you will see the following note: CONDUIT UP AND STUBBED OUT ABOVE CEILING. This empty conduit is for telephone/communications cables that will be installed later by the telephone company.

Reading Notes

The notes are crucial elements of the drawing set. Receptacles, for example, are hard to position precisely based on a scaled drawing alone, and yet the designer may call for exact locations. For example, the designer may want receptacles exactly 6" above the kitchen counter backsplash and centered on the sink.

8.3.2 Busways

The office/warehouse project uses three types of busways: two types of lighting busways and one power busway. Only the power busway is shown on the power plan; the lighting busways will appear on the lighting plan.

Figure 32 shows two runs of busways: one running the length of the building on the south end (top wall on drawing) and one running the length of the north wall. The symbol list in *Figure 31* shows this busway to be designated by two parallel lines with a series of X's inside. The symbol list further describes the busway as General Electric Type DK-100. These busways are fed from the main distribution panel (circuits MDP-1 and MDP-2) through GE No. DHIBBC41 tap boxes.

The NEC® defines a busway as a grounded metal enclosure containing factory-mounted, bare or insulated conductors, which are usually copper or aluminum bars, rods, or tubes.

The relationship of the busway and hangers to the building construction should be checked prior to commencing the installation so that any problems due to space conflicts, inadequate or inappropriate supporting structure, openings through walls, etc. are worked out in advance so as not to incur lost time.

For example, the drawings and specifications may call for the busway to be suspended from brackets clamped or welded to steel columns. However, the spacing of the columns may be such that additional supplementary hanger rods suspended from the ceiling or roof structure may be necessary for the adequate support of the busway. To offer more assistance to workers on the office/warehouse project, the engineer may also provide an additional drawing that shows how the busway is to be mounted.

Other details that appear on the floor plan in *Figure 33* include the general arrangement of the underfloor duct system, junction boxes and feeder conduit for the underfloor duct system, and plan views of the service and telephone equipment, along with duplex receptacle outlets. A note on the drawing requires all receptacles in the toilets to be provided with ground fault circuit interrupter (GFCI) protection. The letters EWC next to the receptacle in the vestibule designates this receptacle for use with an electric water cooler.

9.0.0 ◆ SPECIAL ELECTRICAL SYSTEMS PLANS

As mentioned in previous sections, an electrical plan can consist of multiple sheets that detail the installation of circuits for electric power, HVAC, and lighting, as well as any special electrical systems that may include environmental control, fire alarm, security, video, electrical instrumentation, or other systems as required. In other cases, special electrical systems may be covered in plan drawings generated by subcontractors that are experienced and qualified in the design and installation of the specific systems.

Examples of special electrical system plans prepared by a subcontractor for an elementary school project are shown in *Figure 34*. Included are partial sheets of plans for the intercom, clocks, and sound systems along with a portion of the cafeteria sound system riser sheet. Riser diagrams are discussed in the next section. Other sheets cover the fire alarm system. The contents of the sheets have been abbreviated and rearranged to allow presentation in this module. Note that many of the drawing elements discussed previously are present in these plans including title blocks, symbol legends, and key plans.

10.0.0 ◆ ELECTRICAL DETAILS AND DIAGRAMS

Electrical diagrams are drawings that are intended to show electrical components and their related connections. They show the electrical association of the different components, but they are seldom, if ever, drawn to scale.

10.1.0 Riser Diagrams

One-line (single-line) **block diagrams** or riser diagrams are used extensively to show the arrangement of equipment and wiring. The **power-riser** diagram in *Figure 35*, for example, was used on the office/warehouse building under discussion and is typical of such drawings. The drawing shows all pieces of electrical equipment as well as the connecting lines used to indicate service-entrance conductors and feeders. Notes are used to identify the equipment, indicate the size of conduit necessary for each feeder, and show the number, size, and type of conductors in each conduit. *Figure 36* shows a building riser diagram for an environmental control system.

A panelboard schedule (*Figure 37*) is included with the power-riser diagram to indicate the exact components contained in each panelboard. This panelboard schedule is for the main distribution panel. On the actual drawings, schedules would also be shown for the other two panels (PNL A and PNL B).

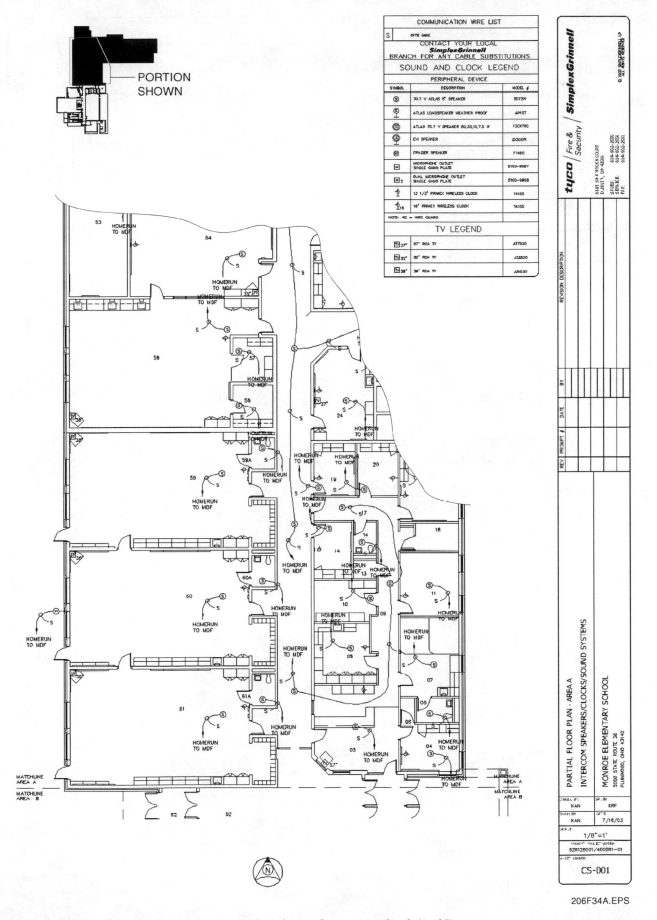

Figure 34 ◆ Examples of special electrical systems plans for an elementary school. (1 of 5)

206F34A.EPS

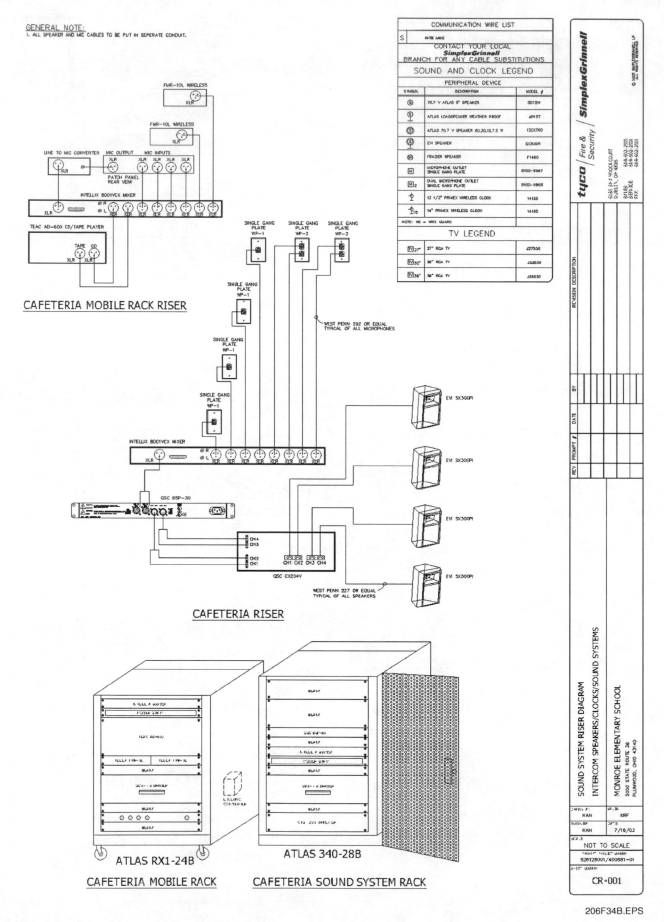

Figure 34 ◆ Examples of special electrical systems plans for an elementary school. (2 of 5)

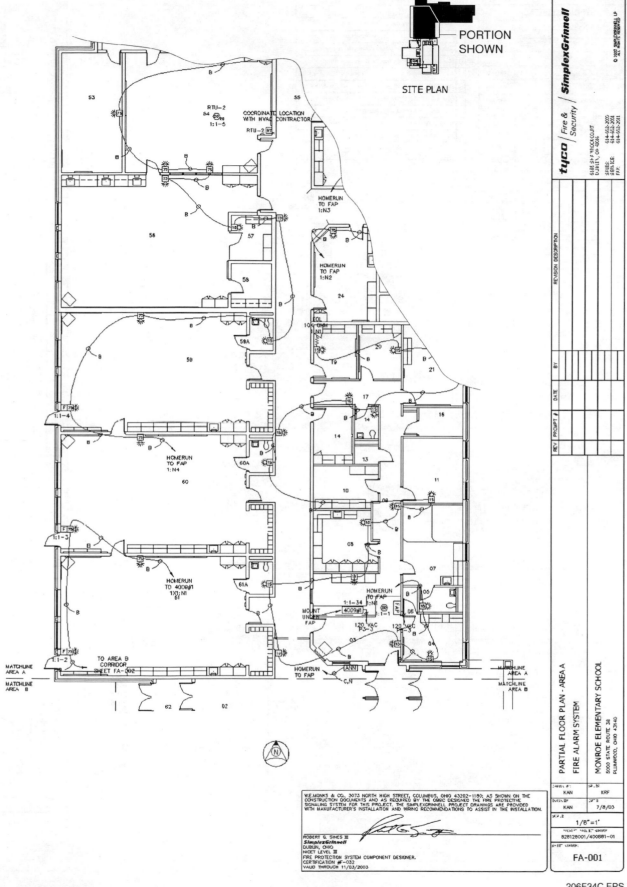

Figure 34 ◆ Examples of special electrical systems plans for an elementary school. (3 of 5)

206F34C.EPS

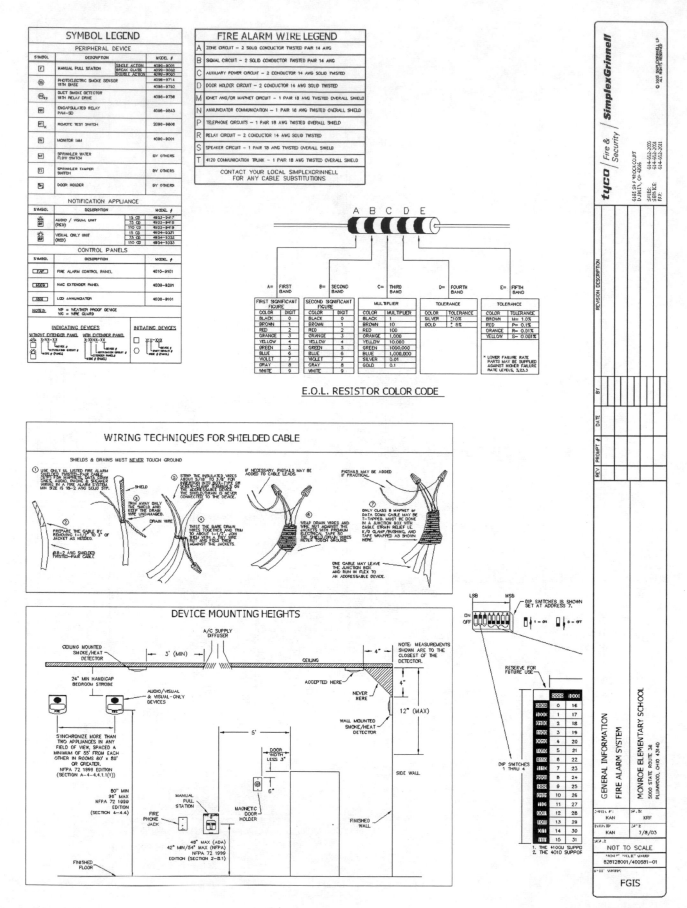

Figure 34 ◆ Examples of special electrical systems plans for an elementary school. (4 of 5)

206F34D.EPS

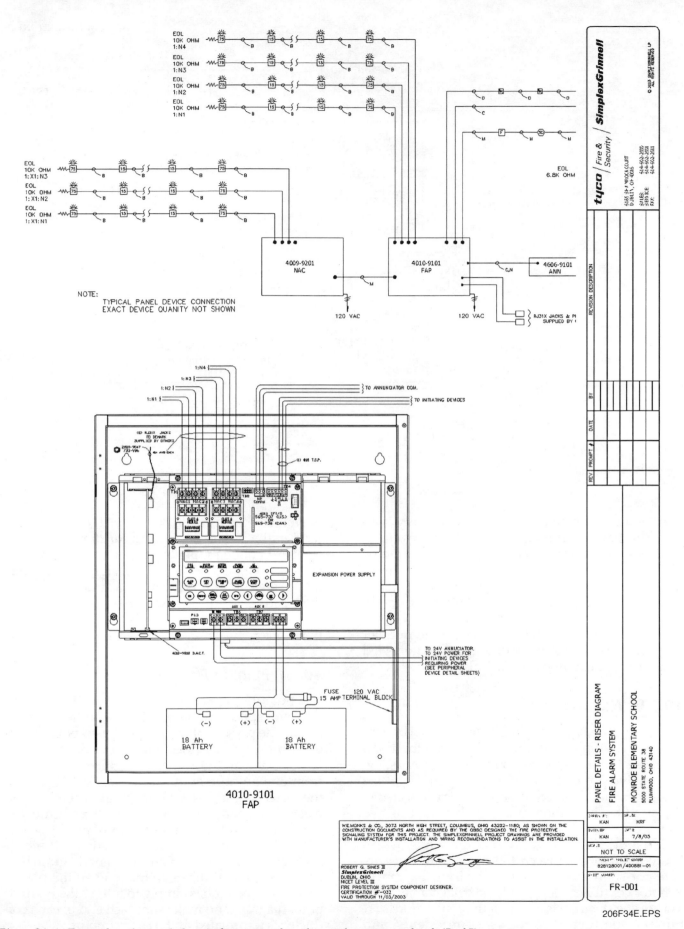

Figure 34 ◆ Examples of special electrical systems plans for an elementary school. (5 of 5)

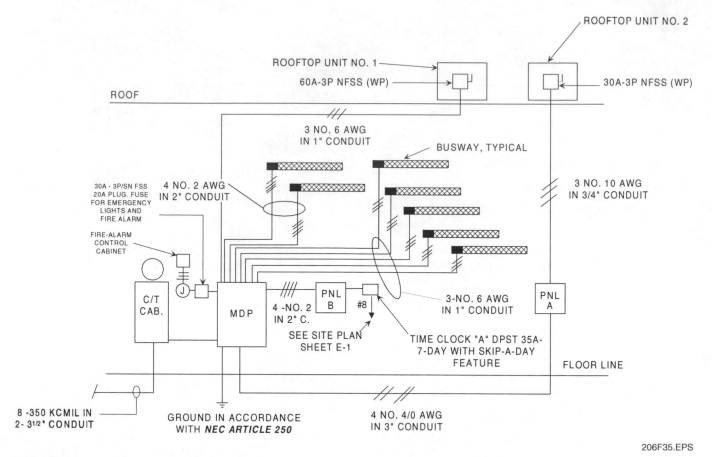

ROOFTOP UNIT NO. 2

ROOFTOP UNIT NO. 1

60A-3P NFSS (WP)

30A-3P NFSS (WP)

ROOF

3 NO. 6 AWG
IN 1" CONDUIT

BUSWAY, TYPICAL

3 NO. 10 AWG
IN 3/4" CONDUIT

30A - 3P/SN FSS
20A PLUG. FUSE
FOR EMERGENCY
LIGHTS AND
FIRE ALARM

4 NO. 2 AWG
IN 2" CONDUIT

FIRE-ALARM
CONTROL
CABINET

C/T
CAB.

J

MDP

4 -NO. 2
IN 2" C.

PNL
B

#8

3-NO. 6 AWG
IN 1" CONDUIT

PNL
A

SEE SITE PLAN
SHEET E-1

TIME CLOCK "A" DPST 35A-
7-DAY WITH SKIP-A-DAY
FEATURE

FLOOR LINE

8 -350 KCMIL IN
2- 3¹/₂" CONDUIT

GROUND IN ACCORDANCE
WITH *NEC ARTICLE 250*

4 NO. 4/0 AWG
IN 3" CONDUIT

206F35.EPS

Figure 35 ◆ Typical power-riser diagram.

In general, panelboard schedules usually indicate the panel number, type of cabinet (either flush- or surface-mounted), panel mains (ampere and voltage rating), phase (single- or three-phase), and number of wires. A four-wire panel, for example, indicates that a solid neutral exists in the panel. Branches indicate the type of overcurrent protection; that is, they indicate the number of poles, trip rating, and frame size. The items fed by each overcurrent device are also indicated.

10.2.0 Wiring Diagrams

There are two types of circuit diagrams. The first type is the wiring diagram, which is used for installation and troubleshooting. Wiring diagrams show the relative position of various components of the equipment and how each conductor is connected in the circuit. These diagrams are classified in two ways: as internal diagrams, which show the wiring inside a device, and as external diagrams, which show the wiring between the devices in a system. *Figure 38* shows a simple internal wiring diagram of a motor controller, along with external connections. The wiring diagram is a map of the connections, terminals, conductors, devices, components, and

equipment in the circuit or system. It is the final tool used in the installation process or the roadmap for troubleshooting the circuit or system. There are several different methods used by designers to illustrate the wiring connection layout of a system, including the point-to-point method, cable method, baseline method, and lineless method. The following subsections look more closely into these methods.

10.2.1 Point-to-Point Method

The point-to-point wiring diagram method (*Figure 39*) saves space and helps alleviate clutter on a diagram. The person following a point-to-point wiring diagram generally does not have a description of the equipment, but rather only traces a conductor from one point to another, making sure it follows the correct path. This diagram is very useful in troubleshooting when the exact wiring path must be determined, but it is more typically used to terminate wiring during installation. Often technicians responsible for using a point-to-point diagram have no knowledge of the circuitry or even the process involved. Their job is simply to follow the point-to-point diagram and make sure each end of each conductor is terminated at the correct point.

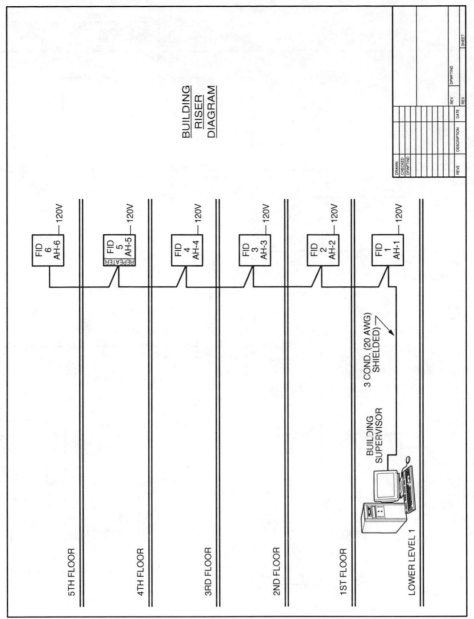

Figure 36 ◆ Building riser diagram for an environmental control system.

PANELBOARD SCHEDULE										
PANEL No.	CABINET TYPE	PANEL MAINS			BRANCHES					ITEMS FED OR REMARKS
		AMPS	VOLTS	PHASE	1P	2P	3P	PROT.	FRAME	
MDP	SURFACE	600A	120/208	3φ,4-W	-	-	1	225A	25,000	PANEL "A"
					-	-	1	100A	18,000	PANEL "B"
					-	-	1	100A		POWER BUSWAY
					-	-	1	60A		LIGHTING BUSWAY
					-	-	1	70A		ROOFTOP UNIT #1
					-	-	1	70A	▼	SPARE
					-	-	1	600A	42,000	MAIN CIRCUIT BRKR

206F37.EPS

Figure 37 ◆ Typical panelboard schedule.

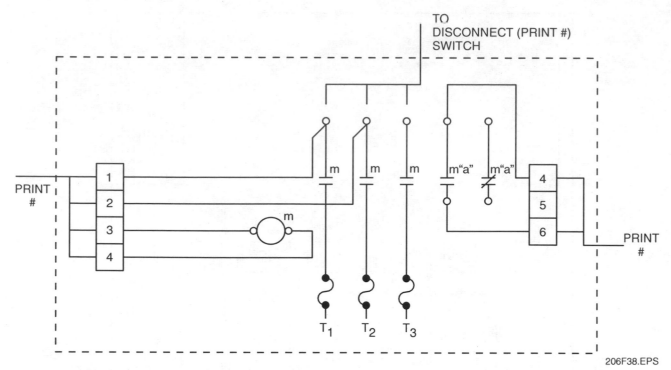

Figure 38 ◆ Wiring diagram of a motor controller.

206F38.EPS

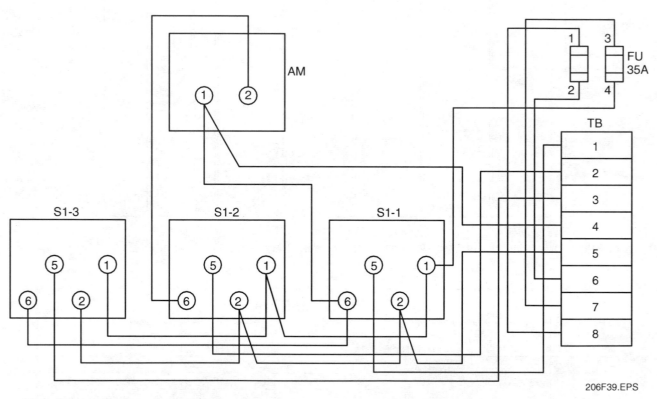

Figure 39 ◆ Point-to-point connection diagram.

206F39.EPS

10.2.2 Cable Method

Unlike the point-to-point wiring diagram, the cable wiring diagram method does not show each conductor from end-to-end. It only shows the cable or cable bundle that the conductor or conductors are in (*Figure 40*). This method requires much less space than the point-to-point method and allows much more white space on the drawing, making it appear less cluttered. One disadvantage of using this method is that the technician or installer does not know the exact path that a conductor takes

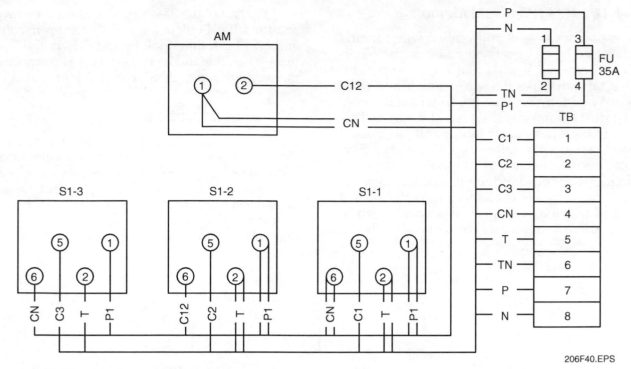

Figure 40 ◆ Cable method connection diagram.

from one point to the other. He or she only knows that it enters a cable or cable bundle and exits at one or more locations. This diagram requires much more attention to detail and often requires a higher level of conductor termination skills. It is a very common method used in wiring diagram layouts. Its primary function is to serve as an installation tool, not as a troubleshooting tool.

10.2.3 Baseline Method

The baseline method is a variation of the cable method. It is used on complex diagrams. In the baseline diagram, one line is used to represent many wires. The use of the baselines decreases the number of lines in the diagram. *Figure 41* is a baseline connection diagram.

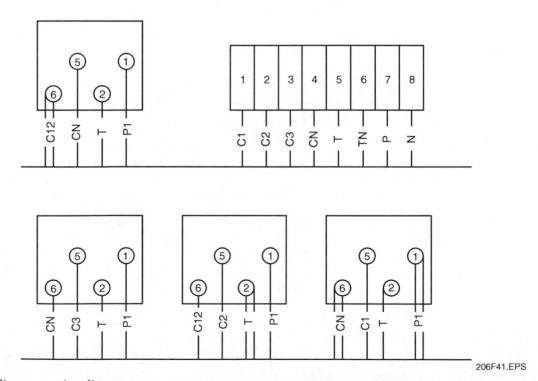

Figure 41 ◆ Baseline connection diagram.

10.2.4 Lineless (Wireless) Method

Lineless diagrams, sometimes called wire running lists, are used by designers on systems that generally have terminal boards that are clearly numbered. A table that lists the wire number, as well as the board and terminal number to which it is connected, must accompany this type of diagram. This is a fairly simple diagram to follow and can be used by less experienced installers as long as they follow the exact cross-reference of the table. Completed terminations should be double-checked to make sure the correct wire is connected to the correct board and terminal. Using the example shown in *Figure 42*, try to find the wire connection numbered 4 for the end of the 35-amp fuse, using the drawing and the table. Look in the table in the From column and find FU4, then look beside FU4 in the To column. The table indicates that FU4 is connected to three different points: S1-1,1; S1-2,1; and S1-3,1.

10.3.0 Schematics

The second type of circuit diagram is the schematic diagram (*Figure 43*), which represents circuit components using symbols. It shows how components are connected together electrically, but typically does not show complete point-to-point wiring. As an electronic systems technician, your schematic diagram reading will most likely be done in the

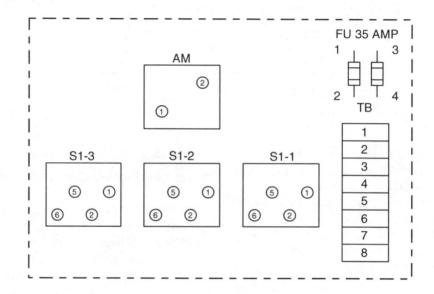

WIRE RUNNING LIST

WIRE CODE	FROM	TO	WIRE SIZE
C1	TB1	S1-1, 5	
C2	TB2	S1-2, 5	
C3	TB3	S1-3, 5	
CN	TB4	AM1, S1-1, 6; S3-1, 6	AS REQUIRED
T	TB5	S1-1, 2; S1-2, 2; S1-3, 2	
TN	TB6	FU2	
P	TB7	FU3	
N	TB8	FU1	
C12	AM2	S1-2, 6	
P1	FU4	S1-1, 1; S1-2, 1; S1-3, 1	

206F42.EPS

Figure 42 ◆ Lineless connection diagram.

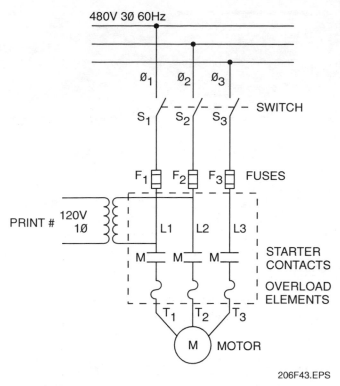

480V 3Ø 60Hz

Figure 43 ◆ Schematic diagram example.

course of troubleshooting. An effective way to interpret a schematic diagram is to start with a load and work back to the source. A load can be a relay coil, motor, light bulb, or alarm annunciator; it can be any electrical device that consumes energy.

In this example, if a motor does not start when it should, the problem could be a burned-out motor or a malfunction in the control circuit for the motor. The obvious and easiest thing to do is to replace the fuses. If that does not correct the problem, the next step is to trace the circuit from the motor back to the voltage source looking for a defective switching device or wire connection. In order to do this effectively, you must understand the logic and interdependencies designed into the circuit. For example, *Figure 44* shows a schematic diagram of a simple lighting circuit. This type of diagram is called a ladder diagram because the voltage source is represented by uprights, and the load lines are like the rungs of a ladder. This method of diagramming clearly shows controls that affect each load, as well as the interdependencies of the loads. Let's assume this ladder represents the control circuit for parking

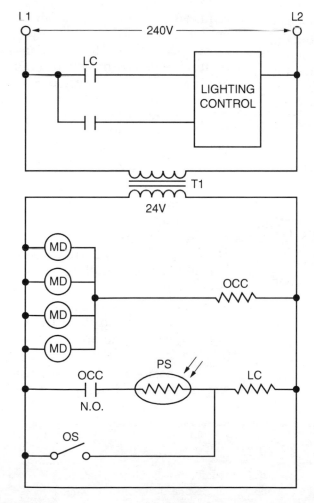

LEGEND

LC = Lighting Contactor
MD = Motion Detector
OCC = Occupied Relay
OS = Override Switch
PS = Photocell Switch

Figure 44 ◆ Ladder diagram.

Testing Lights
What simple design change to the circuit in *Figure 44* would make it easier to test the lights?

lot lights and that there are three conditions for operating the lights:

1. The lights should be enabled whenever it is dark.
2. In order to save utility costs, the lights will only come on when the building is occupied.
3. A means of bypassing the automatic controls is required.

The lights themselves are activated by the closure of lighting contactor (LC1) in a 240-volt circuit. However, the contactor is controlled by a 24-volt circuit that obtains its supply voltage from step-down transformer T1. This is a very common design approach. The lower voltage is safer and the 24-volt circuit can use less expensive wiring and components than a 240-volt circuit.

A photoelectric relay is in series with the contactor coil. Motion sensors are used to prevent the light from being turned on when the building is unoccupied. Working back from the contactor coil, you can see that a photo switch must be activated in order to energize the lighting contactor. This satisfies condition 1.

The OCC relay is energized by motion detectors scattered around the building. Because the normally open contacts of the OCC relay are in series with the photo switch, the circuit is not complete until a motion detector energizes the OCC relay while the photo switch is activated. That is, it must be after dark and someone must be in the building. This satisfies condition 2.

A single-pole switch is wired into the contactor coil circuit so the lights can be manually turned on. This satisfies condition 3.

10.4.0 Drawing Details

A detail drawing is a drawing of a separate item or portion of an electrical system, giving a com-plete and exact description of its use and all the details needed to show exactly what is required for its installation. For example, the power plan for the office/warehouse has a sectional cut through the busduct. This is a good example of where an extra, detailed drawing is desirable.

A set of electrical drawings will sometimes require large-scale drawings of certain areas that are not indicated with sufficient clarity on the small-scale drawings. For example, the site plan may show exterior pole-mounted lighting fixtures that are to be installed by the contractor.

11.0.0 ◆ WRITTEN SPECIFICATIONS

The written specifications for a building or project are the written descriptions of work and duties required of the owner, architect, and consulting engineer. Together with the working drawings, these specifications form the basis of the contract requirements for the construction of the building or project. Those who use the construction draw-ings and specifications must always be alert to discrepancies between the working drawings and the written specifications. These are some situa-tions where discrepancies may occur:

• Architects or engineers use standard or proto-type specifications and attempt to apply them without any modification to specific working drawings.
• Previously prepared standard drawings are changed or amended by reference in the speci-fications only and the drawings themselves are not changed.
• Items are duplicated in both the drawings and specifications, but an item is subsequently amended in one and overlooked in the other contract document.

In such instances, the person in charge of the pro-ject has the responsibility to ascertain whether the

Understanding Contact Symbols
When a drawing shows normally open or normally closed contacts, the word *normally* refers to the condition of the contacts in their de-energized or shelf state.

Although most of the work of an EST involves the use of electrical drawings and schedules, other drawings in the set are important as well:

- The site plan often shows the locations of underground utilities and wall penetrations.
- Elevation drawings show exterior finish and trim.
- Section drawings (at right) show the construction of walls, floors, and ceilings. This is extremely important to installers who are creating pathways for conduit and cables because it provides guidance on where and how to make penetrations.
- Mechanical drawings show the location of heating and air conditioning equipment and ductwork, which can serve as obstacles to cabling pathways.
- Floor plans show the interior layout of the structure and are a valuable tool in planning an installation.
- Schedules, along with the other drawings in the set, are important when doing equipment and material takeoffs.
- The electrical plan usually includes details of the electronic systems and cabling (see the following page).

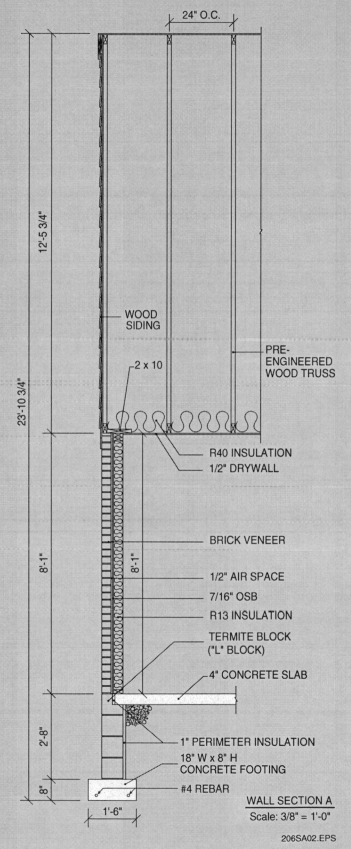

WALL SECTION A
Scale: 3/8" = 1'-0"

206SA02.EPS

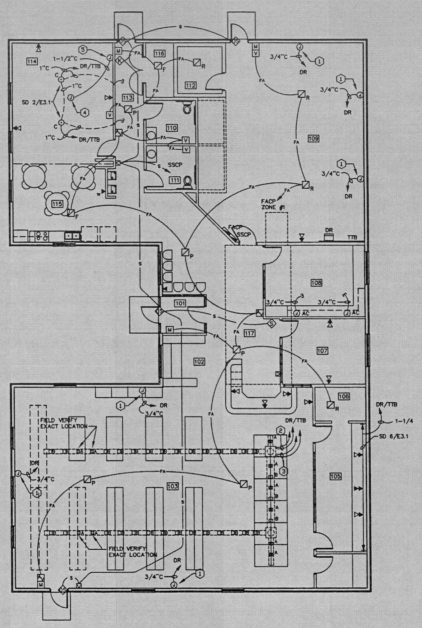

$$\diamondsuit \begin{matrix} 3 \\ \text{E2.1} \end{matrix} \quad \underline{\text{SYSTEMS PLAN}}$$

1/8" = 1' - 0"

NORTH

NOTES: DETAIL 3/E2.1

① SINGLE GANG ROUGH-IN WITH PULL STRING AND BLANK COVER AT 18"AFF, FOR FUTURE.

② TWO 1-1/4"C FOR VOICE CABLES. ONE TO BE SPARE WITH PULL STRING ONLY, FOR FUTURE USE.

③ TWO 1-1/4"C FOR DATA CABLES. ONE TO BE SPARE WITH PULL STRING ONLY, FOR FUTURE USE.

④ 6"X6"X4" JUNCTION WITH BLANK COVER AND PULL STRING IN ACCESSIBLE CEILING SPACE FOR FUTURE PROJECTOR.

⑤ 4-11/16" ROUGH-IN WITH SINGLE GANG COVER FOR FUTURE USE.

206SA03.EPS

drawings or the specifications take precedence. Such questions must be resolved, preferably before the work begins, to avoid added costs to the owner, architect/engineer, or contractor.

11.1.0 How Specifications are Written

Writing accurate and complete specifications for building construction is a serious responsibility for those who design the buildings because the specifications, combined with the working drawings, govern practically all important decisions that are made during the construction span of every project. Compiling and writing these specifications is not a simple task, even for those who have had considerable experience in preparing such documents. A set of written specifications for a single project will usually contain thousands of products, parts, and components, and the methods of installing them, all of which must be covered in either the drawings and/or specifications. No one can memorize all of the necessary items required to describe accurately the various areas of construction. One must rely upon reference materials such as manufacturer's data, catalogs, checklists, and, most of all, a high-quality master specification.

11.2.0 Format of Specifications

For convenience in writing, speed in estimating, and ease of reference, the most suitable organization of the specifications is a series of sections dealing with the construction requirements, products, and activities, and is easily understandable by the different trades. Those people who use the specifications must be able to find all the information they need without spending too much time looking for it.

The most commonly used specification-writing format used in North America is the *MasterFormat*™. This standard was developed jointly by the Construction Specifications Institute (CSI) and Construction Specifications Canada (CSC). For many years prior to 2004, the organization of construction specifications and suppliers catalogs has been based on a standard with 16 sections, otherwise known as divisions, where the divisions and their subsections were individually identified by a five-digit numbering system. The first two digits represented the division number and the next three individual numbers represented successively lower levels of breakdown. For example, the number 13213 represents division 13, subsection 2, sub-subsection 1 and sub-sub-subsection 3. In this older version of the standard, electrical systems, including any electronic or special electrical systems, were lumped together under Division 16 – *Electrical*. Today, specifications conforming to the 16 division format may still be in use.

In 2004, the *MasterFormat*™ standard underwent a major change. What had been 16 divisions was expanded to four major groupings and 49 divisions with some divisions reserved for future expansion (*Figure 45*). The first 14 divisions are essentially the same as the old format. Subjects under the old Division 15 – *Mechanical* have been relocated to new divisions 22 and 23. The basic subjects under old Division 16 – *Electrical* have been relocated to new divisions 26 and 27. In addition, the numbering system was changed to 6 digits to allow for more subsections in each division, which allowed for finer definition. In the new numbering system, the first two digits represent the division number. The next two digits represent subsections of the division and the two remaining digits represent the third level sub-subsection numbers. The fourth level, if required, is a decimal and number added to the end of the last two digits. For example, the number 132013.04 represents division 13, subsection 20, sub-subsection 13 and sub-sub-subsection 04. Under the new standard, the Facility Service Subgroup contains the divisions that are most important to the EST. These include the following divisions:

- *Division 25 – Integrated Automation*
- *Division 26 – Electrical*
- *Division 27 – Communications*
- *Division 28 – Electronic Safety and Security*

Appendix B contains a detailed breakdown of these divisions.

Specifications

Written specifications supplement the related working drawings in that they contain details not shown on the drawings. Specifications define and clarify the scope of the job. They describe the specific types and characteristics of the components that are to be used on the job and the methods for installing some of them. Many components are identified specifically by the manufacturer's model and part numbers. This type of information is used to purchase the various items of hardware needed to accomplish the installation in accordance with the contractual requirements.

Division Numbers and Titles

PROCUREMENT AND CONTRACTING REQUIREMENTS GROUP
Division 00   Procurement and Contracting Requirements

SPECIFICATIONS GROUP

GENERAL REQUIREMENTS SUBGROUP
Division 01   General Requirements

FACILITY CONSTRUCTION SUBGROUP
Division 02   Existing Conditions
Division 03   Concrete
Division 04   Masonry
Division 05   Metals
Division 06   Wood, Plastics, and Composites
Division 07   Thermal and Moisture Protection
Division 08   Openings
Division 09   Finishes
Division 10   Specialties
Division 11   Equipment
Division 12   Furnishings
Division 13   Special Construction
Division 14   Conveying Equipment
Division 15   Reserved
Division 16   Reserved
Division 17   Reserved
Division 18   Reserved
Division 19   Reserved

FACILITY SERVICES SUBGROUP
Division 20   Reserved
Division 21   Fire Suppression
Division 22   Plumbing
Division 23   Heating, Ventilating, and Air Conditioning
Division 24   Reserved
Division 25   Integrated Automation
Division 26   Electrical
Division 27   Communications
Division 28   Electronic Safety and Security
Division 29   Reserved

SITE AND INFRASTRUCTURE SUBGROUP
Division 30   Reserved
Division 31   Earthwork
Division 32   Exterior Improvements
Division 33   Utilities
Division 34   Transportation
Division 35   Waterway and Marine Construction
Division 36   Reserved
Division 37   Reserved
Division 38   Reserved
Division 39   Reserved

PROCESS EQUIPMENT SUBGROUP
Division 40   Process Integration
Division 41   Material Processing and Handling Equipment
Division 42   Process Heating, Cooling, and Drying Equipment
Division 43   Process Gas and Liquid Handling, Purification, and Storage Equipment
Division 44   Pollution Control Equipment
Division 45   Industry-Specific Manufacturing Equipment
Division 46   Reserved
Division 47   Reserved
Division 48   Electrical Power Generation
Division 49   Reserved

Div Numbers - 1

206F45.EPS

Figure 45 ◆ 2004 *MasterFormat*™.

ELECTRONIC SYSTEMS TECHNICIAN LEVEL TWO — TRAINEE MODULE 33206-05

Summary

In this module, the symbols and conventions used on architectural and engineering drawings are discussed. As an electronic systems technician, you need to know how to recognize the basic symbols used on electrical drawings and other drawings used in the building construction industry. You should also know where to find the meaning of symbols that you do not immediately recognize. Schedules, diagrams, and specifications often provide detailed information that is not included on the working drawings.

Reading architectural and engineering drawings takes practice and study. Now that you have the basic skills, take the time to master them.

Review Questions

1. The location of a building on the property is shown on the _____.
   a. elevation drawings
   b. floor plan
   c. site plan
   d. section drawing

2. The drawing shown in *Figure 1* is a(n) _____.
   a. sectional view
   b. floor plan
   c. elevation drawing
   d. site plan

3. Government regulations require that electrical drawings from engineering firms be complete, precise, and easy to read.
   a. True
   b. False

4. A 24 × 36 drawing equates to letter size _____.
   a. A
   b. B
   c. C
   d. D

5. A dashed line on an electrical diagram represents _____.
   a. a busway
   b. wiring concealed in the floor
   c. wiring concealed in a wall
   d. a branch circuit homerun

6. The purpose of a key plan on a drawing sheet is to _____.
   a. show specific dimensions and other pertinent information concerning a particular piece of equipment
   b. provide a detailed view taken from an area of a drawing
   c. identify that part of the project to which that drawing sheet applies
   d. show the location of a building on a building site

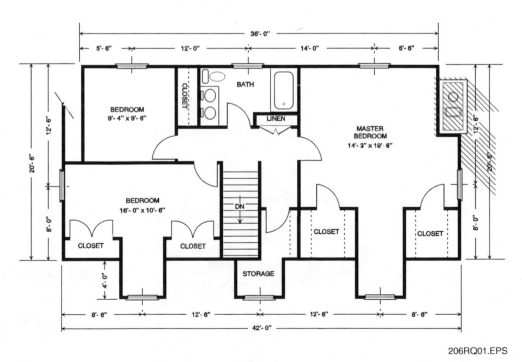

206RQ01.EPS

Figure 1

INTRODUCTION TO ELECTRICAL BLUEPRINTS

7. A drawing that shows the relative position of various components of electrical equipment and how each conductor is connected in the circuit is a _____.
   a. plan drawing
   b. one-line diagram
   c. wiring diagram
   d. panelboard schedule

8. In the CSI format for written specifications, most of the work concerning an EST can be found in the _____ subgroup.
   a. general requirements
   b. site and infrastructure
   c. facility construction
   d. facility services

9. In the CSI format for written specifications, integrated automation systems are covered under Division _____.
   a. 25
   b. 26
   c. 27
   d. 28

10. In the CSI format for written specifications, communications systems are covered under Division _____.
    a. 25
    b. 26
    c. 27
    d. 28

Trade Terms Introduced in This Module

Architectural drawings: Working drawings consisting of plans, elevations, details, and other information necessary for the construction of a building. Architectural drawings usually include:

- A site (plot) plan indicating the location of the building on the property
- Floor plans showing the walls and partitions for each floor or level
- Elevations of all exterior faces of the building
- Several vertical cross sections to indicate clearly the various floor levels and details of the footings, foundations, walls, floors, ceilings, and roof construction
- Large-scale detail drawings showing such construction details as may be required

Block diagram: A single-line diagram used to show electrical equipment and related connections. See power-riser diagram.

Blueprint: An exact copy or reproduction of an original drawing.

Detail drawing: An enlarged, detailed view taken from an area of a drawing and shown in a separate view.

Dimensions: Sizes or measurements printed on a drawing.

Electrical drawing: A means of conveying a large amount of exact, detailed information in an abbreviated language. Consists of lines, symbols, dimensions, and notations to accurately convey an engineer's designs to electricians and electronic systems technicians who install the electrical system on a job.

Elevation drawing: An architectural drawing showing height, but not depth; usually the front, rear, and sides of a building or object.

Floor plan: A drawing of a building as if a horizontal cut were made through a building at about window level, and the top portion removed. The floor plan is what would appear if the remaining structure were viewed from above.

One-line diagram: A drawing that shows, by means of lines and symbols, the path of an electrical circuit or system of circuits along with the various circuit components. Also called a single-line diagram.

Plan view: A drawing made as though the viewer were looking straight down (from above) on an object.

Power-riser diagram: A single-line block diagram used to indicate the electric service equipment, service conductors and feeders, and subpanels. Notes are used on power-riser diagrams to identify the equipment; indicate the size of conduit; show the number, size, and type of conductors; and list related materials. A panelboard schedule is usually included with power-riser diagrams to indicate the exact components (panel type and size), along with fuses, circuit breakers, etc., contained in each panelboard.

Scale: On a drawing, the size relationship between an object's actual size and the size it is drawn. Scale also refers to the measuring tool used to determine this relationship.

Schedule: A systematic method of presenting equipment lists on a drawing in tabular form.

Schematic diagram: A detailed diagram showing complicated circuits, such as control circuits.

Sectional view: A cutaway drawing that shows the inside of an object or building.

Shop drawing: A drawing that is usually developed by manufacturers, fabricators, or contractors to show specific dimensions and other pertinent information concerning a particular piece of equipment and its installation methods.

Site plan: A drawing showing the location of a building or buildings on the building site. Such drawings frequently show topographical lines, electrical and communication lines, water and sewer lines, sidewalks, driveways, and similar information.

Written specifications: A written description of what is required by the owner, architect, and engineer in the way of materials and workmanship. Together with working drawings, the specifications form the basis of the contract requirements for construction.

Metric Conversion Chart

| inches | | m m | inches | | m m | inches | | m m | inches | | m m |
fractions	decimals		fractions	decimals		fractions	decimals		fractions	decimals	
—	.0004	.01	25/32	.781	19.844	—	2.165	55.	3-11/16	3.6875	93.663
—	.004	.10	—	.7874	20.	2-3/16	2.1875	55.563	—	3.7008	94.
—	.01	.25	51/64	.797	20.241	—	2.2047	56.	3-23/32	3.719	94.456
1/64	.0156	.397	13/16	.8125	20.638	2-7/32	2.219	56.356	—	3.7401	95.
—	.0197	.50	—	.8268	21.	—	2.244	57.	3-3/4	3.750	95.250
—	.0295	.75	53/64	.828	21.034	2-1/4	2.250	57.150	—	3.7795	96.
1/32	.03125	.794	27/32	.844	21.431	2-9/32	2.281	57.944	3-25/32	3.781	96.044
—	.0394	1.	55/64	.859	21.828	—	2.2835	58.	3-13/16	3.8125	96.838
3/64	.0469	1.191	—	.8661	22.	2-5/16	2.312	58.738	—	3.8189	97.
—	.059	1.5	7/8	.875	22.225	—	2.3228	59.	3-27/32	3.844	97.631
1/16	.062	1.588	57/64	.8906	22.622	2-11/32	2.344	59.531	—	3.8583	98.
5/64	.0781	1.984	—	.9055	23.	—	2.3622	60.	3-7/8	3.875	98.425
—	.0787	2.	29/32	.9062	23.019	2-3/8	2.375	60.325	—	3.8976	99.
3/32	.094	2.381	59/64	.922	23.416	—	2.4016	61.	3-29/32	3.9062	99.219
—	.0984	2.5	15/16	.9375	23.813	2-13/32	2.406	61.119	—	3.9370	100.
7/64	.109	2.778	—	.9449	24.	2-7/16	2.438	61.913	3-15/16	3.9375	100.013
—	.1181	3.	61/64	.953	24.209	—	2.4409	62.	3-31/32	3.969	100.806
1/8	.125	3.175	31/32	.969	24.606	2-15/32	2.469	62.706	—	3.9764	101.
—	.1378	3.5	—	.9843	25.	—	2.4803	63.	4	4.000	101.600
9/64	.141	3.572	63/64	.9844	25.003	2-1/2	2.500	63.500	4-1/16	4.062	103.188
5/32	.156	3.969	1	1.000	25.400	—	2.5197	64.	4-1/8	4.125	104.775
—	.1575	4.	—	1.0236	26.	2-17/32	2.531	64.294	—	4.1338	105.
11/64	.172	4.366	1-1/32	1.0312	26.194	—	2.559	65.	4-3/16	4.1875	106.363
—	.177	4.5	1-1/16	1.062	26.988	2-9/16	2.562	65.088	4-1/4	4.250	107.950
3/16	.1875	4.763	—	1.063	27.	2-19/32	2.594	65.881	4-5/16	4.312	109.538
—	.1969	5.	1-3/32	1.094	27.781	—	2.5984	66.	—	4.3307	110.
13/64	.203	5.159	—	1.1024	28.	2-5/8	2.625	66.675	4-3/8	4.375	111.125
—	.2165	5.5	1-1/8	1.125	28.575	—	2.638	67.	4-7/16	4.438	112.713
7/32	.219	5.556	—	1.1417	29.	2-21/32	2.656	67.469	4-1/2	4.500	114.300
15/64	.234	5.953	1-5/32	1.156	29.369	—	2.6772	68.	—	4.5275	115.
—	.2362	6.	—	1.1811	30.	2-11/16	2.6875	68.263	4-9/16	4.562	115.888
1/4	.250	6.350	1-3/16	1.1875	30.163	—	2.7165	69.	4-5/8	4.625	117.475
—	.2559	6.5	1-7/32	1.219	30.956	2-23/32	2.719	69.056	4-11/16	4.6875	119.063
17/64	.2656	6.747	—	1.2205	31.	2-3/4	2.750	69.850	—	4.7244	120.
—	.2756	7.	1-1/4	1.250	31.750	—	2.7559	70.	4-3/4	4.750	120.650
9/32	.281	7.144	—	1.2598	32.	2-25/32	2.781	70.6439	4-13/16	4.8125	122.238
—	.2953	7.5	1-9/32	1.281	32.544	—	2.7953	71.	4-7/8	4.875	123.825
19/64	.297	7.541	—	1.2992	33.	2-13/16	2.8125	71.4376	—	4.9212	125.
5/16	.312	7.938	1-5/16	1.312	33.338	—	2.8346	72.	4-15/16	4.9375	125.413
—	.315	8.	—	1.3386	34.	2-27/32	2.844	72.2314	5	5.000	127.000
21/64	.328	8.334	1-11/32	1.344	34.131	—	2.8740	73.	—	5.1181	130.
—	.335	8.5	1-3/8	1.375	34.925	2-7/8	2.875	73.025	5-1/4	5.250	133.350
11/32	.344	8.731	—	1.3779	35.	2-29/32	2.9062	73.819	5-1/2	5.500	139.700
—	.3543	9.	1-13/32	1.406	35.719	—	2.9134	74.	—	5.5118	140.
23/64	.359	9.128	—	1.4173	36.	2-15/16	2.9375	74.613	5-3/4	5.750	146.050
—	.374	9.5	1-7/16	1.438	36.513	—	2.9527	75.	—	5.9055	150.
3/8	.375	9.525	—	1.4567	37.	2-31/32	2.969	75.406	6	6.000	152.400
25/64	.391	9.922	1-15/32	1.469	37.306	—	2.9921	76.	6-1/4	6.250	158.750
—	.3937	10.	—	1.4961	38.	3	3.000	76.200	—	6.2992	160.
13/32	.406	10.319	1-1/2	1.500	38.100	3-1/32	3.0312	76.994	6-1/2	6.500	165.100
—	.413	10.5	1-17/32	1.531	38.894	—	3.0315	77.	—	6.6929	170.
27/64	.422	10.716	—	1.5354	39.	3-1/16	3.062	77.788	6-3/4	6.750	171.450
—	.4331	11.	1-9/16	1.562	39.688	—	3.0709	78.	7	7.000	177.800
7/16	.438	11.113	—	1.5748	40.	3-3/32	3.094	78.581	—	7.0866	180.
29/64	.453	11.509	1-19/32	1.594	40.481	—	3.1102	79.	—	7.4803	190.
15/32	.469	11.906	—	1.6142	41.	3-1/8	3.125	79.375	7-1/2	7.500	190.500
—	.4724	12.	1-5/8	1.625	41.275	—	3.1496	80.	—	7.8740	200.
31/64	.484	12.303	—	1.6535	42.	3-5/32	3.156	80.169	8	8.000	203.200
—	.492	12.5	1-21/32	1.6562	42.069	3-3/16	3.1875	80.963	—	8.2677	210.
1/2	.500	12.700	1-11/16	1.6875	42.863	—	3.1890	81.	8-1/2	8.500	215.900
—	.5118	13.	—	1.6929	43.	3-7/32	3.219	81.756	—	8.6614	220.
33/64	.5156	13.097	1-23/32	1.719	43.656	—	3.2283	82.	9	9.000	228.600
17/32	.531	13.494	—	1.7323	44.	3-1/4	3.250	82.550	—	9.0551	230.
35/64	.547	13.891	1-3/4	1.750	44.450	—	3.2677	83.	—	9.4488	240.
—	.5512	14.	—	1.7717	45.	3-9/32	3.281	83.344	9-1/2	9.500	241.300
9/16	.563	14.288	1-25/32	1.781	45.244	—	3.3071	84.	—	9.8425	250.
—	.571	14.5	—	1.8110	46.	3-5/16	3.312	84.1377	10	10.000	254.001
37/64	.578	14.684	1-13/16	1.8125	46.038	3-11/32	3.344	84.9314	—	10.2362	260.
—	.5906	15.	1-27/32	1.844	46.831	—	3.3464	85.	—	10.6299	270.
19/32	.594	15.081	—	1.8504	47.	3-3/8	3.375	85.725	11	11.000	279.401
39/64	.609	15.478	1-7/8	1.875	47.625	—	3.3858	86.	—	11.0236	280.
5/8	.625	15.875	—	1.8898	48.	3-13/32	3.406	86.519	—	11.4173	290.
—	.6299	16.	1-29/32	1.9062	48.419	—	3.4252	87.	—	11.8110	300.
41/64	.6406	16.272	—	1.9291	49.	3-7/16	3.438	87.313	12	12.000	304.801
—	.6496	16.5	1-15/16	1.9375	49.213	—	3.4646	88.	13	13.000	330.201
21/32	.656	16.669	—	1.9685	50.	3-15/32	3.469	88.106	—	13.7795	350.
—	.6693	17.	1-31/32	1.969	50.006	3-1/2	3.500	88.900	14	14.000	355.601
43/64	.672	17.066	2	2.000	50.800	—	3.5039	89.	15	15.000	381.001
11/16	.6875	17.463	—	2.0079	51.	3-17/32	3.531	89.694	—	15.7480	400.
45/64	.703	17.859	2-1/32	2.03125	51.594	—	3.5433	90.	16	16.000	406.401
—	.7087	18.	—	2.0472	52.	3-9/16	3.562	90.4877	17	17.000	431.801
23/32	.719	18.256	2-1/16	2.062	52.388	—	3.5827	91.	—	17.7165	450.
—	.7283	18.5	—	2.0866	53.	3-19/32	3.594	91.281	18	18.000	457.201
47/64	.734	18.653	2-3/32	2.094	53.181	—	3.622	92.	19	19.000	482.601
—	.7480	19.	2-1/8	2.125	53.975	3-5/8	3.625	92.075	—	19.6850	500.
3/4	.750	19.050	—	2.126	54.	3-21/32	3.656	92.869	20	20.000	508.001
49/64	.7656	19.447	2-5/32	2.156	54.769	—	3.6614	93.			

206A01.EPS

Breakdown of Certain MasterFormat™ Divisions

DIVISION 25 – INTEGRATED AUTOMATION

25 00 00　INTEGRATED AUTOMATION

25 01 00　Operation and Maintenance of Integrated Automation
- 25 01 10　　Operation and Maintenance of Integrated Automation Network Equipment
- 25 01 20　　Operation and Maintenance of Integrated Equipment
- 25 01 30　　Operation and Maintenance of Integrated Automation Instrumentation and Terminal Devices
- 25 01 90　　Diagnostic Systems for Integrated Automation

25 05 00　Common Work Results for Integrated Automation
- 25 05 13　　Conductors and Cables for Integrated Automation
- 25 05 26　　Grounding and Bonding for Integrated Automation
- 25 05 28　　Pathways for Integrated Automation
  - 25 05 28.29　Hangers and Supports for Integrated Automation
  - 25 05 28.33　Conduits and Backboxes for Integrated Automation
  - 25 05 28.36　Cable Trays for Integrated Automation
  - 25 05 28.39　Surface Raceways for Integrated Automation
- 25 05 48　　Vibration and Seismic Controls for Integrated Automation
- 25 05 53　　Identification for Integrated Automation

25 06 00　Schedules for Integrated Automation
- 25 06 11　　Schedules for Integrated Automation Network
- 25 06 12　　Schedules for Integrated Automation Network Gateways
- 25 06 13　　Schedules for Integrated Automation Control and Monitoring Network
- 25 06 14　　Schedules for Integrated Automation Local Control Units
- 25 06 30　　Schedules for Integrated Automation Instrumentation and Terminal Devices

25 08 00　Commissioning of Integrated Automation

25 10 00　INTEGRATED AUTOMATION NETWORK EQUIPMENT

25 11 00　Integrated Automation Network Devices
- 25 11 13　　Integrated Automation Network Servers
- 25 11 16　　Integrated Automation Network Routers, Bridges, Switches, Hubs, and Modems
- 25 11 19　　Integrated Automation Network Operator Workstations

25 12 00　Integrated Automation Network Gateways
- 25 12 13　　Hardwired Integration Network Gateways
- 25 12 16　　Direct-Protocol Integration Network Gateways
- 25 12 19　　Neutral-Protocol Integration Network Gateways
- 25 12 23　　Client-Server Information/Database Integration Network Gateways

25 13 00　Integrated Automation Control and Monitoring Network
- 25 13 13　　Integrated Automation Control and Monitoring Network Supervisory Control
- 25 13 16　　Integrated Automation Control and Monitoring Network Integration Panels
- 25 13 19　　Integrated Automation Control and Monitoring Network Interoperability

25 14 00　Integrated Automation Local Control Units
- 25 14 13　　Integrated Automation Remote Control Panels
- 25 14 16　　Integrated Automation Application-Specific Control Panels
- 25 14 19　　Integrated Automation Terminal Control Units
- 25 14 23　　Integrated Automation Field Equipment Panels

25 15 00　Integrated Automation Software
- 25 15 13　　Integrated Automation Software for Network Gateways
- 25 15 16　　Integrated Automation Software for Control and Monitoring Networks

25 - 1

| 25 15 19 | Integrated Automation Software for Local Control Units |

25 20 00   *Reserved*

25 30 00   **INTEGRATED AUTOMATION INSTRUMENTATION AND TERMINAL DEVICES**

25 31 00	**Integrated Automation Instrumentation and Terminal Devices for Facility Equipment**
25 32 00	**Integrated Automation Instrumentation and Terminal Devices for Conveying Equipment**
25 33 00	**Integrated Automation Instrumentation and Terminal Devices for Fire-Suppression Systems**
25 34 00	**Integrated Automation Instrumentation and Terminal Devices for Plumbing**
25 35 00	**Integrated Automation Instrumentation and Terminal Devices for HVAC**
25 35 13	Integrated Automation Actuators and Operators
25 35 16	Integrated Automation Sensors and Transmitters
25 35 19	Integrated Automation Control Valves
25 35 23	Integrated Automation Control Dampers
25 35 26	Integrated Automation Compressed Air Supply
25 36 00	**Integrated Automation Instrumentation and Terminal Devices for Electrical Systems**
25 36 13	Integrated Automation Power Meters
25 36 16	Integrated Automation KW Transducers
25 36 19	Integrated Automation Current Sensors
25 36 23	Integrated Automation Battery Monitors
25 36 26	Integrated Automation Lighting Relays
25 36 29	Integrated Automation UPS Monitors
25 37 00	**Integrated Automation Instrumentation and Terminal Devices for Communications Systems**
25 38 00	**Integrated Automation Instrumentation and Terminal Devices for Electronic Safety and Security Systems**

25 40 00   *Reserved*

25 50 00   **INTEGRATED AUTOMATION FACILITY CONTROLS**

25 51 00	**Integrated Automation Control of Facility Equipment**
25 52 00	**Integrated Automation Control of Conveying Equipment**
25 53 00	**Integrated Automation Control of Fire-Suppression Systems**
25 54 00	**Integrated Automation Control of Plumbing**
25 55 00	**Integrated Automation Control of HVAC**
25 56 00	**Integrated Automation Control of Electrical Systems**
25 57 00	**Integrated Automation Control of Communications Systems**
25 58 00	**Integrated Automation Control of Electronic Safety and Security Systems**

25 60 00   *Reserved*

25 70 00   *Reserved*

25 80 00   *Reserved*

25 90 00   **INTEGRATED AUTOMATION CONTROL SEQUENCES**

| 25 91 00 | **Integrated Automation Control Sequences for Facility Equipment** |

25 - 2

206A03.EPS

25 92 00	**Integrated Automation Control Sequences for Conveying Equipment**
25 93 00	**Integrated Automation Control Sequences for Fire-Suppression Systems**
25 94 00	**Integrated Automation Control Sequences for Plumbing**
25 95 00	**Integrated Automation Control Sequences for HVAC**
25 96 00	**Integrated Automation Control Sequences for Electrical Systems**
25 97 00	**Integrated Automation Control Sequences for Communications Systems**
25 98 00	**Integrated Automation Control Sequences for Electronic Safety and Security Systems**

206A04.EPS

DIVISION 26 – ELECTRICAL

26 00 00   ELECTRICAL

26 01 00   Operation and Maintenance of Electrical Systems
26 01 10   Operation and Maintenance of Medium-Voltage Electrical Distribution
26 01 20   Operation and Maintenance of Low-Voltage Electrical Distribution
26 01 26   Maintenance Testing of Electrical Systems
26 01 30   Operation and Maintenance of Facility Electrical Power Generating and Storing Equipment
26 01 40   Operation and Maintenance of Electrical and Cathodic Protection Systems
26 01 50   Operation and Maintenance of Lighting
   26 01 50.51   Luminaire Relamping
   26 01 50.81   Luminaire Replacement

26 05 00   Common Work Results for Electrical
26 05 13   Medium-Voltage Cables
   26 05 13.13   Medium-Voltage Open Conductors
   26 05 13.16   Medium-Voltage, Single- and Multi-Conductor Cables
26 05 19   Low-Voltage Electrical Power Conductors and Cables
   26 05 19.13   Undercarpet Electrical Power Cables
26 05 23   Control-Voltage Electrical Power Cables
26 05 26   Grounding and Bonding for Electrical Systems
26 05 29   Hangers and Supports for Electrical Systems
26 05 33   Raceway and Boxes for Electrical Systems
26 05 36   Cable Trays for Electrical Systems
26 05 39   Underfloor Raceways for Electrical Systems
26 05 43   Underground Ducts and Raceways for Electrical Systems
26 05 46   Utility Poles for Electrical Systems
26 05 48   Vibration and Seismic Controls for Electrical Systems
26 05 53   Identification for Electrical Systems
26 05 73   Overcurrent Protective Device Coordination Study

26 06 00   Schedules for Electrical
26 06 10   Schedules for Medium-Voltage Electrical Distribution
26 06 20   Schedules for Low-Voltage Electrical Distribution
   26 06 20.13   Electrical Switchboard Schedule
   26 06 20.16   Electrical Panelboard Schedule
   26 06 20.19   Electrical Motor-Control Center Schedule
   26 06 20.23   Electrical Circuit Schedule
   26 06 20.26   Wiring Device Schedule
26 06 30   Schedules for Facility Electrical Power Generating and Storing Equipment
26 06 40   Schedules for Electrical and Cathodic Protection Systems
26 06 50   Schedules for Lighting
   26 06 50.13   Lighting Panelboard Schedule
   26 06 50.16   Lighting Fixture Schedule

26 08 00   Commissioning of Electrical Systems
26 09 00   Instrumentation and Control for Electrical Systems
26 09 13   Electrical Power Monitoring and Control
26 09 23   Lighting Control Devices
26 09 26   Lighting Control Panelboards
26 09 33   Central Dimming Controls

206A05.EPS

26 - 2

206A06.EPS

26 27 23	Indoor Service Poles	
26 27 26	Wiring Devices	
26 27 73	Door Chimes	
26 28 00	**Low-Voltage Circuit Protective Devices**	
26 28 13	Fuses	
26 28 16	Enclosed Switches and Circuit Breakers	
26 29 00	**Low-Voltage Controllers**	
26 29 13	Enclosed Controllers	
	26 29 13.13	Across-the-Line Motor Controllers
	26 29 13.16	Reduced-Voltage Motor Controllers
26 29 23	Variable-Frequency Motor Controllers	

26 30 00   FACILITY ELECTRICAL POWER GENERATING AND STORING EQUIPMENT

26 31 00	**Photovoltaic Collectors**	
26 32 00	**Packaged Generator Assemblies**	
26 32 13	Engine Generators	
	26 32 13.13	Diesel-Engine-Driven Generator Sets
	26 32 13.16	Gas-Engine-Driven Generator Sets
26 32 16	Steam-Turbine Generators	
26 32 19	Hydro-Turbine Generators	
26 32 23	Wind Energy Equipment	
26 32 26	Frequency Changers	
26 32 29	Rotary Converters	
26 32 33	Rotary Uninterruptible Power Units	
26 33 00	**Battery Equipment**	
26 33 13	Batteries	
26 33 16	Battery Racks	
26 33 19	Battery Units	
26 33 23	Central Battery Equipment	
26 33 33	Static Power Converters	
26 33 43	Battery Chargers	
26 33 46	Battery Monitoring	
26 33 53	Static Uninterruptible Power Supply	
26 35 00	**Power Filters and Conditioners**	
26 35 13	Capacitors	
26 35 16	Chokes and Inductors	
26 35 23	Electromagnetic-Interference Filters	
26 35 26	Harmonic Filters	
26 35 33	Power Factor Correction Equipment	
26 35 36	Slip Controllers	
26 35 43	Static-Frequency Converters	
26 35 46	Radio-Frequency-Interference Filters	
26 35 53	Voltage Regulators	
26 36 00	**Transfer Switches**	
26 36 13	Manual Transfer Switches	
26 36 23	Automatic Transfer Switches	

26 40 00   ELECTRICAL AND CATHODIC PROTECTION

26 41 00	**Facility Lightning Protection**	
26 41 13	Lightning Protection for Structures	

206A07.EPS

26 - 4

DIVISION 27 – COMMUNICATIONS

27 00 00   COMMUNICATIONS
  27 01 00   Operation and Maintenance of Communications Systems
    27 01 10   Operation and Maintenance of Structured Cabling and Enclosures
    27 01 20   Operation and Maintenance of Data Communications
    27 01 30   Operation and Maintenance of Voice Communications
    27 01 40   Operation and Maintenance of Audio-Video Communications
    27 01 50   Operation and Maintenance of Distributed Communications and Monitoring
  27 05 00   Common Work Results for Communications
    27 05 13   Communications Services
      27 05 13.13   Dialtone Services
      27 05 13.23   T1 Services
      27 05 13.33   DSL Services
      27 05 13.43   Cable Services
      27 05 13.53   Satellite Services
    27 05 26   Grounding and Bonding for Communications Systems
    27 05 28   Pathways for Communications Systems
      27 05 28.29   Hangers and Supports for Communications Systems
      27 05 28.33   Conduits and Backboxes for Communications Systems
      27 05 28.36   Cable Trays for Communications Systems
      27 05 28.39   Surface Raceways for Communications Systems
    27 05 43   Underground Ducts and Raceways for Communications Systems
    27 05 46   Utility Poles for Communications Systems
    27 05 48   Vibration and Seismic Controls for Communications Systems
    27 05 53   Identification for Communications Systems
  27 06 00   Schedules for Communications
    27 06 10   Schedules for Structured Cabling and Enclosures
    27 06 20   Schedules for Data Communications
    27 06 30   Schedules for Voice Communications
    27 06 40   Schedules for Audio-Video Communications
    27 06 50   Schedules for Distributed Communications and Monitoring
  27 08 00   Commissioning of Communications

27 10 00   STRUCTURED CABLING
  27 11 00   Communications Equipment Room Fittings
    27 11 13   Communications Entrance Protection
    27 11 16   Communications Cabinets, Racks, Frames and Enclosures
    27 11 19   Communications Termination Blocks and Patch Panels
    27 11 23   Communications Cable Management and Ladder Rack
    27 11 26   Communications Rack Mounted Power Protection and Power Strips
  27 13 00   Communications Backbone Cabling
    27 13 13   Communications Copper Backbone Cabling
      27 13 13.13   Communications Copper Cable Splicing and Terminations
    27 13 23   Communications Optical Fiber Backbone Cabling
      27 13 23.13   Communications Optical Fiber Splicing and Terminations
    27 13 33   Communications Coaxial Backbone Cabling
      27 13 33.13   Communications Coaxial Splicing and Terminations
    27 13 43   Communications Services Cabling

27 - 1

27 - 2

| 27 25 37 | Virtual Private Network Software |
| 27 25 39 | Internet Conferencing Software |

27 26 00 Data Communications Programming and Integration Services
27 26 13	Web Development
27 26 16	Database Development
27 26 19	Application Development
27 26 23	Network Integration Requirements
27 26 26	Data Communications Integration Requirements

27 30 00   VOICE COMMUNICATIONS

27 31 00 Voice Communications Switching and Routing Equipment
| 27 31 13 | PBX/ Key Systems |
| 27 31 23 | Internet Protocol Voice Switches |

27 32 00 Voice Communications Telephone Sets, Facsimiles and Modems
27 32 13	Telephone Sets
27 3216	Wireless Transceivers
27 32 23	Elevator Telephones
27 32 26	Ring-Down Emergency Telephones
27 32 29	Facsimiles and Modems
27 32 36	TTY Equipment

27 33 00 Voice Communications Messaging
27 33 16	Voice Mail and Auto Attendant
27 33 23	Interactive Voice Response
27 33 26	Facsimile Servers

27 34 00 Call Accounting
| 27 34 13 | Toll Fraud Equipment and Software |
| 27 34 16 | Telemanagement Software |

27 35 00 Call Management
27 35 13	Digital Voice Announcers
27 35 16	Automatic Call Distributors
27 35 19	Call Status and Management Displays
27 35 23	Dedicated 911 Systems

27 40 00   AUDIO-VIDEO COMMUNICATIONS

27 41 00 Audio-Video Systems
27 41 13	Architecturally Integrated Audio-Video Equipment
27 41 16	Integrated Audio-Video Systems and Equipment
27 41 16.25	Integrated Audio-Video Systems and Equipment for Restaurants and Bars
27 41 16.28	Integrated Audio-Video Systems and Equipment for Conference Rooms
27 41 16.29	Integrated Audio-Video Systems and Equipment for Board Rooms
27 41 16.51	Integrated Audio-Video Systems and Equipment for Classrooms
27 41 16.61	Integrated Audio-Video Systems and Equipment for Theaters
27 41 16.62	Integrated Audio-Video Systems and Equipment for Auditoriums
27 41 16.63	Integrated Audio-Video Systems and Equipment for Stadiums and Arenas
27 41 19	Portable Audio-Video Equipment
27 41 23	Audio-Video Accessories

27 42 00 Electronic Digital Systems
27 42 13	Point of Sale Systems
27 42 16	Transportation Information Display Systems
27 42 19	Public Information Systems

27 - 3

27 50 00   DISTRIBUTED COMMUNICATIONS AND MONITORING SYSTEMS

27 51 00   Distributed Audio-Video Communications Systems

27 51 13   Paging Systems

    27 51 13.13   Overhead Paging Systems

27 51 16   Public Address and Mass Notification Systems

27 51 19   Sound Masking Systems

27 51 23   Intercommunications and Program Systems

    27 51 23.20   Commercial Intercommunications and Program Systems

    27 51 23.30   Residential Intercommunications and Program Systems

    27 51 23.50   Educational Intercommunications and Program Systems

    27 51 23.63   Detention Intercommunications and Program Systems

    27 51 23.70   Healthcare Intercommunications and Program Systems

27 52 00   Healthcare Communications and Monitoring Systems

27 52 13   Patient Monitoring and Telemetry Systems

27 52 16   Telemedicine Systems

27 52 19   Healthcare Imaging Systems

27 52 23   Nurse Call/Code Blue Systems

27 53 00   Distributed Systems

27 53 13   Clock Systems

27 53 16   Infrared and Radio Frequency Tracking Systems

27 53 19   Internal Cellular, Paging, and Antenna Systems

27 60 00   Reserved

27 70 00   Reserved

27 80 00   Reserved

27 90 00   Reserved

27 - 4

DIVISION 28 – ELECTRONIC SAFETY AND SECURITY

28 00 00   ELECTRONIC SAFETY AND SECURITY
  28 01 00   Operation and Maintenance of Electronic Safety and Security
    28 01 10   Operation and Maintenance of Electronic Access Control and Intrusion Detection
      28 01 10.51   Maintenance and Administration of Electronic Access Control and Intrusion Detection
      28 01 10.71   Revisions and Upgrades of Electronic Access Control and Intrusion Detection
    28 01 20   Operation and Maintenance of Electronic Surveillance
    28 01 30   Operation and Maintenance of Electronic Detection and Alarm
      28 01 30.51   Maintenance and Administration of Electronic Detection and Alarm
      28 01 30.71   Revisions and Upgrades of Electronic Detection and Alarm
    28 01 40   Operation and Maintenance of Electronic Monitoring and Control
      28 01 40.51   Maintenance and Administration of Electronic Monitoring and Control
      28 01 40.71   Revisions and Upgrades of Electronic Monitoring and Control
  28 05 00   Common Work Results for Electronic Safety and Security
    28 05 13   Conductors and Cables for Electronic Safety and Security
      28 05 13.13   CCTV Communications Conductors and Cables
      28 05 13.16   Access Control Communications Conductors and Cables
      28 05 13.13   Intrusion Detection Communications Conductors and Cables
      28 05 13.13   Fire Alarm Communications Conductors and Cables
    28 05 26   Grounding and Bonding for Electronic Safety and Security
    28 05 28   Pathways for Electronic Safety and Security
      28 05 28.29   Hangers and Supports for Electronic Safety and Security
      28 05 28.33   Conduits and Backboxes for Electronic Safety and Security
      28 05 28.36   Cable Trays for Electronic Safety and Security
      28 05 28.39   Surface Raceways for Electronic Safety and Security
    28 05 48   Vibration and Seismic Controls for Electronic Safety and Security
    28 05 53   Identification for Electronic Safety and Security
  28 06 00   Schedules for Electronic Safety and Security
    28 06 10   Schedules for Electronic Access Control and Intrusion Detection
    28 06 20   Schedules for Electronic Surveillance
    28 06 30   Schedules for Electronic Detection and Alarm
    28 06 40   Schedules for Electronic Monitoring and Control
  28 08 00   Commissioning of Electronic Safety and Security

28 10 00   ELECTRONIC ACCESS CONTROL AND INTRUSION DETECTION
  28 13 00   Access Control
    28 13 13   Access Control Global Applications
    28 13 16   Access Control Systems and Database Management
    28 13 19   Access Control Systems Infrastructure
    28 13 26   Access Control Remote Devices
    28 13 33   Access Control Interfaces
      28 13 33.16   Access Control Interfaces to Access Control Hardware
      28 13 33.26   Access Control Interfaces to Intrusion Detection
      28 13 33.33   Access Control Interfaces to Video Surveillance
      28 13 33.36   Access Control Interfaces to Fire Alarm
    28 13 43   Access Control Identification Management Systems
    28 13 53   Security Access Detection

28 - 1

28 13 53.13	Security Access Metal Detectors
28 13 53.16	Security Access X-Ray Equipment
28 13 53.29	Security Access Sniffing Equipment
28 13 53.23	Security Access Explosive Detection Equipment

28 16 00　Intrusion Detection

28 16 13	Intrusion Detection Control, GUI, and Logic Systems
28 16 16	Intrusion Detection Systems Infrastructure
28 16 19	Intrusion Detection Remote Devices and Sensors
28 16 33	Intrusion Detection Interfaces
28 16 33.13	Intrusion Detection Interfaces to Remote Monitoring
28 16 33.16	Intrusion Detection Interfaces to Access Control Hardware
28 16 33.23	Intrusion Detection Interfaces to Access Control System
28 16 33.33	Intrusion Detection Interfaces to Video Surveillance
28 16 33.36	Intrusion Detection Interfaces to Fire Alarm
28 16 43	Perimeter Security Systems
28 16 46	Intrusion Detection Vehicle Control Systems

28 20 00　ELECTRONIC SURVEILLANCE

28 23 00　Video Surveillance

28 23 13	Video Surveillance Control and Management Systems
28 23 16	Video Surveillance Monitoring and Supervisory Interfaces
28 23 19	Digital Video Recorders and Analog Recording Devices
28 23 23	Video Surveillance Systems Infrastructure
28 23 26	Video Surveillance Remote Positioning Equipment
28 23 29	Video Surveillance Remote Devices and Sensors

28 26 00　Electronic Personal Protection Systems

28 26 13	Electronic Personal Safety Detection Systems
28 26 16	Electronic Personal Safety Alarm Annunciation and Control Systems
28 26 19	Electronic Personal Safety Interfaces to Remote Monitoring
28 26 23	Electronic Personal Safety Emergency Aid Devices

28 30 00　ELECTRONIC DETECTION AND ALARM

28 31 00　Fire Detection and Alarm

28 31 13	Fire Detection and Alarm Control, GUI, and Logic Systems
28 31 23	Fire Detection and Alarm Annunciation Panels and Fire Stations
28 31 33	Fire Detection and Alarm Interfaces
28 31 33.13	Fire Detection and Alarm Interfaces to Remote Monitoring
28 31 33.16	Fire Detection and Alarm Interfaces to Access Control Hardware
28 31 33.23	Fire Detection and Alarm Interfaces to Access Control System
28 31 33.26	Fire Detection and Alarm Interfaces to Intrusion Detection
28 31 33.33	Fire Detection and Alarm Interfaces to Video Surveillance
28 31 33.43	Fire Detection and Alarm Interfaces to Elevator Control
28 31 43	Fire Detection Sensors
28 31 46	Smoke Detection Sensors
28 31 49	Carbon-Monoxide Detection Sensors
28 31 53	Fire Alarm Initiating Devices
28 31 33.13	Fire Alarm Pull Stations
28 31 33.23	Fire Alarm Level Detectors Switches
28 31 33.33	Fire Alarm Flow Switches
28 31 33.43	Fire Alarm Pressure Sensors
28 31 63	Fire Alarm Integrated Audio Visual Evacuation Systems
28 31 63.13	Fire Alarm Horns and Strobes

28 32 00　Radiation Detection and Alarm

28 - 2

28 32 13	Radiation Detection and Alarm Control, GUI, and Logic Systems
28 32 23	Radiation Detection and Alarm Integrated Audio Evacuation Systems
23 32 33	Radiation and Alarm Detection Sensors
28 32 43	Radiation and Alarm Dosimeters

28 33 00   Fuel-Gas Detection and Alarm

28 33 13	Fuel-Gas Detection and Alarm Control, GUI, and Logic Systems
28 33 23	Fuel-Gas Detection and Alarm Integrated Audio Evacuation Systems
28 33 33	Fuel-Gas Detection Sensors

28 34 00   Fuel-Oil Detection and Alarm

28 34 13	Fuel-Oil Detection and Alarm Control, GUI, and Logic Systems
28 34 23	Fuel-Oil Detection and Alarm Integrated Audio Evacuation Systems
28 34 33	Fuel-Oil Detection Sensors

28 35 00   Refrigerant Detection and Alarm

28 35 13	Refrigerant Detection and Alarm Control, GUI, and Logic Systems
28 35 23	Refrigerant Detection and Alarm Integrated Audio Evacuation Systems
28 35 33	Refrigerant Detection Sensors

28 40 00   ELECTRONIC MONITORING AND CONTROL

28 46 00   Electronic Detention Monitoring and Control Systems

28 46 13	Hard-Wired Detention Monitoring and Control Systems
28 46 16	Relay-Logic Detention Monitoring and Control Systems
28 46 19	PLC Electronic Detention Monitoring and Control Systems
28 46 23	Computer-Based Detention Monitoring and Control Systems
28 46 26	Discreet-Logic Detention Monitoring and Control Systems
28 46 29	Discreet-Distributed Intelligence Detention Monitoring and Control Systems

28 50 00	*Reserved*
28 60 00	*Reserved*
28 70 00	*Reserved*
28 80 00	*Reserved*
28 90 00	*Reserved*

28 - 3

DIVISION 31 – EARTHWORK

31 - 1

Additional Resources

This module is intended to be a thorough resource for task training. The following reference works are suggested for further study. These are optional materials for continued education rather than for task training.

American Electrician's Handbook, 2002. Terrell Croft, Wilfred Summers. New York, NY: McGraw-Hill.

National Electrical Code® Handbook, Latest Edition. Quincy, MA: National Fire Protection Association.

Figure Credits

Carrier Corporation	206F36
John Traister	206F08–206F14, 206F16–206F18, 206F20, 206F26, 206F28–206F33, 206F35, 206F37
Ritterbush-Ellig-Hulsing, P.C.	206SA03
SoundCom Systems	206F34
Walter Johnson	205SA01
Construction Specifications Institute (CSI) and Construction Specifications Canada (CSC)*	206F45, Appendix B

* The Numbers and Titles used in this textbook are from the 2004 *MasterFormat*™, which is published by the Construction Specifications Institute (CSI) and Construction Specifications Canada (CSC), and used with permission from CSI, 2004. For those interested in a more in-depth explanation of the 2004 *MasterFormat*™ and its use in the construction industry, visit www.csinet.org/masterformat or contact:

The Construction Specifications Institute (CSI)
99 Canal Center Plaza, Suite 300
Alexandria, VA 22314
800-689-2900; 703-684-0300
CSINet URL: http://www.csinet.org

NCCER makes every effort to keep these textbooks up-to-date and free of technical errors. We appreciate your help in this process. If you have an idea for improving this textbook, or if you find an error, a typographical mistake, or an inaccuracy in NCCER's Contren® textbooks, please write us, using this form or a photocopy. Be sure to include the exact module number, page number, a detailed description, and the correction, if applicable. Your input will be brought to the attention of the Technical Review Committee. Thank you for your assistance.

Instructors – If you found that additional materials were necessary in order to teach this module effectively, please let us know so that we may include them in the Equipment/Materials list in the Annotated Instructor's Guide.

Write: Product Development and Revision
National Center for Construction Education and Research
P.O. Box 141104, Gainesville, FL 32614-1104

Fax: 352-334-0932

E-mail: curriculum@nccer.org

Craft _____ Module Name _____

Copyright Date _____ Module Number _____ Page Number(s) _____

Description _____

_____

_____

_____

(Optional) Correction _____

_____

_____

(Optional) Your Name and Address _____

_____

_____

Switching Devices and Timers

COURSE MAP

This course map shows all of the modules in the second level of the *Electronic Systems Technician* curriculum. The suggested training order begins at the bottom and proceeds up. Skill levels increase as you advance on the course map. The local Training Program Sponsor may adjust the training order.

ELECTRONIC SYSTEMS TECHNICIAN LEVEL TWO

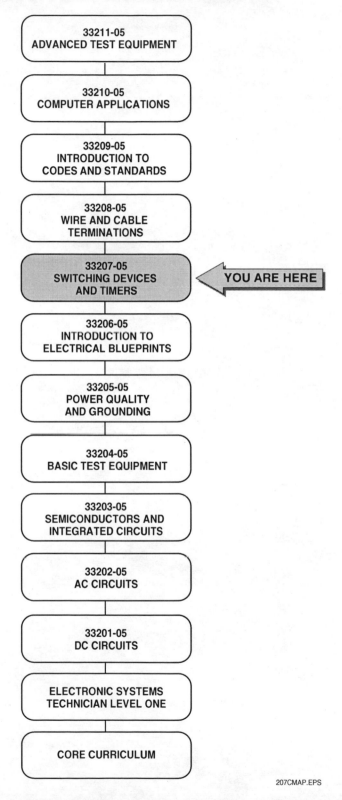

207CMAP.EPS

Figures

Table

Switching Devices and Timers

Objectives

When you have completed this module, you will be able to do the following:

1. Identify and describe the operation of commonly used types of switches.
2. Classify switches based on schematic diagram symbols, according to the number of poles and throws.
3. Identify and describe the operation of photoelectric devices.
4. Describe the operation of a solar cell.
5. Describe the uses and operation of relays.
6. Explain the differences between electromechanical and solid-state relays.
7. Identify and describe the operation of various types of timing devices.
8. Install and program a timer.

Prerequisites

Before you begin this module, it is recommended that you successfully complete *Core Curriculum; Electronic Systems Technician Level One*; and *Electronic Systems Technician Level Two*, Modules 33201-05 through 33206-05.

Required Trainee Materials

1. Pencil and paper
2. Appropriate personal protective equipment

1.0.0 ◆ INTRODUCTION

The switch is one of the most common electrical devices in commercial and industrial installations. Virtually every system and piece of electrical equipment has some type of switch. Photoelectric devices, such as motion sensors and photocells, are also becoming very popular.

There are many types and name brands of switches and photoelectric devices. You are encouraged to learn the principles of operation of the most popular types and apply this knowledge to other devices encountered in the field.

The aim of this module is to describe the characteristics and operation of the most popular types of switches and photoelectric devices so that proper equipment can be selected and installed in commercial and industrial applications.

A relay is an electrical or mechanical device used for remote or automatic control of other devices or systems.

There are many relay types, sizes, and applications. The aim of this module is to describe the basic operating characteristics and applications of some of the more common types of relays used in industry. This information helps the technician to make an educated choice when selecting and installing such relays.

2.0.0 ◆ SWITCHES

An electrical switch is a device that is manually or automatically operated to make or break an electrical **circuit** operating within the rated current and voltage of the switch. Standard electrical switches should not be used to interrupt a circuit under **short circuit** or **overload** conditions. Fuses or circuit breakers should be used to prevent such conditions.

 NOTE
When selecting and installing switches, make sure that each switch is rated for the voltage and current of its circuit.

Switches

It is common to think of switches in terms of manually operated devices used to turn lights on and off. In fact, most switches used in electronic systems are automatically operated by actuators that sense heat, light, pressure, motion, flow, liquid level, or other environmental conditions.

2.1.0 Switch Classifications

Switches can be classified in many different ways:

- According to the number of poles
  - Single-pole
  - Double-pole
  - Multiple-pole
- According to the number of positions (throws)
  - Single-throw
  - Double-throw
  - Multiple-throw
- According to the method of insulation (arc control)
  - Air
  - Oil
  - Gas
- According to the method of operation
  - Electrical
  - Mechanical
  - Motion
  - Light
  - Pneumatic
  - Hydraulic
  - Manual
  - Magnetic
- According to function
  - Breaking a circuit
  - Making a circuit

2.1.1 Switch Contacts

Switches contain contacts that are made of conducting material, such as copper. Current flow is established when the contacts touch each other. Current flow is shut off when the contacts are separated. As previously explained, there are many methods used to make and break electrical contact. A spring mechanism is often used to hold contacts securely in place when the switch is closed. Switches incorporating solid-state technology do not have moving parts, but this module focuses mainly on mechanical-type switches.

Two common types of contacts are knife-blade and butt contacts (*Figure 1*). Many times the contacts

are coated with a conductive material to prevent **oxidation** and to ensure a low-resistance contact point.

Switch contacts that are burnt or charred will need to be replaced to maintain a low-resistance contact point.

2.1.2 Pole of a Switch

The pole of a switch is where the movable contacts are attached. The poles, in conjunction with the contacts, are used to make and break the electrical circuit. The poles are electrically insulated and isolated from one another.

A single-pole switch will make and break only one conductor or leg of a circuit, whereas a two-pole switch will make and break two legs of a circuit, and so on (*Figure 2*). The dotted line between the double-pole and triple-pole contacts indicates that the poles operate at the same time. Switch poles are often abbreviated as shown in *Figure 2*.

BUTT CONTACTS

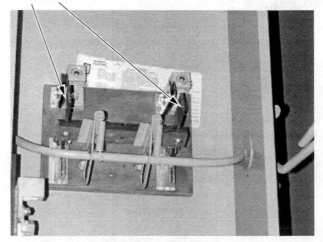

KNIFE BLADES

KNIFE-BLADE CONTACTS

207F01.EPS

Figure 1 ◆ Switch contacts.

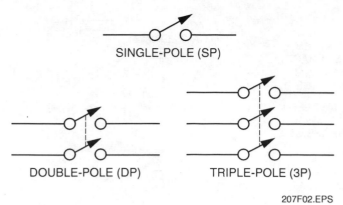

Figure 2 ◆ Poles of a switch.

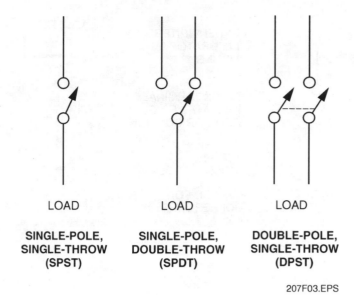

Figure 3 ◆ Single- and double-throw switches.

2.1.3 Closed Positions or Throws of a Switch

As shown in *Figure 3*, a single-throw switch will only make a closed circuit when the switch is in one position. A double-throw switch will make a closed circuit when placed in either of two positions. The poles and throws of a switch are often abbreviated as shown in *Figure 3*. When more than two closed positions are needed, special multiple throw switches should be employed.

2.1.4 Typical Switch Wiring

The following explanations use an electrical light switch and light bulb analogy to illustrate the wiring of various switch poles and throw configurations when applying a signal or power to a load:

- *Single-pole, single-throw switch* – Toggle switches or wall switches can be used to control lights and fractional horsepower motors. The most common switch is the single-pole, single-throw switch used to control lights or equipment from one location. This switch has two brass-colored screw terminals for wire connections (*Figure 4*).

- *Single-pole, double-throw switch* – The three-way switch is a single-pole, double-throw switch that is commonly used in pairs to control a light or piece of equipment from two locations. The common toggle type used in lighting has three screw terminals for wire connections: one black or copper, which is the common, or pole, and two brass or silver colored, which are the throws, as shown in *Figure 5*.

- *Double-pole, single-throw switch* – The double-pole, single-throw switch shown in *Figure 6* is normally used to control 240-volt equipment. This switch has four brass-colored screw terminals for wire connections. The terminals on the left side of the switch are connected through one set of contacts, and the terminals on the right side of the switch are connected through another set of contacts.

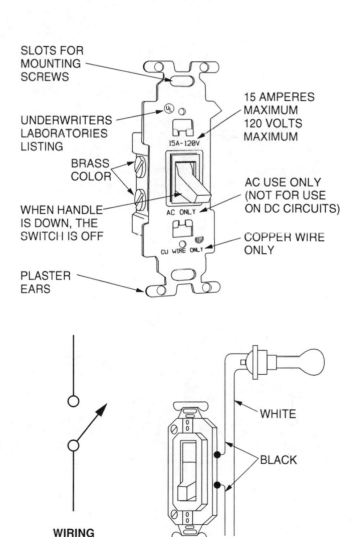

Figure 4 ◆ Single-pole, single-throw switch.

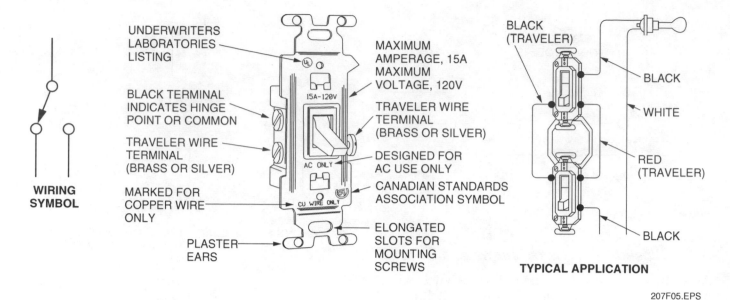

WIRING SYMBOL

UNDERWRITERS LABORATORIES LISTING

BLACK TERMINAL INDICATES HINGE POINT OR COMMON

TRAVELER WIRE TERMINAL (BRASS OR SILVER)

MARKED FOR COPPER WIRE ONLY

PLASTER EARS

MAXIMUM AMPERAGE, 15A MAXIMUM VOLTAGE, 120V

15A–120V

TRAVELER WIRE TERMINAL (BRASS OR SILVER)

AC ONLY

DESIGNED FOR AC USE ONLY

CANADIAN STANDARDS ASSOCIATION SYMBOL

CU WIRE ONLY

ELONGATED SLOTS FOR MOUNTING SCREWS

BLACK (TRAVELER)

BLACK

WHITE

RED (TRAVELER)

BLACK

TYPICAL APPLICATION

207F05.EPS

Figure 5 ◆ Three-way switch (single-pole, double-throw).

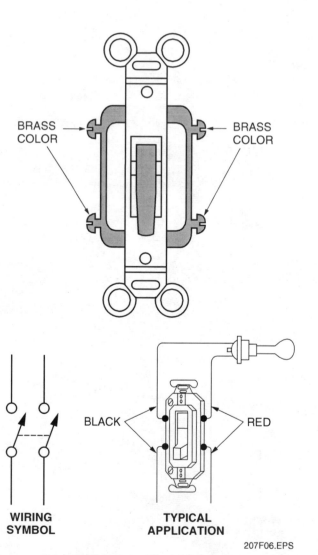

BRASS COLOR

BRASS COLOR

WIRING SYMBOL

BLACK

RED

TYPICAL APPLICATION

207F06.EPS

Figure 6 ◆ Double-pole, single-throw switch.

2.2.0 Switch Descriptions

Switches are available in a variety of types, styles, current/voltage ratings, and pole/contact configurations. Some of the more common types are covered in the following sections.

2.2.1 Panel-Mounted Switches

Figure 7 shows some of the most common types of switches. Except for rotary or joystick switches that can have many poles and multiple throws, most pushbutton, rocker, and snap-action switches have the same pole and throw designations as toggle switches. Switches are also available with center-off positions, momentary-contact positions, and make-before-break contacts.

2.2.2 Float-Level Switches

A float switch contains a floating device that is placed in contact with the fluid to be controlled. Float switches are often used to start and stop motor-driven pumps used to empty and fill tanks. Float switch contacts may be either normally open (N.O.) or normally closed (N.C.), and they must not be allowed to contact the process fluid.

Rod-operated float switches (*Figure 8*) are commonly used to open or close solenoid valves to control fluids.

HEAVY-DUTY
PUSHBUTTON

HEAVY-DUTY
ROTARY

ROCKER

TOGGLE

4-POSITION
JOYSTICK

LIGHTED
PUSHBUTTON

PUSHBUTTON
(USED WITH BOOT COVER)

207F07.EPS

Figure 7 ◆ Panel-mounted switches.

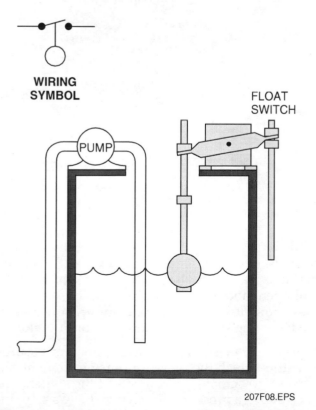

WIRING
SYMBOL

FLOAT
SWITCH

PUMP

207F08.EPS

Figure 8 ◆ Rod-operated float valve with wiring symbol.

Level switches (*Figure 9*) are hermetically sealed or isolated from the fluid being monitored and are designed to be mounted inside or through the exterior of fluid tanks. *Figure 10* shows an explosion-proof, exterior-mounted switch that uses pressure against a diaphragm to detect the level of liquids or dry bin materials in a storage vessel. Two or more switches must be positioned at various levels on a tank to detect high, low, and (if desired) one or more intermediate levels.

2.2.3 Pressure Switches

Pressure switches are used to control the pressure of liquids and gases by starting and stopping motors and compressors when the pressure in the system reaches a preset level. Air compressors, for

> **NOTE**
>
> When selecting pressure switches, pay particular attention to the pressure rating of the switch and the operating pressure of the system. Never exceed the switch ratings.

MAGNETIC FLOAT SWITCH (SEALED)

OPTICAL LIQUID LEVEL SWITCH (SEALED)

HORIZONTAL HERMETICALLY SEALED MAGNETIC SWITCH AND FLOAT

VERTICAL HERMETICALLY SEALED MAGNETIC SWITCH AND FLOAT

207F09.EPS

Figure 9 ◆ Isolated or hermetically sealed level switches.

207F10.EPS

Figure 10 ◆ Explosion-proof, pressure-activated liquid or dry material level switch.

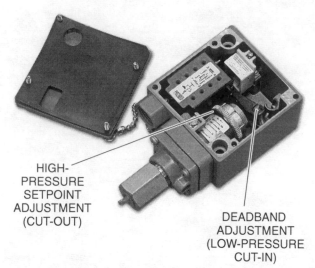

HIGH-PRESSURE SETPOINT ADJUSTMENT (CUT-OUT)

DEADBAND ADJUSTMENT (LOW-PRESSURE CUT-IN)

MACHINE PRESSURE SWITCH

HIGH-PRESSURE SETPOINT ADJUSTMENT (CUT-OUT)

DEADBAND ADJUSTMENT (LOW-PRESSURE CUT-IN)

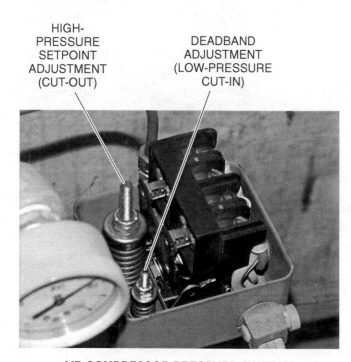

AIR COMPRESSOR PRESSURE SWITCH

207F11.EPS

Figure 11 ◆ Pressure switches.

example, are started directly or indirectly on a call for more air by a pressure switch (*Figure 11*).

Pressure switches use mechanical motion from pressure changes to operate one or more sets of contacts. A typical pressure switch may use a bellows, diaphragm, or **bourdon tube** as the pressure-sensing element. Most pressure switches have an adjustment screw to adjust the high-pressure setpoint, while other, more complex switches have an adjustable **deadband** range that allows the low-pressure setpoint to be adjusted, as shown in *Figures 11* and *12*.

Many pressure switches come with two sets of contacts: one N.O. set and one N.C. set. When the pressure reaches the setpoint, one set of contacts will open, and the other will close. In selecting a pressure switch for a particular application, each of the following should be taken into consideration:

- *Setpoint range* – This is the span of pressures within which the pressure-sensing element can be set to actuate the contacts of the switch. For example, a pressure switch may have a setpoint range of 20 to 100 psi.

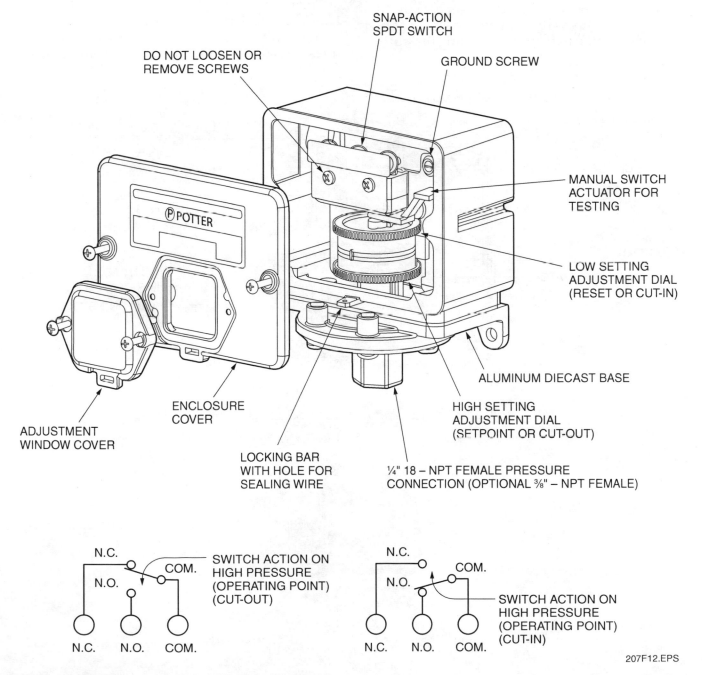

Figure 12 ◆ Pressure switch with calibrated adjustments.

207F12.EPS

- *Deadband adjustment* – This is the span of pressure between the setpoint limit (which actuates the contacts) and the **reset point** (which resets the contacts to their normal position). For example, a pressure switch may have a setpoint range between 20 and 100 psi with a deadband of only 5 to 15 psi between a high-pressure setpoint and a low-pressure setpoint (reset).

- *Rating of the switch and contacts* – The rating includes the type of switch used, such as single-pole, single-throw, or single-pole, double-throw. It also includes the amount of current and voltage that the contacts can switch safely.

- *Accuracy* – The accuracy of the switch refers to the ability of the switch to actuate repeatedly at the setpoint. This value is typically stated as a percentage of the maximum operating pressure of the switch.

Another application of a pressure switch is the detection of high and low liquid or gas flow in pressurized lines. This is accomplished with a differential pressure switch (*Figure 13*). As shown in *Figure 14*, the pressure drop across a calibrated orifice in the pressure line provides an indication of the flow rate. The high-pressure setpoint and deadband of the pressure switch are set so that an alarm sounds or the flow is shut down if the flow rises above (or falls below) the desired rate.

2.2.4 Limit Switches

Limit switches react to physical changes. There are many types of limit switches. One type contains a metal element that reacts mechanically to changes in heat. It will open a circuit if temperatures exceed preset limits. Another common type

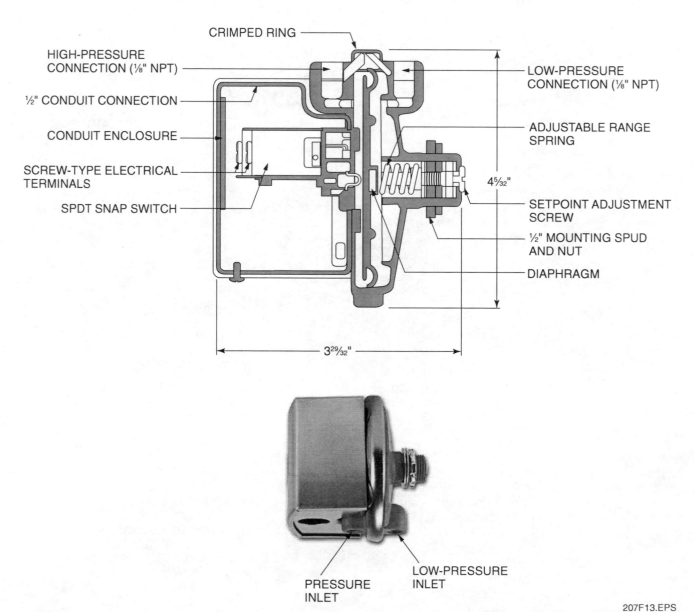

Figure 13 ◆ Differential pressure switch.

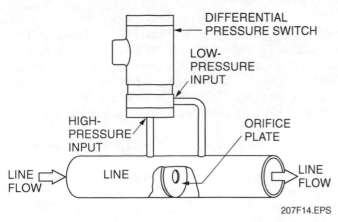

Figure 14 ◆ Differential pressure switch application.

of limit switch contains a mechanical actuator that activates a switch when it reaches a preset travel limit. Another type, similar to the limit switch that stops a garage door at either end of its travel, is actuated when its actuator is touched by a dog mounted on a chain or rod.

In general, the operation of a limit switch begins when the moving machine or part strikes an operating lever, which activates the limit switch. *Figure 15* shows a limit switch with its operating lever.

Limit switches can be used either as control devices for regular operation or as emergency

switches to prevent the improper functioning of equipment. The contacts may be normally open (N.O.) or normally closed (N.C.), and they may be momentary (spring-returned) or maintained-contact types.

The installation of limit switches involves selecting the best actuator, or operating lever, and then mounting the limit switch in the correct position in relationship to the moving part. It is important that the limit switch not be operated beyond the manufacturer's recommended specifications of travel. If overtravel of the moving part will affect the switch or cause the switch to operate beyond its travel range, an adequate mechanical stop, as shown in *Figure 16*, should be used to prevent damage to the switch.

Various types of snap-action switches (*Figure 17*) are used as limit switches. They are also used as equipment and instrument panel access-safety interlocks. These types of switches are characterized by very short travel-activation distances.

2.2.5 *Electronic Switches*

Silicon-controlled rectifiers (SCRs) are electronic devices that normally operate in the ON or OFF state very much like a switch does. SCRs are used extensively in motor controls and power supplies (*Figure 18*).

The basic purpose of the SCR is to switch on and off large amounts of power. It performs this function with no moving parts to wear out and no contacts that require replacing or maintaining.

The SCR can be used in AC and DC circuits, but it will only allow current to pass in the forward

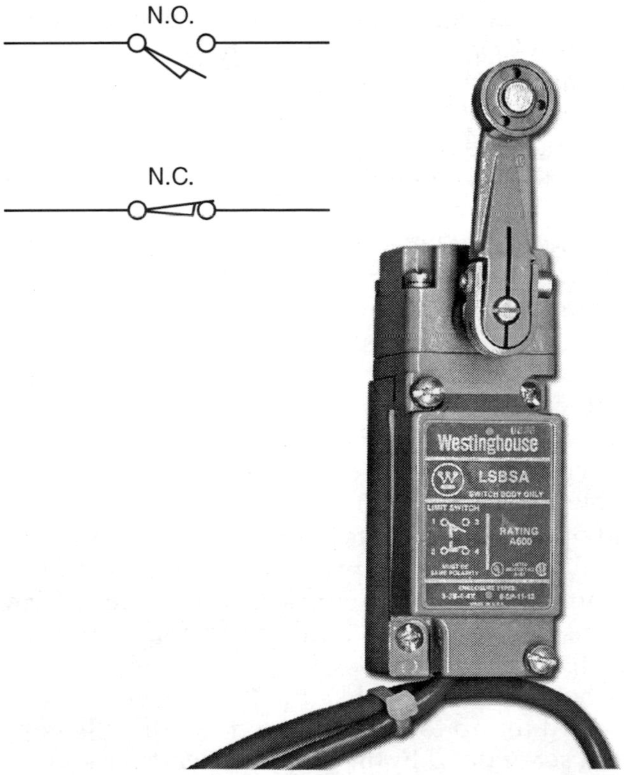

N.O.

N.C.

Figure 15 ◆ Machine limit switch.

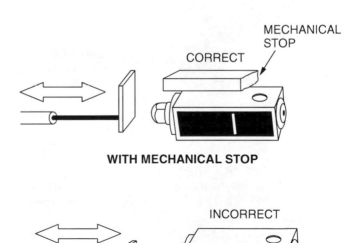

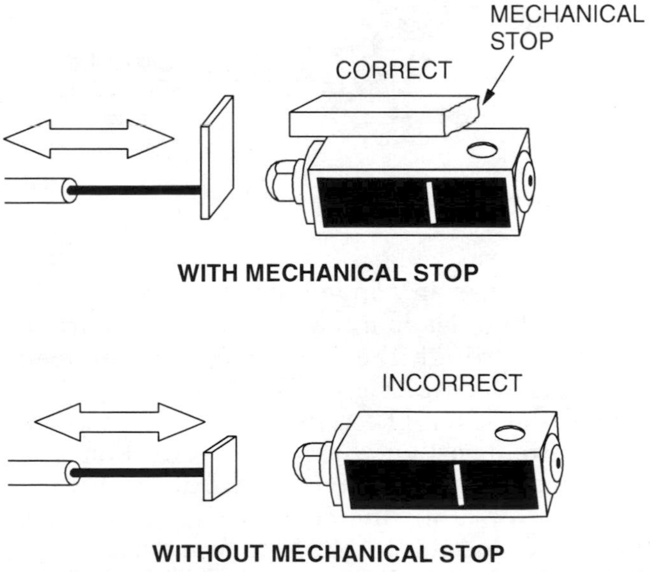

Figure 16 ◆ Limit-switch protection.

Figure 17 ◆ Snap-action switches used as limit switches and interlock switches.

207F17.EPS

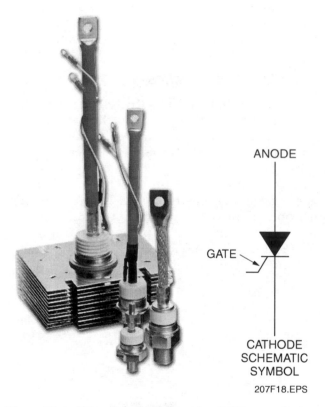

Figure 18 ◆ Silicon-controlled rectifier (SCR).

207F18.EPS

ANODE

GATE

CATHODE
SCHEMATIC
SYMBOL

direction. It has an **anode** (positive) lead, a **cathode** (negative) lead, and a gate lead. No current will flow through the SCR until the gate lead receives an electrical signal. Once the gate receives a signal, the SCR is like a closed switch, and it conducts current easily in the forward direction. The gate signal can now be removed, but conduction will not stop until the current tries to reverse direction or the current drops to near zero.

Once conduction has stopped, the SCR will be like an open switch, and no current will flow. A new gate signal must be applied before the SCR will conduct again. The SCR is similar to a **diode** in that it will only allow current to flow in the forward direction and never in the reverse direction (cathode to anode). A diode, however, is not a switch, has no gate, and acts as an electrical check valve that continually allows forward current flow only.

3.0.0 ◆ PHOTOELECTRIC DEVICES

Light-sensitive devices, sometimes called photoelectric transducers, alter their electric characteristics when exposed to visible or **infrared radiation (IR)** light. Light-sensitive devices include photocells, solar cells, motion detectors, and phototransistors.

Such light-sensitive devices can trigger many different kinds of circuits for the control of alarms, lights, motors, relays, and other actuators.

Some of the more common photoelectric devices and their principles of operation are discussed in this section.

3.1.0 Photocell Switches

Photocells are also called many other names, including photoconductive cells, light-dependent resistors (LDRs), and photoresistors. Basically, photocells are variable resistors in which the resistance values depend on the amount of light that falls on the device.

The resistance of the photoconductive material shown in *Figure 19* varies inversely with the amount of light that shines on it. In other words, as more light hits the photocell, the lower the resistance of the photoconductive material becomes. The photoconductive material is usually made of cadmium sulfide (CdS) or cadmium selenide (CdSe) and is placed directly between the two terminals of the photocell. In bright light, the resistance between the terminals can be as low as 50 ohms; in darkness, it can be as high as 5 megohms.

The most common use for photocells is the photocell switch used to control lights between dusk and dawn. Essentially, the photocell switch, shown in *Figure 20*, acts as a single-pole, single-throw (SPST) switch that closes when it gets dark and opens when it gets light. Unlike the common SPST toggle switch previously discussed, the photocell must receive power to operate, which means that it requires both conductors of the circuit (L1 and L2).

Photocell switches like the one in *Figure 20* can be rated for up to 300 watts and can directly control a set of flood lights rated 300 watts or less.

A photoelectric detector (also known as an electric eye) is an active infrared sensor that transmits

Photoelectric Detectors

Photoelectric detectors are used in some smoke alarms. In beam-type detectors, the presence of smoke blocks the light path, which results in an alarm. In the light-scattering type of smoke detector, smoke reflects light onto a photo sensor, which then activates the alarm.

207SA01.EPS

infrared light. This infrared light beam is received by a photocell. If an intruder interrupts the light beam, the detector is activated.

Photoelectric detectors are often used to cover long perimeter areas, but they can also be used to form a multi-beam wall for space detection (*Figure 21*).

TYPICAL WIRING

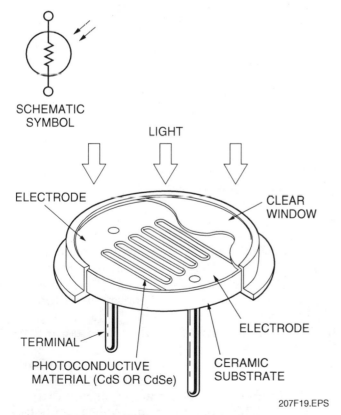

Figure 19 ◆ Cutaway view of a photocell or light-dependent resistor (LDR).

207F19.EPS

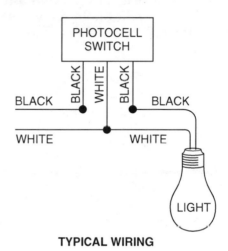

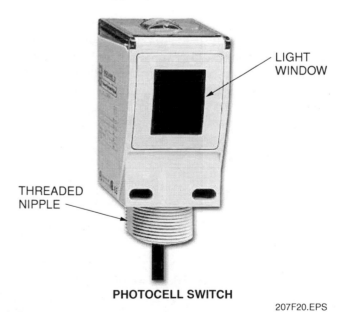

PHOTOCELL SWITCH

207F20.EPS

Figure 20 ◆ Photocell switch and typical application.

Photoelectric detectors are designed to detect an interruption of a concentrated beam of infrared light that is pulsed at varying frequencies. An intruder would have to duplicate the exact pulse rate of the light in order to defeat the system by projecting light on the photocell. This makes photoelectric detectors difficult to defeat, although concealing the transmitter and receiver also contributes greatly to a successful detection.

Typical photo sensors used for direct scanning are shown in *Figure 22*. Most of these sensors use IR only or visible red light beams to avoid interference from ambient area lighting.

Reflective scan is another popular scanning method used to detect moving objects. In reflective scan, the light source and the photoreceiver are mounted in the same housing and on the same side of the object to be detected (*Figure 23*). This method is used when there is limited space or when mounting restrictions prevent aiming the light beam directly at the photoreceiver. As shown in *Figure 23*, the reflected light is absorbed by the scanner/receiver.

3.2.0 Solar Cells

Solar cells, also called photovoltaic detectors, generate a DC voltage proportional to the amount of light that is absorbed by the cell.

NOTE

Keep sensor and reflector lenses clean to ensure proper operation of detectors. Avoid abrasive cleaners or cloth that may scratch the lenses.

A typical single cell (*Figure 24*) only generates about one-half volt DC. Therefore, cells are often arranged in series and parallel combinations to obtain the desired output voltage and current levels. Cells can be joined together to make a solar module, a solar panel, or a solar array. *Figure 24* shows the components of a photovoltaic array

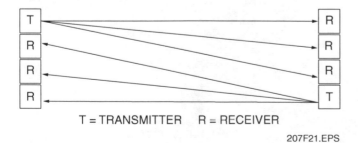

T = TRANSMITTER   R = RECEIVER

207F21.EPS

Figure 21 ◆ Multi-beam photoelectric wall.

IN-LINE UNITS

RIGHT-ANGLE UNITS

207F22.EPS

Figure 22 ◆ Photo sensors used for direct scanning.

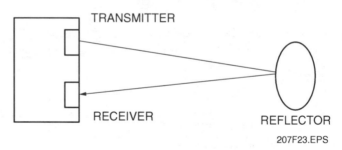

207F23.EPS

Figure 23 ◆ Bounce back active infrared detector.

system that could be used to charge a battery bank or power a piece of DC equipment.

3.3.0 Infrared Devices

Any object that is not at a temperature of **absolute zero** (–273°C or –460°F) will emit invisible electromagnetic energy in the form of IR. A detector can pick up this transmitted energy and produce an electrical signal proportional to the amount of IR detected. This signal can then be used to control other devices.

IR detectors are commonly used as motion/people detectors, TV remote control detectors, and IR scanners that detect hot spots in electrical equipment. The most common application is the motion/people detector.

SOLAR PANELS

MODULE

SINGLE CELL

207F24.EPS

Figure 24 ◆ Photovoltaic array components.

3.3.1 Motion Detectors

Motion detectors are typically used in security systems. The security system may be an elaborate zone-controlled system that incorporates motion detectors, door switches, smoke detectors, and moisture sensors, or it may be as simple as a set of motion detector floodlights mounted above a garage. The basic operation of the detector is the same in all cases.

Figure 25 shows a set of floodlights controlled by an IR motion detector switch. Many times, a photocell switch is added to the light controller, which automatically turns the lights off during daylight hours. Also, the light controller usually comes equipped with a manual override method that can be used to turn the lights on at any time.

207F25.EPS

Figure 25 ◆ Motion detector floodlights.

When the light controller contains both the photocell switch and an IR detector switch, it operates as if there were three SPST (single-pole, single-throw) switches involved (*Figure 26*). The photocell switch and the motion detector switch are in series, so both must be closed (there must be motion, and it must be dark) in order for the light to come on. The third switch is the manual bypass switch, which is in parallel with the series motion detector and photocell switches.

3.4.0 Fiber-Optic Switching Devices

Fiber optics transmit light and can be used in all types of photoelectric devices. In one application, fiber-optic sensors use an emitter to send the light, a receiver to receive the light sent by the emitter, and a flexible cable packed with tiny fibers to transmit the light. There may be separate fiber cables for the emitter and the receiver, depending

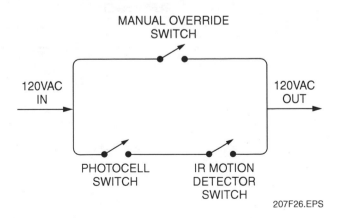

207F26.EPS

Figure 26 ◆ Single-line diagram of a motion detector with a photocell light controller.

on the sensor, or there may be a single cable with two internal fiber cables. When a single cable is used, the emitter and receiver use various methods to distribute emitter and receiver fibers within a cable. Glass fibers are used when the emitter source is infrared light, while plastic fibers are used when the emitter source is visible light.

Figure 27 illustrates three different methods for using fiber optics in photoelectric sensing. In *Figure 27A*, the method illustrated is referred to as thru-beam, in which light is emitted and received with individual cables. The target either breaks this light transmission or allows it to pass from emitter to receiver. *Figure 27B* and *Figure 27C* show the retro-reflective and the diffuse methods in which the light is emitted and received with the same cable. In the retro-reflective method, a reflective material is installed to bounce the emitted light back to the receiver, as was the case in the infrared installations previously discussed. The diffuse method uses the target to diffuse or reflect the emitted light from the cable.

4.0.0 ◆ PROXIMITY SENSORS

Proximity sensors are used to detect the presence or absence of an object without ever touching it. They are becoming increasingly popular for industrial applications because they are versatile, safe, reliable, and can be used in place of mechanical limit switches. A typical proximity sensor is shown in *Figure 28*.

The most popular type of proximity sensor is the line-powered inductive type (*Figure 29*), which operates on the **eddy current** killed **oscillator** (ECKO) principle. The oscillator consists of an inductor/capacitor (LC) tuned tank circuit which generates an AC magnetic flux as it oscillates and emits this field from the face of the switch. When a metal object is placed near the magnetic lines of flux, eddy currents are induced in the metal object, which in turn disrupt the tuning of the LC tank circuit. The oscillations die out or are killed. When the oscillations die, the integrator signals the trigger to switch the output of the proximity sensor.

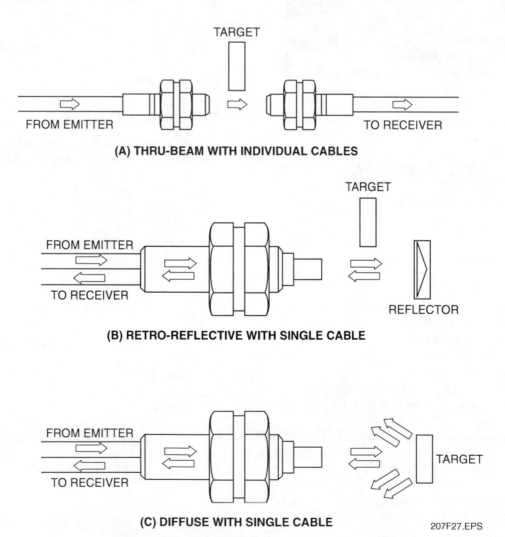

TARGET

FROM EMITTER   TO RECEIVER

(A) THRU-BEAM WITH INDIVIDUAL CABLES

TARGET

FROM EMITTER

TO RECEIVER

REFLECTOR

(B) RETRO-REFLECTIVE WITH SINGLE CABLE

FROM EMITTER

TO RECEIVER

TARGET

(C) DIFFUSE WITH SINGLE CABLE

207F27.EPS

Figure 27 ◆ Fiber-optic applications.

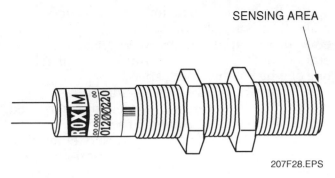

SENSING AREA

207F28.EPS

Figure 28 ◆ Proximity sensor.

5.0.0 ◆ ELECTRICAL RELAYS

Most electrical relays can be classified as either electromechanical or solid state. Electromechanical relays (EMRs) consist of devices with contacts that are closed by some type of magnetic effect. **Solid-state relays (SSRs),** by contrast, have no mechanical contacts and switch entirely by electronic means. A third category that is sometimes used is a type of solid-state relay called the hybrid relay. Hybrid relays are a combination of electromechanical and solid-state technology. A major advantage of a relay is that it is able to act as a voltage amplifier. That is, the relay can be operated by a low voltage in order to control a larger voltage.

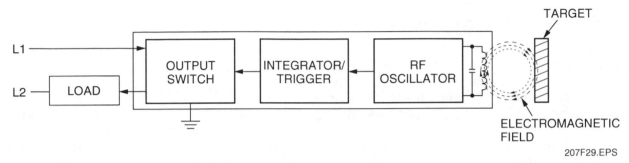

207F29.EPS

Figure 29 ◆ Line-powered inductive proximity sensor.

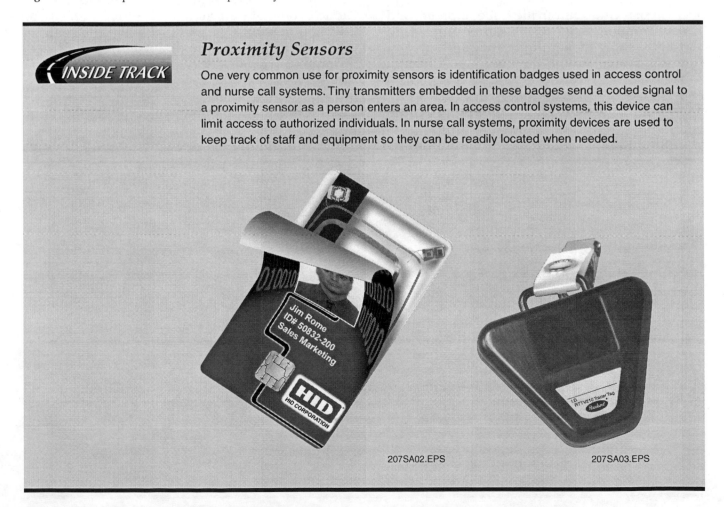

Proximity Sensors

One very common use for proximity sensors is identification badges used in access control and nurse call systems. Tiny transmitters embedded in these badges send a coded signal to a proximity sensor as a person enters an area. In access control systems, this device can limit access to authorized individuals. In nurse call systems, proximity devices are used to keep track of staff and equipment so they can be readily located when needed.

207SA02.EPS   207SA03.EPS

5.1.0 Electromechanical Relays

Electromechanical relays (EMRs) generally can be subdivided into three classifications: reed relays, control relays, and general-purpose relays. The major differences between these three types of relays are their costs, life expectancies, and intended uses in circuits.

Reed relays are used in applications requiring low and stable contact resistance, low capacitance, high insulation resistance, long life, and small size. These include automatic test equipment and instrumentation. Reed relays can be fitted with coaxial shielding for high-frequency applications. They are available with the very high isolation voltage ratings typically required for medical applications. Also, their low cost and versatility make them suitable for many security and general-purpose applications. Reed relays offer several advantages over electromechanical relays, one of which is switching speed.

Reed switches are often misidentified as reed relays. The difference between the two reed devices is discussed in the next section.

Control relays are used in a variety of applications covering all aspects of electrical and electronic systems. *Figure 30* shows the structure of a typical electromagnetic control relay.

Plug-in relays are available with relatively low coil voltage ratings such as 24VAC and 120VAC. DC coil voltage relays are also available. The greatest benefit of this type of relay is that all the wiring stays in place with the socket if the relay requires replacement. It is relatively inexpensive and is often referred to as a throw-away relay.

5.1.1 Reed Relays and Switches

Reed switches are magnetically operated components with contacts hermetically sealed in a glass capsule. Positioning a permanent magnet next to the switch or placing the switch in or near a DC electromagnet causes the contact reeds to flex and touch, completing a circuit. Protective inert gas, or a vacuum within the capsule, keeps the contacts clean and protected for the life of the device. These very small, inexpensive devices provide millions of switching cycles in normal circuits.

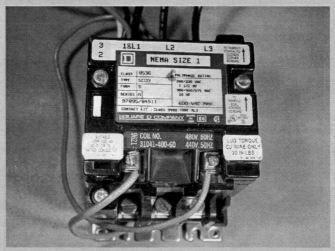

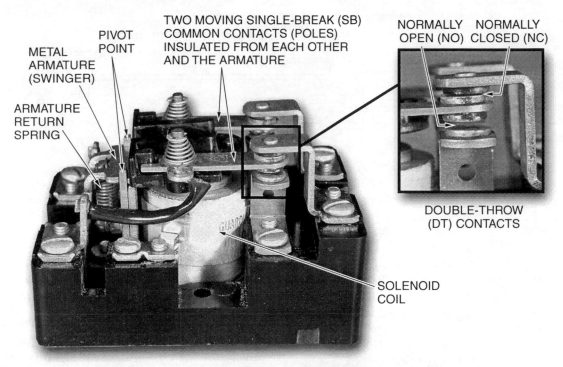

METAL ARMATURE (SWINGER)

PIVOT POINT

TWO MOVING SINGLE-BREAK (SB) COMMON CONTACTS (POLES) INSULATED FROM EACH OTHER AND THE ARMATURE

NORMALLY OPEN (NO)   NORMALLY CLOSED (NC)

ARMATURE RETURN SPRING

DOUBLE-THROW (DT) CONTACTS

SOLENOID COIL

(A) DOUBLE-POLE (DP), DOUBLE-THROW (DT), OPEN-FRAME CONTROL RELAY

DOUBLE-THROW (DT) CONTACTS

FOUR POLES

(B) FOUR-POLE (4P), DOUBLE-THROW (DT), GENERAL-PURPOSE RELAY

207F30.EPS

Figure 30 ◆ Electromechanical relays.

Applications include alarms, relays, and proximity sensors. Any time an electric current must be turned on or off without mechanically touching the switch, the reed switch is a good choice. Reed switches are often referred to as proximity switches because the switch is not required to actually contact the magnetic device or to be a component of it in order to operate.

Figure 31 illustrates a reed switch with a single set of normally open contacts. Reed switches are available with combinations of normally open (N.O.) and normally closed (N.C.) sets of contacts.

Reed relays are similar in design to reed switches with the exception that they contain their own coil or electromagnet. Like all electromechanical relays, they require an electrical signal to activate the coil. Reed relays extend the contact terminals and the coil terminals to the outside of the sealed unit for connections.

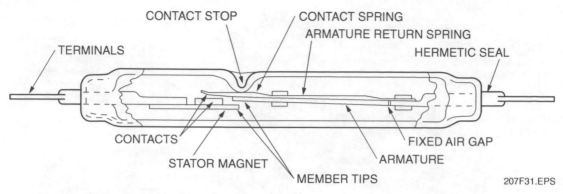

CONTACT STOP **CONTACT SPRING**
ARMATURE RETURN SPRING
TERMINALS **HERMETIC SEAL**

CONTACTS **FIXED AIR GAP**
STATOR MAGNET **ARMATURE**
MEMBER TIPS 207F31.EPS

Figure 31 ◆ Reed switch.

Figure 32 shows a typical reed relay and a comparable sealed, electromechanical, microminiature relay with terminal schematics for each. An M.S. on a schematic denotes a magnetic shield, while an E.S.S. denotes an electrostatic shield. As with the reed switches, reed and microminiature relays are also available with combinations of normally open (N.O.) and normally closed (N.C.) sets of contacts. Reed relays are typically rated by the following factors:

• Minimum coil operating voltage

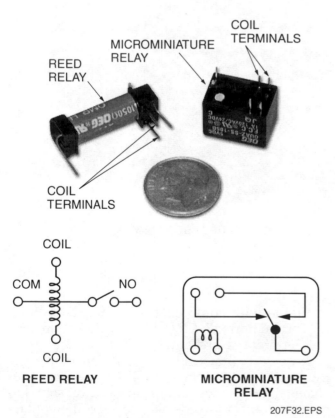

REED RELAY   **MICROMINIATURE RELAY**

207F32.EPS

Figure 32 ◆ Reed relay and microminiature relay with terminal schematics.

• Release voltage, which is the level of voltage at which the coil drops out and returns to its rest state
• Contact voltage and amperage ratings

Reed relay applications in the instrumentation industry include control of valves and solenoid operation, where contact arcing must be isolated for hazardous locations. Reed relays are also used in applications in which repetitive and rapid switching is required.

5.1.2 General-Purpose Relays

Figure 33 illustrates several different styles of general-purpose relays. These relays are designed for commercial and industrial application where economy and fast replacement are high priorities. Most general-purpose relays have a plug-in design that makes for quick replacement and simple troubleshooting.

Refer to *Figure 34*. A general-purpose relay consists of an electromagnet or coil, contact springs, a contact **armature**, and the mounting. When the coil is energized, the flow of current through the coil creates a magnetic field, which pulls the armature downward to close or open a set of contacts.

The general-purpose relay is available for both AC and DC operations. These relays are available with coils that can open or close the contacts by applying voltages from millivolts to several hundred volts. Relays with coil voltages of 6V, 12V, 24V, 48V, 115V, and 230V designs are the most common. Today, manufacturers offer a number of general-purpose relays that require as little as 4 milliamperes at 5VDC or 22 milliamperes at 12VDC for use in controlling solid-state circuits.

When the solenoid coil is specified for use with an AC control signal, a shading coil (aluminum or copper) at the top of the core (*Figure 35*) is used to create a weak, out-of-phase, auxiliary magnetic field. As the main field collapses when the AC current

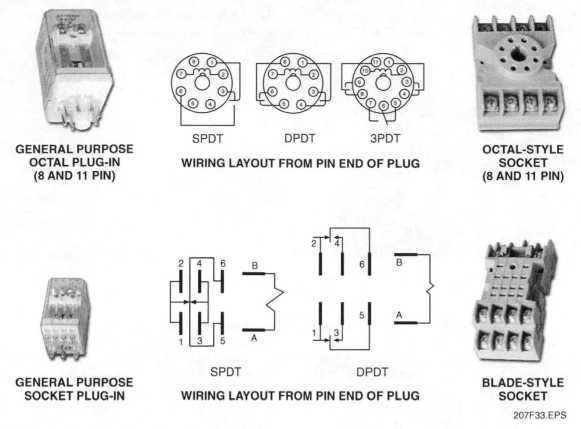

Figure 33 ◆ Examples of general-purpose relays.

periodically drops to zero, the weak field generated by the shading coil is strong enough to keep the armature in contact with the core and prevent the relay from chattering at a 120Hz rate. If the shading coil is loose or missing, the relay will produce excessive noise and be subject to abnormal wear and coil heat buildup. Light-duty relays use a flexible, thin-gauge copper spring stock for the moving contact arm. The armature is designed to swing over a slightly greater distance than necessary in both directions. This overswing, known as armature over-travel, causes the flexible moving arm to bend slightly in the normally open and closed positions. This eliminates any contact-pressure mating or bounce problems. In control relays that use an inflexible moving contact arm (*Figure 35*), a contact-pressure spring is used to allow flexible positioning of the moving contact arm to eliminate contact-pressure mating and bounce problems. As in light-duty relays, this is accomplished through armature over-travel. A degree of contact cleaning is also accomplished by armature over-travel. As the contacts mate, the over-travel causes the contact surfaces to wipe slightly. This action removes oxides from the mating surfaces of the contacts and helps keep contact resistance low.

Like all relays, general-purpose relays are rated and selected according to their maximum voltage rating, contact current rating, and the coil voltage rating. The maximum voltage rating is the maximum voltage that the relay can handle safely. The insulation, contacts, and other design elements of the relay are rated according to a safe maximum voltage. This is the maximum voltage rating of the

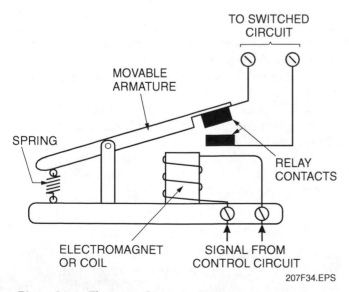

Figure 34 ◆ Electromechanical relay operation.

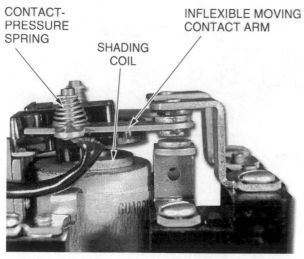

CONTACT-PRESSURE SPRING

SHADING COIL

INFLEXIBLE MOVING CONTACT ARM

207F35.EPS

Figure 35 ◆ Control relay.

Table 1 Relay and Contact Pole Designations

Designation	Meaning
ST	Single-Throw
DT	Double-Throw
N.O.	Normally Open
N.C.	Normally Closed
SB	Single-Break
DB	Double-Break (industrial relays)
SP	Single-Pole
DP	Double-Pole
3P	Three-Pole
4P	Four-Pole
5P	Five-Pole
6P	Six-Pole
(N)P	N = numeric number of poles

relay. The contact current rating represents the maximum current that the contacts may handle without severely arcing, burning, melting, or otherwise damaging the contact surfaces. The coil voltage represents the nominal voltage level needed to operate the coil properly. In latching relays that are constructed with two individual coils, the latching voltage represents the minimum momentary voltage required to operate the relay, while the release voltage is the minimum momentary voltage required to return the relay to its rest state. These are usually low DC voltage levels and are found in low-voltage electronic circuits.

Relays are available in a wide range of contact configurations. Terms typically associated with relay contacts include poles, throws, and breaks. *Figure 36* and *Table 1* help define these terms. Pole describes the number of completely isolated circuits that can pass through the switch at one time. Isolated means that maximum rated voltage of the same polarity may be applied to each pole of a given switch without danger of shorting between poles or

contacts. A double-pole switch can carry current through two circuits simultaneously, with each circuit isolated from the other. With double-pole switches, the two circuits are mechanically connected so that they open or close at the same time, while still being electrically insulated from each other. This mechanical connection is represented in the symbol by a dashed line connecting the poles.

Throws are the number of different closed contact positions per pole that are available on the switch. In other words, throw denotes the total number of different circuits that each individual pole is capable of controlling. The number of throws is independent of the number of poles. It is possible to have a single-throw switch with one, two, or more poles, as shown in *Figure 36*.

Breaks are the number of separate contacts the switch uses to open or close each individual circuit. If the switch breaks the electrical circuit in one place, then it is a single-break switch. If the switch breaks the electrical circuit in two places, then it is a double-break switch.

To simplify the listing of contact and switching arrangements, general-purpose relays carry **NARM** (National Association of Relay Manufacturers) code numbers. The numerals are used as abbreviations of the switching arrangements and are listed in *Table 2*. *Figure 37* illustrates the basic contact forms that are available. These forms are also designated by NARM.

5.1.3 Relay Holding Circuits

Some relays are energized by the momentary closure of a switch, such as a pushbutton switch or a cam-operated switch. A holding, or latching, circuit is needed in order to keep the relay energized

Low-Voltage Relays

Low-voltage systems controller relays are often used to provide one-way communications between a systems controller and low-voltage interface controls provided with devices such as screens, drapes, shades, door locks, and gates. Low-voltage control system relays can also be used to drive high-voltage relays used in lighting and HVAC systems.

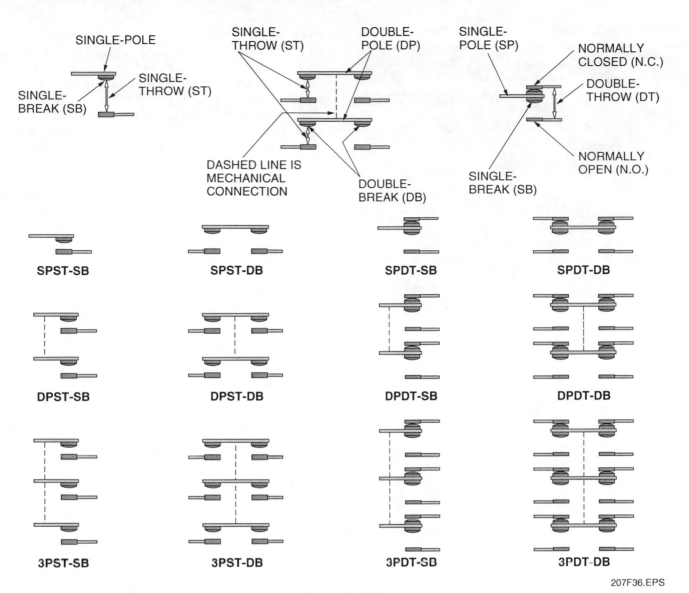

SINGLE-POLE

SINGLE-
BREAK (SB)

SINGLE-
THROW (ST)

SINGLE-
THROW (ST)

DOUBLE-
POLE (DP)

DASHED LINE IS
MECHANICAL
CONNECTION

DOUBLE-
BREAK (DB)

SINGLE-
POLE (SP)

NORMALLY
CLOSED (N.C.)

DOUBLE-
THROW (DT)

SINGLE-
BREAK (SB)

NORMALLY
OPEN (N.O.)

SPST-SB   SPST-DB   SPDT-SB   SPDT-DB

DPST-SB   DPST-DB   DPDT-SB   DPDT-DB

3PST-SB   3PST-DB   3PDT-SB   3PDT-DB

207F36.EPS

Figure 36 ◆ Relay contact arrangements.

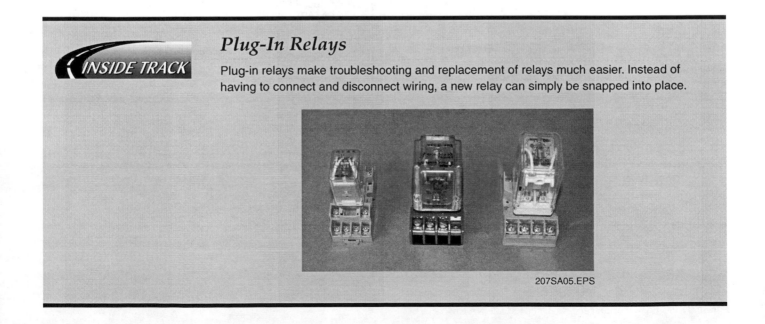

Plug-In Relays

INSIDE TRACK

Plug-in relays make troubleshooting and replacement of relays much easier. Instead of having to connect and disconnect wiring, a new relay can simply be snapped into place.

207SA05.EPS

Table 2 Relay Switching Arrangment Code Number

Contact Code and NARM Designator	
1-SPST-NO	15-4PST-NO 16-4PST-NC
2-SPST-NC	17-4PDT
3-SPST-NO-DM	18-5PST-NO
4-SPST-NC-DB	19-5PST-NC
5-SPDT	20-5PDT
6-SPDT-DB	21-6PST-NO
7-DPST-NO	22-6PST-NC
8-DPST-NC	23-6PDT
9-DPST-NO-DB	24-7PST-NO
10-DPST-NC-DB	25-7PST-NC
11-DPDT	26-7PDT
12-3PST-NO	27-8PST-NO
13-3PST-NC	28-8PST-NC
14-3PDT	29-8PDT

once the momentary contact is broken. This is often accomplished by dedicating one set of the relay's contacts to act as holding contacts. *Figure 38* shows an example. The coil of relay R-1 is energized when the N.O. pushbutton is pressed. At that time, the two sets of normally open relay contacts close. Contacts R1-2 light the lamp when they close. Contacts R1-1 act as holding contacts. Because these contacts are connected to the power source, they provide an energizing path to the relay coil. These contacts keep the coil energized until the supply voltage to the coil is interrupted by momentarily depressing the N.C. pushbutton.

Mechanically held relays and **contactors** are also used in some applications. The action is accomplished through the use of two coils and a latching mechanism. Energizing one coil (called the latch coil) through a momentary signal causes

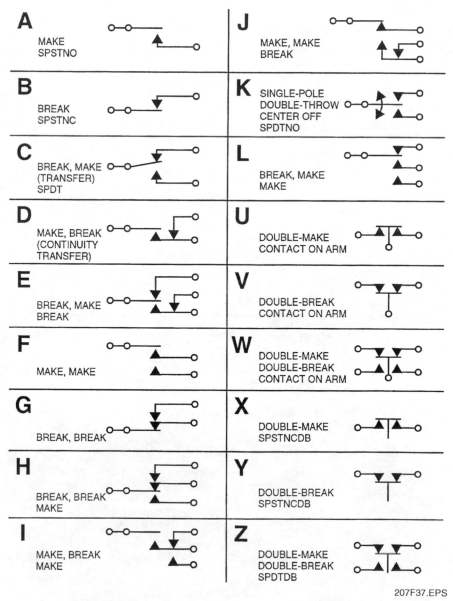

Figure 37 ◆ Basic contact forms accepted by NARM.

207F37.EPS

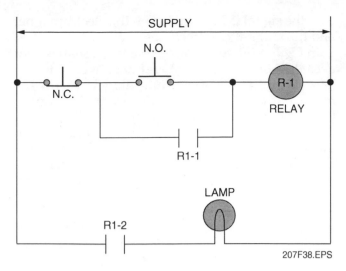

Figure 38 ◆ Relay holding circuit.

Relay Operations

Binding at the hinges or excessive armature spring force may cause a relay that normally has a snap action to make and break sluggishly. This condition is often encountered in relays that are rarely operated. Relays should be operated and observed from time to time to make sure their parts are working freely with proper clearances and spring actions.

the contacts to switch, and a mechanical latch holds the contacts in this position even though the initiating signal is removed. The coil is then de-energized. To restore the contacts to their initial position, a second coil (called the unlatch coil) is momentarily energized.

Mechanically held relays and contactors are used where the hum of an electrically held device would be objectionable, such as in auditoriums, hospitals, and churches.

5.1.4 Magnetic Relay Testing

The relay case generally contains a schematic of the relay showing the pins assigned to the coil and each set of contacts. This schematic shows which contacts are normally open and which are normally closed. There are two aspects to relay testing. One is the test to identify pins when they cannot be read from the relay itself. The other type of test is a verification that the relay is working properly.

The first step in identifying relay contacts is to locate the coil terminals. This is done by placing ohmmeter leads on two terminals until a resistance reading is obtained. This reading is typically in the 50- to 120-ohm range. A normally open contact pair will read infinite resistance, indicating an open circuit. A normally closed contact pair will read zero or very small resistance.

If a relay is suspected of being defective, check the contacts by applying an energizing voltage to the relay coil and then using a test light to check the contacts. You should hear an audible click when the contacts transfer. If the contacts are normally open, the light should go on when the relay is energized and be off when it is de-energized. The opposite is true for normally closed contacts.

5.2.0 Solid-State Relays

Solid-state relays (SSRs) are becoming more popular because, in addition to their size advantage, they offer long operating life and no contact bounce. They offer higher reliability than EMRs because there are no moving parts. Current leakage and the inability to carry higher loads are some disadvantages associated with SSRs. The devices are also sensitive to over-voltage spikes. SSRs are used in circuit designs that aren't sensitive to minor leakage. In addition to instrumentation applications, SSRs are used in automatic test equipment, data acquisition systems, mobile communications, and security systems, which require high reliability and long life.

The SSR business is growing at the expense of mechanical relays, particularly for applications that need millions of cycles of repetitive operation. SSRs are a good choice for customers seeking to control relays with less power. *Figure 39* shows an example of one type of SSR.

5.2.1 Comparison of Electromechanical and Solid-State Relays

Both EMRs and SSRs are designed to provide a common switching function, but they accomplish

207F39.EPS

Figure 39 ◆ Solid-state relay.

the end result in very different ways. As previously noted, EMRs provide electrical switching by using electromagnetic energy to move a set or sets of contacts, creating or breaking the path for the electrical circuit or circuits passing through them. The SSR, on the other hand, depends on electronic devices such as silicon-controlled rectifiers (SCRs) and triacs to create switching by triggering solid-state, semiconductor devices to pass or block the flow of electrical current without using mechanical contacts. Simply put, the EMR provides or breaks a mechanical path for the flow of current, while the SSR allows or prevents current flow through it based on an electrical trigger or exciter that it receives. *Figure 40* shows a basic schematic for both an EMR and an SSR.

In *Figure 40A*, the electromagnet is controlled by the control signal applied to it. When this signal is at the appropriate level to cause the electromagnet to operate, the magnetic field created by the electromagnet will attract the opposite pole of the armature, pulling it down until it rests against the core of the electromagnet. This movement causes the movable contacts to mate with the stationary contacts. In this example, two breaks are used to make the circuit. The function of the spring is to cause the contacts to separate whenever the signal is removed from the electromagnet and the magnetic field ceases to exist.

In *Figure 40B*, the control voltage (signal) is connected to the positive (+) and negative (–) terminals. The **LED** (light-emitting diode) accepts the relay control voltage and converts this energy to light energy. This light is collected by a photo detector that controls the gate firing circuit. When the control voltage or signal specified for this relay is reached, the gate circuit fires and in turn triggers the triac, allowing current to flow through the triac. Removal or reduction of control voltage reduces light output by the LED and stops the triggering of the gate firing circuit. This particular SSR requires a DC signal or control voltage that usually falls in the range of 3 to 32VDC.

Some disadvantages of using SSRs instead of EMRs were discussed earlier. They include the tendency of an SSR to leak current, a limited ability to carry heavy loads, sensitivity to over-voltage spikes, and higher costs compared to electromagnetic relays. However, the advantages of using SSRs instead of EMRs exceed the disadvantages:

- *Improved reliability* – The usual failure modes of conventional relays are contacts welding together or fatigue failure of the contact supports or

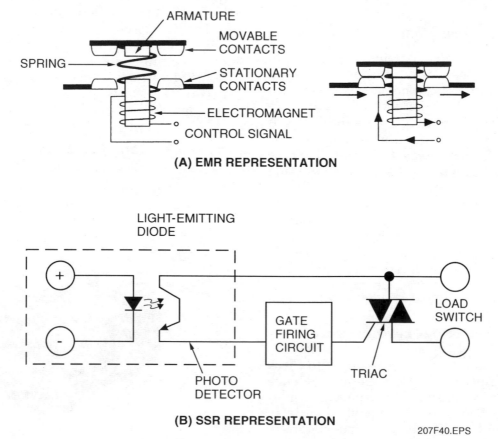

Figure 40 ◆ Comparison between an EMR and an SSR.

207F40.EPS

springs. Because the SSR has no moving parts, these types of failures are eliminated.

- *Noiseless operation* – With no moving parts, operation is noiseless.

- *Voltage drop constant with time* – The voltage drop across the closed contacts of a conventional relay will vary depending on contact force, corrosion, or number of cycles. The SSR does not suffer from these problems.

- *Low input requirements* – The SSR requires much less power to turn on than a conventional relay of similar current-carrying capability.

- *Self-protection* – The thermal shutdown designed into the SSR provides internal protection against overload. Short-circuited loads may result in contact welding in a conventional relay, while the SSR simply turns itself off.

- *Sealed housing and weatherproof connector* – This feature allows the SSR to be installed in locations that would be detrimental to ordinary relays.

5.2.2 Connecting SSRs to Achieve Multiple Outputs

SSRs are made with either single- or multiple-switched outputs. If an application requires that multiple SSR outputs be controlled by one control input signal, a multiple-output SSR would normally be used. However, the inputs of two or more single-output SSRs can be connected in series or in parallel to obtain the required number of switched outputs. *Figure 41* shows examples of

three single-output SSRs controlled by a single switch connected both in parallel and in series to control the application of three-phase power to a load. When connecting SSR relays in series, the DC supply voltage divides equally across the individual SSRs. This reduces the voltage available to each one. For this reason, the supply voltage must be increased proportionally when multiple SSRs are connected in series. For the example shown, the control supply voltage for the three SSRs must be increased to at least three times the minimum operating voltage required for use with a single relay.

5.2.3 SSR Temperature Considerations

The operation of an SSR, or any solid-state device, can be affected by exposure to high temperatures. Manufacturers of SSRs specify the maximum temperature permitted for use with their relays. Typically this is 40°C (104°F). High temperatures at an

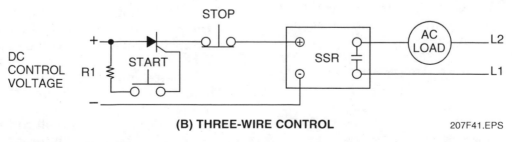

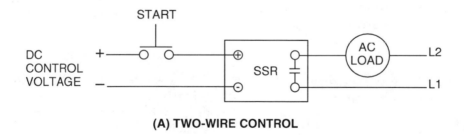

Figure 41 ◆ Two-wire and three-wire SSR control.

207F41.EPS

SSR can result from high ambient temperatures in the surrounding area and in the enclosure in which the SSR is mounted. The power switching devices in an SSR itself also generate heat that contributes to its temperature. The larger the current passing through the relay, the more heat is produced.

Sometimes the temperature of SSRs and other devices is controlled by forced air cooling, but in most cases it is controlled by the use of a heat sink. A heat sink is a metal base, commonly aluminum, used to dissipate the heat of the solid-state components mounted on it. In some SSRs, the package of the device often serves as its heat sink, but when an SSR is used to handle higher power, a separate heat sink may be needed. A heat sink has a low thermal resistance as determined by the size of its surface area and the type of material from which it is made. Thermal resistance is the resistance of a material to the conduction of heat. The lower the thermal resistance number, the greater the ability to dissipate heat. The use of a heat sink enables the SSR to control higher current levels.

When selecting a heat sink for use with a particular SSR, follow the SSR manufacturer's recommendations. When installing a heat sink, follow these guidelines:

- When it is practical, use a heat sink with fins to provide the greatest heat dissipation per unit of surface area.
- Make sure that the mounting surface between the heat sink and the SSR is flat and smooth.
- Use thermal grease or a pad between the SSR and the heat sink surfaces to eliminate any air gaps and enhance the thermal conductivity.
- Make sure all mounting hardware is securely fastened.

5.2.4 Solid-State Relay Overvoltage and Overcurrent Protection

An SSR's power transistors, SCRs, and triacs can be damaged by a shorted load. It is a good practice to install an overcurrent protection fuse to open the output circuit should the load current increase to a higher value than the nominal load current. This fuse should be an ultra-fast fuse designed for use with semiconductor devices.

NOTE

Solid-state relays are much more sensitive to loads than equivalent mechanical devices. Always follow the manufacturer's application instructions.

Fire Alarm System Relays

In a fire alarm system, a relay may be used to seize a phone line. If an alarm is triggered, the communicator sends a signal that energizes the seizure relay. This action disconnects the house phone so it can be used to send an alarm to the monitoring station. Once the signal has been acknowledged by the monitoring station, the seizure relay is de-energized, and the house phone is reconnected.

To protect the power transistors, SCRs, and triacs from overvoltages resulting from transients induced on the load lines, it is recommended to install a peak-voltage clamping device such as a varistor, zener diode, or snubber network in the output circuit.

6.0.0 ◆ TIMERS

Many industrial control applications require timing relays that can provide dependable service and are easily adjustable over the timing ranges. The proper selection of timing relays for a particular application can be made after a study of the service requirements and with a knowledge of the operating characteristics of each available device.

There are two basic categories of timers of concern to the EST: synchronous clock timers and solid-state electronic timers. Although each device works in a different way, all timers have the ability to introduce some degree of time delay into a control circuit.

6.1.0 Synchronous Time Switches

A time switch, also called a time clock, is an electrically operated switch used to control the amount of time and the time of day that power is switched on and off to a piece of equipment. The term synchronous is used to describe such time switches because AC synchronous motors are used for the clock motors. A typical synchronous time switch is shown in *Figure 42*.

To help understand how these controls operate, the technician must first become acquainted with the interior components of a time switch. Descriptions of these components follow.

The clock gear train (*Figure 43*) is heavy and rugged enough to withstand prolonged normal usage. It is designed and built to function properly even under adverse conditions.

The switch mechanism (*Figures 43* and *44*) is made of channeled, spring brass U-beam blades

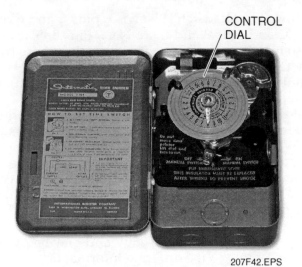

Figure 42 ◆ Synchronous time switch.

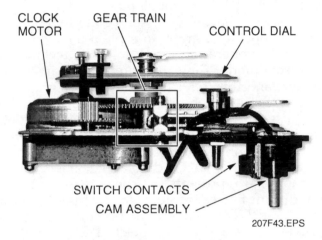

Figure 43 ◆ Clock gear train, motor, and switch mechanism.

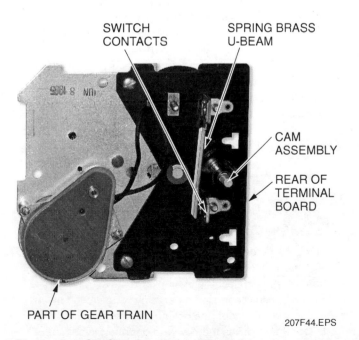

Figure 44 ◆ Single-pole switch mechanism.

operated by a rugged cam to give instant and positive make and break action. The contacts are usually self-cleaning and made of a special alloy to prevent pitting. They are rated to carry an inrush current ten times their normal amperage rating without arcing or sticking. They are supplied as single- or double-pole switches.

Time switches are usually powered by heavy-duty, industrial-type motors. These motors are a synchronous type and are self-lubricating and extremely quiet. They never need service or attention and are practically immune to adverse temperature and humidity conditions.

The terminal board (*Figure 45*) allows easy connection of equipment wiring. Note the connections for the line and load wires. On most two-pole time switches, the line terminals of both poles are located on the left side of the enclosure, and the load terminals are located on the right side of the enclosure.

The control dial is divided accurately into units of time. Most are painted with contrasting colors to facilitate the placement of trippers. The on and off trippers are adjustable to provide the desired on and off periods. These dials provide 24-hour coverage. The trippers actuate a toggle mechanism that turns the contact cam slightly to open and close the contact(s). The trippers are secured in position on the rotating dial by thumbscrews. More on and off trippers can be added for additional on/off cycles. The dial is usually set at the correct time by pulling the spring-loaded dial toward the front of the control and rotating until

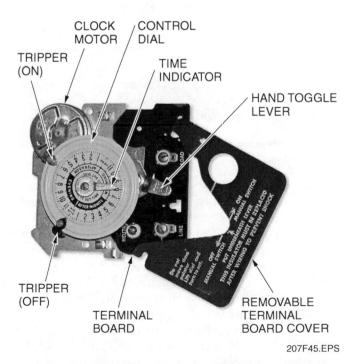

Figure 45 ◆ Terminal board and control dial.

the correct time corresponds to the time indicator. Consult the manufacturer's recommendations before attempting to set the dial.

The toggle lever allows operation of the switch without disturbing the dial settings. The toggle lever can be moved to the on position to determine if the system is functioning properly.

In operation, the time switch motor is wired in parallel with the switching contacts across two phases or from a neutral terminal to a line terminal. The toggle lever also moves from off to on and back when the toggle mechanism is actuated by the on or off trippers. As the motor rotates the dial and trippers, the contacts are opened or closed, depending on the desired operation. Timers provide automatic control over a great variety of functions and equipment. Some of the more common uses include the following:

- Chemical injection
- Water heaters
- Defrost control

6.2.0 Solid-State Timers

Solid-state timers derive their name from the fact that the time delay is provided by solid-state electronic devices enclosed within the timing device. Because solid-state timing devices are almost always replaced and never repaired, this module does not elaborate on how the device electronically provides time delay. In short, most solid-state devices use a resistance/capacitance (RC) network with a transistor and silicon-controlled rectifier (SCR) to provide the time delay and switching characteristics.

NOTE

Make sure that solid-state devices are connected properly before applying power. An improper connection could damage the solid-state device.

Figure 46 shows a typical solid-state timer with the manufacturer's specifications, wiring diagram (pin arrangement), and operation chart. This timer uses an eight-pin plug-in relay socket. This enables the timer to be replaced without disconnecting any wires once the socket has been wired. To read the pin numbers, find the identification notch (*Figure 47*) and count clockwise starting with one. Not all manufacturers mark the numbers on the pins, so remembering this is important.

Using the pin arrangement diagram, it is evident that pin 1 is common to both pins 3 and 4. Pins 1 and 3 are N.O. contacts, and pins 1 and 4 are N.C. contacts. This is called a single-pole, single-

throw (SPST) switch. Likewise, pins 8 and 6 are N.O., and pins 8 and 5 are N.C.. Pins 8, 5, and 6 also make up an SPST switch. Together, these two switches (pins 1, 3, 4, and 8, 5, 6) would make a double-pole, double-throw (DPDT) switch, as listed under the specifications for this timer.

Refer to the specifications in *Figure 46*, and find the current rating for the contacts, 10 amps at 115VAC and 6 amps at 230VAC. This is the amount of current that each contact can safely handle. It determines the maximum size load the timer can turn on or off. It is important not to confuse the rating of the contacts with the coil voltage needed to run the timer. The coil voltage for this model can be anywhere from 12V to 230V and may be AC or DC according to the ordering information, but the contact rating does not change. Thus it is possible that a coil voltage of 12VDC (pins 2 and 7) can control a 120VAC load (through pins 1 and 3).

Most manufacturers of solid-state timers provide an operating graph to explain the logic of a timer. The operating graph tells the technician that this model timer is an on-delay timer. To understand this, refer to the operating graph in *Figure 46*. The graph refers to the condition of the contacts, normal condition or switched, at reset, during timing, and after timing. On the graph, the off level is on the bottom and represents the contacts in their normal condition. The on level is on the top and represents the contacts in switched condition. For example, refer to pins 1 and 3 (N.O.) in *Figure 46*. When the coil voltage is applied (pins 2 and 7), contacts 1 and 3 remain open for the set delay period. After the delay, pins 1 and 3 close, turning on the load, and remain closed for as long as the voltage is applied to the coil. When the coil voltage is removed from the coil, pins 1 and 3 instantly open. Thus the logic of this timer is on-delay and can be used in any on-delay circuit as long as it meets specifications, such as for contact rating and time duration.

The timer in *Figure 46*, like most solid-state timers, has a knob mounted on top to adjust the time delay. This knob can be set between 0 and 10. The numbers 0 to 10 on the timer do not indicate

Timer Relays

Transistor-controlled and sensor-controlled timer relays are also available. A transistor-controlled timer relay is controlled by a transistor switch located in an external electronic device. A sensor-controlled timer relay is controlled by an external proximity or photoelectric sensor in which the timer relay supplies the power to operate the sensor.

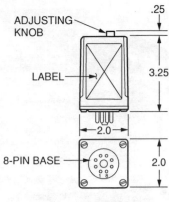

Economical Solid-State DPDT operate delay relay for new or replacement applications, for use in automatic control circuits, machine tool programming, sequencing controls, heating & cooling operations, warm-up delays, etc. A complete range of delay times is available as shown below. Models with fixed delay and models with remote adjustment are also available.

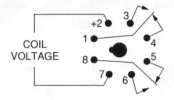

SPECIFICATIONS

- 4 ADJUSTABLE TIMING RANGES
- 100ms RESET TIME
- TEMPERATURE RANGE = −20 TO +60° C
- TRANSIENT PROTECTION = 2,500 VOLTS
- FALSE TRIGGER PROTECTION
- ±2% REPEATABILITY
- MAX. TIME ACCURACY = ±5%
- CONTACTS = DPDT

      10 AMPS @ 115V
      6 AMPS @ 230V

PIN ARRANGEMENT

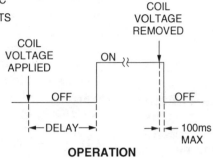

OPERATION

DIMENSIONS

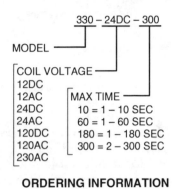

330 – 24DC – 300

MODEL

COIL VOLTAGE
12DC
12AC
24DC   MAX TIME
24AC   10 = 1 – 10 SEC
120DC   60 = 1 – 60 SEC
120AC   180 = 1 – 180 SEC
230AC   300 = 2 – 300 SEC

ORDERING INFORMATION

207F46.EPS

Figure 46 ◆ Solid-state timer specification sheet.

the time delay directly. They represent the percentage of the total time for which the timer can be set. If the total range of the timer is 1 to 60 seconds, a setting of 5 on the timer dial would indicate a 30-second delay. If the total range of the timer was 2 to 300 seconds, a setting of 5 on the timer dial would indicate a 150-second delay. Manufacturers use this system of making and setting the time range when repeatability can be about +0.1 percent. If greater repeatability is required, a digital adjustable switch can be used. The repeatability of this type of timer is generally about +0.005 percent.

6.3.0 Programmable Electronic Time Switches

Electronic time switches provide the same kind of load control as the mechanical synchronous time switches covered previously. However, the control capability can be very extensive and sophisticated in comparison to the mechanical-switch versions. The programmable electronic time switch shown in *Figure 48* can simultaneously control up to 16 separate isolated load circuits, each rated at 20 amps with any voltage up to 277 volts applied. In addition to this 16-load circuit model, this particular manufacturer has 1-, 2-, 4-, and 8-load circuit units available. All units have a user selectable

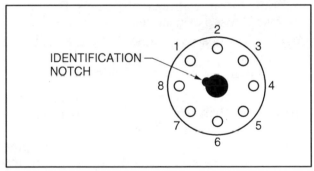

207F47.EPS

Figure 47 ◆ Octal socket pin numbers.

12- or 24-hour clock format with a 7-day (5/2) or full-year programming capability and one-minute time resolution for each load circuit. Full-year control provides up to 99 holidays with variable duration scheduling, automatic daylight adjustment with selectable override, and automatic leap year adjustment. At least 4,000 set points are available along with Fixed, Pulsed, Interval, and Astro switching-time options for flexible load control. The switching times for each load circuit can be programmed in any combination of the following:

- *Fixed* – The selected load circuit is switched on and off at user-selected times.

Figure 48 ◆ Programmable electronic time switch.

207F48.EPS

- *Pulse* – Pulse switching is the same as fixed switching except the on or off operation occurs only for a selectable short duration from 1 to 127 seconds for such things as bell ringing, signal control, or the operation of latching relays.
- *Interval* – Interval switching is the same as pulse switching except for a longer duration, from a minimum of one minute up to 7 days. This option can also be used as user-selectable override.
- *Astro* – This is a varying switching time based on the changing times of sunset and sunrise for a specific time zone and latitude.

Form C load circuit contacts provide a common input and both normally open and normally closed circuit configurations.

Depending on the model of this particular unit, the operating power can be either 24VDC or any selectable voltage up to 277VAC. A non-volatile memory maintains the program for the life of the time switch. A replaceable lithium battery maintains accurate time keeping and calendar information for a minimum of 8 years. To prevent overloads when the power is restored after a power failure, the unit will soft-start all load circuits that are supposed to be on, by staggering the activation of each applicable load circuit by about 15 seconds, beginning with the lowest numbered load circuit. To prevent damage to the control modules in AC-powered units due to lightning strikes or power line disturbances, the AC-powered units have 6kV transient surge protection on the operating power input.

Summary

Understanding the classifications and operation of switches and photoelectric devices is very important to anyone working in the electronics field. Without such knowledge, the selection, installation, and maintenance of electronic equipment and circuits would be difficult, if not impossible.

Switches are classified according to the number of poles, the number of throws, and according to their function. Their operation depends mostly on the type of contacts involved and the method of operating the contacts. Overloading is the most frequent cause of switch failure.

Photoelectric devices are generally classified only according to their function. The operation of these devices usually depends on electronics that are not readily accessible to the user. For photoelectric devices, overloading and improper wiring connections are the most common causes of failure.

Relays and timers play an important role in the operation of today's highly sophisticated electronic systems. Relays and timers make the remote or automatic operation of electrical or mechanical equipment possible. They may be electromechanical or electronic. The application usually determines the type of relay or timer. However, the technician must often decide on such factors as compatibility, accuracy, and operating characteristics.

Electrical relays usually control one or more sets of contacts. These contacts in turn control or send electrical signals to other electrical devices. Electrical relays generally fall into two categories: electromechanical and solid state. Electromechanical relays use a magnetic effect to open or close mechanical contacts, whereas solid-state relays use electronic devices such as SCRs and triacs to simulate open or closed contacts.

Timers are used to introduce delays into controllers and control circuits or to allow for timed operations to occur accurately. Timers are usually synchronous clocks or solid-state relays.

Electric time clocks are normally used to provide automatic operation of electrical equipment by using a synchronous AC motor to turn the dial of a clock. The dial then opens and closes the mechanical contacts, which supply power to the equipment at the proper time of day.

Solid-state timers perform their timing function using solid-state electronics. This method has the advantage of having no moving parts to wear out or replace. Theoretically, solid-state timers should never wear out or need maintenance.

Review Questions

1. Switches are normally used to make or break an electrical circuit that is operating _____.
   a. within the rated voltage of the switch only
   b. under an overload condition
   c. within the rated current of the switch only
   d. within the voltage and current ratings of the switch

2. Switch contacts are often coated with a material that _____.
   a. provides lubrication
   b. increases circuit resistance
   c. decreases contact pressure
   d. prevents oxidation

3. The poles of a switch define the number of _____ in a switch.
   a. contacts
   b. isolated circuits
   c. toggles
   d. throws

4. A double-pole, single-throw switch with four brass-colored screw terminals for wire connections is normally used to control _____ circuits or equipment.
   a. 24V
   b. 100V
   c. 240V
   d. 480V

Refer to *Figure 1* to answer Questions 5 through 7.

Label the switches according to the number of poles and the number of throws.

5. _____ Switch A

6. _____ Switch B

7. _____ Switch C

   a. SPST
   b. DPDT
   c. SPDT
   d. DPST

8. Switches designed to be activated mechanically by the motion of machinery are called _____ switches.
   a. pressure
   b. limit
   c. float
   d. photoelectric

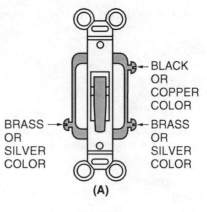

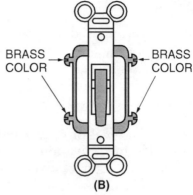

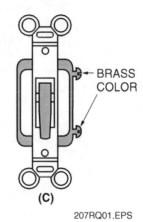

207RQ01.EPS

Figure 1

9. An SCR is an electronic device that _____.
   a. is triggered by infrared radiation
   b. can switch on and off large amounts of power
   c. uses four leads during normal operation
   d. will conduct current in the forward and reverse direction

10. An increase in the amount of light that shines on a photocell _____.
    a. decreases its resistance
    b. increases its resistance
    c. increases its operating voltage
    d. decreases its operating voltage

11. Light is converted into electrical power by a(n) _____.
    a. photocell
    b. proximity sensor
    c. solar cell
    d. SCR

12. A motion detector switch typically uses a(n) _____ sensor to detect motion.
    a. vibration
    b. temperature
    c. infrared
    d. pressure

13. Proximity sensors are used to detect _____.
    a. the presence of smoke
    b. the presence or absence of an object
    c. the sound of a window breaking
    d. a reduction in light level

14. Regardless of the level of power being controlled by a relay, the activation of the relay is accomplished by a(n) _____ power level signal.
    a. higher
    b. inverse
    c. identical
    d. lower

15. Relays that are often referred to as throw-away relays are _____ relays.
    a. plug-in
    b. solid-state
    c. electromechanical
    d. programmable

16. Mechanical latching types of general-purpose relays typically have _____ coils.
    a. DC
    b. interlocking
    c. two
    d. plug-in

17. A disadvantage associated with solid-state relays is _____.
    a. current leakage
    b. contact bounce
    c. a long operating life
    d. a lack of moving parts

18. In a solid-state relay, an LED converts electrical energy into _____.
    a. pneumatic energy
    b. light energy
    c. current
    d. capacitance

19. Clock motors on time switches are typically _____.
    a. DC motors
    b. protected by thermal overload relays
    c. rated at a higher voltage rating than the contacts in the clock
    d. AC synchronous motors

20. Switching functions in a solid-state timer are provided by _____.
    a. SCRs
    b. SSRs
    c. EMRs
    d. LEDs

Trade Terms Introduced in This Module

Absolute zero: A hypothetical temperature at which all molecular movement stops (–273°C or –460°F).

Anode: The positive terminal of an electrical device.

Armature: The part or windings of an electric device at which the voltage is induced.

Bourdon tube: A curved tube that straightens when exposed to a pressure increase. A linkage connects to the end of the tube to provide the mechanical motion required to operate a pressure switch.

Cathode: The negative terminal of an electrical device.

Circuit: The complete path of an electric current.

Contactor: A heavy-duty relay used to switch heavy current, high voltage, or both.

Deadband: The difference between the setpoint and the reset point in calibration.

Diode: An electronic device that allows current to flow in one direction only.

Eddy currents: Currents induced into the core by transformer action.

Infrared radiation (IR): Describes invisible heat waves having wavelengths longer than those of red light.

LED: Light-emitting diode.

NARM: National Association of Relay Manufacturers

Oscillator: An electrical device that generates an electrical frequency.

Overload: A condition in which a circuit operates at greater than rated current.

Oxidation: The pressure that resets the contacts in a pressure switch after the setpoint has been achieved.

Additional Resources

This module is intended to present thorough resources for task training. The following reference works are suggested for both instructors and motivated participants interested in further study. These are optional materials for continued education rather than for task training.

American Electricians' Handbook, 2002. Terrell Croft and Wilford I. Summers. New York, NY; McGraw-Hill.

National Electrical Code® Handbook, Latest Edition. Quincy, MA: National Fire Protection Association.

Figure Credits

Courtesy of Honeywell International, Inc.	207F17
Cutler-Hammer	207F22
Dwyer Instruments, Inc.	207F09, 207F10, 207F13 (photo), 207F14, 207F28
HID Corporation	207SA02
Intermatic, Inc.	207F49
Potter Electric Signal Company	207F12
National Burglar and Fire Alarm Association	207F21, 207F23
Powerex, Inc.	207F18
Rauland-Borg Corporation	207SA03
Reproduced by written permission from RadioShack Corporation	207F07
Siemens Energy and Automation	207F40A
Solar World, www.solarworld.com	207F24
Reprinted courtesy of Square D/Schneider Electric	207F20 (photo)
System Sensor	207SA01
Topaz Publications, Inc.	207F01, 207F11, 207F15, 207F25, 207F30, 207SA04, 207F32, 207F33 (photo), 207F35, 207SA05, 207F42–207F46 (photo)
Tyco Electronics	207F39

CONTREN® LEARNING SERIES — USER UPDATE

NCCER makes every effort to keep these textbooks up-to-date and free of technical errors. We appreciate your help in this process. If you have an idea for improving this textbook, or if you find an error, a typographical mistake, or an inaccuracy in NCCER's Contren® textbooks, please write us, using this form or a photocopy. Be sure to include the exact module number, page number, a detailed description, and the correction, if applicable. Your input will be brought to the attention of the Technical Review Committee. Thank you for your assistance.

Instructors – If you found that additional materials were necessary in order to teach this module effectively, please let us know so that we may include them in the Equipment/Materials list in the Annotated Instructor's Guide.

Write: Product Development and Revision
National Center for Construction Education and Research
P.O. Box 141104, Gainesville, FL 32614-1104

Fax: 352-334-0932

E-mail: curriculum@nccer.org

Craft _____ Module Name _____

Copyright Date _____ Module Number _____ Page Number(s) _____

Description _____
_____
_____
_____

(Optional) Correction _____
_____
_____

(Optional) Your Name and Address _____
_____
_____

Wire and Cable Terminations

COURSE MAP

This course map shows all of the modules in the second level of the *Electronic Systems Technician* curriculum. The suggested training order begins at the bottom and proceeds up. Skill levels increase as you advance on the course map. The local Training Program Sponsor may adjust the training order.

ELECTRONIC SYSTEMS TECHNICIAN LEVEL TWO

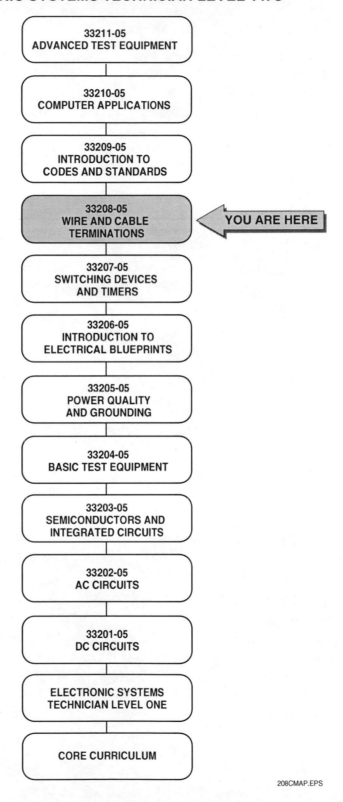

33211-05
ADVANCED TEST EQUIPMENT

33210-05
COMPUTER APPLICATIONS

33209-05
INTRODUCTION TO
CODES AND STANDARDS

33208-05
WIRE AND CABLE
TERMINATIONS

YOU ARE HERE

33207-05
SWITCHING DEVICES
AND TIMERS

33206-05
INTRODUCTION TO
ELECTRICAL BLUEPRINTS

33205-05
POWER QUALITY
AND GROUNDING

33204-05
BASIC TEST EQUIPMENT

33203-05
SEMICONDUCTORS AND
INTEGRATED CIRCUITS

33202-05
AC CIRCUITS

33201-05
DC CIRCUITS

ELECTRONIC SYSTEMS
TECHNICIAN LEVEL ONE

CORE CURRICULUM

208CMAP.EPS

Figures

Tables

Wire and Cable Terminations

Objectives

When you have completed this module, you will be able to do the following:

1. Identify various types of communications connectors.
2. Identify various coaxial connectors.
3. Identify basic fiber-optic connectors.
4. Identify common crimp connectors.
5. Prepare and terminate a cable or wire with various types of connectors.

Prerequisites

Before you begin this module, it is recommended that you successfully complete *Core Curriculum; Electronic Systems Technician Level One;* and *Electronic Systems Technician Level Two*, Modules 33201-05 through 33207-05.

Required Trainee Materials

1. Pencil and paper
2. Appropriate personal protective equipment

1.0.0 ◆ INTRODUCTION

This module provides information on the different types of conductor and cable termination devices used for various low-voltage circuits. It also covers the typical tools and procedures for installing cable or conductors to some of the most commonly used devices, as well as appropriate inspection and testing procedures.

The five major types of terminations covered in this module include the following:

* Coaxial
* Twisted pair
* Solderless (crimp)
* Solder type
* Optical fiber

2.0.0 ◆ COAXIAL CABLE TERMINATIONS

Coaxial cable is used in CATV, video, and computer networks. The coaxial cables in common use contain a solid center conductor surrounded by an insulating material, which is, in turn, surrounded by at least one braided copper shield. A PVC or plenum-rated outer jacket covers the entire cable (*Figure 1*). The types of coaxial cable and their applications were discussed in the Level One module entitled *Low-Voltage Cabling*.

2.1.0 Types of Coaxial Cable Connectors

A number of different types of cable connectors are used on coaxial cables. Coaxial cable connectors are used primarily in CATV and network-powered broadband systems. These connectors are generally available as mechanically crimped or clamped connectors or as soldered connectors, depending on the type. Some are also available as twist-on types, but these are not recommended for professional installations. Video and audio cable connectors are used for distribution of video and audio signals within a confined area.

2.2.0 Coaxial Cable Management

Coaxial cables must be securely supported. Supports should be placed no greater than five feet apart unless project specifications allow greater intervals. The important thing to consider is that

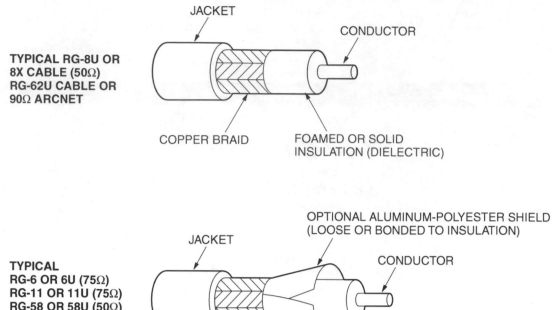

TYPICAL RG-8U OR
8X CABLE (50Ω)
RG-62U CABLE OR
90Ω ARCNET

JACKET

CONDUCTOR

COPPER BRAID

FOAMED OR SOLID
INSULATION (DIELECTRIC)

TYPICAL
RG-6 OR 6U (75Ω)
RG-11 OR 11U (75Ω)
RG-58 OR 58U (50Ω)
THINNET OR RG-59 OR
59U (75Ω)

JACKET

OPTIONAL ALUMINUM-POLYESTER SHIELD
(LOOSE OR BONDED TO INSULATION)

CONDUCTOR

COPPER OR
ALUMINUM BRAID

FOAMED, SOLID, OR GAS
INJECTED INSULATION (DIELECTRIC)

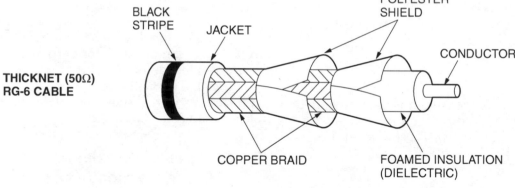

THICKNET (50Ω)
RG-6 CABLE

BLACK
STRIPE

JACKET

ALUMINUM-
POLYESTER
SHIELD

CONDUCTOR

COPPER BRAID

FOAMED INSULATION
(DIELECTRIC)

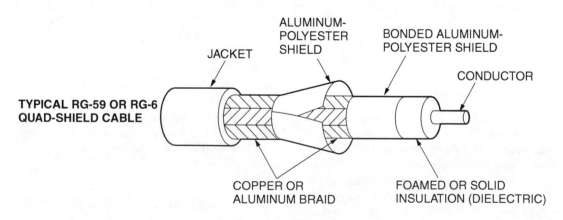

TYPICAL RG-59 OR RG-6
QUAD-SHIELD CABLE

JACKET

ALUMINUM-
POLYESTER
SHIELD

BONDED ALUMINUM-
POLYESTER SHIELD

CONDUCTOR

COPPER OR
ALUMINUM BRAID

FOAMED OR SOLID
INSULATION (DIELECTRIC)

NOTE: RG-59, RG-6, AND RG-11 CABLES ARE ALSO REFERRED TO AS SERIES 59, SERIES 6, AND
SERIES 11, RESPECTIVELY.

208F01.EPS

Figure 1 ◆ Typical shielded coaxial cables.

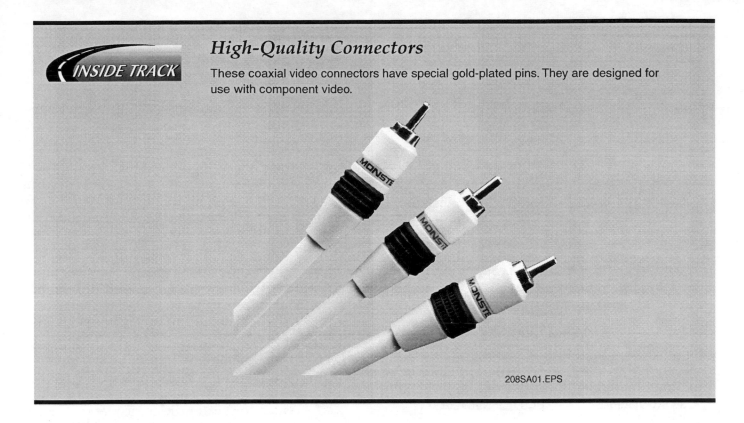

High-Quality Connectors

These coaxial video connectors have special gold-plated pins. They are designed for use with component video.

208SA01.EPS

the performance of coaxial cable can be negatively affected by bending it too tightly. The minimum bend radius stated by the manufacturer must not be exceeded. If the manufacturer does not specify a minimum bend radius, use a minimum radius of ten times the cable diameter. An approved pipe bending tool should be used to bend the cable. The outer conductor or sheath should show no wrinkles after bending.

2.3.0 Termination of Coaxial Cable

Coaxial cable is no longer recognized as an acceptable media under *American National Standards Institute/Telecommunications Industry Association/ Electronic Industries Alliance (ANSI/TIA/EIA)-568B* for new communications/data system installations. However, it is used for other systems such as CATV, satellite, CCTV, video, security, and fire alarm systems that may be integrated with a communications/data system. Cabling installers

should possess a working knowledge of the various common coaxial cables used in the industry and the methods of termination.

The predominant coaxial cables are RG-6, RG-11, RG-58, RG-59, and RG-62. *Table 1* shows the characteristic impedance and general use of each type of cable.

Coaxial cables used in data applications typically have a stranded inner conductor and are usually terminated using BNC connectors. Coaxial cables for video applications have a solid center conductor and may use either a crimp-on, solder-on, or twist-on BNC connector. CATV applications use a solid inner conductor and are terminated with an F connector.

The tools required for proper stripping and terminating coaxial cable are a diagonal wire cutter, adjustable two-step cable stripper, and a ratchet-type crimping tool (*Figure 2*). The ratchet crimping tool is designed specifically for the crimping of certain types of BNC connectors. If the center

Table 1 Coaxial Cable Characteristics

Cable Type	Nominal Impedance	General Usage
RG-58 IEEE 802.3, RG-8	50Ω	10BASE-2 Ethernet
RG-6, RG-11, RG-59 (TIA/SCTE Series 6, 11, and 59)	75Ω	CATV, Video, Antennas
RG-62	93Ω	ARCNet, IBM 3XXX

Workmanship

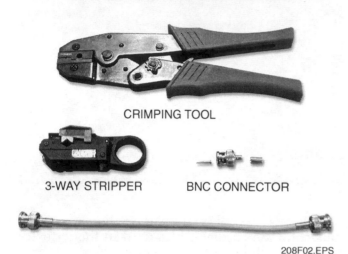

CRIMPING TOOL

3-WAY STRIPPER   BNC CONNECTOR

208F02.EPS

Figure 2 ◆ Coaxial cable termination tools.

pin of the connector is a soldered type, a soldering tool will also be required.

Although other connector systems (screw-on style) are available, captive-pin connectors, which ensure positive retention of the center conductor, are normally used. Always follow the connector manufacturer's instructions when installing a connector.

2.3.1 Terminating Coaxial Cable with a BNC Connector

Figure 2 shows a typical captive-pin BNC connector. Only crimp-type BNC connectors should be used.

Step 1   Make a straight cut on the termination ends of the coaxial cable.

Step 2   Place the connector ferrule over the end of the cable.

Step 3   Adjust the stripping tool (*Figure 3*) to meet the desired cable diameter and stripping requirements. Make sure the stripper is adjusted so that center conductor will not be nicked. The stripping tool should be adjusted to the dimen-

sions required by the specific conductor. Different connectors use different strip lengths. *Figure 3* shows stripping tools designed for use on Series 6 and Series 11 coaxial cable.

Step 4   Insert cable into the stripper by the distance required by the connector manufacturer.

Step 5   Rotate the stripper three to five full turns.

Step 6   Remove the stripper from the cable.

Step 7   Remove the severed sheathing, **shielding**, and **dielectric** material.

Step 8   Inspect the cable for stripping quality, and ensure that the center conductor and the insulation are not nicked or scored, and that any stray strands of the braided/foil shield are pushed away from the center conductor.

Step 9   If the connector center pin is a solder type, tin the center conductor using a soldering tool. Make sure not to melt the center-conductor dielectric material. Soldering procedures are discussed later in this module.

Step 10   Seat the connector center pin on the center conductor.

208F03.EPS

Figure 3 ◆ Coaxial cable stripping tools.

Step 11 If the connector center pin is a solder type, solder the center conductor to the pin; otherwise, crimp the pin to the center conductor using the small diameter die of the crimping tool (*Figure 4*).

Step 12 Make sure the ferrule is on the cable; then insert the sleeve end of the connector over the center conductor and pin.

Step 13 Mate the connector body to the cable by aligning it so that its sleeve fits over the center conductor pin and between the center conductor dielectric and the braid/foil shield. Make sure no strands of the foil or braid shield are touching the center conductor and pin as the sleeve is forced over the dielectric and under the shielding. Braid/foil and sheathing should butt up against back of the connector body when the body is fully seated (*Figure 5*).

Step 14 Slide the connector ferrule up over the cable sheath so that it butts against the back of the connector. If necessary, place the edge of the connector body face down on the edge of a table and place pliers lightly around the cable behind the ferrule. Push down with the pliers to seat the ferrule against the back of the connector (*Figure 6*).

Step 15 Make sure the crimping tool has the correct die for the coaxial cable being terminated (i.e., RG-6, RG-8, RG-58, RG-59, or RG-62, with PVC or plenum-rated sheath).

Step 16 Place the crimp tool over the connector ferrule (*Figure 7*), and squeeze the tool until the die is completely closed.

Step 17 Inspect the connection for neatness (no exposed braiding strands). The connector has to be tight.

Step 18 After both ends of the cable are terminated, use a volt-ohmmeter to check for continuity of the center conductor from pin to pin. Then check for shorts between the center conductor pin and the connector body.

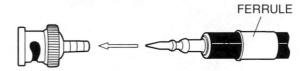

1. SLIDE THE SLEEVE OF THE CONNECTOR BODY ONTO THE DIELECTRIC AND UNDER THE SHIELD.

2. THE BRAID/FOIL PUSHES FLUSH AGAINST THE BACK OF THE CONNECTOR BODY.

208F05.EPS

Figure 5 ◆ Seating the connector body on coaxial cable.

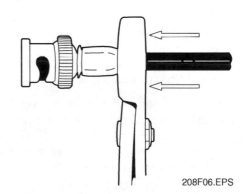

208F06.EPS

Figure 6 ◆ Seating the ferrule with pliers.

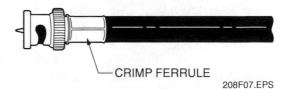

CRIMP FERRULE

208F07.EPS

Figure 7 ◆ Connector body with crimped ferrule.

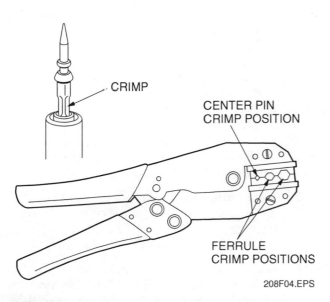

CRIMP

CENTER PIN CRIMP POSITION

FERRULE CRIMP POSITIONS

208F04.EPS

Figure 4 ◆ Crimping the connector pin to the center conductor.

BNC Connectors

Besides BNC cable connectors, BNC connectors are available to support many configurations. This BNC T connector is designed to direct a signal to two locations. Adapters are available to convert from one connector type to another, such as F to BNC. However, the application, bandwidth, and other characteristics must be considered before using such adapters.

208SA02.EPS

2.3.2 Typical Procedure for Terminating a Coaxial Cable with an F Connector

F connectors are the most common type used on coaxial cable. These connectors are used extensively in terminating CATV cables, for example. Although screw-type and crimp-type connectors are available, the crimp-type is preferred because it provides a more secure connection that is less susceptible to interference and intermittent problems. *Figure 8* shows a typical F connector.

Use the following procedure to terminate coaxial cable with an F connector:

Step 1 Make a straight cut on the termination ends of the coaxial cable.

Step 2 If required, adjust the stripping tool to meet the desired cable diameter and stripping requirements. The stripping tool may have to be adjusted to the dimensions required by the specific conductor. Different connectors use different strip lengths.

Step 3 For the crimped connector, insert the cable into the stripper and strip off the sheath, shield(s), and dielectric down to the center conductor for a distance of ⅜" from the end (*Figure 9*). Then strip off an additional ⅓" of sheath and shield(s) down to the dielectric. For the Snap-N-Seal® connector, the stripper will determine the proper strip length.

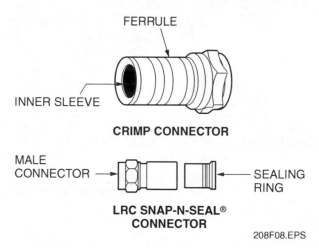

Figure 8 ◆ Typical crimp F connector (rear view) and LRC Snap-N-Seal® F connector.

Step 4 Inspect the cable for stripping quality. Ensure that the center conductor and the insulation are not nicked or scored, and that any stray strands of the braided/foil shield are pushed away from the center-conductor dielectric.

Step 5 Make sure that the appropriate type of F connector is selected for the coax cable to be terminated (i.e., RG-6, RG-6 quad-shield, RG-58, RG-59, or RG-59 quad-shield).

Step 6 For the Snap-N-Seal® connector, slide the seal ring onto the cable. Mate the connector body to the cable by aligning it so that its sleeve fits over the center-conductor dielectric and between the center conduc-

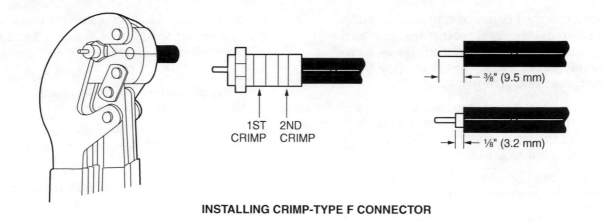

$3/8"$ (9.5 mm)

$1/8"$ (3.2 mm)

INSTALLING CRIMP-TYPE F CONNECTOR

IT-1000 TOOL

1. PUSH DOWN THE WIRE STRIPPING BLADE SO THAT YOU CAN INSERT THE RG-6 OR RG-9 WIRE.

2. BE SURE TO INSERT THE WIRE IN THE CORRECT SIDE. LET THE BLADE CLAMP DOWN ON THE WIRE.

3. THE COAX SHOULD EXTEND JUST A LITTLE BEYOND THE TOOL BODY. TWIST THE TOOL AROUND THE COAX WIRE.

4. HOLD THE WIRE AND PULL THE TOOL SO IT STRIPS OFF THE COAX WIRE SO IT LOOKS LIKE THIS.

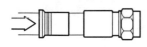

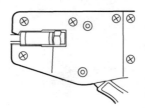

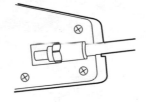

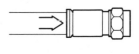

5. PUT THE SNAP-N-SEAL® RING UP THE COAX AS SHOWN. THEN PUSH THE CONNECTOR ONTO THE COAX UNTIL THE FOAM IS EVEN WITH THE BOTTOM OF THE INSIDE OF THE THREADED END OF THE CONNECTOR.

6. MOVE THE HANDLE AWAY FROM THE TOOL AND PUT THE WIRE WITH THE CONNECTOR IN THE TOOL CAVITY. BE SURE THE CENTER CONDUCTOR IS STRAIGHT.

7. PULL THE LEVER TOWARD THE TOOL UNTIL YOU FEEL IT SNAP AS THE RING IS FORCED INTO THE CONNECTOR, MAKING A WEATHERTIGHT SEAL.

8. THE CONNECTOR IS NOW INSTALLED ON THE COAX AND READY TO USE.

INSTALLING SNAP-N-SEAL® TYPE F CONNECTOR

208F09.EPS

Figure 9 ◆ Stripping cable and crimping an F connector.

tor dielectric and the braid/foil shield. Make sure no strands of the foil or braid shield are touching the center conductor as the sleeve is forced over the dielectric and under the shielding. Braid/foil and sheathing must be inserted fully under the ferrule. Rotate the connector body to aid in seating it on the cable. When fully seated, the dielectric showing at the front side of the connector should be flush with the inside surface of the connector.

Step 7 For a crimped connector, make sure the crimping tool has the correct die for the F-type connector being crimped.

WIRE AND CABLE TERMINATIONS

Step 8 For a crimped connector, place the crimp tool over the connector ferrule and squeeze the tool until the die is completely closed (Refer to *Figure 9*). Repeat for the second crimp, if necessary.

Step 9 For the Snap-N-Seal® connector, open the tool and place the connector and sealing ring in the tools. Close the tool until the ring snaps into the connector.

Step 10 Inspect the connection for neatness. The connector has to be tight.

Step 11 After both ends of the cable are terminated, use a volt-ohmmeter to check for continuity of the center conductor from pin to pin. Then check for shorts between the center conductor pin and the connector body.

2.4.0 Coaxial Cable Testing

The two tests commonly performed on coaxial cable are the continuity test and the length test. The continuity test is performed with a multimeter:

Step 1 Start by checking the cable for unwanted signals on the AC and DC voltage scales. If any voltage is detected, remove the source.

Step 2 Use the multimeter on the ohms scale to verify that there is no continuity with the far end of the cable open.

Step 3 Install a shorting plug on the cable, and measure the DC loop resistance.

Step 4 Calculate the required DC loop resistance from the cable manufacturer's data, and compare the two values.

The length of the cable segment is determined using the DC loop resistance reading from the continuity test:

Step 1 Obtain the DC resistance value of the core and shield from the cable manufacturer's data and add these two values.

Step 2 Divide the result by 1,000 to obtain the resistance per foot. (Values are given in resistance per 1,000 feet.)

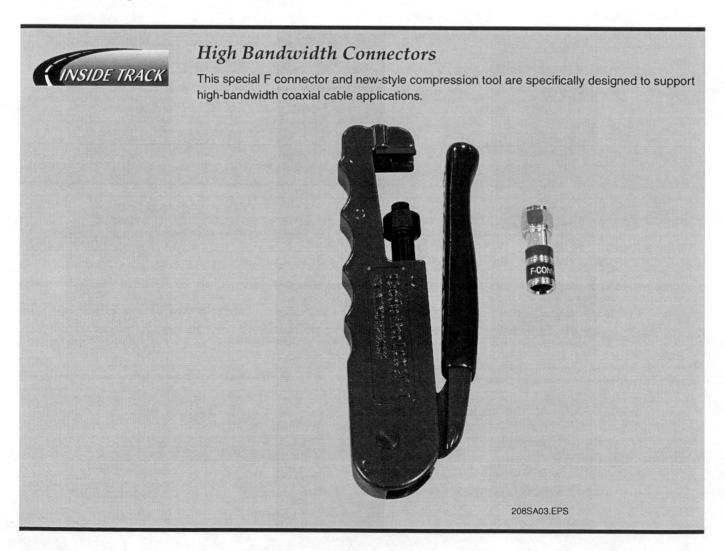

INSIDE TRACK

High Bandwidth Connectors

This special F connector and new-style compression tool are specifically designed to support high-bandwidth coaxial cable applications.

208SA03.EPS

Step 3 Divide the DC loop resistance measure in the continuity test by the DC resistance per foot obtained from the manufacturer's data. The result is the length of the cable in feet.

3.0.0 ◆ TERMINATING UTP CABLE

Category 5 and higher communications/data conductor termination involves organizing cables/conductors by destination, followed by the mounting or installation of the appropriately-sized termination panels or devices. Then the cables/conductors must be formed, supported, and dressed to length, including appropriate slack. Once this has been accomplished, the cables/conductors are properly labeled and then terminated to an appropriate connection device.

Proper cable termination practices are important for the complete and accurate transfer of both analog and digital information signals. **Insulation displacement connection (IDC)** termination is the recommended method of copper termination recognized by *ANSI/TIA/EIA-568B* for UTP cable terminations. This method removes or displaces the conductor insulation as it is seated in the connection. Specific tools designed for making these connections are required. During termination, the cable is pressed between two edges of a metal clip, displacing the insulation and exposing the copper conductor. The copper conductor is held tightly within the metal clip, ensuring a solid connection.

Screw-type terminal faceplates commonly used in voice applications are not recommended by *ANSI/TIA/EIA-568B* for UTP terminations and should not be used.

There are a number of different applications used to terminate cable:

- Termination blocks
- Work area outlets
- Patch cords

The following types of cable are currently recognized by the *ANSI/TIA/EIA-568B* for communications/data use. However, only UTP cable termination will be covered in this module.

- 100Ω unshielded twisted-pair copper cable (UTP)
- 100Ω screened twisted-pair copper cable (ScTP)
- 150Ω shielded twisted-pair copper cable (STP-A)

3.1.0 Types of UTP Connectors

Most communications/data network systems being installed in commercial buildings and some newer residential structures are wired using Category 5e or higher shielded or unshielded twisted-pair cable. In many cases, optical fiber, CATV, video, and/or audio channels are installed as part of the communications/data network. Several manufacturers, such as Siemon, Krone, and BIX, provide a wide variety of equipment for these systems. The equipment supplied by each manufacturer, while functionally equivalent, varies in the devices, meth-

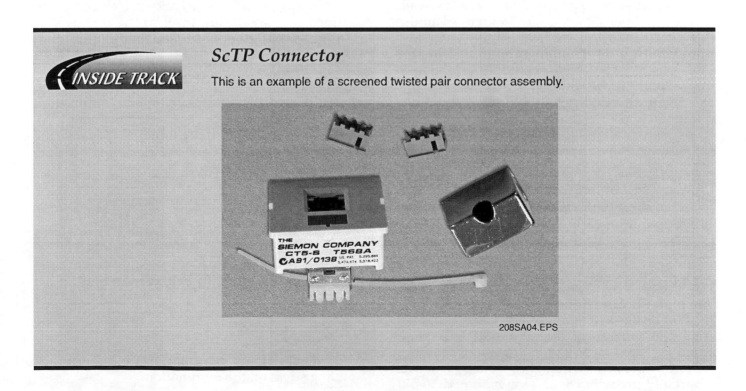

ScTP Connector

This is an example of a screened twisted pair connector assembly.

INSIDE TRACK

208SA04.EPS

ods, and tools used to terminate communication cable wires. Cable manufacturers may require special tools to seat the wiring in their style of insulation displacement connector (IDC) pins used in their station jacks, plugs, and consolidation point or cross-connect punchdown blocks. The IDC pins are special slotted clips that squeeze through the insulation on each wire and make contact with the copper center conductor of the wire as the wire is pushed down through the clip. The use of an IDC eliminates the need to strip insulation from each wire as the wire is connected to the equipment.

Some of the most common equipment used for station jacks, plugs, and punchdown block installations include the following:

- *Couplers/modules* – Typical snap-in station outlet jacks (also referred to as couplers) that are used in horizontal runs to feed equipment are illustrated in *Figure 10*. Most communications/data couplers are six-pin (three-pair) or eight-pin (four-pair) couplers for UTC or ScTP cable. The couplers shown in *Figure 10* require a special tool to seat the wiring in the IDC pins of the coupler. Other couplers or adapters that are available include fiber-optic adapters, video, or audio coaxial cable connections. The couplers/adapters that are shown in *Figure 10* snap into various types of faceplates for recessed boxes.

  Figure 11 shows typical snap-in compact modules, faceplates, and mounting frames that do not require any special tool to seat UTP cable. Like the couplers/adapters shown in *Figure 10*, these compact modules are also available for fiber-optic, CATV, video, or audio applications. These types of modules snap into various types of mounting frames and employ various types of faceplates to trim the modules and mounting boxes.

  Figure 12 shows several typical styles of surface-mount boxes that can accommodate various combinations of one or the other of the couplers/adapters previously described.

- *Consolidation point or cross-connect punchdown blocks* – A number of different punchdown blocks, along with various rack equipment, are available from each manufacturer. Two of the most popular punchdown blocks are known as the 66 block and the 110 block. Both of these blocks are available as rack- or wall-mounted devices and require manufacturer-unique tools to punch down wiring into the IDC pins.

  The 66 block (*Figure 13*) is vertically mounted and comes in sizes ranging from 12- to 100-pair that are rated for field termination of Category 3 or Category 5e wiring. They are also available partially pre-wired with side-mounted 50-pin

(25-pair) receptacles for Category 3 use or modular jacks for Category 5e use. The 66 blocks can be obtained with either bridged or unbridged IDC pins. The unbridged pins can be electrically connected as desired by bridging clips that can also serve as quick disconnects for testing purposes or as service disconnects without disturbing the wiring connected to the block. At least one manufacturer allows the use of bridging clips on Category 5e-rated 66 blocks.

The horizontally mounted 110 block (*Figure 14*) is available in sizes ranging from 25- to 300-pair that can be field terminated for up to Category 5e use or that is available prewired for Category 3 use. The 110 block is also available with 6 to 36 modular jacks or as a 25- to 150-pair disconnect block. The disconnect block has the advantage of allowing the insertion of test plugs or disconnect plugs without disturbing the wiring connected to the block.

- *Plugs, patch cords, and workstation equipment cables* – A variety of plugs that can be used for field terminating of patch cords or workstation equipment cables for up to Category 5e installations are shown in *Figure 15*. When used as a patch cord, normal configuration is a straight-through pinout, with the exception that some network equipment require a crossover pinout to facilitate connection. Most of these plugs require manufacturer unique tools for assembly. Factory-assembled cords and cables (*Figure 16*) are usually recommended for Category 5e or higher installations. These cables and patch cords are available in straight-through or reverse configurations based on use.

3.2.0 UTP Cable Management

It is important to perform preliminary termination procedures before actual cable termination procedures. Proper preliminary procedures improve the quality of the job and decrease the amount of time required for termination.

Preliminary termination procedures involve organizing the cable by destination. Cables should be placed in close proximity to the point of termination and must also be identified properly to make sure that they are terminated at the correct position.

Forming and dressing cables will result in the positioning of the cables in a neat and orderly manner for termination. The length of cable needed to reach the termination location must be determined, taking into account enough slack to re-terminate if necessary and not placing undue pulling stress on the termination.

ANGLED COUPLER
WITH 2 ST ADAPTERS
(2 PORTS)

ANGLED COUPLER
WITH 2 FC ADAPTERS
(2 PORTS)

ANGLED COUPLER
WITH 1 DUPLEX
ST-TO-SC ADAPTER
(2 PORTS)

FLAT CT COUPLER
WITH 2 DUPLEX
SC-TO-ST ADAPTER
(4 PORTS)

TYPICAL FIBER ADAPTER COUPLERS

ANGLED COUPLER
WITH 1 BNC
ADAPTER

ANGLED COUPLER
WITH 1 F-TYPE
ADAPTER

FLAT COUPLER
WITH 2 F-TYPE
ADAPTERS

TYPICAL COAX COUPLERS

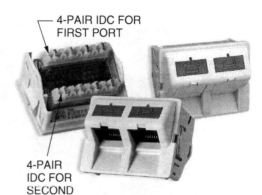

— 4-PAIR IDC FOR
FIRST PORT

4-PAIR
IDC FOR
SECOND
PORT

**DUAL-PORT ANGLED
5e COUPLER
WITH OR WITHOUT
SPRING-LOADED
PROTECTIVE COVER**

  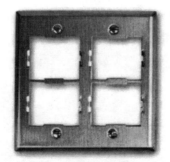

ANGLED COUPLER
WITH 1 RCA
CONNECTOR
W/ SOLDER TAIL

FLAT COUPLER
WITH 1 SVHS
CONNECTOR
W/ SOLDER TAIL

FLAT COUPLER
WITH 2, 2-WAY
BINDING POSTS

TYPICAL AUDIO/VIDEO COUPLERS

**TYPICAL AUDIO/
SPEAKER COUPLER**

SINGLE-GANG STAINLESS
STEEL FACEPLATE FOR
TWO COUPLERS

DOUBLE-GANG STAINLESS
STEEL FACEPLATE FOR
FOUR COUPLERS

TRIPLE-GANG STAINLESS
STEEL FACEPLATE FOR
SIX COUPLERS

TYPICAL PLASTIC OR STAINLESS STEEL FACEPLATES

208F10.EPS

Figure 10 ◆ 5e-rated snap-in couplers/adapters and faceplates.

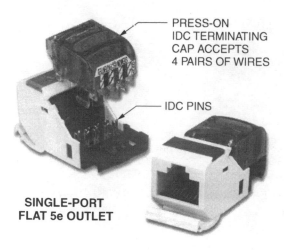
PRESS-ON
IDC TERMINATING
CAP ACCEPTS
4 PAIRS OF WIRES

IDC PINS

**SINGLE-PORT
FLAT 5e OUTLET**

FLAT MODULE WITH
1 SIMPLEX FC ADAPTER
(1 PORT)

ANGLED MODULE
WITH 2 ST ADAPTERS
(2 PORTS)

ANGLED MODULE
WITH 1 DUPLEX SC
ADAPTER (2 PORTS)

ANGLED MODULE
WITH 2 SIMPLEX FC
ADAPTER (2 PORTS)

FIBER-OPTIC FLAT AND ANGLED ADAPTER MODULES

FLAT MODULE
WITH 1 F-TYPE
ADAPTER

FLAT MODULE
WITH 1 BNC
ADAPTER

FLAT MODULE WITH
1 RCA CONNECTOR
WITH SOLDER TAIL

ANGLED MODULE WITH
1 RCA CONNECTOR
WITH SOLDER TAIL

COAX MODULES

SINGLE-GANG
PLASTIC
DUPLEX
FACEPLATE

SINGLE-GANG
PLASTIC
FACEPLATE

DOUBLE-GANG
PLASTIC DUPLEX
FACEPLATE

DOUBLE-GANG
PLASTIC
FACEPLATE

DUPLEX AND DESIGNER FACEPLATES

SIMPLEX
MOUNTING
FRAME,
ACCEPTS 1 FLAT
OR ANGLED
MODULE

DUPLEX
MOUNTING
FRAME,
ACCEPTS
2 ANGLED
MODULES

DUPLEX
MOUNTING
FRAME,
ACCEPTS
4 FLAT
MODULES

DESIGNER MOUNTING
FRAME, ACCEPTS
6 FLAT OR ANGLED
MODULES

MODULAR MOUNTING FRAMES

208F11.EPS

Figure 11 ◆ Typical 5e-rated snap-in compact modules, faceplates, and mounting frames.

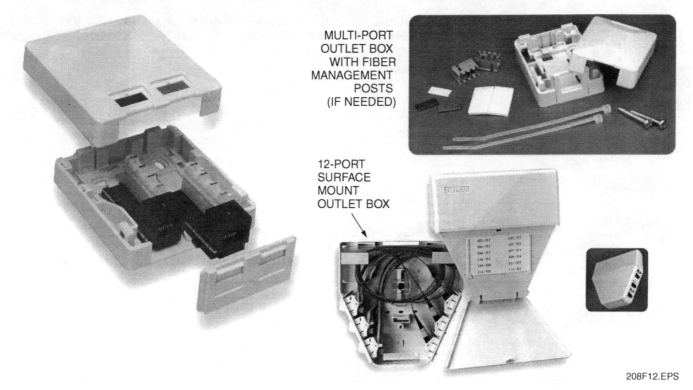

MULTI-PORT
OUTLET BOX
WITH FIBER
MANAGEMENT
POSTS
(IF NEEDED)

12-PORT
SURFACE
MOUNT
OUTLET BOX

208F12.EPS

Figure 12 ◆ Typical surface-mount boxes.

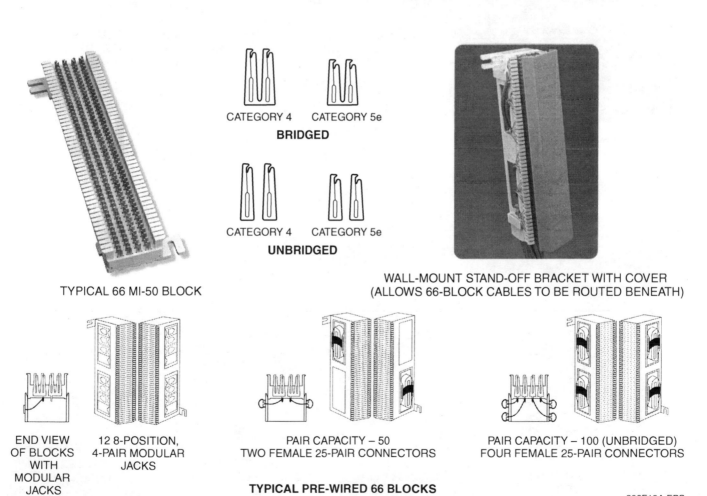

CATEGORY 4   CATEGORY 5e

BRIDGED

CATEGORY 4   CATEGORY 5e

UNBRIDGED

TYPICAL 66 MI-50 BLOCK

WALL-MOUNT STAND-OFF BRACKET WITH COVER
(ALLOWS 66-BLOCK CABLES TO BE ROUTED BENEATH)

END VIEW
OF BLOCKS
WITH
MODULAR
JACKS

12 8-POSITION,
4-PAIR MODULAR
JACKS

PAIR CAPACITY – 50
TWO FEMALE 25-PAIR CONNECTORS

PAIR CAPACITY – 100 (UNBRIDGED)
FOUR FEMALE 25-PAIR CONNECTORS

TYPICAL PRE-WIRED 66 BLOCKS

208F13A.EPS

Figure 13 ◆ Typical type 66 punchdown blocks and accessories (1 of 2).

Proper cable management results in neat and orderly bundles of cables that are formed into a symmetrical pattern. Besides being aesthetically acceptable, proper cable management provides support and mechanical protection of the pairs.

You must remember that termination procedures are not complete until all terminations are properly identified and labeled.

The following are some preliminary termination procedures for Category 5 and higher communications/data cable:

- Check for proper cable routing, wiring scheme, and compatible equipment
- Form and dress cables at consolidation points or cross-connect panels
- Determine the length and slack required for cables
- Use the proper cable management hardware

BRIDGING CLIPS

THESE INDUSTRY STANDARD BRIDGING CLIPS ARE USED TO CONNECT ADJACENT QUICK CLIPS ON 66 BLOCKS. THE CLIPS ARE EASY TO REMOVE FOR ISOLATING AND TESTING INCOMING PAIRS FROM OUTGOING PAIRS AND ARE REUSABLE.

COLORED BRIDGING CLIPS

DESIGNED TO FIT THE 66M TYPE CONNECTING BLOCK, EACH OF THESE PLUG-ON ADAPTERS CONTAINS TWO STANDARD BRIDGING CLIPS, SO THEY ACTUALLY BRIDGE A COMPLETE PAIR WHEN INSTALLED, NOT JUST A SINGLE WIRE. THE PLASTIC HOUSINGS ARE COLOR-CODED AND SERVE TO PROTECT THE QUICK CLIP. TECHNICIANS CAN TEST LINES WITH THE CLIPS IN PLACE BY USING AN IN-LINE TEST PROBE.

RED SPECIAL SERVICE MARKERS

THESE RED PLASTIC MARKERS SLIDE OVER 66 QUICK CLIPS AND TERMINATED WIRES AND ARE IDEAL FOR MAKING SPECIAL CIRCUITS ON BLOCKS.

CAPACITY EXPANDING ADAPTERS

THESE ADAPTERS CREATE ADDITIONAL CAPACITY ON 66 BLOCKS BY PLUGGING DIRECTLY ONTO THE 66 QUICK CLIPS – WITH OR WITHOUT WIRES PUNCHED DOWN. THE ADAPTERS COME WITH EITHER ONE OR TWO ADDITIONAL QUICK CLIPS (CATEGORY 3 USE ONLY) .

ADAPTER WITH
1 DOUBLE QUICK CLIP
(BRIDGED)

ADAPTER WITH
2 DOUBLE QUICK CLIPS
(UNBRIDGED)

ADAPTER WITH
1 SINGLE QUICK CLIP

208F13B.EPS

Figure 13 ◆ Typical type 66 punchdown blocks and accessories (2 of 2).

TYPICAL 110 DISCONNECT BLOCK

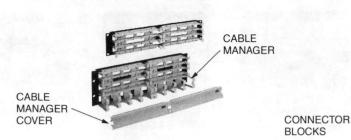

CABLE MANAGER

CABLE MANAGER COVER

TYPICAL RACK-MOUNT 110 BLOCK

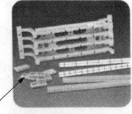

CONNECTOR BLOCKS

TYPICAL WALL-MOUNT 110 BLOCK

50-PIN CONNECTORS

TYPICAL 110 MODULAR JACK BLOCK

How the disconnect block works:

The disconnect block is category 5e compliant in its "normal through" circuit state.

S110* blocks and disconnect modules are mounted on a printed circuit board. "A" side terminations and "B" side terminations are electrically common until a test adapter monitor or disconnect plug is inserted into the disconnect module.

"A" side terminations

Disconnect module

"B" side terminations

TYPICAL PRE-WIRED 110 BLOCK

THE DISCONNECT AND MONITOR TEST ADAPTERS OFFER THE ABILITY TO ISOLATE A CATEGORY 5 CIRCUIT FOR TESTING OR MONITORING WITHOUT REMOVING CABLE TERMINATIONS.

CATEGORY 5 DISCONNECT TEST ADAPTERS

4-PAIR   2-PAIR   1-PAIR

110 DISCONNECT PLUGS

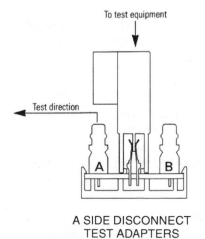

To test equipment

Test direction

A   B

A SIDE DISCONNECT TEST ADAPTERS

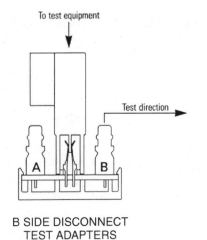

To test equipment

Test direction

A   B

B SIDE DISCONNECT TEST ADAPTERS

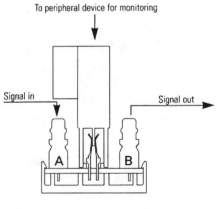

To peripheral device for monitoring

Signal in   Signal out

A   B

DISCONNECT TEST ADAPTERS MONITOR

208F14.EPS

Figure 14 ◆ Typical type 110 punchdown blocks and accessories.

8-POSITION
MODULAR PLUG
WITH 8 CONTACTS

6-POSITION
MODULAR PLUG
WITH 4 CONTACTS

8-POSITION KEYED
MODULAR PLUG
WITH 8 CONTACTS

8-POSITION SHIELDED
MODULAR PLUG
WITH 8 CONTACTS

6-POSITION
MODULAR PLUG
WITH 6 CONTACTS

MODULAR PLUGS

FIELD TERMINATED,
1-PAIR, 110
PATCH PLUG

FIELD TERMINATED,
2-PAIR, 110
PATCH PLUG

UNIVERSAL MODULAR PLUG

PERMITS FIELD-TERMINATION
OF MODULAR CORDS IN 2-, 3-,
OR 4-PAIR INCREMENTS AND
TERMINATES TWISTED PAIR
CABLE WITH 26-22 AWG SOLID
OR 7-STRAND CONDUCTORS.

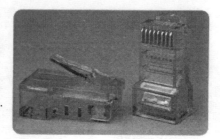

FIELD TERMINATED,
3-PAIR, 110
PATCH PLUG

FIELD TERMINATED,
4-PAIR, 110
PATCH PLUG

PATCH PLUG
BEING INSTALLED
ON A 110 BLOCK

110-BLOCK PATCH PLUGS

208F15.EPS

Figure 15 ◆ Typical plugs used for wire terminations.

3.2.1 Check for Proper Cable Routing, Wiring Scheme, and Compatible Equipment

Use the following procedure to check for proper cable routing, wiring scheme, and compatible equipment:

Step 1 Know the desired wiring scheme for the system installation. If incompatible parts and/or incorrect wiring schemes are used for terminating, unsatisfactory results will occur. Three major wiring schemes are used for copper IDC terminations: T568A, T568B, and Universal Service Order Code (USOC). The T568A and T568B wiring schemes are the only ones compliant with the *ANSI/TIA/EIA-568B* or *570B* standard. Each of the wiring schemes, along with various modular jack, plug, and cord styles, is shown and described in *Figure 17*. The USOC schemes are used for FCC-approved voice communication equipment and systems only.

NOTE

The T568A scheme is preferred and should be used unless the customer specifies otherwise.

Step 2 At work area terminations, check that all cables are available and properly labeled at the wall outlet locations. In the equipment rooms, make sure that cross-connect blocks and/or panels are installed in accordance with the building plans.

Step 3 Make sure that the correct products are available for the application. For example, modular furniture uses a different variety of outlets than those installed in drywalls of offices.

3.2.2 Form and Dress Cables at Consolidation Points or Cross-Connect Panels

Use the following procedure to form and dress cables at consolidation points or cross-connect panels:

DOUBLE-ENDED,
4-PAIR STRANDED
MODULAR CORD

SINGLE-ENDED,
NON-PLENUM OR
PLENUM RATED,
4-PAIR, SOLID
MODULAR CORD

CATEGORY 5e
SCREENED,
DOUBLE-ENDED
4-PAIR STRANDED CORD

SINGLE-ENDED,
4-PAIR SOLID
CABLE ASSEMBLY

110 PATCH CABLE
ASSEMBLIES

110 TO MODULAR PATCH
CABLE ASSEMBLIES

SOLID 110 PATCH
CABLE ASSEMBLIES

SOLID 110 TO MODULAR
PATCH CABLE ASSEMBLIES

25-PAIR CABLE ASSEMBLIES
(CATEGORY 3)

25-PAIR TO MODULAR CABLE
ASSEMBLIES (CATEGORY 3)

TYPICAL 1-, 2-, 3-, OR 4-PAIR
CROSS-CONNECT WIRE

208F16.EPS

Figure 16 ◆ Typical factory-assembled cords and cables.

Step 1 Bring all cables into a layout shaped so that it does not allow the cables to be crossed over each other and is cascaded into a sweeping curve to the destination.

Step 2 Dress the cables by making sure that all cables are parallel to each other. If necessary, smooth them out by hand until they form a neat, orderly bundle. Temporarily secure the bundles with tie wraps.

3.2.3 Determine the Length and Slack Required for Cables

Cables may enter a wiring closet from multiple directions, which frequently results in cables of many different lengths. After the cables are bundled as described previously, the correct amount of slack must be determined as outlined in the following steps. Once the correct slack is determined, the cables should be re-labeled and cut to uniform lengths. Each cable must be carefully re-marked, using the same markings, prior to cutting off the excess cable containing the original markings. Make sure that any new labeling is correct.

Step 1 Make sure that enough cable is present to reach the destination point for termination.

Step 2 For horizontally run cables where the exact location of the termination in the wiring closet is unknown, ensure that enough cable is available to reach the farthest point

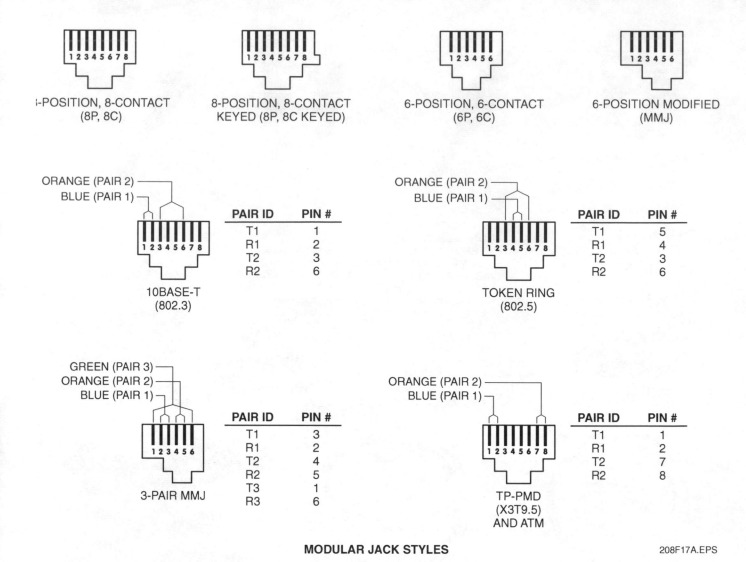

MODULAR JACK STYLES

208F17A.EPS

Figure 17 ◆ Communications/data system wiring schemes (1 of 2).

in the wiring closet plus enough to reach the floor from the horizontal cable management. It is important to remember that if enough cable slack exists, the exact location of the backboard and/or rack termination hardware location can be adjusted to allow for additional equipment and/or hardware. If there is not enough cable slack, then either the location of the termination hardware must be adjusted to fit the horizontal cable, or a new horizontal cable must be re-pulled. By the unplanned relocation of the termination hardware, the organized symmetry of the telecommunications closet can be altered to such a point that future move, add, or change (MAC) activity becomes extremely difficult.

Step 3 For feeder/tie/backbone cables running vertically, enough slack should be available to reach the floor or ceiling, plus the

distance across the backboard. When terminating these types of cable, remember that you are usually dealing with very heavy, high pair-count cables that require a great many termination connections. If an adequate amount of slack is available, situations could arise that could cause a major redesign of the telecommunications closet. As a result, the proper design specifications and symmetry of the system may not be able to be maintained.

3.2.4 Using Proper Cable Management Hardware

There are a number of different types and styles of hardware used to organize and support cables. These include cable management devices, wire ties, hook and loop straps, mushrooms, and D-rings (*Figure 18*). These products are designed to support the in-place cables properly and to

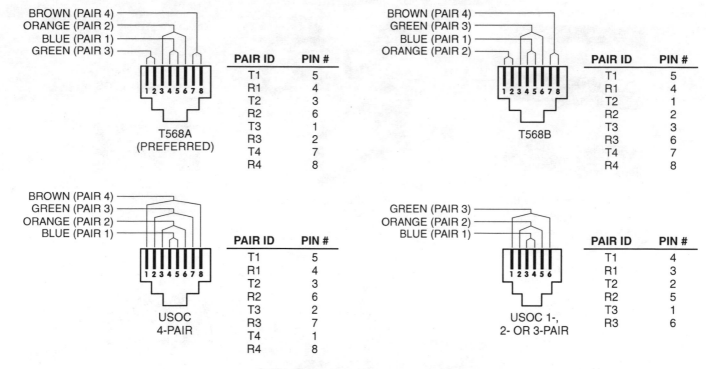

PAIR ID	PIN #
T1	5
R1	4
T2	3
R2	6
T3	1
R3	2
T4	7
R4	8

T568A
(PREFERRED)

PAIR ID	PIN #
T1	5
R1	4
T2	1
R2	2
T3	3
R3	6
T4	7
R4	8

T568B

PAIR ID	PIN #
T1	5
R1	4
T2	3
R2	6
T3	2
R3	7
T4	1
R4	8

USOC
4-PAIR

PAIR ID	PIN #
T1	4
R1	3
T2	2
R2	5
T3	1
R3	6

USOC 1-,
2- OR 3-PAIR

COMMON OUTLET CONFIGURATIONS

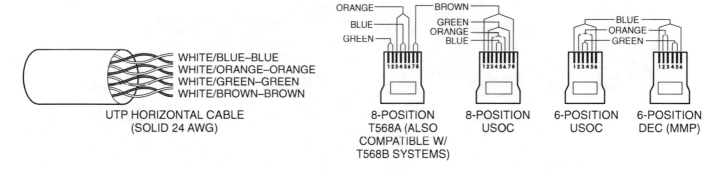

MODULAR PLUG PAIR CONFIGURATIONS

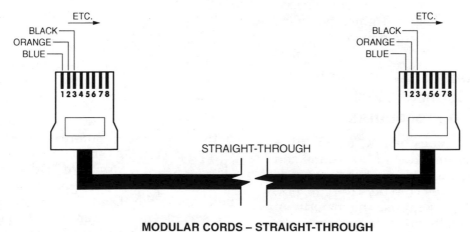

MODULAR CORDS – STRAIGHT-THROUGH

208F17B.EPS

Figure 17 ◆ Communications/data system wiring schemes (2 of 2).

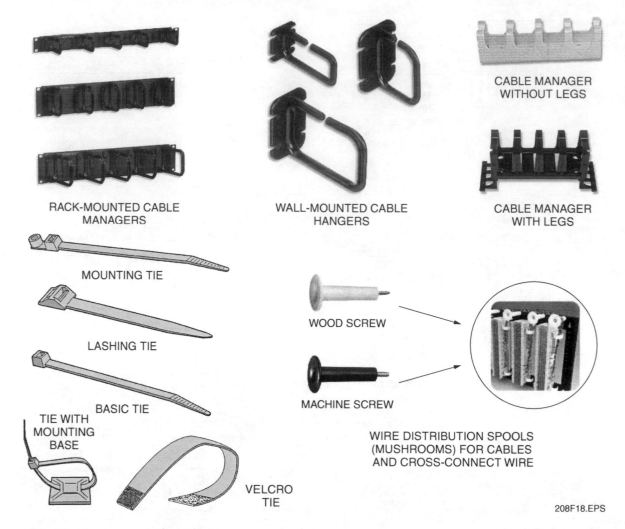

CABLE MANAGER
WITHOUT LEGS

RACK-MOUNTED CABLE
MANAGERS

WALL-MOUNTED CABLE
HANGERS

CABLE MANAGER
WITH LEGS

MOUNTING TIE

LASHING TIE

BASIC TIE

TIE WITH
MOUNTING
BASE

VELCRO
TIE

WOOD SCREW

MACHINE SCREW

WIRE DISTRIBUTION SPOOLS
(MUSHROOMS) FOR CABLES
AND CROSS-CONNECT WIRE

208F18.EPS

Figure 18 ◆ Wire/cable management devices.

relieve tension, and to provide support for future cable that may be installed.

Use the appropriate cable management devices to keep the dressed cables secured. This cable management hardware should be evenly spaced throughout the dressed length. Tighten tie wraps by hand only. Do not overtighten them.

3.3.0 Typical Consolidation Point or Cross-Connect Block Termination Procedures

There are five basic types of IDC termination blocks used in the termination of horizontal and backbone copper cabling. The most common IDC termination blocks are the 66-type, 110-type, BIX, Krone, and LSA. These comprise the majority of the market; however, other devices are available. Most manufacturers provide both rack-mountable and wall-mountable hardware which can house single- or multiple-termination blocks.

To ensure a good connection, follow the hardware manufacturer's specifications closely. Attention must be given to complying with the proper procedures for the following:

- Determining the proper method and length of sheath removal
- The length of pair untwisting that is permitted (*ANSI/TIA/EIA-568B* recommends a maximum of ½" [13 mm] of untwisted pairs, measured from the last twist to the IDC)

A number of manufacturers, including termination equipment manufacturers, offer cable stripping as well as special wire termination tools (*Figure 19*). Each type of IDC termination clip requires a specially designed terminating tool for performing the IDC termination correctly. There are several manufacturers that market termination tools with the ability to interchange the blades for use on several styles of IDC termination blocks. Most punchdown blades are double-ended. One end has the cutoff blade and the other end does not. The end without a cutoff blade is used to punch down wires that will be run to more than one clip, a process known as bridging.

Multi-pair (four- and five-pair) punchdown/ cutoff tools are available that can be used on 110 blocks to seat cabling and cross-connect wire as well as connection blocks. Like the single-wire termination tools, the blade used in the multi-pair tools is doubled-ended so that it can be used to either seat and cut off wires or just seat wires.

> **WARNING!**
> Make sure the brand of tool being used is compatible with the blade. Different blades may look very similar, but there are slight differences in design. Improper matching of the termination tool handle and termination blade can lead to serious personal injury as well as poor IDC terminations.

The usual punchdown order for tie/feed/backbone multi-pair cable (25 pairs or more) or four-pair station cable starts at clip 1 of a punchdown block and progresses toward the higher clip numbers. For four-pair UTP workstation cables, the tip conductor (white) of the blue pair is applied to clip 1, and the ring conductor (blue) is applied to clip 2. The remaining three pairs of wires are applied in the succession of tip and ring, as shown in *Table 2*.

The next four-pair UTP cable is punched down in the same order starting with pin 9.

The punchdown order for tie/feed/backbone multi-pair cable is the same for each of the five-pair groups of wires differentiated by the primary tip colors (*Table 3*).

As an example, each of the five pair groups of a 25-pair cable is punched down in the order shown in *Figure 20* and as partially detailed in *Table 4*. This punchdown order matches the tip and ring connections for a 50-pin Telco connector. If a Telco connector is used on the other end of the tie/feeder/backbone cable, troubleshooting between the block and the connector is much easier.

Always follow the manufacturer's and/or customer's specifications and guidelines regarding all terminations. The following are some examples of recommended termination practices from one manufacturer.

Do:

• Use connecting hardware that is compatible with the installed cable.

TYPE 110 MULTI-PAIR TERMINATION TOOL

SHEATH REMOVAL TOOL

PUNCHDOWN WITH 110/66 BLADE

BIX INSERTION TOOL

KRONE INSERTION TOOL

208F19.EPS

Figure 19 ◆ Typical cable stripping and punchdown tools.

Table 2 Four-Pair Cable Punchdown Order

Four-Pair Cable	Pair Color	Tip or Ring Conductor	Tip/Ring Color	Punchdown Block Clip No.
No. 1	Blue/White (first pair)	Tip	White	1
		Ring	Blue	2
	Orange/White (second pair)	Tip	White	3
		Ring	Orange	4
	Green/White (third pair)	Tip	White	5
		Ring	Green	6
	Brown/White (fourth pair)	Tip	White	7
		Ring	Brown	8
No. 2	Blue/White (first pair)	Tip	White	9
		Ring	Blue	10
	Orange/White (second pair)	Tip	White	11
		Ring	Orange	12
	Green/White (third pair)	Tip	White	13
		Ring	Green	14
	Brown/White (fourth pair)	Tip	White	15
		Ring	Brown	16
Etc.	Repeats for remaining cables	Repeats	Repeats	17 thru 48

- Terminate each horizontal cable on a dedicated telecommunications outlet.
- Maintain the twist of horizontal and backbone cable pairs up to the point of termination.
- Tie and dress horizontal cables neatly and with a minimum bend radius of four times the cable diameter.
- Tie and dress backbone cables or cable bundles neatly and with a minimum bend radius of 10 times the cable or bundle diameter.

Do Not:

- Use connecting hardware that is of a lower category than the cable being used.
- Create multiple appearances of the same cable at several distribution points (called bridged taps).
- Leave any wire pairs untwisted.
- Overtighten cable ties, use staples, or make sharp bends with cables.
- Place cable near equipment that may generate high levels of electromagnetic interference.

Follow these recommended guidelines for UTP termination:

- Pair twists shall be maintained as close as possible to the point of termination.

- Untwisting shall not exceed 1" (25 mm) for Category 4 links and ½" (13 mm) for Category 5 and Category 5e links. Follow manufacturer's guidelines for Category 3 products. If no guidelines exist, then untwisting shall not exceed 3" (75 mm).
- Connecting hardware shall be installed to provide well-organized installation with cable management and in accordance with manufacturer's guidelines.
- Strip back only as much jacket as is required to terminate individual pairs.

Remember to document all termination information properly. This helps to identify the origin, destination, and routing of all cables for future MAC (moves, adds, changes) activities or for troubleshooting.

3.3.1 Typical Type 66 Block Termination Procedures

A typical 50-pair 66 block used for a cross-connect application consist of two vertical columns of internally bridged clips, as shown in *Figure 21*. One block is used to terminate tie/feed/backbone cable and another block is used to terminate horizontal runs to the workstations. The appropriate pairs of the tie/feed/backbone cable are then

Table 3 Typical Color Codes for Wire Pairs and Binder Groups

Pair			Binder Group	
Number	Tip	Ring	Color	Pair Count
1	White	Blue	White-Blue	001 – 025
2	White	Orange	White-Orange	026 – 050
3	White	Green	White-Green	051 – 075
4	White	Brown	White-Brown	076 – 100
5	White	Slate	White-Slate	101 – 125
6	Red	Blue	Red-Blue	126 – 150
7	Red	Orange	Red-Orange	151 – 175
8	Red	Green	Red-Green	176 – 200
9	Red	Brown	Red-Brown	201 – 225
10	Red	Slate	Red-Slate	226 – 250
11	Black	Blue	Black-Blue	251 – 275
12	Black	Orange	Black-Orange	276 – 300
13	Black	Green	Black-Green	301 – 325
14	Black	Brown	Black-Brown	326 – 350
15	Black	Slate	Black-Slate	351 – 375
16	Yellow	Blue	Yellow-Blue	376 – 400
17	Yellow	Orange	Yellow-Orange	401 – 425
18	Yellow	Green	Yellow-Green	426 – 450
19	Yellow	Brown	Yellow-Brown	451 – 475
20	Yellow	Slate	Yellow-Slate	476 – 500
21	Violet	Blue	Violet-Blue	501 – 525
22	Violet	Orange	Violet-Orange	526 – 550
23	Violet	Green	Violet-Green	551 – 575
24	Violet	Brown	Violet-Brown	576 – 600
25	Violet	Slate	No Binder	

208T03.EPS

cross-connected with twisted-pair wire to the appropriate pairs of the horizontal cable runs. The outer clips of each column are used to terminate up to a 25-pair backbone cable or up to six four-pair horizontal cables. The inner (bridged) clips of each column are used to terminate the cross-connect wire that connects the tie/feed/backbone pairs to the assigned workstation pairs.

The following is a typical procedure for installing a 66 block and terminating wires on the block:

Step 1   Mount a standoff bracket (sometimes called a dead bug) firmly on a properly prepared plywood wall using two wood screws, or mount the bracket on a cross-connect frame. Orient the bracket so that the upper mounting tab is to the left.

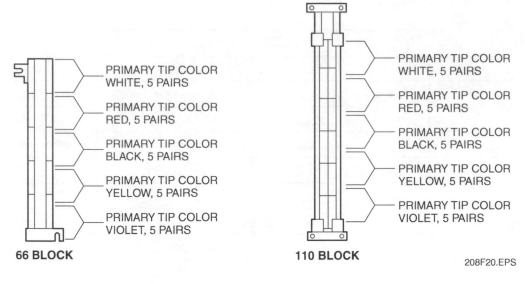

Figure 20 ◆ Typical 25-pair cable, five-pair group punchdown order for 66 and 110 blocks.

Table 4 Partially Detailed Five-Pair Group Punchdown Order for a 25-Pair Cable

Primary Tip Color	Pair Color	Tip or Ring Conductor	Tip/Ring Color	Punchdown Block Clip Pair No.
White (first group)	Blue/White (first pair)	Tip	White	1
		Ring	Blue	2
	Orange/White (second pair)	Tip	White	3
		Ring	Orange	4
	Green/White (third pair)	Tip	White	5
		Ring	Green	6
	Brown/White (fourth pair)	Tip	White	7
		Ring	Brown	8
	Slate/White (fifth pair)	Tip	White	9
		Ring	Slate	10
Red (second group)	Blue/Red (first pair)	Tip	Red	11
		Ring	Blue	12
	Orange/Red (second pair)	Tip	Red	13
		Ring	Orange	14
	Green/Red (third pair)	Tip	Red	15
		Ring	Green	16
	Brown/Red (fourth pair)	Tip	Red	17
		Ring	Brown	18
	Slate/Red (fifth pair)	Tip	Red	19
		Ring	Slate	20
Black, yellow, and violet (third, fourth, and fifth group)	Repeats for black, yellow, and violet primary tip colors	Repeats	Repeats for black, yellow, and violet primary tip colors	21 thru 50

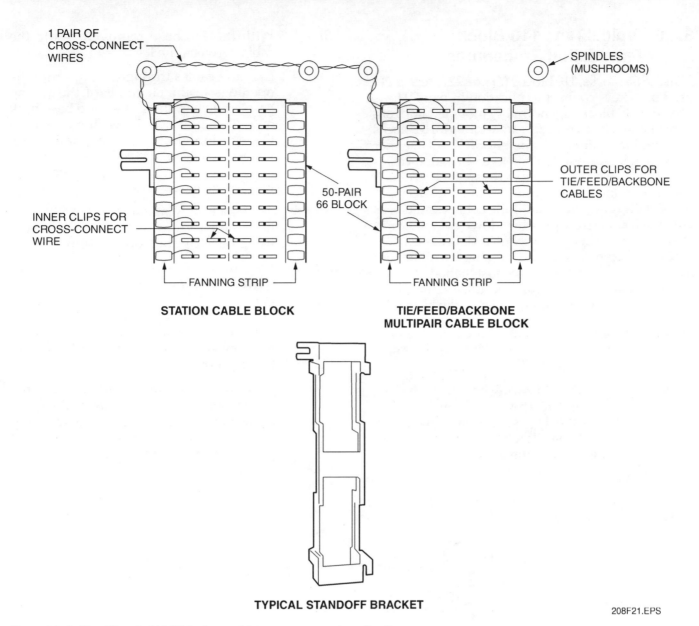

1 PAIR OF CROSS-CONNECT WIRES

SPINDLES (MUSHROOMS)

50-PAIR 66 BLOCK

INNER CLIPS FOR CROSS-CONNECT WIRE

OUTER CLIPS FOR TIE/FEED/BACKBONE CABLES

FANNING STRIP

FANNING STRIP

STATION CABLE BLOCK

TIE/FEED/BACKBONE MULTIPAIR CABLE BLOCK

TYPICAL STANDOFF BRACKET

208F21.EPS

Figure 21 ◆ Two 50-pair 66 MI blocks used in a cross-connect application.

Step 2   Route the horizontal or backbone cables inside the bracket, and feed the cables out either the right or left side of the bracket, as needed.

Step 3   Install the 66 block onto the bracket with the upper mounting tab to the left.

Step 4   Strip back only as much cable jacket as is necessary to terminate the conductors.

Step 5   Route each pair through the slot in the fanning strip in its twisted state. Separate the conductors of each pair inside the fanning strip and place in the appropriate quick clips. Route all wires through the clips in the same direction. Pair twist must be maintained to within ½" (13 mm) of the point of termination for Category

5e installations. Check the conductor sequence at this point and correct any miswires or reversals.

Step 6   Terminate each conductor using an impact tool or equivalent.

NOTE

It is important to keep the tool perpendicular to the block when terminating the conductors. Twisting the tool while terminating can result in bent pins and subsequent damage to the 66 block.

Step 7   Label each block on the fanning strip, designation strips, or cover, as needed.

3.4.0 Typical Type 110 Block Termination Procedures

Like the 66 blocks, 110 blocks (*Figure 22*) are usually used in a cross-connect application in pairs. However, the 110 blocks are normally mounted horizontally and stacked vertically. In addition, 110 blocks are terminated in a completely different manner than 66 blocks. With this block, workstation cables or tie/feed/backbone cable wires are fanned out on the base, seated, and trimmed, but the electrical connection is not made until the connecting blocks with the IDC clips are punched down over the wires after they have been properly placed. Cross-connect wires are then punched down on the top of the connecting blocks to complete the installation.

The following is a typical procedure for installing a 110 block and terminating wires on the block:

Step 1   Mount the 110 wiring base and legs onto a suitable mounting surface with the necessary screws and hardware.

Step 2   Remove the 110 wiring base from the legs by depressing the outer four fanning strips of the 110 wiring base inward to defeat the leg latches. Pull the wiring base away from the legs.

Step 3   With the 110 base removed, route the cable between the legs.

Step 4   Lace the cables through the appropriate openings in the channels of the wiring base, and snap the 110 wiring base back onto the legs. Push the wiring base onto the legs until the latches snap into place. For additional security, the assembly can be fastened together using self-tapping, Phillips-head screws.

Step 5   Strip back only as much cable jacket as necessary to terminate the conductors using the stripping tool or equivalent.

Step 6   Lace the conductor pairs into the 110 wiring base. Pair twist must be maintained to within ½" (13 mm) of the point of termination for Category 5e installations. Ample channel space is provided to allow jacketed cable to continue close to the point of termination.

Step 7   Seat the conductors, and trim off the excess wire with the cutting edge of an impact tool or equivalent. Be sure that the cutting edge is properly oriented prior to trimming the wire.

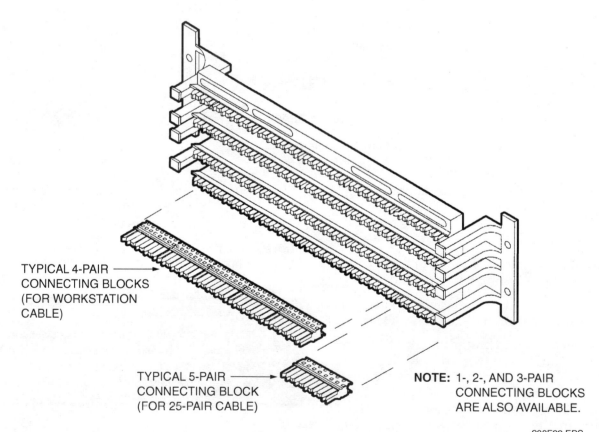

TYPICAL 4-PAIR CONNECTING BLOCKS (FOR WORKSTATION CABLE)

TYPICAL 5-PAIR CONNECTING BLOCK (FOR 25-PAIR CABLE)

NOTE: 1-, 2-, AND 3-PAIR CONNECTING BLOCKS ARE ALSO AVAILABLE.

208F22.EPS

Figure 22 ◆ Typical 110 block.

Step 8 Visually inspect the conductor and cable placement at this point to eliminate any miswires or reversals.

Step 9 Insert a 110 connecting block into the head of an impact tool. Use a three- or four-pair connecting block for UTP and a five-pair connecting block for 25-pair cable as applicable.

Step 10 Carefully align the 110 connecting block over the wiring base, with the blue marking to the left side of the block (gray stripe down), and seat the connecting block.

Step 11 Label the circuits, slide the designation strip into the 110 holder, and snap the holder onto the wiring base. Complete the connections by punching down the cross-connect wire or using 110 patch cables.

Step 12 Remove designation strips prior to removing base from legs. A vertical wire manager can be mounted to the legs of the 110 block to provide vertical cable management for patch cords or cross-connect wire.

3.5.0 Typical Workstation Coupler or Modular Jack Termination

It is recommended that the same manufacturer's workstation devices, patch cords, and patch or cross-connect blocks be used for a given installation. This will minimize the possibility of component mismatches. Besides stripping and punchdown tools, palm guards (*Figure 23*) and electrician's scissors are recommended for termination of four-pair UTP work area couplers or compact modules (jacks).

The following are guidelines for terminating wires in a workstation coupler:

Step 1 Pull cables through the appropriate openings in the faceplate, and mount the faceplate to the electrical box with the screws provided.

> **NOTE**
>
> The faceplate may be mounted before or after cable termination.

Step 2 Strip back as much of the cable jacket as is needed (approximately ½" [40 mm]) using a cable stripper.

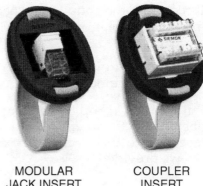

MODULAR JACK INSERT   COUPLER INSERT

208F23.EPS

Figure 23 ◆ Palm guards.

Step 3 Secure the cables to the coupler housing using the cable ties provided. The cables should be in the inside of the coupler housing and the cable jacket should be as close as possible to the cable tie. Do not overtighten the cable ties. For angled couplers, the cable tie anchor points have been modified to optimize cable entrance to the 110-type connector termination blocks depending upon cable orientation.

Step 4 Lace the conductors into the 110-type termination blocks so that the pair colors correspond with those on the blocks. The white conductor of each pair corresponds to the position marked T on the block. Pair twist must be maintained to within ½" (13 mm) of the point of termination for Category 5e installations.

Step 5 Visually inspect the cable conductors and terminate using a 110-type termination tool. For dual flat couplers, a four-pair 110-type impact tool can also be used. The palm guard is recommended to properly secure the module during termination.

Step 6 If the faceplate has not been mounted to the wall, turn the wired coupler 90° and insert through the opening in the faceplate.

Step 7 Insert the coupler into the faceplate. For Category 5e installations, be sure that the minimum cable bend radius behind the coupler is at least 1" (25 mm).

Step 8 Label the faceplate using the designation labels provided, and snap the appropriate icons or tabs into the space above each outlet.

Step 9   If required, the coupler can be removed from the faceplate using a small, flathead screwdriver.

The following are guidelines for terminating wires in a compact 5e module:

Step 1   Pull cables through the appropriate openings in the faceplate, and mount the faceplate to the electrical box with the screws provided.

NOTE

The faceplate may be mounted before or after cable termination.

Step 2   Strip back as much of the cable jacket as needed to terminate the cable using a cable stripper (approximately 1½" [40mm]).

Step 3   Fan out the cable pairs (or individual wires for some non-category 5e terminations) according to the color code label on the clear termination cap. Cable pairs (or individual wires for some non-category 5e terminations) should be fanned out back to the cable jacket.

Step 4   Place the cable pairs (or wires for some non-category 5e terminations) into the four individual channels according to the color code on the termination cap. The cable should be placed so that the cable jacket is adjacent to the termination cap.

Step 5   Lace the individual wires into the proper locations according to the color code label. For Category 5e termination, tip wires (white) should be to the left side of each pair.

NOTE

Cable pairs must be laced into the termination cap for proper termination and transmission performance. For some non-category 5e terminations, such as four-pair USOC, pair groupings may vary.

Step 6   Trim the excess wires as shown using either electrician's scissors or diagonal wire cutters. After trimming, inspect wires to ensure they are still located in the proper positions.

Step 7   Align the rear posts and latches of the termination cap with the corresponding slots in the module.

Step 8   Press down or use the non-cutting side of a single-position Type 110 impact tool placed into the 110-slot on the termination cap to snap-lock the cap into place. Inspect the wires in the viewing window to ensure proper color-coding following termination.

Step 9   If re-termination is required, unlatch the termination cap at screwdriver slots. Do not pry it off. Insert a screwdriver and rotate gently. Pull the cap straight off to remove wires from the terminations, and repeat termination procedure.

3.6.0 Typical Surface-Mount Box Termination Procedure

It is recommended that the same manufacturer's surface-mount boxes or workstation devices, patch cords, and patch or cross-connect blocks be used for a given installation. This will minimize the possibility of component mismatches. The following are instructions for mounting and terminating wires in a typical surface-mount box.

Step 1   Remove the cover of the box by inserting a screwdriver into the cover release slots, beginning with the front slots.

Step 2   Using pliers, remove the appropriate breakout position. For large cable counts or raceway, use secondary breakouts (needle-nose pliers are required). Cable access is provided on three sides and through the bottom of the base.

Step 3   Align the raceway with the opening and mount box using screws, tape or optional magnets. For optional magnets, insert magnets into slots in the base.

Step 4   Route cable(s) through the raceway.

Step 5   Using the provided cable ties, secure the cable(s) to the strain relief anchor points.

Step 6   For optical fiber cable installation, allow enough slack to store about 3' (1 m) of optical fiber cable after termination. Insulation material must be stripped back to the strain relief anchor point. Terminate the fiber and route excess slack around the fiber management posts. Fiber bend radius should be 1.2" (30 mm) or greater.

Step 7   Attach fiber connectors to the bezel.

Step 8   Insert the bezel into the base. Route the remaining fiber so that the minimum bend radius is maintained.

Step 9   Insert the modular jack(s) into the base by sliding the jack assembly into place.

Step 10   Slide the bezel(s) into the base to secure the jacks into position. If all ports are not utilized, optional blanks should be attached to the bezel before sliding the bezel into the base.

Step 11   Strip the jacket from the cable using a cable stripper. Be sure enough slack remains to route excess slack around cable management posts.

Step 12   Position wires in the 110-type termination block slots according to the color-coding. Maintaining pair twists within ½" (13 mm) to the point of termination and cable management with bend radii of no less than 1" (25 mm) are essential to achieving Category 5e transmission performance.

Step 13   Use a single-position, 110-type punch-down tool to terminate 22 to 26 AWG (0.63 mm to 0.40 mm) cable to the 110-type blocks. Ensure that the cutting edge of the tool trims excess wire. Termination caps may also be used to terminate two pairs at a time. For proper terminations, ensure that caps are fully seated. Termination caps may be kept on 110-type blocks for future use.

Step 14   For coaxial, measure cable for proper length, install the connectors, and attach the bezel.

Step 15   Slide the bezel into place, routing cable around the fiber management area.

Step 16   Snap the cover onto base.

Step 17   Snap icons or blank tabs into the space provided above each port. Additional designation areas are provided on the cover for 1½" × ½" (38.1mm × 12.7mm) labels. Remove icon tabs by first placing your thumb or finger over the tab and then prying it out with a pick or a similar instrument.

3.7.0 Modular Plug/Cord Fabrication and Termination Procedures

Normally, field fabrication of cords with modular plugs is not recommended for use as patch cords or equipment connection cords in Category 5 or higher installations. Factory-fabricated, tested, and certified cords with modular plugs (either single-ended or double-ended) should be used to eliminate the possibility of degraded system operation.

If modular cords with plugs are fabricated, the 4-conductor, 6-conductor, or 8-conductor modular plugs (*Figure 24*) that are rated for the category of service desired and available for either flat or round UTP cable are crimped using a special tool applied to the stranded or solid conductors of the UTP cable.

Before the plugs are applied to the cable, it must be determined if the modular cord must be configured in a straight-through or reversed configuration. The T568A configuration is preferred because it is compatible with the USOC one-pair or two-pair color schemes that use T568A UTP cable. *Tables 5* and *6* show the ANSI/TIA/EIA color codes. The USOC wiring scheme for up to three-pair or four-pair patch-cord wiring is shown in *Tables 7* and *8*. The USOC wiring scheme is used only for voice circuits.

The following are instructions for mounting and terminating wires in a typical modular plug:

Step 1   Cut the end of the cable cleanly at a 90-degree angle.

Step 2   Strip and remove only enough cable sheath to allow the conductors to reach past the IDC clips to the end of the plug and still have the jacket under the cable-clamp portion of the plug.

Step 3   Fan out and arrange the conductors in the desired color-code sequence.

Step 4   Insert conductors into the modular plug. Make sure conductors are in the desired order, are flush with the end of the plug, and that the cable sheath is in the cable-clamp portion of the plug.

Step 5   Place the modular plug in the appropriate die of the crimp tool and crimp the IDC clips onto the conductors.

Step 6   Check the plug to ensure that all the conductors are seated properly and are in the correct position.

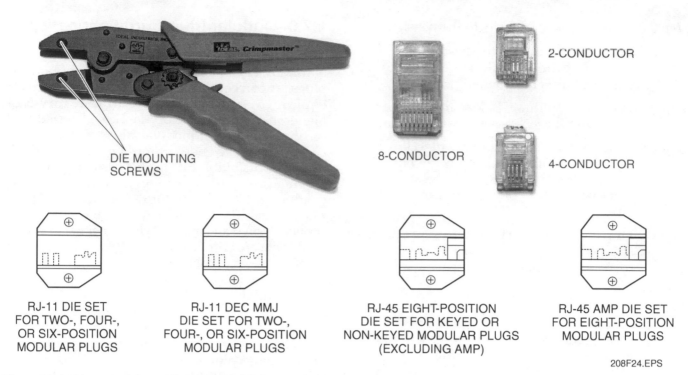

DIE MOUNTING
SCREWS

2-CONDUCTOR

8-CONDUCTOR

4-CONDUCTOR

RJ-11 DIE SET
FOR TWO-, FOUR-,
OR SIX-POSITION
MODULAR PLUGS

RJ-11 DEC MMJ
DIE SET FOR TWO-,
FOUR-, OR SIX-POSITION
MODULAR PLUGS

RJ-45 EIGHT-POSITION
DIE SET FOR KEYED OR
NON-KEYED MODULAR PLUGS
(EXCLUDING AMP)

RJ-45 AMP DIE SET
FOR EIGHT-POSITION
MODULAR PLUGS

208F24.EPS

Figure 24 ◆ Clear plastic modular plugs and crimping tool with interchangeable dies.

Table 5 Straight-Through Patch-Cord Wiring (Typical for Data Circuits)

T568A Configuration (Preferred)			T568B Configuration		
First Modular Plug Pin No.	Conductor Color Code	Second Modular Plug Pin No.	First Modular Plug Pin No.	Conductor Color Code	Second Modular Plug Pin No.
1	W/GN	1	1	W/OR	1
2	GN	2	2	OR	2
3	W/OR	3	3	W/GN	3
4	BL	4	4	BL	4
5	W/BL	5	5	W/BL	5
6	OR	6	6	GN	6
7	W/BR	7	7	W/BR	7
8	BR	8	8	BR	8

Table 6 Data Cross-Over Patch-Cord Wiring

T568A Configuration (Preferred)			T568B Configuration		
First Modular Plug Pin No.	Conductor Color Code	Second Modular Plug Pin No.	First Modular Plug Pin No.	Conductor Color Code	Second Modular Plug Pin No.
1	W/GN	3	1	W/OR	3
2	GN	6	2	OR	6
3	W/OR	1	3	W/GN	1
4	BL	4	4	BL	4
5	W/BL	5	5	W/BL	5
6	OR	2	6	GN	2
7	W/BR	7	7	W/BR	7
8	BR	8	8	BR	8

Table 7 USOC Four-Pair Wiring Scheme Using UTP Cable

USOC Straight-Through Wiring			USOC Reverse (Cross-Over) Wiring		
First Modular Plug Pin No.	Conductor Color Code	Second Modular Plug Pin No.	First Modular Plug Pin No.	Conductor Color Code	Second Modular Plug Pin No.
1	W/BR	1	1	W/BR	8
2	W/GN	2	2	W/GN	7
3	W/OR	3	3	W/OR	6
4	BL	4	4	BL	5
5	W/BL	5	5	W/BL	4
6	OR	6	6	OR	3
7	GN	7	7	GN	2
8	BR	8	8	BR	1

Table 8 USOC Three-Pair Wiring Scheme Using USOC Color Code

USOC Straight-Through Wiring			USOC Reverse (Cross-Over) Wiring		
First Modular Plug Pin No.	Conductor Color Code	Second Modular Plug Pin No.	First Modular Plug Pin No.	Conductor Color Code	Second Modular Plug Pin No.
1	BL	1	1	BL	6
2	BLK	2	2	BLK	5
3	RED	3	3	RED	4
4	GN	4	4	GN	3
5	YEL	5	5	YEL	2
6	WHT	6	6	WHT	1

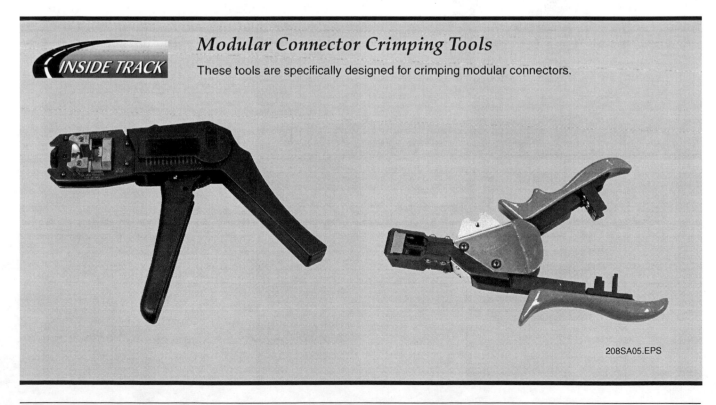

INSIDE TRACK

Modular Connector Crimping Tools

These tools are specifically designed for crimping modular connectors.

208SA05.EPS

3.8.0 Patch Cord and 110 Block Plug Termination Procedures

Patch cords with 110 block plugs can be field-fabricated in most cases for up to Category 5e use. The following are typical instructions for terminating UTP patch cord with 110 block plugs:

Step 1 Cut cable to desired length. Strip the cable jacket at least 1½" (40 mm) from the cable end using stripper cable. Remove a sufficient length of outer jacket to pull the wires into the appropriate wire channel in the cover. Use 24 AWG (0.51 mm) solid or seven-strand twisted-pair conductors. Plugs are also capable of terminating 26 AWG (0.40 mm) solid or seven-strand twisted-pair conductors.

Step 2 Fan out individual pairs according to the desired wiring scheme.

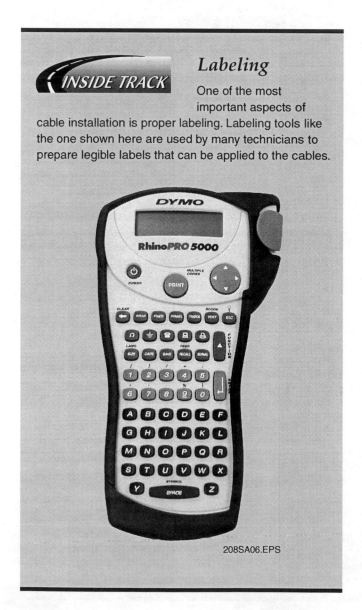

INSIDE TRACK

Labeling

One of the most important aspects of cable installation is proper labeling. Labeling tools like the one shown here are used by many technicians to prepare legible labels that can be applied to the cables.

208SA06.EPS

Step 3 For two-pair, three-pair, and four-pair patch plugs, align the end of the cable jacket over the first rib in the cover. The jacket end should not extend beyond the rib. Use cable tie that is included (except with one-pair) to secure the cable, making sure the cable tie latch is positioned on either side of cable. For one-pair patch plugs, extend the cable jacket over both cover ribs.

Step 4 Insert each wire into the appropriate channel in the cover, maintaining twists as close as possible, ½"(13 mm) or less, to the point of termination. Keep twists symmetrical to avoid a corkscrew effect.

Step 5 Using a diagonal cutting tool, trim the wire ends flush with the front of the housing cover. Wire tips should not protrude past the front surface of the housing cover. Trim any extra pairs back to the cable jacket. Also trim cable tie excess at this time.

Step 6 Align latches on the cover with receptacle holes in the base, and fit the cover IDC slots over contact IDCs.

Step 7 Press the cover and housing base together until all latches are fully engaged. If necessary, squeeze gently with pliers.

Step 8 Snap in the colored icon with the arrow side up. (Colored oval icons are not available for the one-pair patch.)

Step 9 Position the plug with colored icon or top symbol up when mating with the 110 connector (the 110 connector should be orientated with the dark strip down). Do not use if cross-connect wires are terminated to the top of the 110 connectors.

3.9.0 Testing Twisted-Pair Cable

NOTE

The test equipment used to perform the testing covered in this section is covered in the Level Two module, *Advanced Test Equipment*.

The most basic test that is required on every cable is a continuity test, or wire map test. Although this test can be performed with a multimeter, it is inefficient when compared with a pair scanner (wire map tester). This test checks for open circuits, short circuits, and incorrect terminations. When used on ScTP cable, it also checks for drain wire continuity. The pair scanner is connected to one end of the cable, and the tester's remote device is connected to the other. If there is a short, open, or misconnection, the tester will not be able to locate the other end.

NOTE

The ability to pass a tone test does not ensure that there is conductor continuity. In addition to the continuity test, all cables must be tested for length, insertion loss, and near-end crosstalk (NEXT).

Source: Building Industry Consulting Services International, Inc. (BICSI)

Because of the twist in the conductors, the electrical length of the cable will be greater than its physical length. The length of a cable determines the amount of attenuation that cable offers to the signal. Attenuation must be kept within specified parameters. A time-domain reflectometer (TDR) can be used for length testing. The TDR sends a signal that is reflected back from the end of the cable. The time it takes the signal to go down and back is used by the TDR to determine the cable length. Cable length can also be determined using a multimeter and shorting plug in a manner similar to that previously described for coaxial cable.

Certification test procedures can also be used to measure additional parameters required for Category 5e cable.

4.0.0 ◆ SOLDERLESS CONNECTIONS

There are many types of solderless connections. Most involve the use of special crimping tools that are used to crimp the connection device to the conductor. *Figure 25* shows examples of common crimp-type connection devices.

In most cases, crimp-type connectors are furnished with specific barrel/crimp sleeve dimensions for specific types of cables. In addition, manufacturers generally require specific crimping or assembly tools for their crimp-connector types. Make sure to match the cables with the correct connector types and the connector types with the correct assembly tools and procedures. Otherwise, unsatisfactory connections and damaged cable or connectors will result. Most of the connectors are also furnished on factory-assembled patch cords available in various lengths.

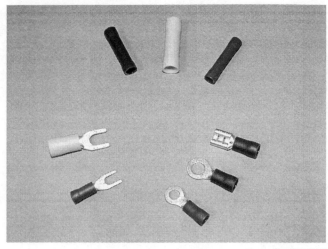
208F25.EPS

Figure 25 ◆ Examples of solderless connection devices.

4.1.0 Crimp Connectors for Screw Terminals

Compression-type connectors used for connecting conductors to screw terminals for low-voltage circuits include those in which hand tools indent or crimp tube-like sleeves that hold one or more conductors. Proper crimping action changes the

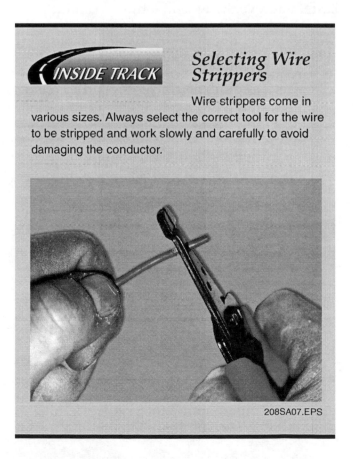

INSIDE TRACK   *Selecting Wire Strippers*

Wire strippers come in various sizes. Always select the correct tool for the wire to be stripped and work slowly and carefully to avoid damaging the conductor.

208SA07.EPS

size and shape of the connector and deforms the conductor strands enough to provide good electrical conductivity and mechanical strength. With the smaller sizes of solid wires, some indenter splice caps require that the wires be twisted together before being placed in the cap and crimped. This will vary with the type of indentations, and the manufacturer's recommendations on the correct method of making the splice should be followed.

One of the disadvantages of indenter or crimp connectors is that special tools are required to make the joint. However, this can be justified by the assurance of a dependable joint and the low cost of the connector.

> **NOTE**
>
> While some types of cable terminations require less installation skill than others, all require a great deal of care and strict compliance with recommendations of the manufacturers of both the cable and terminating means.

Figure 26 shows the basic structure of a crimp connector. The crimp barrel receives and is crimped to the wire. The V's or dimples inside the barrel improve the wire-to-terminal conductivity and also increase termination tensile strength. Most crimp connectors are also available with insulation such as nylon or vinyl covering the barrel to reduce the possibility of shorting to adjacent terminals. The insulation is color-coded according

to the connector's wire range to reduce the problem of wire-to-connector mismatch. An inspection hole is provided at the end of the barrel to allow visual inspection of the wire position. The barrel is connected to the terminal tongue, which physically connects the wire to the termination point, such as a terminal screw. Information about the connector size and conductor range is usually stamped on the tongue by the manufacturer. Tongue styles vary depending on termination requirements. *Figure 27* shows standard tongue styles. The styles most frequently used are the ring tongue and flanged or locking fork. These types are preferred because the terminals will not slip off the terminal screw as the screw is tightened. They are also compatible with most vendor-supplied termination points.

The color-coding of barrel insulation coverings is common among most manufacturers to provide quick identification for installation and as an aid to inspection. Different colors, or a combination of colors, have special meanings. Although manufacturers vary, common or standard colors have been accepted. For example, *Table 9* lists color codes for AMP Special Industries.

Table 9 Color Codes for AMP Special Industries

AWG Wire Size	Color Code
No. 22-16	Red
No. 16-14	Blue
No. 12-10	Yellow

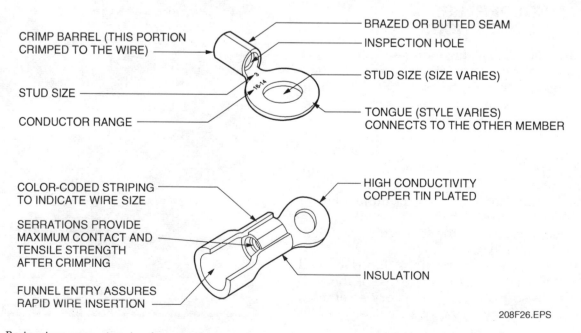

208F26.EPS

Figure 26 ◆ Basic crimp connector structure.

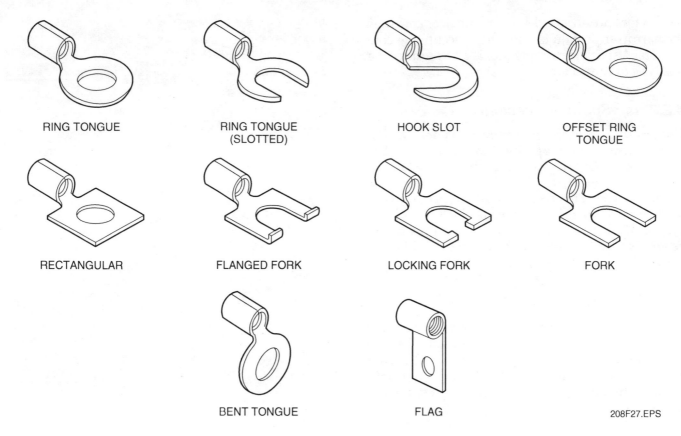

RING TONGUE

RING TONGUE
(SLOTTED)

HOOK SLOT

OFFSET RING
TONGUE

RECTANGULAR

FLANGED FORK

LOCKING FORK

FORK

BENT TONGUE

FLAG

208F27.EPS

Figure 27 ◆ Standard tongue styles of crimped connectors.

Color combinations are sometimes varied to indicate the class or grade rating of an individual **lug** or splice. AMP Special Industries uses a clear plastic or other suitable insulation on the crimp barrel with a colored line to indicate wire size range.

4.2.0 Splice-Type Crimp Connections

A variety of inexpensive devices are used to splice conductors together. Many of these devices require the use of a special crimping tool.

4.2.1 Butt-Splice Connectors

Butt-splice connectors (*Figure 28*) are designed to receive a stripped conductor at either end. The interior of the connector contains a conductive material. After the conductors are inserted into the connector, the center of the connector is crimped with a special crimping tool.

4.2.2 Closed-End Splice Connectors

The closed-end splice connector (*Figure 29*) is also known as a pre-insulated crimp connector. Stripped

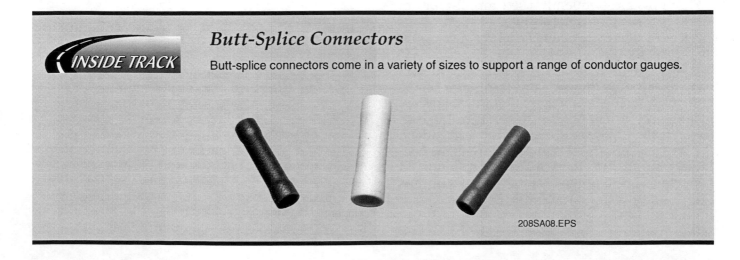

INSIDE TRACK

Butt-Splice Connectors

Butt-splice connectors come in a variety of sizes to support a range of conductor gauges.

208SA08.EPS

conductors are inserted into the device, and then the narrower portion is crimped to secure the conductors. A special crimping tool is also required for this type of connector.

4.2.3 Tap-Splice Connectors

The tap-splice connector (*Figure 30*) is designed to allow a wire to be tapped without cutting it. The connector contains a piece of conductive metal. The two pieces of wire to be spliced are inserted into the connector. When the connector is crimped, the conductive metal penetrates both conductors and acts as a bridge between them. Electrician's pliers can generally be used to crimp the connector, although a special crimping tool is also available.

4.2.4 Molex Connectors

The connector body of a Molex connector (*Figure 31*) is made of nylon or similar material. The two segments are keyed so that they cannot be mated incorrectly. A special round pin is crimped onto each conductor. The pins are then inserted into the connector body.

4.3.0 Wire Nuts

Ever since its invention in 1927, the wire nut, or solderless connector (*Figure 32*), has been a favorite wire connector for use on residential and commercial branch circuit applications. Several varieties of wire nuts are available, but the following are the most common:

• Those for use on wiring systems 300V and under
• Those for use on wiring systems 600V and under (1,000V in lighting fixtures and signs)

Most brands of wire nuts are UL-listed for aluminum-to-copper in dry locations only, aluminum-to-aluminum only, and copper-to-copper only. The maximum temperature rating is 105°C (221°F).

Wire nuts are frequently used for all types of splices in residential and commercial applications and are considered to be the fastest connectors on the market for this type of work.

To use a wire nut, trim the bare conductors using the appropriate tool, and then screw on the wire nut. In doing so, the wire nut draws the conductors and insulation into the shirt of the connector, which increases resistance to flashover. The internal spring is designed to thread the conductors tightly into the wire nut and then hold them with a positive grip. Some types of wire nuts

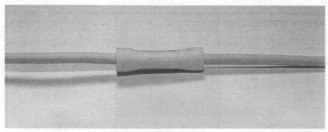

208F28.EPS

Figure 28 ◆ Butt-splice connector.

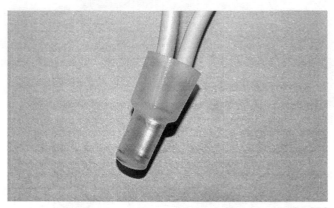

208F29.EPS

Figure 29 ◆ Closed-end splice connector.

208F30.EPS

Figure 30 ◆ Tap-splice connector.

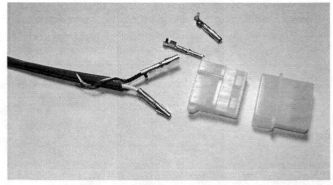

208F31.EPS

Figure 31 ◆ Molex connector.

have thin wings on each side of the connector to facilitate their installation (*Figure 33*). Wire nuts are normally made in sizes to accommodate conductors as small as No. 22 AWG up to as large as No. 10 AWG, with practically any combination of those sizes in between.

Specially designed wire nuts are also made for use in wet locations and/or direct burial applications. These wire nuts have a water-repellent, non-hardening sealant inside the body that protects the conductors against moisture, fungus, and corrosion. The sealant remains in a gel state and will not melt or run out of the wire nut body throughout the life of the connection. Unlike other types of wire nuts, this type can be used one time only. The wire nut can be backed off, eliminating the need to cut the wires for future or retrofit applications, but once removed, it must be discarded.

The general procedure for splicing wires with wire nuts is as follows:

Step 1 Select the proper size wire nut to accommodate the wires. Wire nut packages contain charts that list the allowable combinations of wires by size. Refer to the label on the wire nut box or container for this information (*Figure 34*).

Step 2 Strip the insulation from the ends of the wires. The length of insulation stripped off is typically about ½"; however, it depends on the wire sizes and the wire nut. Follow the manufacturer's directions given on the wire nut package.

Step 3 Stick the ends of the wires into the wire nut and turn clockwise until tight. Note that some manufacturers of wire nuts require that the wires be twisted before the nut is screwed on. Also, some manufacturers recommend using a nut driver to tighten the wire nut. Always follow the manufacturer's instructions.

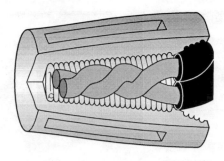

208F32.EPS

Figure 32 ◆ Typical wire nut showing interior arrangement.

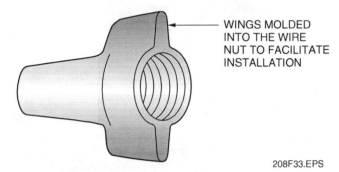

WINGS MOLDED INTO THE WIRE NUT TO FACILITATE INSTALLATION

208F33.EPS

Figure 33 ◆ Some wire nuts have thin wings on each side to facilitate installation.

4.4.0 Cable/Conductor Routing and Inspection Considerations

Cable/conductor routing and inspection are necessary during the job and at the completion of the job. Improper routing and slack can result in future failures of the installed system due to physical damage, shorts, or wire and connector strain. Cable/conductor marking and slack must be sufficient for future system modifications and additions. The cable/conductor routing and inspection considerations include the following:

- *Wire bends* – For all wire and cable routing, maintain a minimum bending radius based on the manufacturer's instructions.

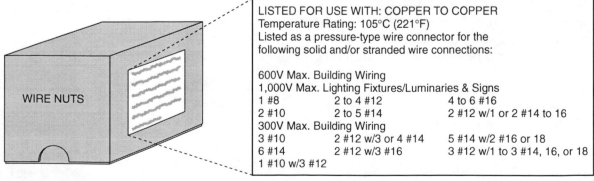

WIRE NUTS

LISTED FOR USE WITH: COPPER TO COPPER
Temperature Rating: 105°C (221°F)
Listed as a pressure-type wire connector for the following solid and/or stranded wire connections:

600V Max. Building Wiring
1,000V Max. Lighting Fixtures/Luminaries & Signs

| 1 #8 | 2 to 4 #12 | 4 to 6 #16 |
| 2 #10 | 2 to 5 #14 | 2 #12 w/1 or 2 #14 to 16 |

300V Max. Building Wiring

3 #10	2 #12 w/3 or 4 #14	5 #14 w/2 #16 or 18
6 #14	2 #12 w/3 #16	3 #12 w/1 to 3 #14, 16, or 18
1 #10 w/3 #12		

208F34.EPS

Figure 34 ◆ Read package label to find allowable wire combinations.

Insulated Spring Connectors

Insulated spring connectors are solderless connectors made in various color-coded sizes that allow for splicing the hundreds of different solid or stranded wire combinations typically encountered in branch circuit and fixture splicing applications.

208SA09.EPS

- *Neatness of routing* – Adequate clearance for movement of mechanical parts should be provided. Enough slack in the cable should be allowed for plugs, terminal boards, and other equipment to be serviced. The cables should be firmly supported to prevent strain on the conductor or the terminals. Strain could cause breaks in the termination and render systems inoperable. The cables should not cross maintenance openings, adjustment openings, or otherwise interfere with normal operation and replacement of components.

- *Protection of wires or cable* – Care must be taken to provide necessary clearance between the conductors and heat-radiating components to prevent deterioration of insulation from heat. All wires should be protected against abrasion. Wire(s) should not be routed over sharp screws, lugs, or terminals, nor should it be bent around sharp openings. Grommets, chase nipples, or some other similar type of protection should be used to protect wire insulation as wire(s) passes through a metal partition.

- *Service loops* – Sufficient slack should be allowed at the termination of each wire to permit one repair, such as cutting off a terminal and re-terminating.

- *Jumper wires* – Bare jumper wires are not permitted. Use only insulated wire for jumpers.

- *Terminal bending limits* – Terminals should not be bent more than 30 degrees above or below the termination point, as shown in *Figure 35*.

- *Quick connectors* – Quick connector considerations include cable retention, strain relief, flexibility, and pull space. Solid wire shall not be used on the removable half of quick connectors.

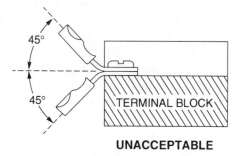

UNACCEPTABLE

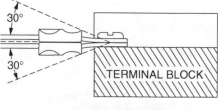

ACCEPTABLE

208F35.EPS

Figure 35 ◆ Terminal bend radius.

• *Inspection* – For wire harness inspections, look for breakouts. This includes checking location, length (stress), and bend radius. Inspect clamping. This includes metal clamps with sleeving or plastic clamps. The maximum sag should be ½" or the cable diameter, whichever is larger. Also, inspect that the wire is not twisted. Inspect cable ties to see that spacing is even.

Inspect for tightness by seeing that the ties do not slip and do not cut or damage any wire. Look at the harness dress and inspect it for neatness, strain relief, visible and legible wire markers and equipment leads, accessible terminals, and closely spaced terminals on sleeving. *Figures 36* through *43* show examples of wire routing and harness dress.

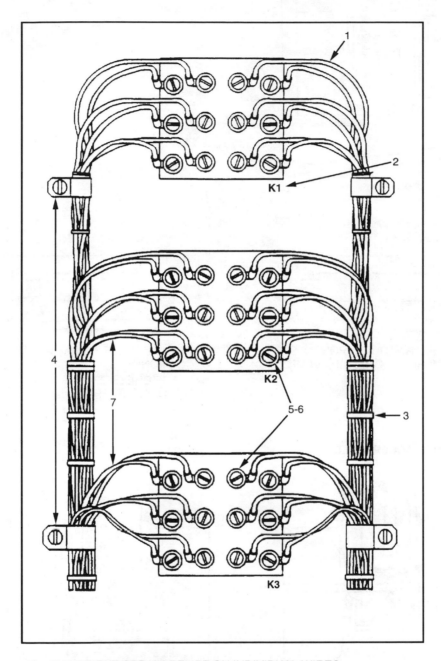

1. SUFFICIENT STRAIN RELIEF ON INDIVIDUAL WIRES
2. REFERENCE DESCRIPTION CLEAR AND EASILY READ
3. PROPER LACING KNOTS OR CABLE TIES USED
4. CABLE PROPERLY CLAMPED
5. TERMINALS ACCESSIBLE
6. SLEEVING USED ON CLOSELY SPACED TERMINALS
7. LEAD DRESS NEAT

208F36.EPS

Figure 36 ◆ Harness dress.

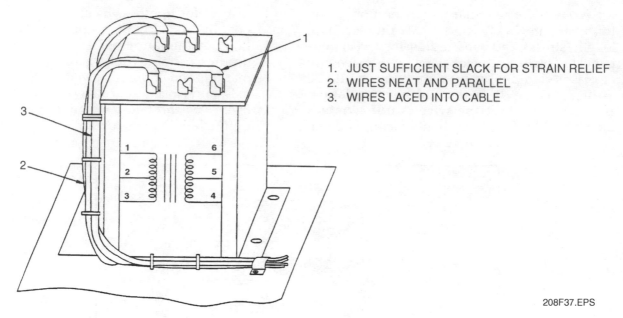

1. JUST SUFFICIENT SLACK FOR STRAIN RELIEF
2. WIRES NEAT AND PARALLEL
3. WIRES LACED INTO CABLE

208F37.EPS

Figure 37 ◆ Wire routing and dress.

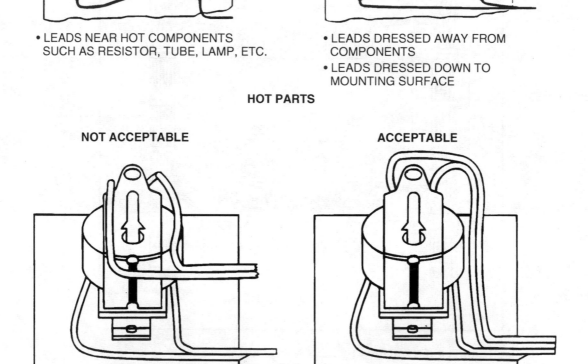

NOT ACCEPTABLE

• LEADS NEAR HOT COMPONENTS SUCH AS RESISTOR, TUBE, LAMP, ETC.

ACCEPTABLE

• LEADS DRESSED AWAY FROM COMPONENTS
• LEADS DRESSED DOWN TO MOUNTING SURFACE

HOT PARTS

NOT ACCEPTABLE

• LEADS NEAR MOVING PARTS

ACCEPTABLE

• LEADS DRESSED SAFELY AWAY FROM ALL MOVING PARTS

MOVING PARTS

208F38.EPS

Figure 38 ◆ Wire routing and dress around hot or moving parts.

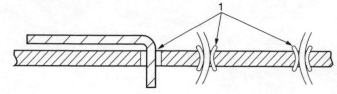 

• NO BREAKS OR DAMAGE TO INSULATION

• WIRE PROTECTED BY GROMMET EYELET, UNIGRIP, OR SIMILAR MATERIAL (SLEEVING NOT ACCEPTABLE)

THROUGH HOLES OF CUTOUTS

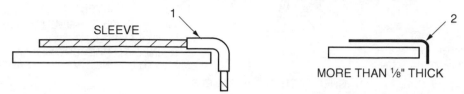

• WIRE PROTECTED WITH UNIGRIP OR OTHER SUITABLE MATERIAL FORMED AROUND SHARP EDGE
• EDGE ROUNDED WITH 1/16" MIN. RADIUS FOR MATERIAL 1/8" THICK OR GREATER

EDGES

NOT ACCEPTABLE	ACCEPTABLE

• BURR ON SHEARED EDGE OF PLATE   • BREAK ON EDGE OF PLATE

ACROSS HOLES, EDGES, AND CUTOUTS

208F39.EPS

Figure 39 ◆ Wire routing – holes, edges, and cutouts.

NOT ACCEPTABLE   **ACCEPTABLE**

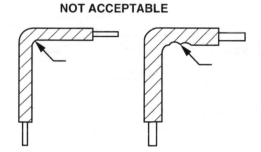   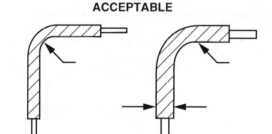

• SHARP BENDS AT CORNER AND RADIUS LESS THAN SHOWN IN TABLE BELOW
• TOOL PINCH MARKS OR DAMAGED INSULATION

• SUFFICIENT BEND RADIUS WITHIN THE SPECIFICATION IN TABLE BELOW
• NO TOOL OR CRIMP MARKS OR DAMAGED INSULATION

WIRE BENDS

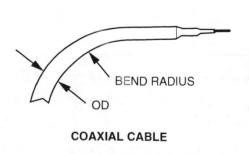

COAXIAL CABLE

CONDUCTOR	MINIMUM BEND RADIUS
INSULATED ELECTRICAL WIRE	2 × OUTSIDE DIAMETER OF WIRE
COAX OR UTP CABLE	4 × OUTSIDE DIAMETER OF WIRE OR CABLE
LARGE WIRE OR CABLE BUNDLES	6 TO 10 × OUTSIDE DIAMETER OR WIRE OR CABLE
FIBER OPTIC CABLE	10 TO 20 × OUTSIDE DIAMETER OF CABLE OR CABLE BUNDLE

TYPICAL MINIMAL BEND RADIUS

208F40.EPS

Figure 40 ◆ Wire/cable routing – minimum bend radius.

NOT ACCEPTABLE	ACCEPTABLE

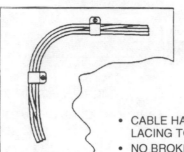

- CABLE HARNESS FORMED BEFORE LACING TO PREVENT STRAINS
- NO BROKEN OR STRAINED WIRES
- CABLE CLAMPED OR OTHERWISE RETAINED TO PREVENT MOVEMENT BY MORE THAN ONE TIME THE DIAMETER

- SHARP BENDS AFTER LACING
- WIRES STRAINED AND BROKEN

HARNESS BENDS FOR PANELS

HARNESS SIZE (DIA. IN INCHES)	SPACING (DIA. IN INCHES)
0.50 AND LARGER	8
0.25 TO 0.50	6
UNDER 0.25	4

SPOT TIE LACING AND CABLE TIE SPACING FOR ALL HARNESSES SHALL BE IN ACCORDANCE WITH THIS TABLE. A SPOT TIE KNOB IS A CLOVE HITCH PLUS A SQUARE KNOT ON TOP.

SPACES OF CABLE TIES OR SPOT TIE LACING

208F41.EPS

Figure 41 ◆ Harness routing – minimum bend radius.

NOT ACCEPTABLE	ACCEPTABLE

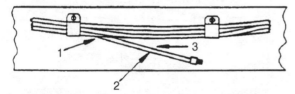 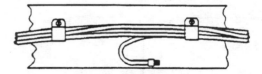

- BREAKOUT IN POOR LOCATION IN RELATION TO TERMINAL
- BREAKOUT WIRE TIGHT
- BEND RADIUS OF BREAKOUT WIRE LESS THAN MINIMUM LIMITS SHOWN IN MINIMUM RADIUS TABLE

- BREAKOUT OPPOSITE TERMINAL
- TIE AT ALL BREAKOUTS
- BREAKOUT WIRE LONG ENOUGH TO ASSURE PROPER STRAIN RELIEF OF WIRE TERMINATIONS
- BEND RADIUS OF BREAKOUT WIRE WITHIN LIMITS SHOWN IN MINIMUM RADIUS TABLE

BREAKOUTS

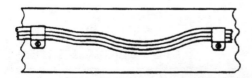 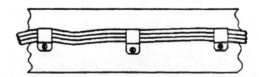

- CABLE SAGS AND CLAMPS NOT POSITIONED ACCORDING TO DRAWING
- METAL CLAMPS USED WITHOUT ADDITIONAL PROTECTION TO CABLE
- WIRES TWISTED IN CABLE

- CABLE STRAIGHT AND PROPERLY CLAMPED PER DRAWING
- APPROVED PLASTIC CLAMPS OR SLEEVING UNDER METAL CLAMP
- WIRES ESSENTIALLY PARALLEL IN CABLE
- MAXIMUM SAG FROM CABLE CENTERLINE OF NO MORE THAN ½" OR THE CABLE DIAMETER, WHICHEVER IS LARGER

CLAMPING

208F42.EPS

Figure 42 ◆ Harness breakouts and clamping.

NOT ACCEPTABLE

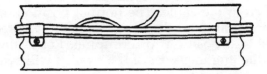

- NO CLAMP OR MAJOR BREAKOUT
- SINGLE TIE IN LACING AT TERMINATION
- WIRES LOOP IN AND OUT OF CABLE

ACCEPTABLE

- MAJOR BREAKOUT MADE AT CLAMPED AREA
- DOUBLE TIES PLUS OVERHEAD KNOT IN LACING AT TERMINATORS
- NO WIRE LOOPS IN AND OUT OF CABLE
- NO TERMINALS, SCREWS, CHIPS, ETC. WITHIN BODY OF CABLE

LACING

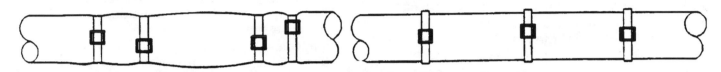

- CABLE TIES TOO TIGHT – MAY CUT INSULATION AND DAMAGE WIRES
- SPACING OF TIES UNEVEN AND NOT PER THE SPACING TABLE

- CABLE TIES PULLED JUST TIGHT ENOUGH TO PREVENT SLIPPAGE
- SPACING EVEN AND WITHIN RECOMMENDED VALUES

PLASTIC CABLE TIES
(NOT RECOMMENDED FOR UTP CATEGORY 5 CABLES)

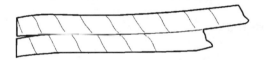

CLAMP-MOUNTED SPIRAL-WRAP TUBING OF VARIOUS SIZES PROVIDES CRUSH-RESISTANCE TO WIRING BUNDLES. FAN-OUT OF WIRES IS ACCOMPLISHED THROUGH THE SPIRAL OPENINGS IN THE WRAP.

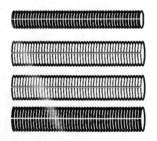

CLAMP-SECURED, SPLIT-LOOM TUBING OF VARIOUS SIZES ALLOWS FAN-OUT OF WIRES AND PROVIDES CRUSH PROTECTION.

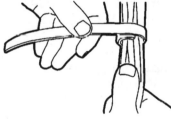

HOOK AND LOOP (VELCRO) WRAP-AROUND CABLE MANAGERS ARE SECURED TO A SINGLE CABLE AND THEN WRAPPED AROUND THE ENTIRE BUNDLE.

HOOK AND LOOP (VELCRO) CABLE ANCHORS HAVE A LARGE HEAD FOR ADDED STRENGTH AND A MOUNTING HOLE FOR SECURING TO A WALL OR RACK.

SPIRAL-WRAP AND SPLIT-LOOM TUBING

VELCRO TIE WRAPS

208F43.EPS

Figure 43 ◆ Harness and cable ties/clamps.

4.5.0 Termination of Conductors/ Cables to Solderless Connectors

The guidelines that apply to the termination of conductors and/or cables to solderless connectors are covered in this section. To ensure an acceptable low-resistance connection, avoid contaminating the termination by cleaning the areas of the cable where it must be cut and stripped. This will prevent pulling compound, oil, grease, or water on the insulation from getting on the exposed conductor.

4.5.1 Conductor Preparation

For cutting cable and wire, a scissor-action type of cutting tool is preferred, if available. Cutting tools with cutting jaws that butt against each other have a tendency to produce a flattened chisel end of the wire, especially when the cutting edges become dull, as shown in *Figure 44*. Chisel point on a wire end makes it difficult, or impossible, to insert the wire into the barrel of a terminal.

The cable jacket should be removed using strippers with an adjustable blade or a die designed for particular size wires to avoid nicking or stretching the wire, or insulation of wires in a multi-conductor cable.

The following points should be observed when stripping wire:

- Nicking, cutting, or scraping of the wire, wire strands, or insulation (inner conductors of multiple conductor cable) must be avoided. This type of damage to the wire indicates that the wrong stripping groove of the stripping tool was used; the tool was improperly set or damaged; the wrong type of stripping blade was used; or the tool was used incorrectly.
- The insulation should be cut cleanly, with no frayed pieces or threads extending past the point of cutoff. Frayed pieces or threads of insulation indicate use of an improper tool or dull cutter blades.

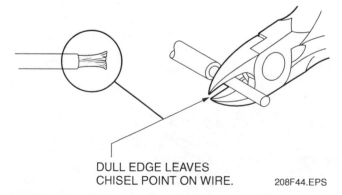

DULL EDGE LEAVES
CHISEL POINT ON WIRE.   208F44.EPS

Figure 44 ◆ Chisel point on a conductor.

- If possible, the wire strands should not be twisted, spread, or disturbed from their normal pitch in the cable. The normal position of the wire strands in the cable, as manufactured, is the best position for crimping. Retwisting or tightening the twist of the strands will eventually result in wire damage.

 WARNING!
Stripping devices that use heated elements can cause severe burns to personnel.

- Stripping devices employing heated elements that burn through the insulation to strip the wire should be used with extreme caution. The conductor, as well as the remaining insulation, can be severely damaged by excess heat or prolonged exposure to heat. Particles of the insulation can melt, flow into the strands of the wire, and prevent establishing a good electrical joint.
- Strip the wire to the proper length required for the terminals being used. The terminal manufacturer will recommend a stripping length. *Figure 45* shows the positioning of the wire in the crimp barrel when stripped to the proper length. Conductor insulation shall be in the belled mouth of the terminal. This relieves stress on the strands or wire and increases strength of the connection. Allowing conductor strands to protrude out of the inspection hole more than 1/32" will interfere with the terminal screw. Cutting the strands too short will reduce the contact surface area.
- Wire that has been stripped should be terminated as soon as possible. The exposed strands will invariably become bent and spread and make subsequent termination operations difficult. Silver-plated strands can become quite tarnished during storage, particularly if ordinary paper, which contains sulphur, is nearby. A minimum amount of handling and storage after stripping will result in better terminations.

4.5.2 Crimping Tools

With a compression-type wire connector, an electrical connection between a wire and a terminal is made by very tightly compressing the crimp barrel with an ordinary pair of pliers. However, such a connection would not necessarily be compressed to the required pressure or in the correct location to ensure a good connection. Therefore, a crimping tool is necessary to produce consistently good connections between terminal and wire.

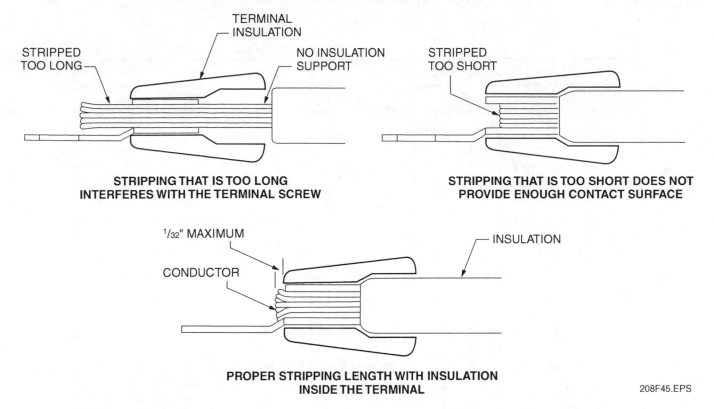

STRIPPED TOO LONG

TERMINAL
INSULATION

NO INSULATION
SUPPORT

**STRIPPING THAT IS TOO LONG
INTERFERES WITH THE TERMINAL SCREW**

STRIPPED TOO SHORT

**STRIPPING THAT IS TOO SHORT DOES NOT
PROVIDE ENOUGH CONTACT SURFACE**

1/32" MAXIMUM

CONDUCTOR

INSULATION

**PROPER STRIPPING LENGTH WITH INSULATION
INSIDE THE TERMINAL**

208F45.EPS

Figure 45 ◆ Proper stripping length.

Figure 46 shows the relationship between the amount of the crimping force and the mechanical and electrical performance. The maximum mechanical strength (A) occurs at a lower crimping force than the maximum electrical performance (B). The point of intersection (C) represents the optimum mechanical and electrical performances, and the corresponding crimping force is selected. Using a crimping die that is too large (undercrimp) has poor electrical performance and a die that is too small (overcrimp) gives a very weak mechanical connection because it literally destroys the crimp barrel and weakens the conductors it is joining.

A simple plier-type crimp tool was the earliest crimping tool developed and continues to be used for repair operations, or where only a few installations are to be made. These tools are similar in construction to ordinary mechanic's pliers except the jaws are specially shaped and the handles are longer, as shown in *Figure 47*.

Since some crimped terminations in the range of 22 to 10 AWG may require a total force of 1,000 pounds or more to complete the crimp, it is necessary to provide a means of increasing the human output force. The force capability of a normal hand for repetitive operations is 75 pounds for adult men and 50 pounds for adult women. The amount of force that the tool multiplies the hand force is termed the **mechanical advantage (MA)** of the tool.

Simple pliers tools are basically constant-MA tools; the MA is the same whether the crimp is being started or finished.

By adjusting linkage or cam mechanisms connecting the handles to the crimp dies, the MA can be varied so that it is low at the start and high at the finish of the crimp stroke. Thus, when the handles start to close from the open position and little or no crimping is being done, the MA is low. As the handles are closed farther, the crimp dies begin to compress the terminal, and the MA increases. In this manner, the MA is patterned to the crimp force requirements and distributed so that a high MA is achieved over the portion of the cycle where it is required.

Figure 48 shows a high MA type of tool that is equipped with a ratchet control. The ratchet mechanism prevents opening the tool and removing the crimped terminal before the handle has been closed all the way and the crimp completed.

The crimp dies of this tool are interchangeable and may contain 2 or 3 positions for crimping different-size terminals. These dies are normally color-coded for use with a color-coded terminal lug for easy identification and to ensure proper crimping force and to minimize overcrimping and undercrimping. The crimp die of the tool determines the completed crimp configuration. There are a variety of configurations in use, such as the simple nest and indenter type die, or the more complicated four-indent die.

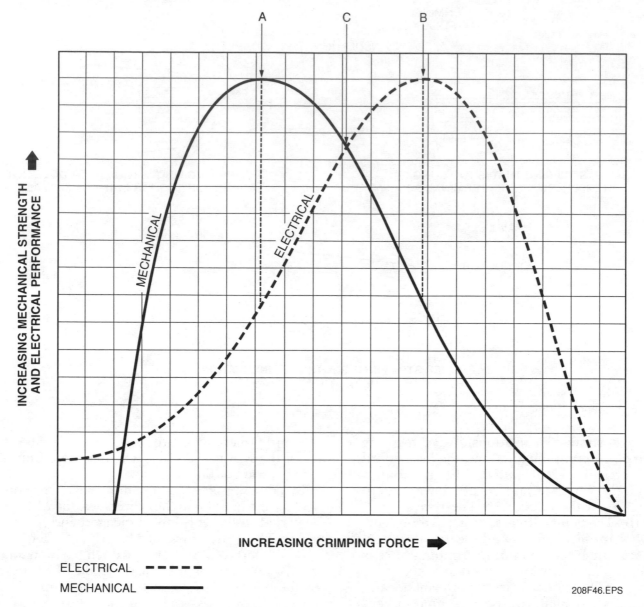

Figure 46 ◆ Mechanical strength versus electrical performance of a crimped connector.

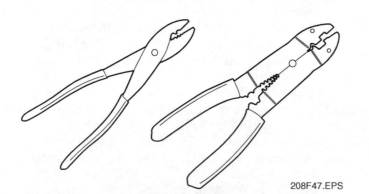

Figure 47 ◆ Commonly used hand crimpers.

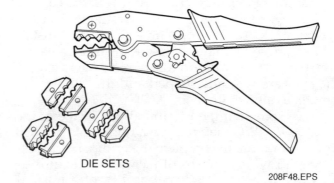

DIE SETS

Figure 48 ◆ Leveraged crimping tool.

Several different configurations may work equally well for some applications, while for others a certain shape is superior. Many considerations affect the determination of crimp die configuration, including the following:

- The type of terminal to be crimped; its size, shape, material, function, and requirements
- The type and size of wires to be accommodated

The major crimp configurations used are shown in *Figure 49* with an explanation of their best application.

The following features are necessary for a good crimping tool:

- The proper location of the terminal in the tool
- The ability to hold the terminal in place while the wire is inserted and the crimp started
- A full cycle control so that the tool will not open and the terminal cannot be removed until the crimp is completed
- The proper size, weight, and shape for efficient operation
- Little or no necessity to adjust or reset the tool parts
- Strong construction and dependable action to provide long, reliable, usage life
- Operation with one hand

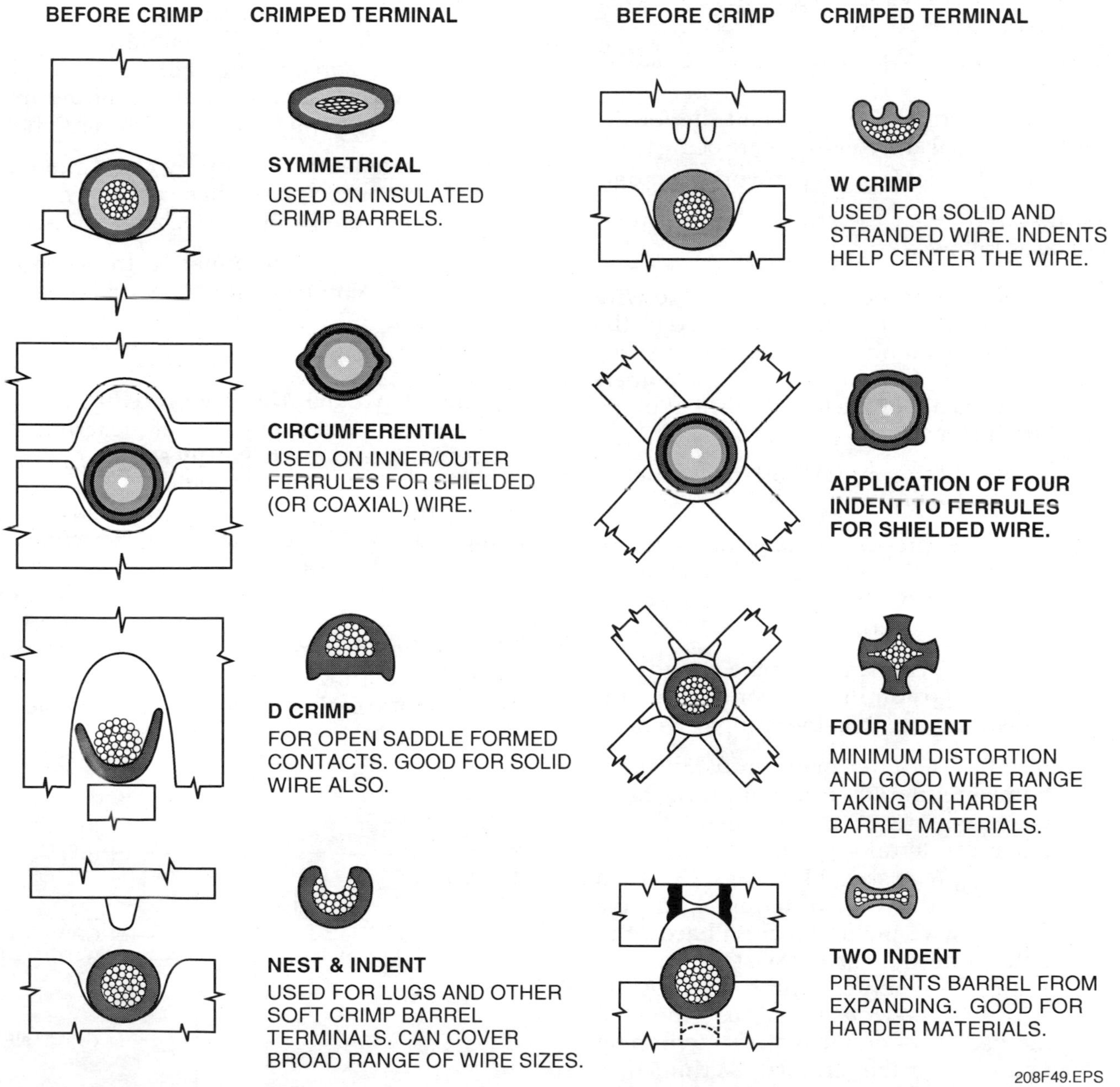

BEFORE CRIMP   **CRIMPED TERMINAL**

SYMMETRICAL
USED ON INSULATED CRIMP BARRELS.

CIRCUMFERENTIAL
USED ON INNER/OUTER FERRULES FOR SHIELDED (OR COAXIAL) WIRE.

D CRIMP
FOR OPEN SADDLE FORMED CONTACTS. GOOD FOR SOLID WIRE ALSO.

NEST & INDENT
USED FOR LUGS AND OTHER SOFT CRIMP BARREL TERMINALS. CAN COVER BROAD RANGE OF WIRE SIZES.

BEFORE CRIMP   **CRIMPED TERMINAL**

W CRIMP
USED FOR SOLID AND STRANDED WIRE. INDENTS HELP CENTER THE WIRE.

APPLICATION OF FOUR INDENT TO FERRULES FOR SHIELDED WIRE.

FOUR INDENT
MINIMUM DISTORTION AND GOOD WIRE RANGE TAKING ON HARDER BARREL MATERIALS.

TWO INDENT
PREVENTS BARREL FROM EXPANDING. GOOD FOR HARDER MATERIALS.

208F49.EPS

Figure 49 ◆ Major crimp configurations.

4.5.3 Crimping Procedure

Prior to making a crimped connection, verify the following items:

- Make sure the size and type of the wire are correct.
- Ensure that the connector and wire materials are compatible. Compare cable and lug designations for material type, such as copper or aluminum. Connectors for this purpose will be properly annotated.

 NOTE

Conductors of dissimilar metals should not be intermixed in a terminal lug or splice connector unless the device is suitable for this purpose.

- Ensure the correct crimp tool and die for the selected terminal and conductor are being used.
- Check that the tool is in proper working order.

The installation of a compression terminal should be as follows:

Step 1 Select the proper terminal for the wire size being terminated, as specified by the terminal manufacturer. This ensures proper fit of the lug and that the **amperage capacity** of the lug equals that of the conductor.

Step 2 Verify that terminal stud size matches terminal screw size.

Step 3 Select the proper crimping tool and die for the terminals being used. This minimizes the possibility of over- or under-crimping the terminal.

Step 4 **Train** the wire strands if the strands are fanned. **Grooming** the conductor provides a proper fit in the crimp barrel.

Step 5 Insert the terminal in the proper die nest. Some insulated terminals now have increased wire ranges that overlap with other-size terminals, such as a terminal covering wire sizes 14 through 18. When crimping this type of terminal, use the middle wire the terminal will accept to determine the correct nest to use. For example, when crimping an 18 AWG wire in a 14-18 terminal, use the nest marked 16-14 on the tool. Overcrimping will occur if the terminal is crimped in the nest marked 18-22. When crimping an uninsulated terminal, the top of the

crimp barrel should be positioned facing the indenter, as shown in *Figure 50*. This will produce the best crimp and allow visual inspection of the crimp after connection to the terminal point. When crimping an insulated terminal, insert the terminal in the color-coded die corresponding to the insulation color.

Step 6 Close the crimp tool handles slightly to secure the terminal while inserting the conductor.

Step 7 Insert the stripped wire into the barrel of the terminal until the insulation butts firmly against the forward stop. Make sure that the insulation is not actually into the barrel. The strands of the conductor should be clearly visible through the inspection opening of the terminal. Make sure all strands of the conductor are inserted into the barrel of the terminal.

Step 8 Complete the crimp by closing the handles until the mechanical cycle has been completed and the ratchet releases.

Step 9 Remove the termination from the tool, and examine it for proper crimping.

4.5.4 Termination Inspection

The final step of the crimping procedure is to perform an inspection to ensure that the termination is electrically and mechanically sound.

Uninsulated terminals should be inspected for correct positioning, centering, and size of the crimp indent. Also, ensure proper terminal size for individual conductors.

Acceptable and unacceptable positioning of a crimp indent is shown in *Figure 51*. The crimp barrel is designed to provide the best mechanical strength when the indent is properly placed on the top (seam) of the barrel. An indent on the side

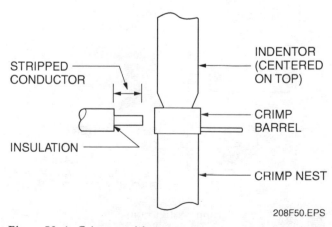

Figure 50 ◆ Crimp position.

can split the terminal seam, thereby reducing both the electrical and mechanical qualities of the termination.

The centering of a crimp indent is very important to the barrel to wire contact area. A poor connection will result if a crimp is placed over the belled mouth or inspection hole of the terminal, as shown in *Figure 52*. A crimp over the belled mouth (if applicable) will compress the insulation of a conductor if the conductor is inserted correctly and will result in poor or no electrical continuity. Crimping over the inspection hole reduces continuity and the holding capabilities of the terminal lug.

Crimping with an incorrect die changes the electrical and mechanical qualities of the termination. The crimp barrel should not be excessively distorted, show any cracks, breaks, or other damage to the base metal.

The previous inspection points were concentrated on the terminal lug and indent location, however, conductor positioning is also very important. Tongue terminals should have the end of the conductor flush with or extending beyond the crimp barrel no more than $\frac{1}{32}$" to ensure proper crimp-to-wire contact area as shown in *Figure 53*. Conductors extending more than $\frac{1}{32}$" past the inspection hole may interfere with the terminal screw. If a conductor is too short and does not reach the inspection hole, then the connection does not provide enough contact surface area; this increases current density and may cause the lug to slip off the conductor.

Terminal lugs having a mouth shall have the conductor's insulation butted against the tapered edge of the crimp barrel as seen in *Figure 53*. Terminal lugs not having a belled mouth shall not have any exposed conductor. Insulation shall be

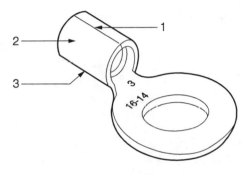

1 – ACCEPTABLE – INDENT ON SEAM (TOP)
2 – UNACCEPTABLE – INDENT ON SIDE
3 – UNACCEPTABLE – INDENI ON BOTTOM

208F51.EPS

Figure 51 ◆ Indent position.

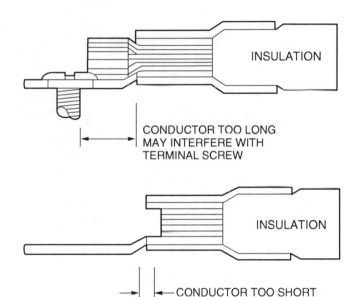

CONDUCTOR TOO LONG MAY INTERFERE WITH TERMINAL SCREW

CONDUCTOR TOO SHORT CANNOT BE SEEN IN INSPECTION HOLE

UNACCEPTABLE

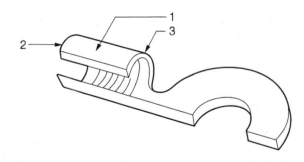

1 – ACCEPTABLE – CENTERED OVER SERRATIONS
2 – UNACCEPTABLE – OVER BELL MOUTH CRIMPS INSULATION
3 – UNACCEPTABLE – OVER INSPECTION HOLE

208F52.EPS

Figure 52 ◆ Crimp centering.

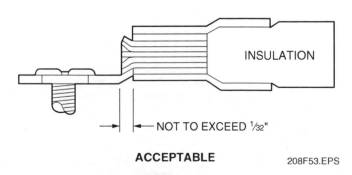

NOT TO EXCEED $\frac{1}{32}$"

ACCEPTABLE   208F53.EPS

Figure 53 ◆ Conductor positioning.

butted to the terminal as seen in *Figure 54*. Any exposed conductor reduces the overall strength of the connection.

Finally, check that the terminal wire sizes stamped on each terminal match the conductor that was crimped. Insulated terminals should be inspected in the same manner as uninsulated terminals; check for proper positioning, centering, and type of crimp. In addition, the terminal insulation should be inspected for breaks, cracks, holes, or any other damage.

Any damage to the insulation is unacceptable. As with uninsulated terminals, insulated terminals shall be inspected to ensure the proper terminal size for each individual conductor. The proper size of terminal and conductor can be checked against the manufacturer's color code. Terminal conductor and stud sizes are usually stamped on the terminal tongue.

The following are some items subject to rejection:

- Wrong-size crimp nest for lug
- Wrong-size lug
- No ring tongue lug used
- Loose connections
- Wire terminated on wrong terminal point per connection drawing

- Conductor strands nicked or broken while stripping insulation
- Too much conductor extended through lug
- All strands not in lug when crimped
- Using the ridge between nests to crimp with
- Incorrect stud size used for a given terminal screw size
- Too much insulation stripped back from lug

4.5.5 Terminal Block Connections

A great variety of terminal blocks is available for use. They range from clamp-type and spring-loaded to the usual screw-type terminal blocks (*Figure 55*).

Wires installed in the spring-loaded terminal blocks are stripped and inserted into holes in the top of the blocks while the spring contact is released with a flat-blade screwdriver. After the wire is inserted, the screwdriver is removed and the wire is clamped in place by the spring contact.

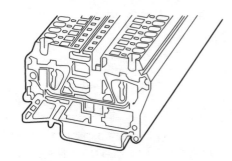

SPRING-LOADED TERMINAL BLOCK

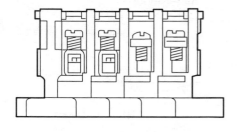

CLAMP-TYPE TERMINAL BLOCK

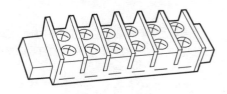

SCREW-TYPE TERMINAL BLOCK

208F55.EPS

Figure 55 ◆ Screw-type terminal blocks.

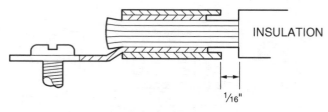

INSULATION

1/16"

UNACCEPTABLE

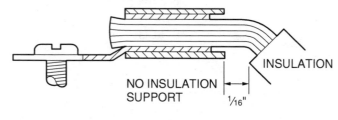

NO INSULATION SUPPORT

INSULATION

1/16"

UNACCEPTABLE

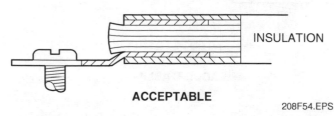

INSULATION

ACCEPTABLE

208F54.EPS

Figure 54 ◆ Conductor support.

Wires installed in a clamp-type terminal block are inserted into a boxed terminal and a screw is tightened to clamp the wire in place. Wires installed on a common screw-type terminal block are installed and curved around the screw or are fitted with a terminal lug that is inserted under the screw. Once the wire or terminal lug is under the head of the screw, the screw is tightened to secure the wire.

When installing wires on a common screw-type terminal block, only two terminals are permitted at any one terminal point. Also, only one flanged fork terminal may be used. When two types of terminals are located at one point, the bottom terminal should be installed upside down, as shown in *Figure 56*. This will provide easier installation and a neater appearance for inspection.

Care should be taken to strip the cable jacket to a point as close as possible to the first termination of the cable, but not to interfere with other cable termination originating from that cable, as seen in *Figure 57*. The cable identification tag should be placed at that point on the cable jacket and located so it can be read easily.

When multiple cables are installed, they should be tied neatly to a support and not block access to the lower terminal blocks or interfere with the connection or disconnection of any wire to the terminal blocks.

Figure 58 shows correct and incorrect termination practices on a typical terminal block.

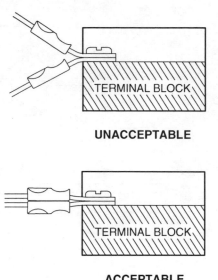

UNACCEPTABLE

ACCEPTABLE

208F56.EPS

Figure 56 ◆ Terminals mounted back-to-back.

When routing individual or multi-conductor cables, the cables should be routed either parallel or at right angles to the frame or wireways. Ensure that wires do not come in contact with or remain in contact with sharp edges. Shielded and coaxial cables should be placed on the outer perimeter of a cable bundle whenever possible. Sharp edges should be avoided and crossover of wire should be kept to a minimum.

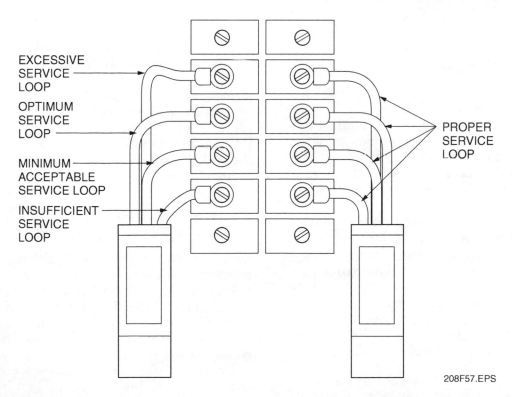

EXCESSIVE SERVICE LOOP

OPTIMUM SERVICE LOOP

MINIMUM ACCEPTABLE SERVICE LOOP

INSUFFICIENT SERVICE LOOP

PROPER SERVICE LOOP

208F57.EPS

Figure 57 ◆ Service loops.

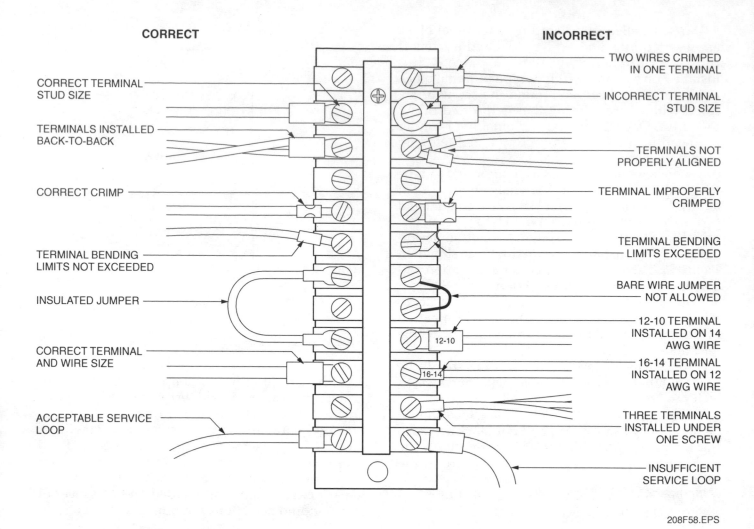

CORRECT

CORRECT TERMINAL STUD SIZE

TERMINALS INSTALLED BACK-TO-BACK

CORRECT CRIMP

TERMINAL BENDING LIMITS NOT EXCEEDED

INSULATED JUMPER

CORRECT TERMINAL AND WIRE SIZE

ACCEPTABLE SERVICE LOOP

INCORRECT

TWO WIRES CRIMPED IN ONE TERMINAL

INCORRECT TERMINAL STUD SIZE

TERMINALS NOT PROPERLY ALIGNED

TERMINAL IMPROPERLY CRIMPED

TERMINAL BENDING LIMITS EXCEEDED

BARE WIRE JUMPER NOT ALLOWED

12-10 TERMINAL INSTALLED ON 14 AWG WIRE

16-14 TERMINAL INSTALLED ON 12 AWG WIRE

THREE TERMINALS INSTALLED UNDER ONE SCREW

INSUFFICIENT SERVICE LOOP

208F58.EPS

Figure 58 ◆ Typical terminal block.

4.6.0 Terminating Typical Shielded Cable

Terminating and grounding other types of shielding, **metallic tape**, and **drain wire** are also important. Depending on the type of installation, shielding may be grounded only at one end of the cable. The ground used for the shield may not be the same potential as the equipment ground. Other circuits may also require terminations at both ends of the cable.

If the shield or drain wire could come in contact with other shields or drain wires, terminal block points, equipment, or equipment cabinets, etc., they should be insulated with insulation tubing or heat-shrinkable tubing to prevent short circuits and circulating currents. Cable that has individually shielded conductors with separate drain wires must have each drain wire and shield tape insulated to eliminate possible circulating currents.

Figure 59 illustrates shield preparation for metallic tape and metallic-braid shielding. While removing the outer jacket, as with coaxial cable, be careful not to damage underlying shielding or individually insulated conductors. For metallic-tape shielding, a bare conductor shall be tack soldered to the metallic shield, as illustrated in *Figure 59*. For metallic-braid shielding, the braid should be pushed back to loosen the braid. After spreading the braid, push the conductors through the braid, tighten the braid, and install a lug (*Figure 60*).

5.0.0 ◆ SOLDER-TYPE CONNECTORS

Some audio connectors are soldered to provide the best possible connection and eliminate interference. Soldering is the process of connecting two pieces of metal together to form a reliable electrical path. The types of solder connectors in common use are the RCA connector and the XLR connector (*Figure 61*).

Occasionally, soldering and desoldering of cables and components is required when installing or making repairs to electrical or electronic equipment.

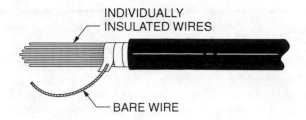

INDIVIDUALLY
INSULATED WIRES

BARE WIRE

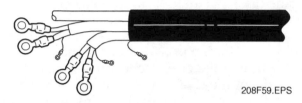

SOLDER
TACK

TO GROUND

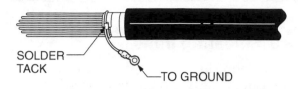

208F59.EPS

Figure 59 ◆ Metallic-tape shielding.

1. PUSH THE METALLIC BRAID BACK.

2. SPREAD THE BRAID AND PULL THE CONDUCTORS
   THROUGH THE BRAID.

3. PULL THE BRAID TO TIGHTEN AND INSTALL A LUG.

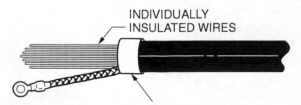

INDIVIDUALLY
INSULATED WIRES

4. SCOTCH TAPE 33, 88 TAPE OR APPROVED HEAT SHRINK
   TUBING CAN BE USED FOR BOTH METALLIC-BRAID
   SHIELDING CABLES.

208F60.EPS

Figure 60 ◆ Metallic-braid shielding.

This section provides some background information about soldering and the gives basic procedures for soldering and desoldering of electronic components and wiring.

5.1.0 Solder

Solder *(Figure 62)* is an alloy made of two or more metals used in various percentages to form a fusible alloy. Solder used in electronic systems is a metal alloy made by combining tin and lead in different proportions. When the proportions are equal, the solder is called a fifty-fifty solder (50 percent tin and 50 percent lead). Similarly, sixty-forty solder consists of 60 percent tin and 40 percent lead. The two most common tin/lead alloys used for soldering electronic equipment and systems are 60/40 and 63/37 (63 percent tin, 37 percent lead). Of the two, the 60/40 is the most widely used. The combination of lead and tin in a solder alloy causes the melting point of the solder to be below that of the melting points of either of its components.

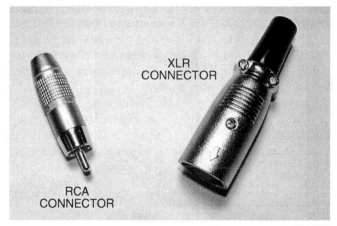

208F61.EPS

Figure 61 ◆ RCA and XLR solder connectors.

208F62.EPS

Figure 62 ◆ A roll of solder.

5.2.0 Soldering Flux

To make a good solder joint, all oxides, corrosion, and tarnish must be removed from the metal surface areas of the items to be soldered. This includes terminals, the stripped end of a wire, or the leads of a component. The removal of surface oxides from metals is accomplished by the use of **soldering flux**. Soldering flux reacts chemically with a metal surface to remove the oxide layer and expose a clean metallic surface that aids in solder bonding.

There are two major categories of soldering flux: inorganic and organic. Inorganic fluxes, commonly called acid fluxes, are not normally used for electrical/electronic soldering because they are too corrosive. They are mainly used in plumbing applications, such as when soldering copper pipe joints. Organic fluxes are of two types: those made from non-rosin materials and those made from rosin-based materials. Non-rosin fluxes also have no or very limited use in electronic soldering. Rosin-type fluxes are made from a variety of natural and modified rosins, such as pine trees. Rosin fluxes are used extensively for soldering in electric/electronic circuits. Rosin flux is sold as a paste. However, most soldering in electronics involves the use of rosin core solder. Rosin-core solder is solder alloy wire that has a central core filled with rosin flux. The rosin flux in the core acts to clean the oxides off the metal parts while they are in the process of being soldered.

5.3.0 Soldering Irons

Most soldering in electrical/electronic equipment is done using a 120V pencil-type electric soldering iron (*Figure 63*). If no electric power is available, butane (gas) or battery-powered soldering equipment can be used. Electric soldering irons consist of a handle, heating element, and the tip. The selection of electric soldering irons and tips is important. The proper heat range or element wattage that is used depends on the size of the connection to be soldered. For example, low-wattage irons are of little use when attempting to solder large joints, such as encountered with large terminals or thick wires. This is because the component being soldered will draw too much heat (sink the heat) away from the tip of the iron, causing the tip to cool down too much. *Table 10* shows the recommended heat ranges for various soldering applications.

208F63.EPS

Figure 63 ◆ Pencil-type soldering iron.

Table 10 Recommended Soldering Iron Heat Range
Versus Application

Heat Range	Application
20–30 watts	Most printed circuits and very small components and terminals
30–40 watts	Small components and terminals and printed circuits with large terminal areas
40–50 watts	Large components and terminals

The method for achieving the required heat range is to use a 40W or 50W element and a variable power source. There is a wide variety of soldering irons available. They range from those with no temperature regulation to those with built-in thermostatic control to ensure that the temperature of the soldering iron tip is maintained at a fixed level. Soldering iron tips are made in different diameters and shapes, and they can be changed to suit the particular soldering job.

Bench-type soldering stations (*Figure 64*) are also available. They typically are a transformer-powered station consisting of a pencil-style soldering iron, non-heat-sinking, heat-resistant soldering iron holder, storage tray for extra tips, and a tip-cleaning cellulose sponge. Most are made to be electrostatic discharge (ESD) safe and safe for soldering heat-sensitive components. Some higher-priced units have the capability of trapping noxious soldering fumes before they have a chance of getting into the atmosphere. Attached to this type of soldering iron is an adjustable tube that can be positioned directly above the soldering tip. The tube is connected to a vacuum pump fitted with special filters within the station. As the fumes rise from the heated solder, they are instantly drawn into the tube where they are captured by the station filter.

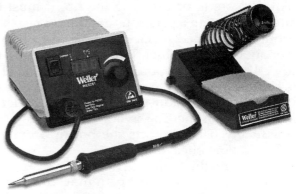

208F64.EPS

Figure 64 ◆ Bench-type soldering station.

5.4.0 The Soldering Process

WARNING!

When soldering, wear proper eye protection. Make sure the area is ventilated and avoid breathing soldering smoke and fumes. The smoke and fumes can irritate mucous membranes and the respiratory system, as well as eyes. If ventilation is insufficient, wear a National Institute of Occupational Safety and Health (NIOSH)-approved respirator with an appropriate filter.

Proper heating of a soldering iron can only be accomplished if the iron is correctly prepared. The tip should be large enough to ensure efficient heat transfer to the area to be soldered or desoldered. However, it must not be so large that it can cause damage to adjacent areas. Smaller tips provide less heat for a given wattage element. Effective wattage may be fine tuned by the proper selection of a tip size. Standard soldering tip shapes include chisel, semi-chisel, bevel, cone, and bevel cone. Select a tip with the largest point face, shortest length, and the largest diameter that will allow safe and adequate access to the work.

5.4.1 Preparing the Soldering Iron

Before attaching the tip to the soldering iron heating element, a small amount of anti-seize compound should be smeared on the tip threads. The tip should be removed and the iron threads cleaned again after about every eight hours of use. Copper tips should be cleaned (dressed) with a smooth file only when cold to prevent oxidation and solder repulsion. When using a plated copper tip, never dress the tip with a file. Clad tips may be cleaned with crocus cloth to remove stubborn oxidation.

 *Soldering Iron Wattage*

A higher wattage for a soldering iron does not mean that the iron is hotter. It means that there is more reserve power for use when soldering larger joints. Thus, the iron can handle heavier-duty soldering applications. This is because the tip of the iron will not cool down as quickly as can happen when soldering with a lower wattage iron.

After cleaning or dressing the tip, it must be **tinned** by flowing a generous amount of new solder onto the shaped surface as soon as the iron is hot enough to melt solder. The iron should then be allowed to idle for about two minutes until it reaches full operating temperature. At this time, additional solder is added to the tip. The solder is left on the tip if the iron is not going to be used immediately. This protects the hot tip from the rapid oxidizing effect of heat.

Before using the soldering iron to make a solder joint, all excess solder must be removed from the tip with a brush, wet cloth, or wet paper towel. If the iron has not been used for some time, fresh solder should be added to the tip prior to wiping off the excess solder. This ensures that any impurities will flow to the surface of the solder and be removed during the wiping action. After the excess solder has been wiped off, the tip should be wiped on a wet cellulose sponge. This provides a clean, dry tip ready for soldering.

5.4.2 Soldering Printed Circuit Board (PC) Mounted Components

When troubleshooting in the field, most faulty printed circuit boards are removed and replaced with a good one, rather than being repaired. However, in some situations it may be necessary to solder components mounted to a printed circuit board. This section gives guidelines for doing so.

Whenever possible, the use of a heat sink is recommended when soldering the leads of all components, especially solid-state components. A heat sink is a device that is clamped on the component lead just above the area to be soldered. It acts to prevent the heat from being transferred up the component lead to the body of the component, thereby protecting the component from heat damage. The useful life of even relatively rugged electronic components can be shortened by heat from soldering. Stainless steel hemostats/forceps-type clamping tools are commonly used as heat sinks. Alligator clips may also be used if space allows. If the component leads do not provide enough access to attach a heat sink, care must be used to prevent heat damage.

Only flux-cored solder should be used as it provides automatic application of flux to the connection while it is being soldered. For large areas or rapidly oxidizing surfaces, a quantity of external rosin flux may also be applied to the joints prior to soldering.

When a dry iron is placed in contact with a metallic surface, the heat causes a rapid oxide layer to form at the junction. This oxide layer is an effective heat insulator and must be removed by the application of flux-cored solder at the junction. The flux will remove the oxide layer, allowing the solder to wet the iron and the surface to be soldered, thereby establishing a **heat transfer bridge**.

After the prepared tip is placed in physical contact with both the component lead and the PC board pad, the solder must be applied to form the solder bond. The iron should not be moved during the soldering operation. Proper cleaning and heating of the joint will allow the solder to flow through the pad hole and form the entire joint with a single application. After the soldering operation is completed, all flux residue should be removed by solvent cleaning with a brush and/or a lint free cloth. Do not allow the solvent to air dry on the board, as it will leave a thin layer of flux residue. Wipe the area dry with a lint-free cloth.

The guidelines for making a good solder joint are as follows:

- Make sure all parts to be soldered are clean and free of dirt and grease.
- Bend the leads of the component to be soldered so that the component is mechanically secured to the PC board. A good mechanical connection is necessary before attempting any soldering.
- Use a small amount of solder to tin the iron tip. This should be done immediately for new tips being used the first time.
- Use a wet cellulose sponge to clean the hot tip of the iron and then add a small amount of solder to the cleansed tip.
- Heat the joint, not the solder. Heat the juncture of the component lead and PC board trace with the iron for about a second. Following this, continue heating the joint with the iron while applying just enough rosin-core solder to the joint to make a good joint. Using too much solder is a waste and can cause short circuits at nearby joints. Using too little solder will cause a poorly formed joint.
- When the joint is soldered, first remove the solder and then the soldering iron tip from the joint.
- Do not move the parts until the solder has cooled naturally.
- When you have finished soldering, the soldered joint should have a nice shiny look; otherwise it is a poor joint.
- Cut the excess length from the component leads. Usually they are cut at the point where the solder has risen up the lead. Note that leads from integrated circuits normally are not cut.

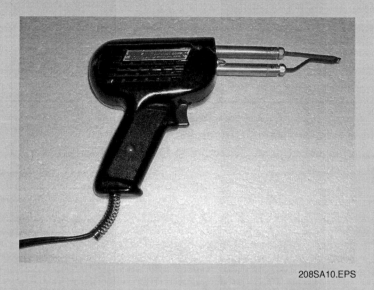
5.4.3 Soldering Wires

Today, most connectors and/or solderless terminals attached to the ends of wires and cables in electronic systems are attached by crimping rather than soldering. Wires terminated using crimp connectors should never be tinned. However, there may be some situations where soldering of wires and cables to non-crimped connectors, or to terminal lugs, is required. When soldering wires, follow the good soldering practices described previously for soldering components on PC boards. Some additional guidelines specific to soldering wires are given here.

Wires to be soldered to connectors should be stripped so that when the wire is placed in the connector solder barrel, there will be a gap of about $\frac{1}{32}$" between the end of the barrel and the end of the wire insulation. This is done to prevent burning the insulation when soldering. It is also done to allow the wire to flex easier at a stress point. Before soldering a copper wire to a connector or terminal lug, make sure that the exposed end is clean and free of oil, grease, and dirt. Following this, tin the exposed end. This is especially important with stranded wire because tinning hold the strands solidly together. Tin the exposed end of the wire for about ½ its length. Note that tinning the exposed wire above this point will cause the wire to be stiff at the point where flexing takes place. This can cause the wire to break. During the

tinning process or when soldering the wire, be sure not to burn the wire insulation.

5.4.4 Desoldering Wires and Components

There will be times when it is necessary to remove solder (desolder) from a joint in order to replace a component, disconnect a wire, or fix an improperly made solder joint. All the guidelines given previously for preparing the soldering iron and making a solder joint also apply when desoldering a joint. However, it should be pointed out that desoldering a joint usually takes somewhat more heat than when soldering it.

Two methods are widely used to desolder a joint. One method uses a desoldering pump, commonly called a solder sucker. The desoldering pump (*Figure 65*) has a heat-proof tip and most are ESD safe. It has a spring-loaded plunger that

when released at the push of a button causes the molten solder from the heated joint to be drawn away from the joint and into the pump. Sometimes it takes more than one attempt to clean all of the solder away from the joint.

The second method uses a woven desoldering braid (*Figure 66*), also called desoldering wick, made of fine strands of oxide-free copper. It has excellent thermal conduction ability and solder retention. When the joint to be desoldered is heated, the desoldering braid works to draw the molten solder up into the braid where it solidifies. Soldering braid is typically packaged on reels and is made in different widths, usually ranging from 0.025" to 0.125". The wider the braid, the more solder it can retain, allowing it to be used when desoldering larger joints. Desoldering braid is recommended when desoldering larger joints that would take several attempts to desolder if using a desoldering pump.

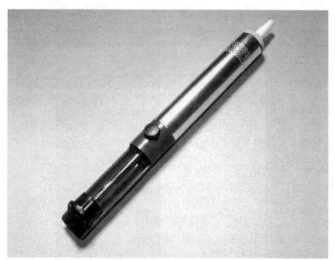

208F65.EPS

Figure 65 ◆ Desoldering pump.

208F66.EPS

Figure 66 ◆ Desoldering a component lead using desoldering braid.

Desoldering using desoldering braid is accomplished as follows:

Step 1   Place the end of the desoldering braid on top of the joint to be desoldered.

> **WARNING!**
> To avoid burning your fingers, hold the braid by its container, or at least 6" from its end.

Step 2   Firmly press a properly heated and tinned soldering iron into the braid and hold it in place.

Step 3   Allow the solder from the heated joint to be drawn into the braid for no more than about ½"; then remove the iron and braid from the joint.

Step 4   Cut the used end of the braid off about ¼" to ⅜" above the point where the solidified solder is visible.

Step 5   If necessary, repeat Steps 1 through 4 until the component or wire is free.

5.5.0 Soldering Safety

Soldering irons, along with the soldering process itself, can cause severe burns. Here are some safety guidelines to follow when soldering.

- Wear safety glasses or goggles.
- Avoid breathing smoke/fumes from soldering. Make sure the smoke/fumes are properly removed by ventilation. If ventilation is inadequate, wear the proper breathing protection.
- Never leave your soldering iron plugged in and unattended.
- To prevent burning your fingers, use needle-nose pliers, heat-resistant gloves, or a small vise to hold small pieces.
- Never set your soldering iron down on anything other than an appropriate soldering iron stand.

Using Desoldering Braid

When using desoldering braid to desolder a component on a PC board, be careful not to allow the solder to cool with the braid adhering to the joint, or you might damage the PC board copper tracks when attempting to pull the braid off the joint.

Desoldering Bulb or Iron

A desoldering bulb or an iron with a built-in desoldering bulb can be used in place of a desoldering pump. Although the tip on the bulb looks like it would melt, it is actually heat-proof.

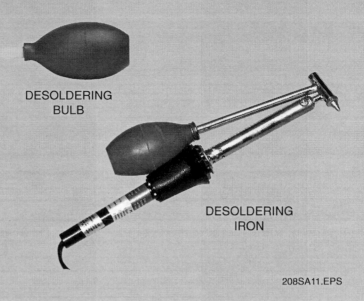

DESOLDERING BULB

DESOLDERING IRON

208SA11.EPS

- Do not shake off excess solder because it can cause serious burns if it contacts your skin.
- Never cut off the grounding prong from a soldering iron plug to make it fit an ungrounded receptacle.
- Replace the soldering iron cord if it is worn or burnt.

5.6.0 Terminating an RCA Connector

The RCA connector is commonly used for consumer-level unbalanced audio. The outer shell of the connector is unscrewed to expose the terminal (*Figure 67*).

Start the process by soldering the cable shield to the shell connection. Then, insert the cable center conductor into the hollow pin and fill with solder. Finally, screw the outer shell onto the connector.

5.7.0 Terminating an XLR Connector

The XLR connector is preferred in professional audio applications. The XLR connector usually has three pins (XLR-3) arranged in a triangular pattern. To terminate a cable, start by removing the set screw from the shell. Then pull out the pin assembly (*Figure 68*). Slide the cable through the outer shell and solder the leads to the pins. There are three pins and a grounding terminal, which is intended to connect the plug to chassis ground.

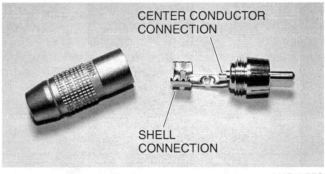

CENTER CONDUCTOR CONNECTION

SHELL CONNECTION

208F67.EPS

Figure 67 ◆ RCA connector with outer shell removed.

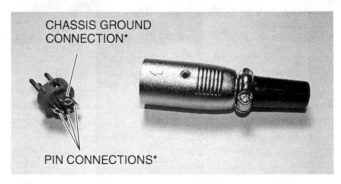

CHASSIS GROUND CONNECTION*

PIN CONNECTIONS*

* The connector pins are numbered. The common arrangement is to solder the black lead to pin 3, the white lead to pin 2, and the shield to pin 1.   *Source: Yamaha Sound Reinforcement Handbook*

208F68.EPS

Figure 68 ◆ XLR connector with outer shell removed.

Connecting the grounding terminal to pin 1 is no longer recommended.

To reassemble the connector, align the keying channel on the shell with the key on the insert. When the connector is reassembled, the pins should not extend beyond the outer shell.

6.0.0 ◆ OPTICAL FIBER CABLE CONNECTORS

Examples of common optical fiber cable connectors are shown in *Figure 69*. Currently, straight tip (ST)-compatible connectors are the most common. However, many installation and application standards now specify the duplex subscriber connector (SC) as the connector of choice because the polarity of the transmit/receive fibers, once established for the installation, will not be a concern for the end user. *ANSI/TIA/EIA-568B* does allow the use of the existing installed base of ST-compatible connectors. In addition, some manufacturers offer duplex ST designs that achieve the same result as the duplex SC connectors. Small form factor (SFF) connectors (*Figure 70*) were developed to save rack space because SC and ST connectors are bulky.

Optical fiber cable may be terminated to these connectors by a number of methods depending on the manufacturer. These methods include the use of an epoxy, a rapid-setting anaerobic adhesive, or a mechanical connection. The mechanical or adhesive methods usually allow termination times of less than two minutes per end. The termination and testing of optical fiber cable is covered in detail in the Level Three module, *Fiber Optics*.

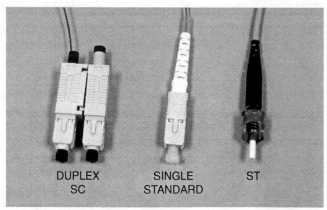

208F69.EPS

Figure 69 ◆ Typical ST and SC optical fiber connectors.

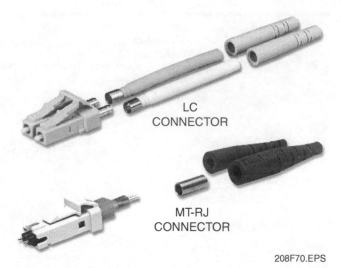

208F70.EPS

Figure 70 ◆ Typical field-installable SFF optical fiber connectors.

7.0.0 ◆ LEGACY COMMUNICATIONS CONNECTORS AND TERMINATIONS

This section covers various types of level 1 and 2 communications connectors and terminations. These connectors and terminations will be encountered in existing buildings that were wired before current standards and methods were implemented.

Old commercial/residential and some new residential telephone installations are generally wired with long-twist, three- or four-wire cable (some without a jacket) that may be initially terminated in one of several ways using the equipment shown in *Figure 71*. In some cases, the telephone equipment is connected directly to a surface-mounted A42 block using spade connectors and the connections are covered with a blank cover. In other cases, the cover may contain a four-wire modular jack (RJ-14C). Sometimes, old style four-prong jacks and plugs were used (*Figure 72*). Some new residential telephone systems are installed and many older telephone systems are extended using modular jacks and four-wire CM-rated type cable (*Figure 73*). These types of installations and equipment are satisfactory for analog telephone and analog fax/data modem use only.

Fiber Termination Tools

A variety of specialized tools and materials are used in terminating optical fiber cable. The tools shown in A are used to strip, cleave, clean, and polish SC and ST connections. A special kit (B) is available for terminating small form factor cables.

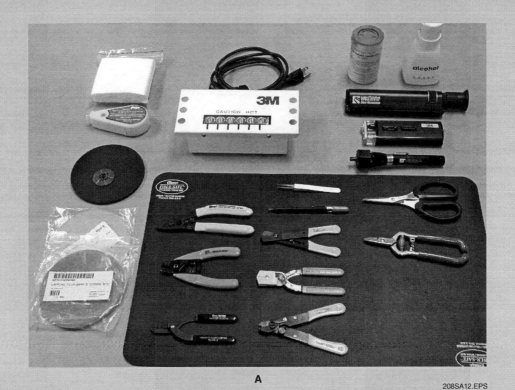

A

208SA12.EPS

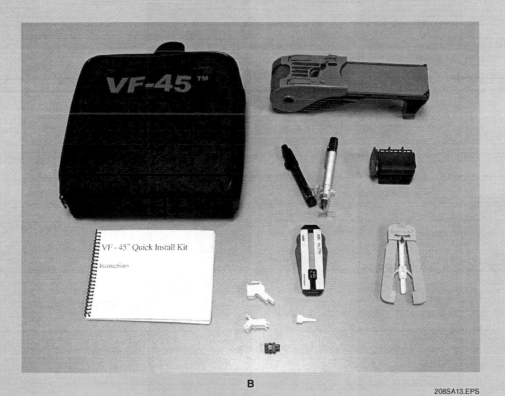

B

208SA13.EPS

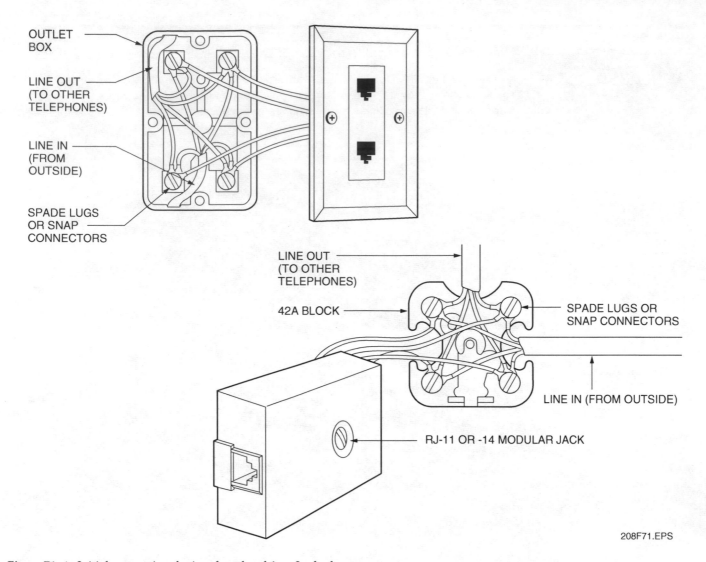

OUTLET BOX

LINE OUT (TO OTHER TELEPHONES)

LINE IN (FROM OUTSIDE)

SPADE LUGS OR SNAP CONNECTORS

LINE OUT (TO OTHER TELEPHONES)

42A BLOCK

SPADE LUGS OR SNAP CONNECTORS

LINE IN (FROM OUTSIDE)

RJ-11 OR -14 MODULAR JACK

208F71.EPS

Figure 71 ◆ Initial connection devices for a level 1 or 2 telephone system.

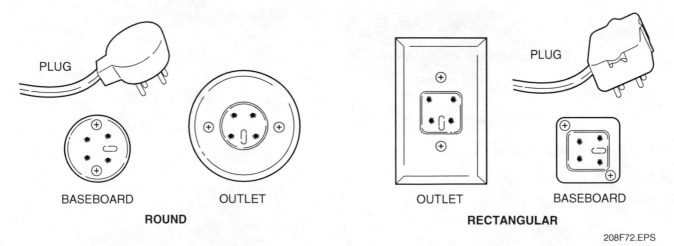

PLUG

BASEBOARD

OUTLET

ROUND

OUTLET

PLUG

BASEBOARD

RECTANGULAR

208F72.EPS

Figure 72 ◆ Old-style jacks and plugs.

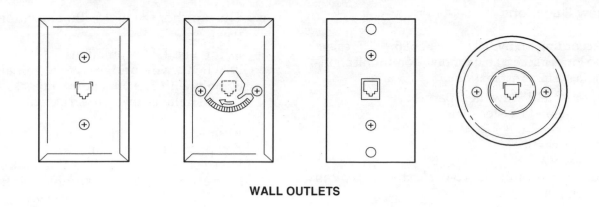

WALL OUTLETS

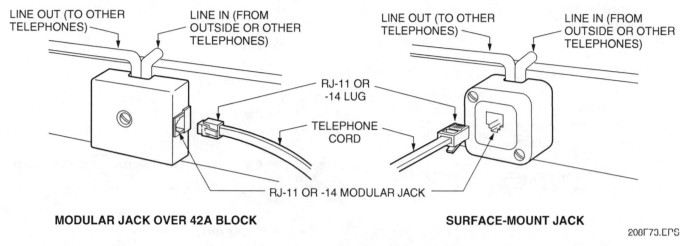

LINE OUT (TO OTHER TELEPHONES)

LINE IN (FROM OUTSIDE OR OTHER TELEPHONES)

RJ-11 OR -14 LUG

TELEPHONE CORD

RJ-11 OR -14 MODULAR JACK

MODULAR JACK OVER 42A BLOCK

LINE OUT (TO OTHER TELEPHONES)

LINE IN (FROM OUTSIDE OR OTHER TELEPHONES)

SURFACE-MOUNT JACK

208Г73.ГРS

Figure 73 ◆ Modular four-wire plugs and recessed or surface-mounted four-wire modular jacks.

Summary

This module has identified a variety of the basic types of terminating devices used for low-voltage cabling. It has covered the basic methods of terminating low-voltage cable to some of the common devices. Low-voltage cable and terminals are commonly joined by solderless connections. Most solderless connections are made with a crimping tool that indents or crimps a connector and conductor together, providing good electrical conductivity and mechanical strength.

Some electronic components, such as audio connectors, are soldered to provide the best possible connection and eliminate interference. This module also describes how to solder and desolder a joint. The electronic systems technician must be aware that the operation of any low-voltage system, including telephone, data, CATV, audio/ video, fire, and security systems, depends primarily on the proper installation and termination of the system cabling using applicable and compatible terminating devices, tools, and testing procedures.

Review Questions

1. According to *ANSI/TIA/EIA-568B*, _____ cable is no longer acceptable for new communications/ data system installations.
   a. unshielded twisted-pair
   b. shielded twisted-pair
   c. coaxial
   d. optical fiber

2. Resistance values for coaxial cable are typically given per _____.
   a. 1,000 yards
   b. one-half mile
   c. square meter
   d. 1,000 feet

3. The abbreviation IDC stands for _____.
   a. industrial device connection
   b. insulation displacement connector
   c. internal device cover
   d. interim digital code

4. The connector wiring scheme shown in *Figure 1* is used for four-pair _____.
   a. token ring
   b. 10base-T
   c. TP-PMD
   d. USOC

5. In a four-pair cable, the first pair in the punch-down order is _____.
   a. blue/white
   b. orange/white
   c. green/white
   d. brown/white

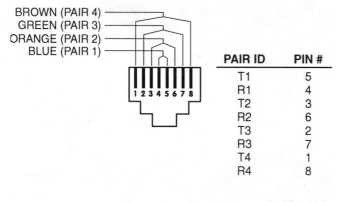

PAIR ID	PIN #
T1	5
R1	4
T2	3
R2	6
T3	2
R3	7
T4	1
R4	8

208RQ01.EPS

Figure 1

6. The punchdown tool without a cutoff blade is used _____.
   a. for all UTP terminations
   b. when a wire bridges two terminals
   c. only with 110 punchdown blocks
   d. when the wire is pre-stripped

7. In telecommunications and data work, it is acceptable to use connecting hardware of a lower category than the cable as long as the connecting hardware is only one generation down.
   a. True
   b. False

8. On a 110 block, electrical connection of the cross-connect wires is made _____.
   a. on base-block IDC clips
   b. before connector blocks are installed
   c. on the top of the connector blocks
   d. on the IDC clips under the connector blocks

9. A palm guard is used when terminating wires on a(n) _____.
   a. A42 block
   b. coupler
   c. modular plug
   d. 66 block

10. Field-fabricated modular cords are not recommended for Category _____ or higher use.
    a. 2
    b. 3
    c. 4
    d. 5

11. The device used to measure the length of copper cable is the _____.
    a. multimeter
    b. TDR
    c. toner
    d. pair tester

12. The continuity test for twisted-pair cable is most efficiently performed with a _____.
    a. volt-ohmmeter
    b. TDR
    c. multimeter
    d. pair scanner

13. The electrical length and the physical length of a twisted-pair cable are the same.
    a. True
    b. False

14. The crimp connector that receives a conductor at either end and is crimped in the center is known as a _____.
    a. closed-end splice connector
    b. butt-splice connector
    c. spade lug
    d. molex connector

15. A circumferential crimping tool is used for _____.
    a. spade lugs
    b. coaxial connectors
    c. ScTP connectors
    d. modular connectors

16. When crimping an uninsulated terminal, the top of the terminal crimp barrel should face _____.
    a. the side of the crimp indenter
    b. away from the crimp indenter
    c. the side of the crimp die
    d. toward the crimp indenter

17. A conductor should extend no more than _____ past the inspection hole on a crimp terminal.
    a. $\frac{1}{32}$"
    b. $\frac{1}{16}$"
    c. $\frac{1}{8}$"
    d. $\frac{1}{4}$"

18. The purpose of soldering flux is to _____.
    a. prevent overheating of the solder joint
    b. remove surface oxides
    c. lubricate the soldering tip
    d. increase the heat of the soldering iron

19. A device commonly used during soldering to protect components from heat damage is a _____.
    a. crocus cloth
    b. solder sucker
    c. stainless steel hemostat
    d. desoldering braid

20. Small form factor connectors were developed primarily to handle increased bandwidth.
    a. True
    b. False

Trade Terms Introduced
in This Module

Amperage capacity: The maximum amount of current that a lug can safely handle at rated voltage.

Coaxial cable: A type of cable that has a center conductor separated from the outer conductor by a dielectric. The center conductor is usually single-strand copper, with the outer conductor made of stranded copper which has been tin plated. This is a very good choice for transmitting very small currents or high frequency signals.

Dielectric: The portion of a coaxial cable that insulates the center conductor.

Drain wire: A wire often attached to a coaxial connector to allow a path to ground from the outer shield.

Grooming: The act of separating the braid in a coaxial conductor. The braid should be separated (untwisted) and pulled back away from the dielectric.

Heat transfer bridge: The application of a small amount of solder to the tip of a soldering iron to facilitate heat transfer between the soldering iron and the joint to be soldered. Also called a solder bridge.

Insulation displacement connector (IDC): The recommended method of copper termination recognized by ANSI/TIA/EIA-568A for UTP cable terminations.

Lug: A term commonly given to crimp type connectors.

Mechanical advantage (MA): The total amount that a crimping tool multiplies the applied hand force.

Metallic tape: The tape that is used to attach a grounding wire to a cable. The tape is metal to make it conductive and allow a good connection to ground.

Service loops: The slack that is left in wire or cable runs to allow for repairs.

Shielding: The metal covering of a cable that reduces the effects of electrical noise.

Soldering flux: A material used to remove surface oxides from metals to be soldered.

Tinned: Applying new solder to the tip of a soldering iron after cleaning or dressing the tip.

Train: The act of smoothing the individual strands in a multi-strand conductor.

Additional Resources

This module is intended to be a thorough resource for task training. The following reference works are suggested for further study. These are optional materials for continued education rather than for task training.

Telecommunications Cabling Installation Manual, Latest Edition. Tampa, FL: BICSI www.bicsi.org.

Telecommunications Distribution Methods Manual, Latest Edition. Tampa, FL: BICSI www.bicsi.org.

Figure Credits

Introduction to Codes and Standards

COURSE MAP

This course map shows all of the modules in the second level of the *Electronic Systems Technician* curriculum. The suggested training order begins at the bottom and proceeds up. Skill levels increase as you advance on the course map. The local Training Program Sponsor may adjust the training order.

ELECTRONIC SYSTEMS TECHNICIAN LEVEL TWO

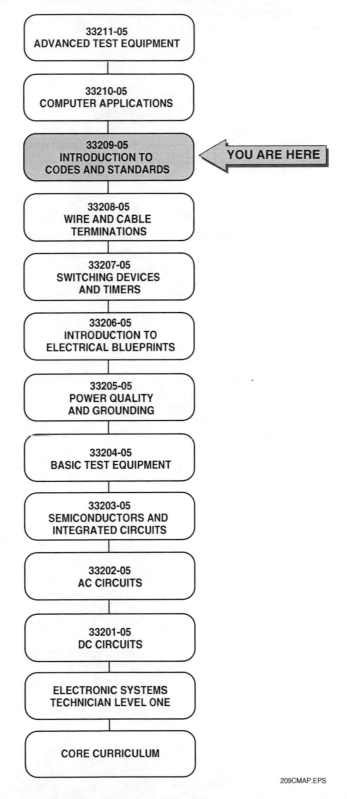

33211-05
ADVANCED TEST EQUIPMENT

33210-05
COMPUTER APPLICATIONS

33209-05
INTRODUCTION TO
CODES AND STANDARDS

YOU ARE HERE

33208-05
WIRE AND CABLE
TERMINATIONS

33207-05
SWITCHING DEVICES
AND TIMERS

33206-05
INTRODUCTION TO
ELECTRICAL BLUEPRINTS

33205-05
POWER QUALITY
AND GROUNDING

33204-05
BASIC TEST EQUIPMENT

33203-05
SEMICONDUCTORS AND
INTEGRATED CIRCUITS

33202-05
AC CIRCUITS

33201-05
DC CIRCUITS

ELECTRONIC SYSTEMS
TECHNICIAN LEVEL ONE

CORE CURRICULUM

209CMAP.EPS

Figures

Tables

Introduction to Codes and Standards

Objectives

When you have completed this module, you will be able to do the following:

1. Identify the difference between codes and standards.
2. Identify trade-relevant codes and standards and their applications.
3. Explain how use and find information in the *National Electrical Code® (NEC®)*.
4. Explain the role of testing laboratories.
5. Use the NEC® to determine the specific requirements for a given telecommunications and/or life safety system application.
6. Use the applicable ANSI/TIA/EIA standards to determine the specific requirements for a given telecommunications and/or life safety system application.

Prerequisites

Before you begin this module, it is recommended that you successfully complete *Core Curriculum*; *Electronic Systems Technician Level One*; and *Electronic Systems Technician Level Two*, Modules 33201-05 through 33206-05.

Required Trainee Materials

1. Pencil and paper
2. Appropriate personal protective equipment
3. Copy of the latest edition of the *National Electrical Code®*

1.0.0 ◆ INTRODUCTION

Most of the work done by ESTs in telecommunications, life safety, security, and other systems is governed by industry standards and building codes. This module introduces and provides a brief overview of the scope and content for the major codes and standards that apply to the industry with an emphasis placed on the *National Electrical Code® (NEC®)*. It is important that you become familiar with the codes and standards that govern your work. During the remainder of your training, you will use these documents extensively and it should become second nature for you to refer to them in the course of your work. A good rule of thumb is: when in doubt, look it up.

2.0.0 ◆ THE PURPOSE OF CODES AND STANDARDS

Building codes and standards regulate all construction work performed in North America. Used together, codes and standards serve to protect life, health, and property and to ensure the quality of construction.

2.1.0 Codes

Building codes and other codes normally address the minimum safety requirements that must be adhered to (*Figure 1*). They typically describe circumstances under which the use of a given type of system, equipment, or component is required, depending on the use and occupancy of a building or structure. Generally, codes are written so

Note: The designations "*National Electrical Code®*" and "NEC®," where used in this document, refer to the *National Electrical Code®*, which is a registered trademark of the National Fire Protection Association, Quincy, MA. *All National Electrical Code® (NEC®) references in this module refer to the 2002 edition of the NEC®.*

- CODES ADDRESS MINIMUM SAFETY REQUIREMENTS.
- CODES DESCRIBE CIRCUMSTANCE(S) UNDER WHICH THE USE OF A GIVEN TYPE OF SYSTEM, EQUIPMENT, OR COMPONENT IS REQUIRED.
- CODES ARE WRITTEN SO THAT THEY CAN BE EASILY ADOPTED INTO LAW.

209F01.EPS

Figure 1 ◆ Codes.

- STANDARDS DETAIL HOW THE PROTECTION SPECIFIED IN CODES IS TO BE ACHIEVED.
- PERFORMANCE STANDARDS SPECIFY FUNCTIONS AND CAPABILITIES OF HARDWARE AND CONDITIONS UNDER WHICH THE EQUIPMENT MUST OPERATE.
- PRESCRIPTIVE STANDARDS SPECIFY PRECISELY WHAT TO DO AND HOW TO DO IT USING APPROVED AND LISTED SYSTEMS AND DEVICES.

209F02.EPS

Figure 2 ◆ Standards.

that they can easily be adopted into law. Some examples of codes developed by the **National Fire Protection Association (NFPA)** and commonly used by ESTs include the following:

- *NFPA 70, The National Electrical Code®*
- *NFPA 72, The National Fire Alarm Code®*
- *NFPA 101, The Life Safety Code®*

Three of the following codes were formerly published and maintained nationally as accepted building codes. Even though they are no longer published, they still may be in force in some state and local jurisdictions.

- *International Building Code®*
- *NFPA 5000, Building and Construction Safety Code™*
- *National Building Code* (no longer published)
- *Uniform Building Code™* (no longer published)
- *Standard Building Code* (no longer published)

There are now only two active building codes: the *International Building Code®* and *NFPA 5000*. State and local governments may also have additional requirements that must be met. More information about these nationally accepted building codes is given later in this module. It is essential that everyone involved in the installation of electronic systems be familiar with the local and national codes in force in the locality where the systems will be installed.

2.2.0 Standards

Standards detail how the protection specified by the code is to be achieved (*Figure 2*). Performance standards specify functions and capabilities of hardware and conditions under which the equipment must operate. Standards should not be confused with technical specifications. For example, NFPA standards require that an audible fire alarm signal be "clearly heard throughout the building," which is usually defined as 15 dBA above ambient

sound levels. It does not specify how many audible alarms to use or where to locate them to achieve the requirement. This kind of information would normally be given in a technical specification.

2.2.1 Types of Standards

There are many types of standards (*Figure 3*). The more common types are defined as follows:

- *International standards* – Mainly used to enhance trade, these standards use a common language and establish equivalent measurements.
- *National standards* – These standards usually define a method of manufacture or design. Two organizations that establish national standards are the **American National Standards Institute (ANSI)** and **National Electrical Manufacturers Association (NEMA)**.
- *Regional and local standards* – Both regional and local standards establish requirements that conform to state and local needs.
- *Safety standards* – These standards establish requirements related to safety issues.
- *Installation standards* – These standards establish requirements related to how the installation of systems and equipment is to be achieved.

TYPES OF STANDARDS

- INTERNATIONAL
- NATIONAL
- REGIONAL AND LOCAL
- SAFETY
- INSTALLATION
- INDUSTRY
- COMPANY
- MANUFACTURER'S INSTRUCTIONS

209F03.EPS

Figure 3 ◆ Types of standards.

- *Industry standards* – Standards developed over the years that form the foundation of how most companies in an industry deal with complexities relating to systems and devices are called industry standards. For example, **Underwriters Laboratories (UL)** has developed standards on most aspects of the telecommunications and electronic security industries, including equipment, installation, and maintenance procedures. To be listed by UL, a company must demonstrate that its products meet the standards for each category.
- *Company standards* – Most companies have developed specific procedures for handling many aspects of your job. These procedures are usually based on applicable codes and standards as well as other good practice methods.
- *Manufacturer's instructions* – Instructions that come with the equipment are called manufacturer's instructions, and they are part of the approval or listing. If the instructions are not followed, the approval or listing is void.

2.2.2 Adopted and Recognized Standards

Adopted standards are standards that have been adopted through statutory or administrative laws in a specific geographic area. Once a document is adopted by reference in a state statute or local ordinance, it has the same force and effect as if printed therein in its entirety.

Recognized standards are standards that are generally accepted as relevant in civil litigation, even in areas where no standards have been formally adopted. Where no applicable code or standard has been officially implemented, it is common for the courts to compare actual circumstances in a case with nationally recognized standards.

The use of standards reduces liability. In civil trials, adopted standards carry the most weight. Where there is no adopted standard, the courts will frequently focus on the issue of compliance with recognized standards. In the case of a conflict between recognized standards and common industry practice, recognized standards will prevail. For example, an alarm system found to be in violation of the recognized standards after a fatal fire may create a liability for the installer.

In a jurisdiction where no particular standard has been legally adopted or no standard is in force, national standards may apply. In addition to the moral and social responsibilities associated with conducting your work correctly, there are added civil liability implications. Complying with or exceeding the recognized codes and standards is your best protection from liability. When faced with justifying an action in a court proceeding, a company is better off if they can cite recognized standards instead of a common company or industry practice.

3.0.0 ◆ DETERMINING WHICH CODES AND STANDARDS TO FOLLOW

There are many types of codes and standards that can apply to a particular system or job. The decision as to which codes and standards apply is made by the Authority Having Jurisdiction (AHJ).

Each job has at least one AHJ. The AHJ is the person or agency that decides the acceptability of the systems and devices being installed in accordance with code requirements.

The AHJ can be a one or all of the following: a building inspector; a fire code inspector; another federal, state, local, or regional authority; a designated insurance company representative; or the property owner. The AHJ determines which rules apply and ultimately resolves any conflicts between any conflicting codes or rules. The AHJ is the organization, office, or individual responsible for approving equipment, an installation, or a procedure.

Deciding which codes and standards apply is not always easy; it requires a familiarity with all of the following:

- Adopted national codes and standards
- Adopted regional codes and standards
- Adopted state codes and standards
- Adopted local codes and standards
- Adopted customer codes and standards
- Codes and standards adopted by your company
- Recognized national codes and standards
- Recognized regional codes and standards
- Recognized state codes and standards
- Recognized local codes and standards

As previously stated, codes and standards establish minimum requirements, thereby allowing work to exceed the requirements and still be considered in compliance. This does not mean that the code or standard requirements can arbitrarily be modified by an individual code enforcement official. Only a person or agency that serves as an AHJ is permitted to modify or supplement the published requirements; however, this should be done in writing through ordinances, statutes, and administrative laws in accordance with applicable legal requirements.

The AHJ is usually authorized to modify the requirements for existing buildings where strict application of the code is not practicable. The AHJ is generally permitted by code to accept an alternate method of compliance if, in the AHJ's judgement, it provides an acceptable level of safety. For example, if an old fire alarm system had 40 manual pull stations mounted at 66 inches above the floor and the current NFPA standard specifies a maximum mounting height of 54 inches, the benefit of bringing the pull stations up to code would not merit the cost.

4.0.0 ◆ WORDS WITH SPECIAL MEANINGS USED IN CODES AND STANDARDS

When interpreting codes and standards, certain words and terms have a special meaning and a major impact on how tasks are to be accomplished. The definition of these words follows:

- *Shall* – This is a mandatory requirement.
- *Should* – This is a recommendation or something that is advised but not required. Note that in some codes and standards the words *may* or *desirable* may be used interchangeably with the word *should*.
- *Approved* – When something is approved, such as an item of equipment, it is sanctioned, certified, and supported.
- *Listed* – When something is listed, such as an item of equipment, the listing person or body has added the equipment to a list, indicating that it shares certain qualities or meets certain tests with other similar listed equipment.
- *Accepted* – When something is accepted, it is believed to satisfy a requirement or standard.
- *Equipment listed for the application* – When used, this term requires that equipment used in a system be properly listed for the application. The key words are *listed for the application*. It is a common misunderstanding for people to assume that, when an item is "listed," it is listed for all applications. Many devices are listed; however, they may not be listed for the specific application. In the listing process, a device becomes listed under one or more of the standards, and should therefore meet the performance requirements of that (those) standard(s) if properly installed, tested, and maintained. It is important to understand which applications each listing is for and to only use products listed for that application.
- *Equivalent* – The "or equivalent" clause in most standards allows the use of systems, methods, or devices of equal or better quality, as long as sufficient documentation is provided to the AHJ to demonstrate the equivalency.

5.0.0 ◆ CODE DEVIATIONS AND CONFLICTS

Code deviations that substitute or provide an alternative feature as accepted by the AHJ are normally considered as conforming to code. Generally, acceptance of any measure not in complete compliance with applicable codes and standards should be specifically stipulated by the AHJ in writing. In the event of conflict between two or more standards, the more stringent requirement should be applied. In one sense, the only true conflict would be a situation in which there is no way to comply with multiple provisions at the same time. For example, if a building code called for manual fire stations to be mounted at 36 to 56 inches above the floor, and another standard specified 42 to 60 inches above the floor, this would appear to be a conflict. Because it is possible to comply with both requirements by installing the manual stations at 42 to 56 inches above the floor, it is not a true conflict.

6.0.0 ◆ *NATIONAL ELECTRICAL CODE®*

NFPA sponsors, controls, and publishes *NFPA 70*, the *National Electrical Code®* (NEC®). It is revised every three years. The NEC® is the most widely used and generally accepted code in the world. It has been translated into several languages. The primary purpose of the NEC® is the practical safeguarding of persons and property from hazards arising from the use of electricity. This purpose is stated in **NEC Section 90.1**. Most federal, state, and local municipalities in the United States have adopted the NEC®, in whole or in part, as their legal electrical code. The NEC® establishes the minimum requirements for the installation of electrical equipment. Some states or localities adopt the minimum requirements established by the NEC®, then add more stringent requirements. Compliance with NEC® requirements increases the safety of electrical installations, which is the reason the NEC® is so widely used. The NEC® is used by many people, such as those listed here:

- Electricians and installers
- Lawyers and insurance companies to determine liability
- Fire marshals and electrical inspectors in loss prevention and safety enforcement
- Designers to ensure a compliant installation

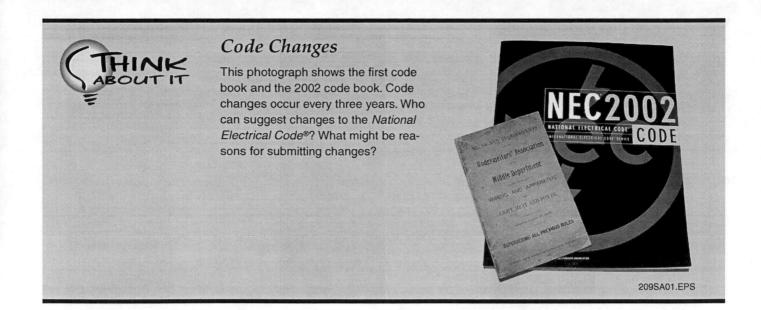

A thorough knowledge of the NEC® is one of the first requirements for becoming a trained electronic systems technician. Although the NEC® itself states "This Code is not intended as a design specification or an instruction manual for untrained persons," it does provide a sound basis for the study of electrical installation procedures, under proper guidance. The NEC® has become the standard reference of the electrical construction industry. Anyone involved in electrical work should obtain an up-to-date copy and refer to it frequently.

Whether you are installing new systems or equipment, or retrofitting existing ones, all electrical work must comply with the current NEC® and all local ordinances. Like most laws, the NEC® is easier to work with once you understand the language and know where to look for the information you need.

6.1.0 The Layout of the NEC®

The NEC® begins with a brief history. The main text of the NEC® is organized into chapters, articles, parts, and sections. Annexes are provided at the back of the book to provide supporting information. *Figure 4* shows how the NEC® is organized.

6.1.1 Types of Rules

There are two basic types of rules in the NEC®: mandatory rules and permissive rules. It is important to understand these rules as they are defined in *NEC Section 90.5*. Mandatory rules contain the words *shall* or *shall not* and must be adhered to. Permissive rules identify actions that are allowed but not required and typically cover options or

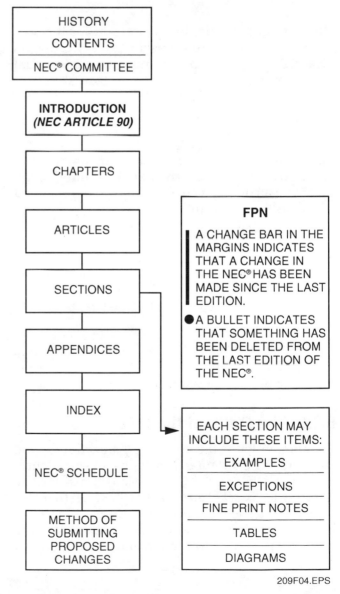

HISTORY

CONTENTS

NEC® COMMITTEE

INTRODUCTION (NEC ARTICLE 90)

CHAPTERS

ARTICLES

SECTIONS

APPENDICES

INDEX

NEC® SCHEDULE

METHOD OF SUBMITTING PROPOSED CHANGES

FPN

| A CHANGE BAR IN THE MARGINS INDICATES THAT A CHANGE IN THE NEC® HAS BEEN MADE SINCE THE LAST EDITION.

● A BULLET INDICATES THAT SOMETHING HAS BEEN DELETED FROM THE LAST EDITION OF THE NEC®.

EACH SECTION MAY INCLUDE THESE ITEMS:

EXAMPLES

EXCEPTIONS

FINE PRINT NOTES

TABLES

DIAGRAMS

209F04.EPS

Figure 4 ◆ The layout of the NEC®.

alternative methods. Be aware that local codes may amend requirements of the NEC®. This means that a city or county may have additional requirements or prohibitions that must be followed in that jurisdiction.

6.1.2 NEC® Introduction

The main body of the text begins with an *Introduction*, also entitled **NEC Article 90**. This introduction gives you an overview of the NEC®. Items included in this section are as follows:

- Purpose of the NEC®
- Scope of the code book
- Code arrangement
- Code enforcement
- Mandatory rules and explanatory material
- Formal interpretation
- Examination of equipment for safety
- Wiring planning
- Metric units of measurement

6.2.0 The Body of the NEC®

The remainder of the book is organized into nine chapters. **NEC Chapter 1** contains a list of definitions used in the NEC®. These definitions are referred to as **NEC Article 100**. **NEC Article 110** gives the general requirements for electrical installations. It is important for you to be familiar with this general information and the definitions.

6.2.1 NEC® Definitions

There are many definitions included in **NEC Article 100**. You should become familiar with the definitions. You should become especially familiar with the following two definitions:

- *Labeled* – "Equipment or materials to which has been attached a label or other identifying mark of an organization that is acceptable to the authority having jurisdiction and concerned with product evaluation, that maintains periodic inspection of production of labeled equipment or materials, and by whose labeling the manufacturer indicates compliance with appropriate standards or performance in a specified manner."

- *Listed* – "Equipment or materials included in a list published by an organization acceptable to the authority having jurisdiction and concerned with product evaluation, that maintains periodic inspection of production of listed equipment or materials, and whose listing states either that the equipment or material meets appropriate designated standards or has been tested and found suitable for use in a specified manner."

Besides installation rules, you will also have to be concerned with the type and quality of materials that are used in electrical wiring systems. Nationally recognized testing laboratories are product safety certification laboratories. Underwriters Laboratories, Inc., also called UL, is one such laboratory. These laboratories establish and operate product safety certification programs to make sure that items produced under the service are safeguarded against reasonably foreseeable risks. Some of these organizations maintain a worldwide network of field representatives who make unannounced visits to manufacturing facilities to counter-check products bearing their seal of approval. The UL label is shown in *Figure 5*.

209F05.EPS

Figure 5 ◆ Underwriters Laboratories, Inc., label.

6.3.0 The Reference Portion of the NEC®

The annexes included with the NEC® provide reference sources. These reference sources can be used to determine the proper application of the NEC® requirements.

6.3.1 Organization of the Chapters

NEC Chapters 1 through 8 each contain numerous articles. Each chapter focuses on a general category of electrical application, such as *NEC Chapter 2, Wiring and Protection*. Each article emphasizes a more specific part of that category, such as *NEC Article 210, Branch Circuits, Part I General Provisions*. Each section gives examples of a specific application of the NEC®, such as *NEC Section 210.4, Multiwire Branch Circuits*. *NEC Chapter 9* contains tables that are used when referenced by any of the articles in *NEC Chapters 1 through 7*. *Annexes A through F* provide informational material and examples that are helpful when applying NEC® requirements.

The chapters of the NEC® are organized into four major categories:

• *NEC Chapters 1, 2, 3, and 4* – The first four chapters present the rules for the design and installation of electrical systems. They generally apply to all electrical installations.

• *NEC Chapters 5, 6, and 7* – These chapters are concerned with special occupancies, equipment, and conditions. Rules in these chapters may modify or amend those in the first four chapters.

• *NEC Chapter 8* – This chapter covers communications systems, such as the telephone and telegraph, as well as radio and television receiving equipment. It may also reference other articles, such as the installation of grounding electrode conductor connections as covered in *NEC Section 250.52*.

• *NEC Chapter 9* – This chapter contains tables that are applicable when referenced by other chapters in the NEC®.

• *Annexes A through F* – Annexes A through F contain helpful information that is not mandatory.
  – *Annex A* contains a list of product safety standards. These standards provide further references for requirements that are in addition to the NEC® requirements for the electrical components mentioned.
  – *Annex B* contains information for determining ampacities of conductors under engineering supervision.
  – *Annex C* contains the conduit fill tables for multiple conductors of the same size and type within the accepted raceways.
  – *Annex D* contains examples of calculations for branch circuits, feeders, and services as well as other loads such as motor circuits.
  – *Annex E* contains information on types of building construction.
  – *Annex F* contains a cross-reference index for the articles that were renumbered in the latest edition of the code.

6.4.0 Text in the NEC®

When you open the NEC®, you will notice several different types of text or printing used. Here is an explanation of each type of text:

• Headings are in bold black letters.

• Italics identify exceptions. Exceptions explain the circumstances under which a specific part of the NEC® does not apply. Exceptions are written in italics under that part of the NEC® to which they pertain.

• **Fine print notes (FPNs)** explain something in an application, suggest other sections to read about the application, or provide tips about the application. These are defined in the text by the term (*FPN*) shown in parentheses before a paragraph in smaller print.

- Figures may be included with explanations to give you a picture of what your application may look like.
- Tables are often included when there is more than one possible application of the NEC®. You would use a table to look up the specifications of your application.

6.5.0 Navigating the NEC®

To locate information for a particular procedure, use the following steps:

Step 1 Familiarize yourself with *NEC Articles 90, 100, and 110,* to gain an understanding of the material covered in the NEC® and the definitions used in it.

Step 2 Turn to the table of contents at the beginning of the NEC®.

Step 3 Locate the chapter that focuses on the desired category.

Step 4 Find the article pertaining to your specific application.

Step 5 Turn to the page indicated. Each application will begin with a bold heading.

> **NOTE**
>
> An index is provided at the end of the NEC®. The index lists specific topics and provides a reference to the location of the material within the NEC®. The index is helpful when you are looking for a specific topic rather than a general category.

Once you are familiar with *NEC Articles 90, 100, and 110,* you can move on to the rest of the NEC®. There are several key sections used often in servicing low-voltage systems. *Table 1* summarizes these sections.

6.5.1 Example of Navigating the NEC®

Suppose you are installing a communications system cable at a building entrance. You know that the metallic members of the cable sheath need to be grounded, but you are not sure of the requirements. To find out this information, use the following procedure:

Step 1 Look in the NEC® table of contents and read down the list until you find an appropriate category. Since you are installing

communications system cables, you would probably first look for the information you need in *NEC Chapter 8, Communications Systems.* Further examination of the table of contents for *NEC Chapter 8* shows that *NEC Article 800, Part IV* covers grounding methods.

Step 2 If the information you are looking for is not readily identified in the table of contents, look in the NEC® index. Follow down the alphabetical listing of subjects until you find an appropriate category for our example, *Grounding.* Listed under *Grounding* you will find the entry *Communication Systems* with references to *NEC Section 800.33* and *NEC Article 800, Part IV.*

Step 3 Referring to *NEC Article 800, Part IV,* you will see that *NEC Section 800.40* covers cable and protector grounding.

Upon reading *NEC Section 800.40* and referenced *NEC Section 800.33,* you know that the metallic members of the cable sheath must be grounded as close as practicable to the point of entrance or they must be interrupted as close to the point of entrance as practicable by an insulating joint or equivalent device. Furthermore, the metallic members of the cable sheath must be grounded using a grounding conductor and electrode that is connected via a bonding jumper to the building or structure power grounding electrode system. These connections must be made as specified in *NEC Section 800.40 (A)* through *(D).*

In addition, the connection made between the communications grounding conductor and electrode shall be made as specified in the applicable sections of *NEC Article 250.*

7.0.0 ◆ CANADIAN ELECTRICAL CODE, PART 1

The Canadian Standards Association (CSA) sponsors, controls, and publishes the *Canadian Electrical Code, Part 1 (CE Code, Part 1).* Like the NEC®, the intent of this code is to establish safety standards for the installation and maintenance of electrical equipment. As with the NEC®, the CE Code may be adopted and enforced by the provincial and territorial regulatory authorities.

Telecommunication and other low-voltage system installers who perform work in Canada must be familiar with the CE Code and comply with its requirements. *Table 2* summarizes several key sections of the CE Code used often in servicing low-voltage systems.

NEC® Reference	Title	Description
NEC Section 90.2	Scope	The Scope provides information about what is covered in the NEC®. This section offers reference to the *National Electrical Safety Code®* (*NESC®*) for industrial or multi-building complexes.
NEC Section 90.3	Code Arrangement	This section explains how the NEC® chapters are positioned. Specifically, *NEC Chapter 8, Communications Systems*, is an independent chapter except where reference is made to other chapters.
NEC Article 100, Part I	Definitions	Definitions are those not commonly defined in English dictionaries. Some terms of interest include *accessible, bonding, explosion-proof apparatus, ground, premises wiring,* and *signaling circuit.*
NEC Section 110.26	Working Space About Electric Equipment (600V, Nominal, or Less)	This section explains the space for working clearances around electrical equipment. This information is useful when placing a terminal in an electrical closet or electronic components on a communications rack.
NEC Article 250	Grounding	This article is referenced from *NEC Article 800.* It contains specific requirements for the communications grounding and bonding network.
NEC Section 300.21	Spread of Fire or Products of Combustion	This section is referenced from various NEC® articles in reference to abandoned cable.
NEC Section 300.22	Wiring in Ducts, Plenums, and Other Air Handling Spaces	This section is referenced from *NEC Article 800 (NEC Section 800.53)* and *NEC Article 820 (NEC Section 820.3).* This article covers the application of listed communication wires, cables, and raceways in ducts, plenums, and other air handling spaces.
NEC Article 500	Hazardous (Classified) Locations	All of *NEC Article 500* is referenced in *NEC Article 800 (NEC Section 800.8).* This article covers hazardous locations such as gasoline stations and industrial complexes. Additionally, healthcare facilities *(NEC Section 517.80)* are of particular importance. Theaters and marinas are also included in this article.
NEC Article 640	Audio Signal Processing, Amplification, and Reproduction Equipment	This article covers audio signal processing, amplification, and reproduction equipment.
NEC Article 725	Class 1, Class 2, and Class 3 Remote-Control, Signaling, and Power-Limited Systems	*NEC Article 725* specifies circuits other than those used specifically for electrical light and power.
NEC Article 760	Fire Alarm Systems	*NEC Article 760* contains requirements for the wiring and equipment used in fire alarm systems.

209T01A.EPS

Table 1 Portions of the NEC® Affecting Telecommunications and Life Safety Systems (2 of 2)

NEC® Reference	Title	Description
NEC Article 770	Optical Fiber Cables and Raceways	*NEC Article 770* pertains to optical fiber cables and raceways. Within this section are the requirements for listing of cable, marking, and installation.
NEC Section 780.6	Cables and Conductors	Referenced from *NEC Article 800 (NEC Section 800.3)*, this section contains requirements for listed hybrid cables used in closed loop and programmed power distribution systems.
NEC Article 800	Communications Systems	*NEC Article 800* contains the requirements for communications circuits.
NEC Article 810	Radio and Television Equipment	*NEC Article 810* contains the requirements for radio and television.
NEC Article 820	Community Antenna Television and Radio Distribution Systems	*NEC Article 820* contains requirements for community antenna television and radio distribution systems.
NEC Article 830	Network-Powered Broadband Communication Circuits	*NEC Article 830* contains the requirements for network-powered broadband communication circuits.

209T01B.EPS

Table 2 Portions of the CE Code Affecting Telecommunications and Life Safety Systems

CE Code Reference	Title	Description
2	General Rules	Provides information on the following: • Permits • Marking of cables • Flame spread requirements for electrical wiring and cables
10	Grounding and Bonding	Contains detailed grounding and bonding information and requirements for using and identifying grounding and bonding conductors
12	Wiring Methods	Involves the requirements for installing wiring systems and outlines the following: • Raceway systems • Boxes • Other system elements
56	Optical Fiber Cables	Contains the requirements for installing optical fiber cables
60	Electrical Communication Systems	Contains the requirements for installing communications circuits

209T02.EPS

8.0.0 ◆ NATIONAL FIRE ALARM CODE® (NFPA 72)

NFPA 72 covers the application, installation, performance, and maintenance of fire alarm systems and their components. The code defines the means of signal initiation, transmission, notification, and annunciation; the levels of performance; and the reliability of the various types of fire alarm systems. It defines the features associated with these systems and also provides the information necessary to modify or upgrade an existing system to meet the requirements of a particular system classification. The intent of the code is simply to define requirements; it is not intended to establish the methods by which the requirements are achieved. The first three chapters contain general information, including administration, referenced publications, and definitions. The remaining eight chapters contain detailed information covering fire alarm systems and their components:

- *Chapter 4, Fundamentals of Fire Alarm Systems* – This chapter covers the fire alarms systems, equipment, and components addressed in the remaining chapters, with the exception of household fire alarm systems, which are covered in Chapter 11.
- *Chapter 5, Initiating Devices* – This chapter defines the requirements for smoke, heat, and fire detectors, as well as alarm initiating devices.
- *Chapter 6, Protected Premises Fire Alarm Systems* – This chapter covers systems, other than household fire protection systems, that are used to protect life and property by detecting and annunciating the existence of heat, fire, or smoke within the protected premises.
- *Chapter 7, Notification Appliances for Fire Alarm Systems* – This chapter contains the requirements for audible and visible devices used to signal a fire alarm.
- *Chapter 8, Supervising Station Fire Alarm Systems* – This chapter covers the requirements for facilities, operation and service, and communications methods used in monitoring protected premises.
- *Chapter 9, Public Fire Alarm Reporting Systems* – This chapter addresses the installation and use of municipal fire alarm systems, including fireboxes and other systems used to initiate fire alarms from the street.
- *Chapter 10, Inspection, Maintenance, and Testing* – This chapter addresses testing and maintenance requirements for protected premises fire alarm systems, supervising station fire alarm reporting systems, public fire alarm reporting systems, and the initiating devices and notification appliances connected to them.
- *Chapter 11, Single- and Multiple-Station Alarms and Household Fire Alarm Systems* – This chapter covers all fire warning equipment used in residential applications.

The code also contains annexes that provide relevant explanatory material, an engineering guide for automatic fire detector spacing, and referenced publications.

9.0.0 ◆ LIFE SAFETY CODE® (NFPA 101®)

NFPA 101® addresses life safety from fire as it applies to both new and existing buildings. It covers those construction, protection, and occupancy features necessary to minimize danger to life from fire, including smoke, fumes, or panic. It identifies the minimum requirements for the design of exit facilities that will permit prompt escape of occupants from buildings or, where desirable, into safe areas within buildings. This code recognizes that life safety is more than a matter of egress and, accordingly, deals with other considerations that are essential to life safety. It does not address all of the general fire prevention or building construction features that are normally a function of fire prevention and building codes.

NFPA 101® is not adopted in all states or local jurisdictions. If it is adopted, there could be conflicts between the various codes and other adopted documents. The AHJ will determine which code takes precedence, although it is usually the most stringent.

The code is divided into 42 chapters. The first 10 chapters contain general information related to the code, as well as material related to building egress and fire safety, including the following:

- General information
- Fundamental requirements
- Definitions
- Classification of occupancy and hazard contents
- Means of egress
- Features of fire protection
- Building service and fire protection equipment

The remaining chapters cover different types of occupancies. The requirements given in these chapters apply to both new and existing occupancies. These chapters cover the requirements for the following types of occupancies:

- Assembly
- Educational
- Healthcare
- Detention and correctional
- Motels and dormitories
- Apartment buildings
- Lodging and rooming houses
- One- and two-family dwellings
- Residential board and care
- Mercantile
- Business
- Industrial
- Storage
- Daycare

Code annexes contain useful explanatory notes and references related to the text contained in the various chapters of the code.

10.0.0 ◆ RELATED NFPA CODES

Other codes published by NFPA that apply to low-voltage installations are briefly described in this section. These codes may be cited in project specifications; therefore, all installers should be familiar with their content.

10.1.0 *Standard for the Installation of Sprinkler Systems (NFPA 13)*

NFPA 13 covers the proper design and installation of sprinkler systems for all types of fire hazards. Coverage includes the following items:

- Design considerations
- Water supplies
- Equipment requirements
- System flow rates
- Sprinkler location and position
- General storage and rack storage of materials
- Specifically identified criteria for special occupancy hazards
- Minimum sizes for sprinklers used in storage applications
- Separation requirements between early suppression fast response and other types of sprinkler systems
- Rules for protecting sprinkler systems against seismic events.

Related standard *NFPA 13D* covers the installation of sprinkler systems in one- and two-family dwellings and manufactured homes. *NFPA 13R* covers the installation of sprinkler systems in residential occupancies up to and including four stories in height.

The fire sprinkler standards, and others (such as *NFPA 20, Standard for the Installation of Stationary Fire Pumps for Fire Protection*) contain requirements for alarms and signaling. Where a structure also requires a fire alarm system in accordance with *NFPA 72*, the more stringent provisions within each standard must be applied in the event of a conflict.

10.2.0 *Standard for the Protection of Information Technology Equipment (NFPA 75)*

NFPA 75 covers the requirements for computer installations needing fire protection and special building construction, rooms, areas, or operating environments. Application is based on risk considerations such as the business interruption aspects of the function, as in computers used in the stock market, or the fire threat to the installation.

10.3.0 *Standard for the Installation of Lightning Protection Systems (NFPA 780)*

NFPA 780 describes the requirements for the protection of people, buildings, special occupancies, heavy-duty stacks, structures containing flammable liquids and gases, and other entities against lightning damage.

11.0.0 ◆ NATIONAL BUILDING CODES

Building codes that are national in scope provide minimum standards to guard the life and safety of the public by regulating and controlling the design, construction, and quality of materials used in modern construction. They have also come to govern the use and occupancy, location of a type of building, and the on-going maintenance of all buildings and facilities. Once adopted by a local jurisdiction, these national building codes then become law. It is common for localities to change or add new requirements to any national code requirements adopted in order to meet more stringent requirements and/or local needs. The provisions of the national building codes apply to the construction, alteration, movement, demolition, repair, structural maintenance, and use of any building or structure within the local jurisdiction.

The national building codes are the legal instruments that enforce public safety in construction of human habitation and assembly structures. They are used not only in the construction industry but also by the insurance industry for compensation appraisals and claims adjustments, and by the legal industry for court litigation.

Up until 2000, there were three model building codes. The three code writing groups, Building Officials and Code Administrators (BOCA), International Conference of Building Officials (ICBO), and Southern Building Code Congress International (SBCCI), combined into one organization called the International Code Council (ICC) with the purpose of writing one nationally accepted family of building and fire codes. The first edition of the *International Building Code®* was published in 2000 and the second edition in 2003. It is intended to continue on a three-year cycle.

In 2002, the NFPA published its own building code, *NFPA 5000®*. There are now two nationally recognized codes competing for adoption by the 50 states.

The format and chapter organization of the two codes differ, but the content and subjects covered are generally the same. Both codes cover all types of occupancies from single-family residences to high-rise office buildings, as well as industrial facilities. They also cover structures, building materials, and building systems, including life safety systems.

Some major cities and local jurisdictions will have their own independent codes. As an EST, you must know the requirements within the jurisdiction where the work is to be performed.

12.0.0 ◆ ANSI/TIA/EIA TELECOMMUNICATIONS RELATED STANDARDS

The American National Standards Institute (ANSI), **Electronic Industries Alliance (EIA)**, and **Telecommunications Industry Association (TIA)** jointly publish standards for the manufacturing, installation, and performance of voice, data, video, audio, security, environmental control, and other electronic and communications equipment and systems. Five ANSI/TIA/EIA standards listed here govern telecommunications cabling in buildings. Each standard covers a specific part of building cabling. They address the required cable, hardware, equipment, design, and installation practices. In addition, each ANSI/TIA/EIA standard lists related standards and other reference materials that deal with the same topics.

- *ANSI/TIA/EIA-568-B, Commercial Building Telecommunications Cabling Standard*
- *ANSI/TIA/EIA-569-B, Commercial Building Standard for Telecommunications Pathways and Spaces*
- *ANSI/TIA/EIA-570-B, Residential Telecommunications Infrastructure*
- *ANSI/TIA/EIA-606, Administration Standard for the Telecommunications Infrastructure of Commercial Buildings*
- *ANSI/TIA/EIA-607, Commercial Building Grounding and Bonding Requirements for Telecommunications*
- *ANSI/TIA/EIA-942, Telecommunications Infrastructure Standard for Data Centers*

12.1.0 *Commercial Building Telecommunications Cabling Standard (ANSI/TIA/EIA-568-B)*

ANSI/TIA/EIA-568-B covers telecommunications cabling in commercial buildings. The standard provides specifications for a generic building cabling system that can be created and used with a variety of products from many different manufacturers. In addition to design specifications, the standard includes performance specifications for cables and components used in commercial building cabling.

The B revision to *ANSI/TIA/EIA-568* was published in 2001. This revision, in addition to incorporating outstanding technical service bulletins, addenda, and interim standards that were added to the previous edition (*568-A*), resulted in a major

---

Association Origins

Both TIA and EIA originated in 1924. TIA started as a small group organizing a telephone industry trade show. That group later became a committee of the United States Independent Telephone Association. It remained that way until 1979, when it split off as a separate association—The United States Telecommunications Suppliers Association (USTSA)—to organize telecom exhibits and trade shows. TIA was formed in 1988 by merging USTSA with the EIA Information and Telecommunications Technical Group. EIA originated in 1924 as the Radio Manufacturers Association.

Source: *Telecommunications Industry Association*

---

reorganization of the standard into three separate technical standards.

- *ANSI/TIA/EIA-568-B.1, General Requirements –* This section covers commercial building cabling for telecommunications products and services. It specifies a telecommunications cabling system for commercial buildings that will support a multi-product, multi-vendor environment.
- *ANSI/TIA/EIA-568-B.2, 100-Ohm Balanced Twisted-Pair Cabling Standard –* This standard covers copper components. It specifies cabling components, transmissions, and cabling models, as well as the measurement procedures needed for verification of balanced twisted pair cabling.
- *ANSI/TIA/EIA-568-B.3, Optical Fiber Cabling Components Standard –* This standard covers fiber components. It specifies components such as cables and connectors, along with the transmission requirements, for an optical fiber cabling system.

A number of addenda have been added since the 2001 publication of the standard. These include the following addenda, which had been issued by the end of 2004.

- *ANSI/TIA/EIA-568-B.1-1 –* This addendum covers bend radii for 4-pair UTP and 4-pair ScTP patch cables.
- *ANSI/TIA/EIA-568-B.1-2 –* This addendum provides additional requirements for grounding and bonding of installed balanced ScTP cables and hardware in commercial buildings.
- *ANSI/TIA/EIA-568-B.1-3 –* This addendum applies to supportable distances and channel attenuation for optical fiber cable.
- *ANSI/TIA/EIA-568-B.1-4 –* This addendum provides recognition for Category 6 and 850 nm laser-optimized 50/125 μm multimode optical fiber cable.
- *ANSI/TIA/EIA-568-B.1-5 –* This addendum covers cabling requirements for telecommunications enclosures.
- *ANSI/TIA/EIA-568-B.2-1 –* This addendum specifies requirements for 100-ohm Category 6 cables and connectors.

- *ANSI/TIA/EIA-568-B.2-2 –* This addendum makes corrections to *ANSI/TIA/EIA-568-B.2*.
- *ANSI/TIA/EIA-568-B.2-3 –* This addendum provides additional considerations for insertion loss and return loss pass/fail criteria.
- *ANSI/TIA/EIA-568-B.2-4 –* This addendum specifies solderless connection reliability for copper connections used in commercial buildings.
- *ANSI/TIA/EIA-568-B.2-5 –* This addendum corrects references in *ANSI/TIA/EIA-568-B.2*.
- *ANSI/TIA/EIA-568-B.2-6 –* This addendum refines and enhances measurement methods specified for Category 6 cable.
- *ANSI/TIA/EIA-568-B.3-1 –* This addendum provides transmission performance standards for 50/125 μm optical fiber cables.

12.2.0 *Commercial Building Standard for Telecommunications Pathways and Spaces (ANSI/TIA/EIA-569-B)*

ANSI/TIA/EIA-569-B provides specifications for designing and constructing the pathways and spaces for telecommunications cabling in commercial buildings. The standard is especially useful for writing construction bids and contracts.

The standard's specifications are based on the belief that the building, the telecommunications needs of its occupants, and the telecommunications technology available may change several times during the life of a building. This standard covers all low-voltage systems that carry information inside a building, including telecommunications pathways and spaces.

ANSI/TIA/EIA-569-B was released in 2004. It has the following structure:

1. *Scope*
2. *Definition of Terms, Acronyms and Abbreviations, Units of Measure, and Symbols*
3. *Telecommunications Diversity*
4. *Entrance Facilities*
5. *Access Provider Spaces and Service Provider Spaces*
6. *Multi-Tenant Building Spaces*

Using Up-to-Date Standards

If your work requires you to refer to standards, you must make sure that you have access to the current version of the standard as well as any addenda to the standard. Addenda, technical service bulletins, and interim standards supplement, clarify, and correct the basic standard. Look through the list of addenda to *ANSI/TIA/EIA-568-B*. You should be able to see how the content of these addenda might have an impact on telecommunications installations.

7. *Building Spaces*
8. *Tenant Building Pathways*

In addition to setting specifications, the standard offers a good working definition of each type of space or pathway discussed. The standard also contains many illustrations of pathways and equipment.

ANSI/TIA/EIA-569-B also includes four annexes:

- *Annex A – Firestopping*
- *Annex B – Additional Section Information*
- *Annex C – Noise Reduction Guidelines*
- *Annex D – Bibliography and References*

12.3.0 *Residential Telecommunications Infrastructure (ANSI/TIA/EIA-570-B)*

ANSI/TIA/EIA-570-B provides standardized requirements for residential telecommunications cabling. The standard's specifications are based on the facilities that are necessary for existing and emerging telecommunications services. Within the standard, services are correlated to grades of cabling for residential units. The cabling requirements covered include support for voice, data, and video; security systems; whole-home audio cabling; and control systems. Cable and connector requirements cover UTP cabling; 75-ohm coaxial cabling; optical fiber cabling; and multi-conductor cabling. Installation requirements are provided for all these types as well as special cabling for security, control, and audio systems. Revision B of the standard was released in 2004. It contains eight sections:

1. *Scope*
2. *Definition of Terms, Acronyms and Abbreviations, Units of Measure, and Symbols*
3. *Single-Dwelling Residence Infrastructure*
4. *Multi-Dwelling/Campus Infrastructure*
5. *Cable and Connecting Hardware*
6. *Installation Requirements*
7. *Field Test Requirements*
8. *Administration*

ANSI/TIA/EIA-570-B also includes four informative annexes:

- *Annex A – Cabling Residential Buildings*
- *Annex B – Installation Guide*
- *Annex C – Typical Applications that Interface to Residential Cabling*
- *Annex D – Bibliography and References*

12.4.0 *Administration Standard for the Telecommunications Infrastructure of Commercial Buildings (ANSI/TIA/EIA-606)*

ANSI/TIA/EIA-606 describes the requirements for the keeping of records and the information that must be available in order to administer the telecommunications system in a commercial building properly. To ensure that administration records are accurate and up-to-date, cabling installers must understand the following concepts:

- What information the records must contain
- How the information should be recorded
- What devices and structures must be labeled
- How the devices and structures should be labeled

The standard requires that an administration system track all aspects of the telecommunications system, including cables, terminations, termination hardware, patching and cross-connect facilities, splices, conduit and other pathways, bonding and grounding, telecommunications rooms, equipment rooms, and other telecommunications spaces.

Each component of the telecommunications system must have a unique identification code. This code is used for the component's record, for the record of connected or related components, and for any drawings where the components appear. If the component requires labeling, the code must be printed on the component's label or on the component itself. Details about identification codes appear in *Section 4.2, Identifiers.*

Section 4.3, Records explains what information must be recorded. Examples of records appear throughout the standard. These examples show both the required information and optional information that would make the record more complete and provide linkage between the components.

Although not part of the standard, *Appendix C* contains a list of common telecommunications symbols used in representing elements of the telecommunications infrastructure.

12.5.0 *Commercial Building Grounding and Bonding Requirements for Telecommunications (ANSI/TIA/EIA-607)*

ANSI/TIA/EIA-607 specifies the grounding and bonding requirements for the telecommunications system in a commercial building. It also specifies the interconnections between telecommunications grounding and other grounding. This standard specifies the following:

- Ground reference for telecommunications systems in entrance facilities, telecommunications rooms, and equipment rooms
- Bonding and connecting of conductors, cable shields, pathways, and hardware in entrance facilities, telecommunications rooms, and equipment rooms

ANSI/TIA/EIA-607 also describes in detail the physical requirements for the five major components of a telecommunications grounding and bonding system:

- Telecommunications bonding backbone (TBB)
- Telecommunications main grounding busbar (TMGB)
- Telecommunications grounding busbar (TGB)
- Telecommunications bonding backbone interconnecting bonding conductor (TBBIBC)
- Bonding conductor for telecommunications

The interconnections between these components and their relationship to other building grounds are explained in detail in the standard. The explanations include installation considerations.

It is specified that all bonding conductors must be a 6 AWG or larger insulated copper wire marked by a distinctive green color. The main bonding conductor must also be labeled with a warning, which asks that any disconnection of this bonding conductor be reported immediately to the building's telecommunications manager.

13.0.0 ◆ RELATED STANDARDS

There are related standards applicable to low-voltage systems published by the **Institute of Electrical and Electronic Engineers (IEEE)**. Some commonly used standards are briefly described here.

13.1.0 *CSMA/CD Access Method (IEEE Standard 802.3)*

This standard covers the telecommunications and information exchange between systems and local and metropolitan area networks. Covered are specific requirements for carrier sense multiple access with collision detection (CSMA/CD) access methods and physical layer specifications. Several supplements to this specification are also available to cover the following subjects: frame extensions for virtual bridged local area networks; physical layer parameters and specifications for 1000MB/s operation over four-pair of Category 5 balanced copper cabling, type 1000BASE-T; and link aggregation.

13.2.0 *Token Ring Access Method (IEEE Standard 802.5)*

This standard covers the telecommunications and information exchange between systems and local and metropolitan area networks. Covered are specific requirements for token ring access methods and physical layer specifications. Supplements to this specification are also available to cover the following subjects: recommended practice for dual ring operation with wrapback reconfiguration and 100Mbit/s dedicated ring operation over two-pair cabling.

13.3.0 *High-Performance Serial Bus (IEEE Standard 1394)*

This specification covers the requirements for the integration of a high-speed serial bus with IEEE standard 32-bit and 64-bit parallel buses, as well as such nonbus interconnects as the *IEEE Standard 1596*, scalable coherent interface. It is intended to provide a low-cost interconnect between cards on the same backplane, cards on other backplanes, and external peripherals. Specifications covered include summary descriptions, cable specifications, backplane specifications, link layer specifications, transaction layer specifications, and serial bus management specifications.

14.0.0 ◆ TESTING LABORATORIES

Testing laboratories are an integral part of the development of codes and standards. The NFPA and other organizations rely on testing laboratories to conduct research into electrical equipment and its safety. These laboratories perform extensive testing of new products to make sure that they are built to NEC® standards for electrical and fire safety. They receive statistics and reports from agencies all over the United States concerning electrical shocks and fires and their causes. Upon seeing developing trends concerning the association of certain equipment and dangerous situations or circumstances, this equipment is specifically targeted for research. All the reports from these laboratories are used in the generation of changes or revisions to the NEC® and other codes.

Underwriters Laboratories, Inc.

The Chicago World's Fair was opened in 1893 and thanks to Edison's introduction of the electric light bulb, the World's Fair lit up the world. But all was not perfect—wires soon sputtered and crackled, and, ironically, the Palace of Electricity caught fire. The fair's insurance company called in a troubleshooting engineer, who, after careful inspection, found faulty and brittle insulation, worn out and deteriorated wiring, bare wires, and overloaded circuits.

He called for standards in the electrical industry, and then set up a testing laboratory above a Chicago firehouse to do just that. Hence, Underwriters Laboratories, Inc., an independent testing organization, was born. Underwriters Laboratories, Inc. (UL), is now an internationally recognized authority on product safety testing and safety certification and standards development.

14.1.0 Nationally Recognized Testing Laboratories

Nationally Recognized Testing Laboratories (NRTLs) are product safety certification laboratories. They establish and operate product safety certification programs to make sure that items produced under the service are safeguarded against reasonably foreseeable risks. NRTLs maintain a worldwide network of field representatives who make unannounced visits to factories to check products bearing their safety marks.

Two such nationally recognized testing laboratories are the UL and the Canadian Standards Association (CSA). The UL is a not-for-profit organization whose principle business is to evaluate U.S. products to determine their compliance with defined standards for safety. UL evaluates electrical and mechanical products, building materials, construction systems, fire protection equipment, and marine products. UL's standards for safety establish the basis for testing products and levels of acceptability. Once a product has successfully passed the testing process, it may be listed and have the UL label affixed to it. The CSA performs product evaluations and tests on Canadian products similar to those performed by the UL. Through international agreements, both organizations are authorized to test and certify specific types of products for use in both U.S. and Canadian markets.

14.2.0 National Electrical Manufacturers Association

The National Electrical Manufacturers Association (NEMA) was founded in 1926. It is made up of companies that manufacture equipment used for generation, transmission, distribution, control, and utilization of electric power. The objectives of NEMA are to maintain and improve the quality and reliability of products; to ensure safety standards in the manufacture and use of products; and to develop product standards covering such matters as naming, rating, performance, testing, and dimensions. NEMA participates in developing the NEC® and the *National Electrical Safety Code*® and advocates their acceptance by state and local authorities.

Summary

In addition to learning the technical aspects of your profession, you must become familiar with the codes and standards that govern your work. Most of the work done by ESTs in telecommunications, life safety, security, and other systems is governed by industry standards and building codes. The NEC® specifies the minimum provisions necessary for protecting people and property from hazards arising from the use of electricity and electrical equipment. As an EST, you must be aware of how to use and apply the NEC® on your job. Using the NEC® correctly will help you to install and maintain the electrical equipment and systems you will encounter.

As an EST, you will be the last person on the design/installation team to see that systems are installed correctly and the devices operate as intended. Your knowledge of the relevant codes and standards, and their requirements, will enable you and your company to provide the customer with an approved, high-quality system.

Review Questions

1. The NEC® is published by the _____.
   a. National Fire Protection Association
   b. National Electrical Code Association
   c. International Code Council
   d. American National Standards Institute

2. A mandatory rule contains the words _____ and must be adhered to.
   a. shall or shall not
   b. accepted or not accepted
   c. should or should not
   d. approved or not approved

3. If there is a conflict between two codes or standards that apply to a job, the best approach is to _____
   a. apply the most demanding standard
   b. apply the least demanding standard
   c. call the organizations that publish the codes
   d. use the average of the two requirements from both standards

4. The NEC® provides the _____ requirements for the installation of electrical systems.
   a. most stringent
   b. complete
   c. design specification
   d. minimum

5. *NEC Article 110* covers _____.
   a. branch circuits
   b. definitions
   c. general requirements for electrical installations
   d. wiring design and protection

6. Hazardous locations are covered in _____.
   a. *NEC Article 250*
   b. *NEC Article 500*
   c. *NEC Article 725*
   d. *NEC Article 760*

7. Which of the following codes is the *National Fire Alarm Code*®?
   a. *NFPA 13*
   b. *NFPA 72*
   c. *NFPA 75*
   d. *NFPA 101*

8. Regarding building codes, _____.
   a. national codes always take precedence over local codes
   b. local codes are the same as national codes, so there is no conflict
   c. local codes may contain requirements that are not covered in national codes
   d. they are for information only and do not have to be rigidly followed

9. The ANSI/TIA/EIA standard that governs commercial pathways and spaces is _____.
   a. *ANSI/TIA/EIA-568-B*
   b. *ANSI/TIA/EIA-569-B*
   c. *ANSI/TIA/EIA-570-B*
   d. *ANSI/TIA/EIA-606*

10. The function of Underwriters Laboratories, Inc., is to _____.
    a. ensure electronic systems against damage or failure
    b. check installed systems to make sure they work correctly
    c. develop product standards for electronic systems and equipment
    d. conduct research into electrical equipment and its safety

Trade Terms Introduced in This Module

American National Standards Institute (ANSI): An international organization that serves in the capacity as administrator and coordinator of the U.S. private sector voluntary standardization system. ANSI does not itself develop standards, rather it facilitates development by establishing consensus among qualified groups.

Electronic Industries Alliance (EIA): An international organization whose goal is to enhance the competitiveness of American electronics producers. The organization supports functions for its members in areas of market statistics, technical standards, government relations, and public affairs.

Institute of Electrical and Electronic Engineers (IEEE): An international organization that produces about 30 percent of the world's published literature and standards pertaining to electrical engineering, computers, and control technology.

National Electrical Manufacturers Association (NEMA): A group made up of companies that manufacture equipment used for the generation, transmission, distribution, control, and utilization of electric power. The objectives of NEMA are to maintain and improve the quality and reliability of products; to ensure safety standards in the manufacture and use of products; and to develop product standards covering such matters as naming, rating, performance, testing, and dimensions.

National Fire Protection Association (NFPA): A nonprofit organization that publishes codes and standards that are nationally recognized by state and local governments as adoptable into law. Some of these codes and standards include the *National Electrical Code*®, *Life Safety Code*®, *Uniform Fire Code*™, *National Fuel Gas Code*, *National Fire Alarm Code*®, and the *Building and Construction Safety Code*™. The organization's mission is to reduce the burden of fire and other hazards on the quality of life by providing and advocating codes and standards, research, training, and education.

Nationally Recognized Testing Laboratories (NRTL): A national group of product safety certification laboratories. They establish and operate product safety certification programs to ensure that items produced under the service are safeguarded against reasonably foreseeable risks.

Telecommunications Industry Association (TIA): An international trade organization that provides communications and information technology products, materials, systems, distribution services, and professional services. It represents suppliers of communications and information technology products on public policy, standards, and market-development issues.

Underwriters Laboratories (UL): A group of laboratories that tests devices and materials for compliance with the standards of construction and performance established by the laboratory and with regards to the suitability for installation in accordance with the appropriate standards of the National Board of Fire Underwriters.

Additional Resources

This module is intended to be a thorough resource for task training. The following reference works are suggested for further study. These are optional materials for continued education rather than for task training.

ANSI/TIA/EIA Standards, Latest Edition, available from Global Engineering Documents, Englewood, CO.
   ANSI/TIA/EIA-568-B, Commercial Building Telecommunications Cabling Standard Plus Addenda 1 through 3, and Telecommunications Systems Bulletins TSB-67, TSB-72, and TSB-75
   ANSI/TIA/EIA-569-B, Commercial Building Standards for Telecommunications Pathways and Spaces
   ANSI/TIA/EIA-570-B, Residential Telecommunications Cabling Standard
   ANSI/TIA/EIA-606, Administration Standard for the Telecommunications Infrastructure of Commercial Buildings
   ANSI/TIA/EIA-607, Commercial Building Grounding and Bonding Requirements for Telecommunications

IEEE Standards, Latest Edition. Piscataway, NJ: Institute of Electrical and Electronic Engineers.
   Standard 802.3, CSMA/CD Access Method
   Standard 802.5, Token Ring Access Method
   Standard 1394, Standard for a High Performance Serial Bus

International Building Code, Falls Church, VA: International Code Council.

NFPA Codes, Latest Edition. Quincy, MA: National Fire Protection Association.
   NFPA 5000, Building Construction and Safety Code™
   NFPA 13, Installation of Sprinkler Systems
   NFPA 13D, Installation of Sprinkler Systems in One- and Two-Family Dwellings and Manufactured Homes
   NFPA 13R, Installation of Sprinkler Systems in Residential Occupancies up to and Including Four Stories in Height
   NFPA 70, National Electrical Code®
   NFPA 72, National Fire Alarm Code®
   NFPA 75, Protection of Information Technology Equipment
   NFPA 77, Recommended Practice on Static Electricity
   NFPA 101, Life Safety Code®
   NFPA 780, Standard For Installation Of Lightning Protection Systems

SCTE Standards, Latest Edition. Exton, PA: Society of Cable Telecommunications Engineers.
   ANSI/SCTE 74 2003 (formerly IPS SP 001), Specification for Braided 75 Ohm Flexible RF Coaxial Drop Cable

Websites:
www.bicsi.org
www.nfpa.org
www.tiaonline.org

Figure Credits

Tim Ely   209SA01

Computer Applications

COURSE MAP

This course map shows all of the modules in the second level of the *Electronic Systems Technician* curriculum. The suggested training order begins at the bottom and proceeds up. Skill levels increase as you advance on the course map. The local Training Program Sponsor may adjust the training order.

ELECTRONIC SYSTEMS TECHNICIAN LEVEL TWO

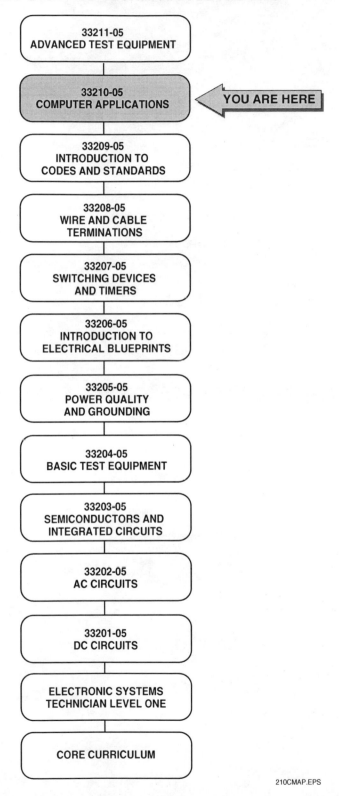

33211-05
ADVANCED TEST EQUIPMENT

33210-05
COMPUTER APPLICATIONS ◁ YOU ARE HERE

33209-05
INTRODUCTION TO
CODES AND STANDARDS

33208-05
WIRE AND CABLE
TERMINATIONS

33207-05
SWITCHING DEVICES
AND TIMERS

33206-05
INTRODUCTION TO
ELECTRICAL BLUEPRINTS

33205-05
POWER QUALITY
AND GROUNDING

33204-05
BASIC TEST EQUIPMENT

33203-05
SEMICONDUCTORS AND
INTEGRATED CIRCUITS

33202-05
AC CIRCUITS

33201-05
DC CIRCUITS

ELECTRONIC SYSTEMS
TECHNICIAN LEVEL ONE

CORE CURRICULUM

210CMAP.EPS

MODULE 33210-05 CONTENTS

Figures

Tables

Computer Applications

Objectives

When you have completed this module, you will be able to do the following:

1. Define terms commonly used in discussing computers and networks.
2. Identify the components of a personal computer and discuss the function of each.
3. Upload and download files to a security, lighting control, or fire system.
4. Build and test a null modem cable.
5. Assemble a personal computer.
6. Back up system configuration files.
7. Load application software on a computer and use the software to perform a task.
8. Explain the function of each level of the open systems interconnection (OSI) reference model for data communication.
9. Describe the characteristics of and uses for various types of data transmission media.
10. Describe the function of the Internet as it relates to network protocols.

Prerequisites

Before you begin this module, it is recommended that you successfully complete *Core Curriculum*; *Electronic Systems Technician Level One*; and *Electronic Systems Technician Level Two*, Modules 33201-05 through 33209-05.

Required Trainee Materials

1. Pencil and paper
2. Appropriate personal protective equipment

1.0.0 ◆ INTRODUCTION

Computers and computer-controlled equipment play an important role in everyone's life. In the world of electronic systems, computers and **microprocessors** play a significant role:

- They enable communication among various systems and devices within a building, and from building to building.
- They manage building facilities such as air conditioning, lighting, and **access control (AC)**.
- They collect and provide information to building owners and facility managers.
- They enable building owners and facility managers to program control devices such as thermostats and alarm systems.

Many special terms and abbreviations are used in the computer world. Here are some terms you may encounter in reading or talking about computers. In addition to the list below, there is a list of common acronyms in the *Appendix*.

- *Bandwidth* – The speed at which data travels in transmission lines. Bandwidth is stated in cycles per second, which in this case represents the number of bits of data per second. A typical telephone modem has a bandwidth of 56,000 bits per second (56 Kbps), while a high-speed connection such as video cable has a bandwidth of 1.5 million bits per second. A file transferred on the 1.5 MHz line would move about 27 times faster than the same data transferred over a 56k line. A speed of 1.5 MHz is required to receive high-quality video.
- *Basic input/output system (BIOS)* – The **basic input/output system (BIOS)** is the first set of instructions to run when a computer is started (booted) up. The BIOS is stored in ROM located on the system board.

- *Binary digit (bit)* – The smallest unit of information in a computer system.
- *Byte* – One character, such as a number. It consists of 8 bits. A kilobyte (KB) is 1,024 bytes; a megabyte is approximately one million bytes; a gigabyte is approximately one billion bytes; and a terabyte is approximately one trillion bytes. These terms are used to define storage capacity.
- *Bus* – The internal wiring pathway between the internal elements of a computer. Buses are defined in terms of their width, which means how many bits of data they can carry, which in turn affects the processing speed of the computer. There are various types of buses such as ISA and PCI. The PCI bus is the faster of the two, and is able to handle 32-bit and 64-bit data. The ISA bus, in contrast, handles 16-bit data.
- *Cache* – Cache is a type of memory in which data is stored, or stockpiled, ahead of time so it is available for use when needed. Having key instructions and information readily available in cache allows the computer to work faster because it does not have to go to the hard drive or other device to search for it.
- *DIP switch* – One of a set of tiny switches, located on a circuit board. DIP switches are used to configure the processor to perform certain functions. They are often used to select options.
- *Digital subscriber line (DSL)* – DSL is a method of providing high-speed communication over telephone lines. It is one of several such methods.
- *Handshake* – The process by which two computers initially establish communication. During the handshake, the computers determine if a connection is possible, then establish the best mode for the transmission.
- *Integrated drive electronics (IDE)* – IDE is a high-speed interface **protocol** associated with hard drives, compact disc-read only memory (CD-ROM), and digital versatile disc-read only memory (DVD-ROM).
- *Integrated services digital network (ISDN)* – ISDN is a high-speed telecommunications connection. It has a bandwidth of 64 Kbps, as compared with DSL, which transfers at about 150 Kbps.
- *Network interface card (NIC)* – A network interface card is a special printed circuit card or adapter that enables a workstation to operate in a computer network.
- *Parallel I/O* – I/O stands for input/output. A parallel I/O is one in which all data bits being transferred from one device to another are sent simultaneously on separate wires. Printers are typically connected to parallel I/O connections on computers.
- *Partition* – A section of a hard drive allocated to a specific function. A computer user might put application software on one partition and data files on another.
- *Plug and play* – A special process in which the BIOS recognizes peripheral devices such as printers, scanners, and drives, and automatically configures the system to interface with them. Before plug and play, the operator had to specifically configure the computer to handle each device as it was added.
- *Random access memory (RAM)* – RAM is memory that temporarily stores information in a computer. The data stored in RAM is erased when the computer is shut off. RAM consists of memory chips located on a memory board connected to the main circuit board (motherboard). The memory modules are known as SIMMs (single in-line memory module), which holds nine memory chips, or DIMMs (dual in-line memory modules), which hold 18 memory chips. The amount of RAM a computer has determines what applications it can use and how many applications it can have running at once. There are two types of RAM: DRAM (D is for dynamic), is the most common. It must be refreshed often by the computer to retain the information stored in it. SRAM (S is for static) retains information without being refreshed. It is more expensive, however, so it is only used where necessary, such as in video and cache applications. Synchronous DRAM is a newer, faster version of DRAM.
- *Read-only memory (ROM)* – ROM chips are pre-programmed with instructions or information for the computer in which they are used. One important allocation of ROM in a PC is the storage of the BIOS, which contains the boot-up instructions for the PC.
- *Small computer system interface (SCSI)* – SCSI (pronounced scuzzy) is an interface specification for connecting peripheral devices to a computer. It supports several high-speed devices through a single 50-pin or 68-pin cable.
- *Serial input/output (serial I/O)* – A method of transferring data between two devices one bit at a time. Modems and some printers are connected to a serial port.
- *T-1 line* – A high-speed communication line used for data transfer. It consists of 24 64 Kbps channels, which can be used separately, combined into clusters, or combined into a single connection that will provide a 1.5 Mbps data transfer rate.

- *T-3 line* – A very high-speed communication line consisting of 43 64 Kbps channels, which can be combined into a single connection that will provide a 43 Mbps transfer rate.
- *Virtual memory* – Hard disk space allocated to augment RAM. It is not as fast as RAM because of the disk access time, but there are some uses for which it is suitable. One of these is to serve as RAM when the PC is running multiple programs that require more RAM than the computer has available.

2.0.0 ◆ MICROPROCESSORS

Microminiaturization enables tiny devices smaller than the tip of your little finger to perform work that, in the early days of computers, used to take a roomful of electronic equipment to do. The **semiconductor** makes microminiaturization possible.

Semiconductors are materials in which the capacity to conduct electricity can be controlled by varying the voltage applied. In this case, we are talking about low-level DC voltages in the range of 5V to 15V. Heat, light, and pressure are also used to control current flow in semiconductors. Silicon and germanium are the two most widely used semiconductor materials. Components made of semiconductors, such as the diode shown in *Figure 1*, are known as solid-state devices.

An integrated circuit chip (*Figure 2*) is a tiny wafer of semiconductor material containing microminiature electronic circuits designed to perform a specific function or functions. To get a perspective on what microminiature means, think about a multi-function digital wristwatch. All the complex timekeeping, calendar, and display functions are contained on a single integrated circuit chip that you might have trouble finding if you looked inside the watch.

Transistors (*Figure 3*) are the most common semiconductor devices. A transistor consists of a wafer of semiconductor material sandwiched

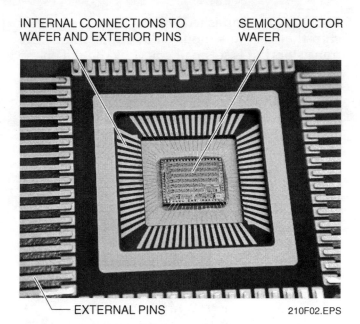

INTERNAL CONNECTIONS TO WAFER AND EXTERIOR PINS   SEMICONDUCTOR WAFER

EXTERNAL PINS   210F02.EPS

Figure 2 ◆ Interior of an integrated circuit.

between two layers of a different semiconductor material. Current flow is controlled by the voltage applied at the junctions of the dissimilar materials. Years ago you would find them soldered onto printed circuit boards, which we will discuss shortly. You might have been able to hold half a dozen of these early transistors in the palm of your hand. It would take hundreds of them, maybe thousands, to perform some of the func-

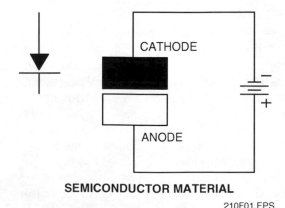

SEMICONDUCTOR MATERIAL

210F01.EPS

Figure 1 ◆ Basic semiconductor.

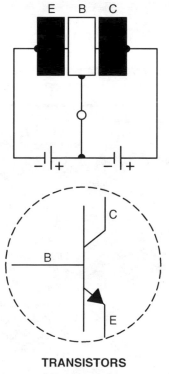

TRANSISTORS

210F03.EPS

Figure 3 ◆ Basic transistor circuit.

tions now accomplished with an integrated circuit chip that contains thousands of microminiature transistors but is so small you could hold it on the tip of your finger.

Special integrated circuit chips known as microprocessors (*Figure 4*) are used to make decisions and perform complex tasks such as calculations. They are the brains of the familiar personal computer. One such chip does all the processing work for a powerful multimedia computer.

Earlier, we mentioned printed circuit boards, which are commonly known as PC boards. (Note that PC is used as the abbreviation for both printed circuit and personal computer.) All the processing functions performed by a personal computer are housed on a PC board known as the motherboard. Components such as capacitors, microprocessors, and integrated circuit chips are mounted on one side of the board. The other side of the board, known as the foil side, has copper foil bonded to it. Instead of being connected by wire, the compo-

nents are connected by the pathways. *Figure 5* shows a graphics board commonly used in PCs. In the manufacturing process, the desired circuit is printed on the foil, and the copper is then chemically etched away from the areas that are not printed. The remaining copper then acts as the conductor between the components on the board. The use of very low (microampere range) currents in modern electronic circuits allows the use of very thin foil conductors.

The PC boards you encounter in the electronic systems you will install and service may range in size from very small (1 or 2 components) to very large (50 to 100 components). Regardless of the size, the maintenance approach is the same for all of them. If there is a problem with the board or module, you will replace the entire thing rather than trying to diagnose the problem to a specific component. Except in rare instances, PC boards are relatively inexpensive, and it is more economical to throw one away and replace it with a new

210F04.EPS

Figure 4 ◆ Microprocessor chips.

210F05.EPS

Figure 5 ◆ PC graphics board.

Maintenance Philosophy

The age of the microminiature circuit and the use of printed circuit boards brought about a major change in the way technicians conducted troubleshooting. In earlier times, the technician would focus on isolating and replacing a defective electronic component such as a transistor. If a circuit board was replaced in the field, it would first be brought to the shop, where further troubleshooting would be performed using complex test equipment. Today, it is cheaper in most cases to just throw the board away, because the cost of the board is less than the cost of the labor and equipment needed to troubleshoot it. Moreover, integrated circuits are so sensitive that troubleshooting and repair can introduce new faults. If a specialized circuit board is costly enough to warrant test and repair, it is usually returned to the manufacturer.

The benefit of this approach is that a failed system can be brought back on line quickly by simply replacing the circuit board. The other side of that coin is that there is a tendency to replace the board without determining what caused the failure. Even if the board has failed, the failure could easily have been caused by a problem outside of the board. In such cases, the board will likely fail again. The result will be a costly return visit and an unhappy customer.

one than it is to spend hours trying to isolate the problem to a component. Because of the delicacy of the integrated circuits, the heat from a normal soldering tool, or the static electricity from your body can cause permanent damage to the components, so in trying to repair the board, you could do more harm than good.

You must verify that the board has failed before you replace it with a new one. Otherwise, the replacement board might fail, or the initial problem will still exist. These days, most microprocessor-controlled systems have their own built-in diagnostic feature that will tell you whether the PC board, or some other device, has failed.

3.0.0 ◆ MAINFRAME COMPUTERS

Computers fall into two major classifications: mainframe computers and personal computers (PCs). A mainframe computer (*Figure 6*) is intended for enterprise-wide applications where a large amount of processing is required. In fact, mainframes are now commonly referred to as enterprise servers. The enterprise server shown in *Figure 6* is over six feet tall. It contains more than 100 1.2 GHz processors and has more than 500 gigabytes of memory. Large businesses and government entities use enterprise servers to handle their accounting and payrolls, keep track of inventory, and manage the flow of information and products. Enterprise servers may be accessed by many people using personal computers or "dumb" terminals that consist of a monitor and keyboard. Clients of mainframe computers are linked to it in a network by cabling. External links are provided by telecommunications systems. Networks will be covered later in the module.

4.0.0 ◆ PERSONAL COMPUTERS

Personal computers have become so commonplace, that it is now unusual to meet someone who does not use one. People use them at home to play video games, obtain movies and music, corre-

210F06.EPS

Figure 6 ◆ Mainframe computer.

spond with friends and family, and do their shopping. At work, people use them to control their environments, manage their schedules, access information, and send information and correspondence around the world.

There are two basic types of personal computers: The Apple® Macintosh® and the IBM® PC®. The latter, although originally developed by IBM®, has been cloned by many companies and is the standard for business computing and most home computers. The Macintosh® (or Mac®) was the first personal computer to use a mouse and has typically been favored by graphic designers and desktop publishers.

Figure 7 shows a PC in the desktop configuration—the central processing unit (CPU) is in a horizontal case with the monitor resting on it. Desktop PCs tend to be aimed at the low-cost home market, where expansion is not important. They have a small footprint and thus do not take up much space.

If a more powerful computer capable of expansion is needed, a tower configuration is chosen for the CPU, as shown in *Figure 8*. It has more internal expansion slots to accommodate additional special-purpose PC boards, more space for built-in storage drives, and more connectors (ports) to hook up peripherals such as scanners, printers, and game devices.

The PC components shown in Figure 8 are described as follows:

• *Central processing unit (CPU)* – This is the main box. It contains the processing circuits and

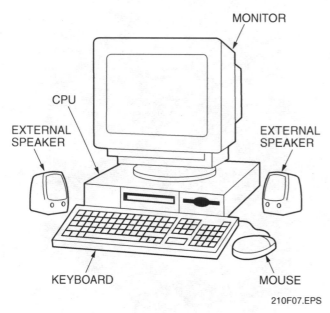

Figure 7 ◆ Personal computer (desktop configuration).

other devices. It will be described further when we take a look inside.

• *Diskette (floppy) drive* – The floppy drive supports a 3½" magnetic diskette that will accommodate about 1.44 megabytes of data. It is therefore a convenient medium for storing and transferring text files outside the computer, but it is not very useful for graphics files, which consume a lot more storage space than text. For many years, the floppy disk was a major means of transferring files from one computer to another, and a floppy disk drive was standard on every PC. With the advent of other read-write devices with much greater capacity, as well as the internet, the use of floppies has diminished.

• *Compact disk-read-only memory (CD-ROM) drive* – The CD is portable and able to contain a large amount of data (640 megabytes). The CD itself is inexpensive (less than $1), but it requires a special recording device to write information from the computer to it, so it is more often used for distribution than as a storage device. A regular CD can only be written to once, but there are special rewritable CDs that can be erased and reused more than once.

• *Digital versatile disk (DVD) drive* – The DVD is newer than the CD-ROM and has a much greater information capacity. It has four different storage modes, the lowest of which can store 4.7 gigabytes of data, which is about seven times the capacity of a CD-ROM. In its highest capacity mode (both sides, two layers per side), it can store more than 17 gigabytes.

• *Monitor* – The monitor is the display device. It receives information from the video card inside the CPU.

• *Keyboard* – The keyboard allows the operator to enter alphabet characters and numbers. It also

Figure 8 ◆ Multimedia computer (tower configuration).

contains function and control keys that are used by computer programs to perform special functions. The use of function keys is less common in a mouse-driven system, where interaction with the computer is done by clicking on graphic objects rather than entering keystrokes. However, they are still functional and can be used in place of the mouse for many tasks.

- *Speakers* – With the advent of multimedia, audio was introduced to the PC. Now, it is unusual to find a PC without a set of speakers. Speakers with a built-in amplifier provide the best quality sound. Powered speakers have their own power source; the sound level can be adjusted using a volume control on the main speaker, rather than doing it in the operating system. Speakers require an audio card or special audio circuits on the motherboard.

- *Mouse* – The mouse is so-called because of its shape and its long, curly tail. Trackballs are other common input devices (*Figure 9*).

 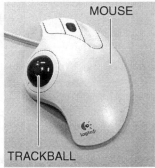

MOUSE   **MOUSE WITH TRACKBALL**

210F09.EPS

Figure 9 ◆ Computer input devices.

Touch Screens

Touch screens are popular computer input devices that you may find on nurse call, intrusion detection, and access control systems. The touch screen takes the place of the keyboard and mouse by providing selections that the operator touches with a finger or stylus. The touch screen overlays the computer monitor. The personal digital assistant (PDA) is one common touch screen device.

Typically, an electric current flows through the touch sensor, which overlays the computer screen. Touching the screen at a particular point changes the current. A special controller and dedicated software translate the change into a signal that the computer is able to process. The change is comparable to clicking a mouse cursor at a particular point on the computer monitor.

210SA01.EPS

4.1.0 Monitors

The monitor that appeared in the first PCs was a **monochrome** device capable of displaying only text and line drawings (simple plots, etc). Over the years since then, the monitor has evolved through several generations of devices capable of displaying color graphics, starting with the color graphic adapter (CGA) standard. CGA was followed by the enhanced graphic adapter (EGA) and video graphics array (VGA), which emerged in 1987. These were followed with super VGA (SVGA) and Video Electronic Standards Association (VESA)-compliant monitors, which brought the screen image to new quality levels.

The quality of a monitor is measured in terms of its resolution and the number of colors it is capable of presenting. Resolution is measured in terms of **pixels** (short for picture element), which are tiny dots that make up the image on the screen. The more pixels, the higher the quality of the image. Your computer must be configured to display the resolution of the subject image. This is done in Windows® by going to the Display function on the Control Panel and resizing the window for a different resolution. Otherwise, the image on the screen will be undersized or oversized. For example, if your monitor is designed for 1280×1024 pixel resolution, and you are using an 800×600 resolution, the image quality will be lower than it could be. This adjustment can be done from a system tray icon at the bottom of the screen on some computers.

The number of colors (color depth) is another important measure of image quality. Early CGA monitors were capable of displaying 16 colors (4-bit color) with a resolution of 320×200 pixels. VGA brought that capability up to 256 colors (8-bit color) and 640×480 resolution, and VESA-compliant designs yielded 1280×1024 resolution with 16.8 million colors (24-bit color).

Monitors generally range in size from 14" to 21", with 15" to 17" being commonplace in homes and businesses. The size refers to the distance from one corner of the screen to the other; the actual display area will be smaller. Although monitors typically look like the ones depicted earlier, flat screen displays are available and have come into general use.

Flat Panel Monitors

Most computer users are familiar with the bulky computer monitors that use cathode ray tube (CRT) display technology. Such monitors can take up two to three cubic feet of desk space and weigh 30 or 40 pounds. In order for laptop computers to be viable, a new technology was needed. This led to the development to the liquid crystal display (LCD), which became the standard for portable computers, and eventually led to the flat panel display now commonly used with desktop computers.

210SA02.EPS

4.2.0 Connections

On most PCs, the connection devices are located at the rear of the unit (*Figure 10*). In the computer world, connectors are called ports.

- *Parallel ports* – A PC will have one or two parallel ports, which are designed to mate with 25-pin male connectors. They are designed to interface with peripheral devices such as printers, scanners, and tape drives. The computer designates parallel ports as LPT. If the PC has two parallel ports, they will be designated LPT 1 and LPT 2.

- *Serial ports* – A serial port is usually a 9-pin connector, but some are 25-pin connectors. They can be distinguished from other ports because they are male connectors. The serial port is used to connect such components as a mouse or external modem to the PC. Serial ports are often called COM ports.

- *Monitor port* – The monitor port is a 15-pin HD15 connector. The cable that carries video from the CPU to the monitor is plugged in here. A digital video interface (DVI) connector may also be provided here.

- *Keyboard port* – The keyboard is usually connected to the CPU with a round 9-pin AT or PS2

connector. The connector is keyed to prevent the cable from being connected incorrectly.

- *Universal serial bus (USB) ports* – A USB port allows a user to connect a wide variety of peripheral devices to the PC. One of the problems with earlier PCs is that they did not have enough ports to connect all the external peripherals a person might want to use. If you wanted to use a document scanner, for example, you might have to disconnect the printer to obtain a connection. On many PCs, the USB ports are located on the front of the CPU to allow easy swapping of peripheral devices. There are two versions of USB. Version 1.1 was designed for high-bandwidth devices such as scanners, digital cameras, and webcams. Version 2.0 supports higher-bandwidth devices, as well as low-bandwidth devices such as the keyboard and mouse. In most cases, Version 2.0 devices are backwards compatible with Version 1.1 ports.

- *Ethernet port* – **Ethernet** is a networking protocol that allows connected devices such as computers and printers to communicate with each other in a network. It is the predominant networking protocol. Ethernet is implemented through use of an Ethernet board installed in each networked device.

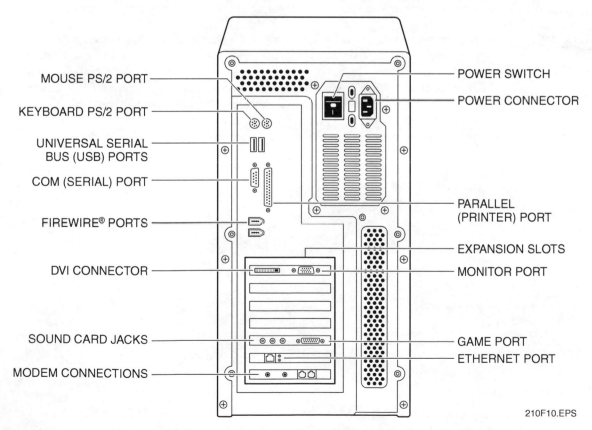

MOUSE PS/2 PORT
KEYBOARD PS/2 PORT
UNIVERSAL SERIAL BUS (USB) PORTS
COM (SERIAL) PORT
FIREWIRE® PORTS
DVI CONNECTOR
SOUND CARD JACKS
MODEM CONNECTIONS

POWER SWITCH
POWER CONNECTOR
PARALLEL (PRINTER) PORT
EXPANSION SLOTS
MONITOR PORT
GAME PORT
ETHERNET PORT

210F10.EPS

Figure 10 ◆ Rear view of a tower configuration PC.

- *FireWire® port* – FireWire® is a data transfer standard specifically designed for high-speed transfer of video and other media. It is similar to USB, but much faster.
- *Game port* – The game port is a 15-pin connector used to connect a joystick or similar device used for games and simulations.

NOTE

Some PC manufacturers color-code the PC connectors and the connectors on the mating cables to eliminate the guesswork about how to connect peripheral devices to the computer. With this scheme, someone unfamiliar with computers can hook one up in minutes.

- *Phone jack* – This jack is used to connect the modem to the phone system. It is the same type of jack used in standard telephone circuits.
- *Mouse port* – Some versions of the mouse are connected with a 6-pin or PS2 connector. Other versions are connected to a serial port.

4.2.1 Null Modem

A null modem is a cable or adapter used to connect two computers together through their serial (COM) ports. A null modem cannot exceed 25' in length.

When two computers are connected, one of the computers will be the transmitter and the other the receiver at any given time. If you run a straight cable from one computer to another, data will not transfer. In a null modem, the wiring is arranged so that the input and output pins are switched, which permits the signals to be sent to the proper pins. *Table 1* compares the pin-to-pin connections for a straight cable to those of three types of null modems.

4.3.0 Inside the PC

Figure 11 shows the major components located on the inside of the PC.

The system board, commonly known as the motherboard, is the heart of the PC. It contains the CPU, or microprocessor, which performs calculations and manages the flow of information through the system. The speed at which a CPU processes

USB

The universal serial bus has become the preferred connection for many computer peripherals. One common device that uses the USB port is the media reader for digital cameras. The removable camera storage media is simply inserted into the reader, and the digital photos can then be opened by, and transferred to, the PC. Digital cameras use a variety of so-called flash media devices, including the SmartMedia® card, thumb drive, various memory sticks, and the CompactFlash® card. The media reader shown here is designed to read the SmartMedia® card (shown) as well as the CompactFlash® card.

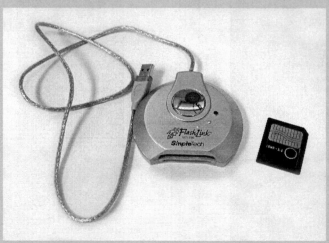

210SA03.EPS

Table 1 Comparison of Pin Connections for Serial Cable and Null Modems

Straight Cable		Null Modems					
9-Pin to 25-Pin		9-Pin to 9-Pin		25-Pin to 25-Pin		9-Pin to 25-Pin	
1	8	1, 6	4	1	1	1, 6	20
2	3	2	3	2	3	2	2
3	2	3	2	3	2	3	3
4	20	4	1, 6	4	5	4	6, 8
5	7	5	5	5	4	5	7
6	6	7	8	6, 8	20	7	5
7	4	8	7	7	7	8	4
8	5	9	9	20	6, 8	9	22
9	22			22	22		

information is stated in megahertz (MHz) or gigahertz (GHz). Early Pentium PCs (around 1995) had a speed of 60 MHz to 75 MHz. By 2000, 1 GHz was common. The added speed not only made data and graphic processing much faster, it opened the way for PCs to become true multimedia devices, delivering high-quality video, audio, and animation.

One function of the motherboard is to act as a base for other boards, known as expansion boards. Among these expansion boards are the hard disk

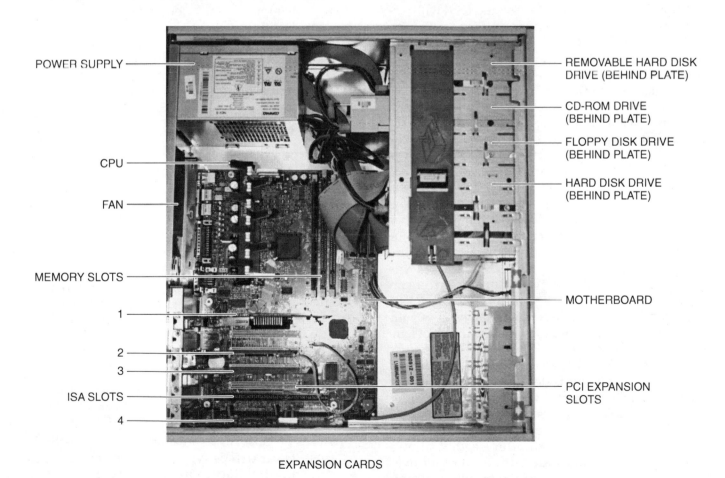

EXPANSION CARDS
1. VIDEO GRAPHICS CARD
2. NETWORK INTERFACE CARD
3. SPECIAL PURPOSE VIDEO CARD
4. AUDIO CARD

210F11.EPS

Figure 11 ◆ Interior of a tower configuration PC.

controller, which allows the computer to communicate with its hard disk drives; the video board, which drives the monitor; and the modem. Inside the CPU, the motherboard is positioned so that its expansion slots are near the rear of the CPU. The expansion boards contain connectors that protrude from openings in the rear of the CPU. These connectors are the ports we discussed earlier.

Until 2004, the 32-bit processor was the standard for CPUs. This technology evolved from the 8-bit and 16-bit standards that existed in the 1980s. The bit designation refers to the width of the data stream that the processor can handle and the amount of random access memory (RAM) it is able to use. With 64-bit processing, the amount of RAM the processor can use will rise exponentially. This added capacity will not mean much in the short term for most PC users. It will, however, dramatically improve processing speed and capacity for heavy-duty applications such as CAD programs, databases, and math processors.

The sound (audio) card is an expansion board that drives the speakers. The sound card contains connectors that are accessible at the back of the CPU. In addition to the speaker jack, the sound card will usually have jacks for other purposes, including an audio input jack that allows a tape or CD player to play sound through the computer, an audio output jack that allows connection to an audio amplifier, and a microphone jack that allows users to record audio. People can use the PC to communicate with each other over phone lines. A full duplex capability means that two parties can talk at once. Half-duplex means they need to take turns talking.

To obtain CD-quality sound, a card with a 16-bit sampling size and a 44.1 kHz sampling rate is required.

A video card contains video memory chips that store information before sending it to the monitor and microprocessors to interpret data from the CPU into displayed images.

A large, high-resolution graphic image can contain many megabytes of data. If your computer is not set up to handle such graphics, it will really slow your computer down. In order to eliminate this problem, newer computers use AGP (accelerated graphics port) video cards and special video RAM to display graphics. The AGP bus is designed to communicate directly with the computer's main memory in order to display complex graphics rapidly.

The term modem is an abbreviation for modulator-demodulator. It is the device that allows computers to communicate with each other over telephone lines. Its purpose is to translate incoming data from the analog form used on phone lines to the digital form required for the computer. It does the opposite for outgoing data. The modem must be plugged into a phone jack in order to work.

Modems communicate with other modems over phone lines. The speed at which the data is transferred is determined by the slowest of the two modems. For example, if a 56 Kbps modem is communicating with a 33.6 Kbps modem, the data will transfer at the 33.6 Kbps rate.

When you log on to the internet, your modem contacts a modem at the site of your ISP (internet service provider). The ISP then reaches out to other computers on the internet to access the website you select. The ISP uses high-speed transmission lines and special modems to make these connections rapidly. Businesses that transfer large amounts of information from one site to another also use high-speed lines to move the data faster.

The hard drive is the main storage device for the computer. It stores the software programs as well as the files created using those programs. The hard drive is covered in more detail in the next section.

The power supply converts the 120VAC source to DC power needed to operate the solid-state electronics. The power supply will include a ventilating fan to dissipate the heat generated by the power supply and other devices. A PC will have one or more ventilating fans, depending on the amount and types of devices it contains.

4.4.0 Hard Drive

All computers contain some type of fixed internal storage device, usually a hard drive (*Figure 12*). The computer programs and accumulated data are stored on the hard drive. The computer's operating system is designed to get data and programs from the hard drive when needed and to store the information back to the hard drive when work on a file is complete. While the computer is processing the work, the information and programs are transferred into random access memory (RAM), which is the memory located on chips inside the computer.

The efficiency of a hard drive is determined by two critical factors:

- The rotational speed of the disk, measured in revolutions per minute (rpm). The faster it rotates, the faster it can find and retrieve information.
- Access (seek) time, measured in milliseconds (ms). Hard drives typically have access times ranging from 8 to 15 milliseconds. The shorter the time, the faster and more expensive the drive.

much faster than looking on the hard drive. Cache memory is one of the features that has significantly enhanced computer processing speeds.

4.4.1 Data Compression

When files are transferred on a network, the transfer time is a function of the file size. The bigger the file, the longer it takes to upload or download. One of the ways to speed up file transfer is to compress the data. Special software programs are available for this purpose.

Another benefit of compression is that it allows data to fit on smaller storage devices. For example, using compression software, you might be able to put several megabytes of data on a 1.44-meg floppy disk.

These programs work on the principle that most data files contain repeating patterns or nonessential characters. For example, not all characters require a full eight-bit word, but the computer will assign them eight bits anyway. Compression programs use compression algorithms to eliminate non-essential bits.

210F12.EPS

Figure 12 ◆ Hard drive.

Computers also use cache memory to improve access times. Cache memory is a special faster memory where the computer stores recently used information. When the computer wants information, it first looks in the cache, which it can do

Hard Drive Operation

If you are familiar with the phonograph, you have a rough idea of how a hard drive works. The storage device is a hard magnetic disk. The data is read onto the disk and extracted from the disk by arms that contain read-write heads. The heads are millionths of an inch from the disk, but do not actually touch it. The disk spins at a very high speed—3,200 rpm is considered slow; 10,000 rpm is a fast drive. As the disk spins, the arms move at a high speed to place or extract data as commanded by the operating system. The disk is divided into tracks and sectors by the operating system in order to simplify data access.

210SA04.EPS

One problem with compression is that anyone using the file must have a software program that allows them to decompress it.

4.4.2 Disk Fragmentation

Ideally, files should occupy consecutive spaces on a disk in order to make it easier for the computer to access the file. Windows® is not very good at doing this. When storing a file, Windows® will begin storing it on the first unused cluster of file space it finds. If the file is bigger than that cluster, it will fill up the cluster, then go looking for more free space. This is known as fragmentation.

When files are fragmented, it takes longer for the system to load them because the drive head must return to the beginning of the drive to identify each location, then go and get the information from it. There are defragmentation routines available in the operating system to eliminate this problem. These programs rearrange files on the disk so that they are stored as a contiguous stream of data on the disk.

4.4.3 Viruses

Viruses are self-replicating programs, sometimes called bugs, that can alter or destroy computer files and, in some cases, render a computer useless. Their effects can range from placing an unpleasant message on a computer screen to wiping out the content of the hard drive. Viruses are usually transmitted from system to system by innocent parties who may think they are merely transferring files to friends or co-workers. Once they are inside your computer, viruses attach themselves to executable files. They may lie dormant for a long time, then at some prearranged date and time, begin their destructive work on every computer that has received the innocent carrier file.

To counteract the spread of viruses, your computer needs a special software program designed to detect the presence of viruses and remove them. These programs are readily available at stores, through catalogs, and via internet download. It is important to check every diskette you put into your system and every file you download for the presence of viruses. Please note that the virus detection software needs to be updated frequently in order to detect and counteract new viruses. There are some things you can do to minimize the impact in case your computer is ever infected:

- Keep a set of emergency startup disks available.
- Back up your data files frequently to a storage device outside your computer.
- If your computer is ever infected, restore the data files (not the software programs) using the backup. Restore the programs from the program disks. If you did not receive a set of original diskettes or CDs for the software applications on your computer when you bought it, then you should make your own backup set as soon as you activate your computer for the first time.

Protecting Computers

When a computer accesses the Internet, it is just one of millions of computers connected together in a giant worldwide network. While this connection provides the user with access to people and information, it also has a downside because other users can reach in while you are reaching out. If an outside user can access one computer in a network, it gives them access to the entire network. They can, for example, log on to a network and use it for their own purposes. They could also access an e-mail server to send viruses or spam e-mail to everyone in every address book on the network.

For that reason, companies that have computer networks should also have a firewall that controls outside access. A firewall is typically a software package, but can also be implemented in hardware using a router or a separate gateway computer that screens all incoming access to the network. Virus protection is also important. Unscrupulous individuals known as hackers develop harmful software routines just for fun and transmit them to unwitting users. If you open one of these files, it will invade your operating system and cause problems for you and others. One common type of virus is the Trojan Horse. It lodges in your hard drive and waits there until a designated day and time at which it will perform whatever mischief it was designed to do. The companies that market virus software build defenses against known viruses and constantly search for new viruses so they can update the virus protection for their clients.

4.5.0 Removable Storage Media

Any PC or laptop computer will have one or more drives or ports through which software programs and data can be imported to (moved into) or exported from (moved out of) the computer. *Figure 13* shows some of the storage media used for this purpose. Currently, the four most popular are the DVD, CD, flash memory card, and USB flash drive. With the exception of magnetic tape cartridges used for mass storage, the other devices shown are older and are becoming obsolete.

4.5.1 DVD and CD Media

The standard 4.75-inch (12-cm) diameter DVD or CD drive is the most commonly installed removable-media device in modern computers. Currently, a DVD-ROM or a recordable DVD has the most storage capacity of any removable media. A prerecorded DVD-ROM looks just like a standard music CD; however, it cannot be written to, it is only used for importing data into a computer. A recordable DVD is very different from a recordable CD, not only in capacity, but also in the

Figure 13 ◆ Data storage devices.

DVD-R

USB FLASH DRIVE

FLASH MEMORY CARD

CD-RW

100 MB ZIP® DISK

1.44 MB DISKETTE

MAGNETIC TAPE CARTRIDGE

210F13.EPS

Inside Track

High-Definition Video DVD Media

Current DVD drives will probably be replaced in the near future by drives using one or both of two new recording formats. Like the two formats used for current DVD media, the new formats, called HD-DVD and Blue-Ray® DVD, are supported by two different manufacturer consortiums. These discs will be single- or double-sided and single- or dual-layer, but will have file capacities far greater than current discs. In most cases, the new drives will also support the current DVD and CD formats.

method of recording. Depending on the DVD recording format, a typical single-sided DVD disc can store up to 4.7 gigabytes (GB) of data. A gigabyte is a billion bytes. A double-sided disc can store up to 9.4 GB. Newer, dual-layer (DL), single-sided or double-sided discs can store up to 8.5 GB or 17 GB respectively.

Today, there are two DVD recording formats in use, each sponsored by a different consortium of equipment manufacturers. Each has some advantages over the other depending on the end use of the DVD disc. One format consists of the DVD-R (write once), DVD-RW (rewritable up to 1,000 times), and DVD-RAM (rewritable up to 100,000 times) discs sponsored by the DVD Forum. However, the double-sided DVD-RAM with 5.2 GB of storage is sealed in a cartridge and can only be used in a special DVD-RAM drive.

The other format consists of DVD+R and DVD+RW (rewritable up to 1,000 times) discs sponsored by the DVD+RW Alliance. Either format requires specific read/write drives, called burners, with a disc-compatible writing speed. Multi-format drives with selectable writing speeds are available that can read and write R or RW discs in either format.

Mini DVD-R media with a capacity of 1.4 GB is also available and can be used in standard or multiple format DVD-R/-RW drives. The CD is similar to the DVD except that it has much less storage capacity; typically 700 MB (megabytes). It is furnished as a prerecorded CD-ROM or as a recordable CD-R or CD-RW for use with an appropriate CD drive. Many DVD drives can also read and write on CD-R and CD-RW media. Mini CD-R media with a capacity of 120 MB is also available.

4.5.2 Flash Memory and USB Flash Drive Media

Many new computers are being furnished with flash memory card slots used for downloading digital pictures. However, the storage capacity of some of these cards is also useful for transferring files between computers. The cards that are available with large capacities are the CompactFlash® (up to 8 GB), Secure Digital® (up to 1 GB), and various manufacturers' memory sticks (up to 2 GB). The other popular file transfer media are various USB flash drives with capacities up to 4 GB, sometimes called thumb, jump, or pen drives. The term drive is misleading because they are solid-state memory devices without any moving parts. These drives plug directly into and are powered from a USB port and are available as USB 1.1 or 2.0 devices. The USB 2.0 devices are usually backward compatible with USB 1.1 ports.

4.5.3 Other Media

The diskette, also known as a floppy disk, was commonly used in older computers. It has a low capacity of about 1.44 megabytes and is useful for transferring small text files between computers. It is inexpensive and is quickly becoming obsolete.

Magnetic tape drives are very common where large amounts of information must be transferred to other computers or stored outside the computer. Tapes with a capacity of up to 200 GB are available. They are used as temporary backup media in computer networks to store data in case of network failures.

There are also other types of removable storage media, including ZIP® disks, Microdrives®, optical cartridges, and removable hard drives that are not used as much or are becoming obsolete.

4.6.0 Servers

A server is a computer that performs a specific task, such as managing a computer network (network server) or managing printers for a network (print server).

Any computer can be set up to act as a server. However, PCs configured and sold as servers usually contain high-reliability components to reduce the risk of a failure that could disable the network. Servers also have larger, faster hard drives. Some companies use arrays of hard drives known as redundant array of inexpensive disks (RAID) drives. A RAID system, such as the eight-bay array shown in *Figure 14*, uses multiple hard drives integrated into a system that places parts of each file

210F14.EPS

Figure 14 ◆ Eight-bay RAID drive.

onto separate disks. That way, if one disk fails, the file can easily be reassembled.

The backup system will also be connected to the server. Magnetic tape drives are often used as backups because of their capacity. Backups are extremely important to a computer network. If the server hard drive crashes, all the information on it could be lost. There are services that can recover information from a failed hard drive, but it is expensive and time-consuming. Backup software that automatically reaches out to every computer on the network and copies all the new or revised files to tape are readily available. In most networks, this type of backup is performed every day. The server is often programmed to do the backup automatically during off-shift hours.

In a client/server network (*Figure 15*), the files are all stored on the server. Client computers access the server to obtain files they need, then return them to the server when they are finished. The server hard drive will contain the special software used to manage the network. Each client computer will have a client version of the software so that the computers can communicate. The network computer will also act as the printer server so that one workgroup printer can serve several clients.

In a peer-to-peer network (*Figure 16*), individuals on the network store their files on their own computers, but others on the network can access these files. It is a good way for people in a small office to share files and resources such as printers, without the need for a dedicated server. One of the members of the network can act as the printer server if a workgroup printer is used. There are software programs specifically designed to manage peer-to-peer networks. One advantage of a peer-to-peer network is that the network can continue to operate if one of the computers fails. In a client/server network, a server failure can shut everyone down.

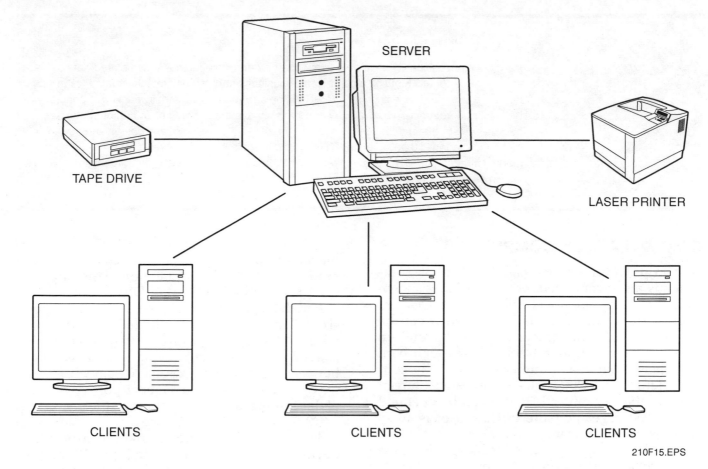

SERVER

TAPE DRIVE

LASER PRINTER

CLIENTS   CLIENTS   CLIENTS

210F15.EPS

Figure 15 ◆ Client/server network.

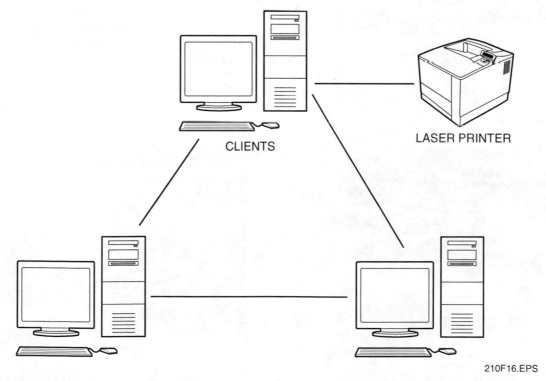

CLIENTS

LASER PRINTER

210F16.EPS

Figure 16 ◆ Peer-to-peer network.

5.0.0 ◆ LAPTOP COMPUTERS

A laptop computer, also called notebook computer, is a self-contained, portable PC. As shown in *Figure 17*, the monitor is built into the upper section of the case which is hinged so it can be opened and closed. Laptops come with a built-in input device such as a trackball or touch pad. Peripherals, such as a mouse, are usually connected through a USB port. A significant feature of the portable computer is that it can be operated from a rechargeable battery if the user is away from a 120VAC power source.

Portable computers usually have a PC card (formerly known as PCMCIA) slot that can accept a removable storage device or a card that allows the laptop to perform special functions such as a modem or network connection. Most are also equipped with a CD-ROM or DVD drive.

Portable computers are available in a range of sizes. The larger the unit, the more it is likely to weigh. A large laptop with a 14" display can weigh about 7 pounds, which does not sound like much, but can begin to feel extremely heavy after you have carried it through a major airport. At the time of this writing, a lightweight unit typically weighs 2 to 3 pounds.

There are also various smaller computing devices. The tablet PC shown in *Figure 18* has an 8" touch screen and uses the tablet PC version of the Windows® XP operating system. It can accept handwritten input to the screen. An even smaller device, also shown in *Figure 18*, is the pocket PC. This PC has a 3.5-inch touch screen and runs the Windows® Mobile operating system.

A laptop computer uses a liquid crystal display (LCD) screen, which uses very little power. This is an important feature for a battery-powered computer.

210F17.EPS

Figure 17 ◆ Laptop computer.

TABLET PC   POCKET PC

210F18.EPS

Figure 18 ◆ Tablet and pocket PCs.

Laptop Input Devices

In order to make the laptop self-contained, developers needed to find a way to replace the mouse with an on-board input device. Three different types of devices emerged: the track pointer, the trackball, and the track pad.

TRACK
POINTER

210SA05.EPS

TRACKBALL

210SA06.EPS

TRACK
PAD

210SA07.EPS

Laptop vs. Desktop

A laptop computer is a self-contained portable package that includes the CPU as well as the monitor. The laptop is designed to run from a DC voltage source. The self-contained rechargeable battery will keep it running for several hours. It can also be powered from an AC power source using an adapter that is supplied with the laptop. While a desktop machine may be equipped with several on-board drives for removable storage, the laptop is likely to have only one. A combination CD-RW/DVD drive is common. Other devices such as floppy drives can be attached through USB ports as needed. Docking stations are used to convert laptops into quasi-desktop computers. The docking station contains the same types of connectors found on the back of a desktop unit. The docking station also allows for the use of a standard monitor, which generally is larger and has a higher screen resolution than a laptop monitor.

6.0.0 ◆ COMPUTER PROGRAMS

A computer is just a box of electronic equipment. It must be instructed what to do and when and how to do it. That is the job of computer programs, commonly known as software, and pre-programmed memory devices containing fixed code that is known as **firmware**. Software is a set of executable instructions designed for a specific purpose. Software comes in two varieties: operating systems, which control all the routine tasks a computer is expected to perform, such as moving files and controlling expansion cards, regardless of which application software it is running; and application software, which performs specific kinds of work, such as word processing. Application software includes off-the-shelf software such as word processing and spreadsheet programs that are sold in stores and through catalogs, and custom (proprietary) software, which is tailor-made for the user by software programmers.

Firmware refers to instructions that are burned into the processor's memory during manufacturing. Programmable read-only memory (PROM) and erasable PROM (E-PROM) are terms often associated with firmware. E-PROM chips are commonly used to give an electronic processing device its personality. In some high-performance cars, for example, the computer chip can be replaced to change the operating characteristics. Unlike software, firmware-driven devices generally have a specific, limited purpose and accommodate a very limited set of input instructions. Firmware in a digital watch, for example, allows the user to change the date and time and set alarms. A programmable thermostat is another firmware-driven device. It displays a limited amount of information and allows the user to select and execute a limited set of options. One of the reasons that devices like

this can be very small is that they do not require hard drives to store software programs and data. Nor do they require input devices to transfer data and software.

6.1.0 Operating Systems

The operating system (also known as the platform) of a computer performs a variety of functions, including:

• Controlling the relationships between the hardware components of the system
• Running application software
• Managing the storage of and access to files

Without an operating system, a computer user would have to be able to write complex instructions in computer language to tell the computer what to do each time a file needed to be accessed, stored, or printed. There are several common operating systems, including:

The MS-DOS® (Microsoft® Disk Operating System) operating system responds to text commands entered on the keyboard. In order to use DOS, the computer operator needs to memorize or keep a handy list of 30 or 40 commands, such as DEL for delete, REN for rename, CHDIR for change directory, and so on. DOS was released by Microsoft® in the 1980s. Within applications programs, many functions were performed with combinations of function keys and other special keys. It was the predominant operating system on IBM®-compatible PCs until the early 1990s, when the first Windows® operating system became accepted. DOS underlies the first few iterations of Windows®, but the presence of DOS is masked by the graphic interface.

When Windows® was released, programs that previously ran under DOS, such as word processing and spreadsheets, had to be rewritten in order to use the graphic-based interactions provided by Windows®. Computers that run Windows® can also run many old DOS applications, but the graphic interface of Windows® will not be available in that program.

Windows® has been the dominant PC operating system since the 1990s. Windows®, like MS-DOS®, was developed by Microsoft®. Windows® has a graphic user interface, or GUI (pronounced gooey). Unlike MS-DOS®, in which a text command must be typed for each task, a Windows® user merely selects a graphic icon or text instruction on the screen, positions the cursor on it with the input device, such as a mouse or trackball, and clicks on it. Most of the same functions can be accomplished with function keys or combinations of keys if an operator prefers to use the keyboard. Different versions of Windows® operate in different ways, and a new version is released about every two years. It is therefore not practical to try to describe Windows® operations in any detail in this module.

One of the most important features of Windows® is its multitasking capability. Multitasking means that the computer is able to have several software programs running at once and to have several files open at the same time. The user can move from one program to another or cut and paste information from one file to another with a few mouse clicks.

Windows® Server is a more powerful version of Windows®, and is used primarily on machines that are connected into computer networks. It is designed for applications that require the sharing of files as well as physical resources such as printers and tape backups. Windows® Server can act as the network server software in a client/server application where the client workstations have PC operating systems such as Windows® or MAC-OS®, and only the server has Windows® Server installed. Windows® Server can be installed on workstations as well. Because Windows® Server is very powerful, many business applications are designed specifically to run on a Server platform and will not work on the other Windows® platforms. Some programs designed for the other Windows® version will run on a Server platform; however, not all will. Most DOS programs will not run on a Server platform. Windows® Server provides better security, better network management, greater speed, and the ability to support multiple users with less effect on performance.

UNIX® was developed as an operating system for mainframe and minicomputers in the 1970s. UNIX® is a high-performance, multitasking platform that provides excellent security features. Until recently, it was used primarily for larger systems, mission-critical enterprise databases and business applications, and for network servers. In recent years, it has become more popular as an operating system for individual workstations, and even some personal computers, due to its better performance and security. Much of the internet is based on UNIX® and UNIX®-like operating systems.

LINUX® is a freely distributed version of UNIX® that runs on a variety of hardware platforms. It is maintained by a group of software developers committed to open-source software. Because it is free, runs on many platforms, and can be adapted to suit the needs of a company or individual, it became very popular in the late 1990s. Many internet service providers use LINUX® because of its high reliability and versatility.

6.1.1 Handling and Storing Information

The information in a computer is organized into a hierarchy. Each storage device has a top level, or root, directory. This directory can contain files or other directories, called subdirectories. These subdirectories can contain other subdirectories and files, and so on. You can create directories, organize them in a systematic way, and then place files into the directories.

The main directory of a drive is called the root directory, and is similar to a file cabinet. It can contain more directories, which are similar to file drawers. Each directory can have more directories in it, like folders in a drawer. Files are like the papers in a folder. To refer to a particular file, you use its name preceded by the names of the directories it is in. This is called the path. For example, to access some manufacturing data in a file on your local hard drive, you might type C:\MANUFACTURING\DATA\FILEX.DAT. C: is usually the identifier for the local hard drive, manufacturing is the first directory, data is another directory inside it, and FILEX.dat is the actual file name.

In Windows®, directories are called folders. To locate a file in Windows®, you would use your mouse to double-click on an icon representing the hard drive. This will display the files and folders on the drive. You then double-click on a folder to see a list of the files and folders it contains. Finally, you can double-click the specific file you want or start a new one.

6.2.0 Application Software

Some of the most common types of application software are described as follows:

- *Word processing* – Word processing programs permit the user to create many types of professional-looking documents, including letters, manuals, reports, newsletters, and flyers. Most word processors will allow you to add illustrations to your text. Word processing programs make it easy to revise your text and to add, delete, or move text.

- *Database* – A database program helps you organize and sort large amounts of information. Databases are especially useful in finding information. For example, in a properly constructed database containing names, addresses, and other information, you could query the database to obtain a list of all the people who live on a particular street.

- *Spreadsheet* – A spreadsheet program is used primarily to organize numbers. They are popular for financial management for both personal and business use. One of the main benefits of spreadsheets is that they allow the user to analyze the information in a variety of ways and print out a variety of charts and graphs. Another advantage of a spreadsheet program is that it can automatically total numbers and recalculate the totals if any number in the spreadsheet is changed. This saves a lot of manual calculation.

- *Utilities* – Utility software is designed primarily to make a computer work better. There are many kinds of utilities. Some of the functions performed by different types of utilities are:

  - Recovering lost files from damaged or defective storage media such as hard disks
  - Protecting computers against viruses
  - Allowing the computer to act as a fax
  - Compressing files to make them easier to store or transfer and decompressing them when needed
  - Providing security to prevent unauthorized access to the computer or to certain files stored on the computer

NOTE

The subject of networks involves a great deal of new terminology, including numerous acronyms. For convenience, a list of commonly used acronyms is included in the *Appendix* at the back of this module.

7.0.0 ◆ NETWORKS

During the early days of computers, large rooms were required to house huge data processing machines. They consumed large amounts of power and produced enough heat to tax even the best air conditioning systems. These machines were extremely expensive, slow, and cumbersome as compared with today's computers.

As computers began to be connected together to form networks, a problem became apparent. Each processor had to use much of its valuable time in communication processing instead of main data processing, which was an inefficient use of the central processor. Front-end processors (FEPs) were developed to take over the communication processing task.

Cluster controllers were developed so that a single telephone line could be shared by simultaneous users in one location. Simple traffic rules for connecting systems, called protocols, were developed and implemented in hardware and software. There was, however, no standardization for inter-operability between manufacturers. In fact, different companies began developing networks independently. For example, Datapoint Corporation developed a standard called ARCnet® at about the same time Xerox® developed Ethernet. A bit later, IBM® developed a third standard called **token ring**. These companies each wanted their product to become the overall standard, so inter-operability was not encouraged. To help prevent this, organizations such as the IEEE supported the International Standards Organization (ISO) objective of worldwide interoperability.

Modern systems move data quickly between dissimilar computers, and information is processed without interfering with the computer's ability to do other work. The capabilities of today's information exchange systems have grown to the point that networks can easily cross national, commercial, and governmental boundaries.

In a relatively short time, networks progressed from university experiments to an essential component of modern business. Few technologies are changing as rapidly as the computer industry, with new technology being unveiled almost daily.

7.1.0 Information Exchange

The first step in understanding how information is transferred between computers is to understand the information format. Computers store information as binary digits called bits which are either 1s or 0s. Digital signals consist of two states, on and off, which can be used to represent these 1s and 0s (see *Figure 19*). These signals can be transferred between computers via cable, allowing computers to transfer information and commands.

The devices sending and receiving the information must be compatible with the transmission medium. A device is compatible when it can reproduce the correct 1s and 0s from the signal on the highway in the receive mode and conversely produce the correct signal when in the transmit mode.

The transmitted information originates in computers in the form of binary-coded data. For example, each letter in the alphabet is assigned its own binary code. Binary code is made up of bits, and these bits are arranged in such a way that they form a binary code.

Bits are transmitted electrically by voltages. The 1 is given a positive voltage (usually 5 VDC) and the 0 is given a value of 0 VDC.

The rate at which the data is transmitted is often stated in bits per second (bps) or bytes per second (Bps). A byte is an eight-bit word. Sometimes data rates are referred to as baud rates, but this is actually not the correct way to identify data rates. The most common receiving and transmitting device is the modem (short for MOdulate/ DEModulate), which translates digital information for transfer over phone lines. Typical data rates for modems are 33.6 kilobits per second (Kbps) and 56 Kbps.

Some other acronyms for data rates are as follows:

- *KBps* – Kilobytes per second (thousands)
- *Mbps* – Megabits per second (millions)
- *MBps* – Megabytes per second (millions)
- *Gbps* – Gigabits per second (billions)
- *GBps* – Gigabytes per second (billions)

A network is a number of computers and peripheral devices connected together or connected to some other control device, such as a **programmable logic controller (PLC)**. The most common types of networks are the **local area network (LAN)** and the **wide area network (WAN)**. A LAN is a network that allows computers and peripheral equipment to communicate with each other within a limited area. A WAN also allows communication between computers and peripherals, but the area is not limited. In most cases, LAN data speeds are greater than WAN data speeds. An example of a LAN application is an office where all of the computers are linked to a server using twisted-pair wire or fiber-optic cable. The server is a computer that contains an array of hard drives and memory that is used by all of the computers in the office. The Internet is an example of a WAN.

Some of the benefits of a LAN are as follows:

- Reduction/control of cabling cost
- User sharing of programs, data, printers, and communication links
- Access to multiple databases
- Access to remote systems within the local network
- Use of lower-cost, diskless workstations
- May allow computers with different operating systems to access common data files
- Possibility of linking multi-vendor machines
- Simplified software license management
- Improved staff communication
- Improved data integrity and security

Some of the advantages of WANs are as follows:

- Access to databases at great distances
- Access to remote systems
- Advertising via the internet
- Long distance communication with e-mail

7.2.0 Local Area Networks

Three basic elements make up a local area network (LAN): transfer medium, hardware, and software.

A transfer medium, usually twisted-pair wires or fiber-optic cable, is used to send information between network elements (*Figure 20*). Twisted-pair

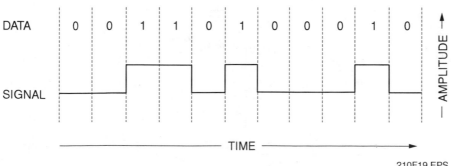

Figure 19 ◆ Binary data format.

210F19.EPS

wires are normally 22 or 24 AWG and are used everywhere in commercial buildings. The twist is engineered to cancel out electrical noise from motors and other inductive sources. These wires are often unshielded, which means that they have no metallic sheath around them. This type of wire is known as unshielded twisted pair (UTP). Pairs with shielding are called screened twisted pair (ScTP). Shielding protects wires from most sources of electromagnetic interference (EMI) by grounding the interference before it gets to the wires.

Fiber-optic cable is used for greater data transfer speed. Fiber-optic cables can routinely handle data rates up to 5 Gbps. Fiber-optic cable is immune to EMI and runs virtually error free. An infrared laser provides a light source that is invisible to the eye. At the other end of the fiber, the light signal is converted back into an electrical signal by a light-sensing diode, which varies its conduction based on the amount of light received.

7.3.0 Transfer Medium

The most basic element of the network is the cable, also known as the transfer medium. It is the most basic of network elements, but it is also the most critical. More networks suffer from a poor choice or poor installation of transfer medium than from any other problem. Choosing the wrong medium may prevent the network from ever functioning properly or, even worse, the network may suddenly deteriorate after working for a certain period of time.

Often, vendors and engineers will recommend fiber optics as the remedy for all network problems. However, fiber optics is not without some problems of its own, with cost being a considerable one. Keep in mind that all types of media have their own strengths, weaknesses, and peculiarities. Because this module covers **real-time** networks, it will focus on the three types of media prevalent in applications today: twisted-pair

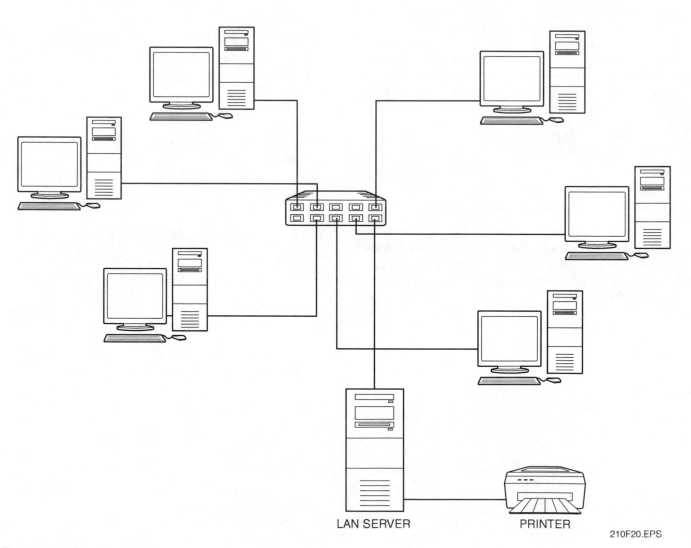

LAN SERVER   PRINTER

210F20.EPS

Figure 20 ◆ Transfer medium.

wiring, coaxial cable, and fiber-optic cable. Wireless data transmission will be discussed briefly, but a thorough treatment is beyond the scope of this module.

7.3.1 Copper Media

Twisted-pair wiring is one of the most common forms of wiring and is probably used more often than any other because of its use in telephone systems. Most commercial buildings or factories have miles of twisted-pair wiring already installed, either as a result of installations by telephone companies or as wiring for other devices within the buildings. Unfortunately, much of this wiring is not suitable for local area networks. This does not mean that it is not suitable for data transmission. In fact, much of this type of wiring is well suited to serial data communication at low to moderate speeds.

The connection of serial devices, such as display terminals or low-speed, asynchronous links between computers, is as much a part of a LAN as the computers. It may be accomplished using the installed UTP wiring, an example of which is shown in *Figure 21*. However, this wiring may not be suited to serve as a backbone or horizontal branch circuit, as not all existing UTP is suited for LAN use. Unless the installed cable matches the specification in the standards or the network vendor's specification, it is unwise to use pre-existing UTP in a network.

There are a number of methods for testing UTP cable to determine its suitability for LAN use. The two best tools are a **time-domain reflectometer (TDR)** and a hand-held cable tester. The cable tester is a digital testing device that employs TDR technology, but displays the information in a plain English readout rather than as a graphic display.

When installing new UTP wiring, it is wise to consider alternate uses of the wire. UTP wiring can be classified by the application for which it is intended. *Table 2* lists the categories of UTP wiring

encountered when considering options for new installations.

In addition to the UTP classifications listed in *Table 2*, standards for Category 7 were being developed at the time this material was written. Category 7 describes a fully shielded twisted-pair cable with a bandwidth in the 600 MHz range. Categories 5 and above are capable of sustained 100 Mbps data transmission.

An excellent reference for designing new wiring plans for commercial buildings is the *Electronics Industries Association (EIA) ANSI/EIA/TIA Standard 568, Commercial Building Telecommunications Wiring Standard*. *Table 3* lists the EIA parameters recommended for UTP cable.

Screened twisted-pair (ScTP) wiring, as shown in *Figure 22*, is also included in the *EIA/TIA 568* standard. ScTP cable is often used in areas susceptible to electromagnetic interference. While UTP is suitable for the office environment, ScTP is more suited to the industrial environment.

Table 2 UTP Classifications

Category	Classification	Typical Usage
3	Digital voice data	Asynch, synch, LANs
5	High-grade data	Very high-speed LANs (10 to 100 Mbps)
5e	High-grade data	Very high-speed LANs (10 to 100 Mbps)
6	High-grade data	Very high-speed LANs (1 to 10 GHz)

Table 3 EIA Parameters for UTP Cable

Parameter	Recommended	Allowable
Pairs	4	
AWG	24	22
Conductor diameter	1.22 mm	
Cable diameter	6.35 mm	
Tensile BP	40.82 kg	
Bend radius	25.4 mm	
Impedance		
64kHz	125 ohms	±15 percent
1 to 6MHz	100 ohms	±15 percent
Crosstalk		
1MHz	41 dB/305 m	
16MHz	23 dB/305 m	
Attenuation		
64MHz	2.8 dB/305 m	
1MHz	7.8 dB/305 m	
16MHz	40 dB/305 m	

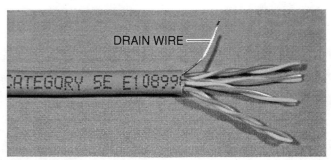

210F21.EPS

Figure 21 ◆ Unshielded twisted-pair wiring.

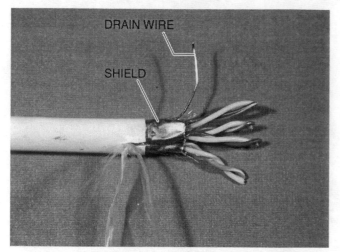

Figure 22 ◆ Screened twisted-pair wiring.

Shielding has both advantages and disadvantages. Shielded cable is more resistant to the types of interference found in many industrial settings and is less likely to radiate emissions caused by high-speed digital data traveling down the cable. The advantages, however, may not outweigh the biggest disadvantage—cable cost. Both the cable and the connector must be shielded. Shielded connectors are up to twenty times more expensive than unshielded ones, and the cable is two to three times as expensive as equivalent unshielded cable.

7.3.2 Fiber-Optic Media

Most vendors recommend fiber-optic cable for industrial networks. The advantages of fiber-optic cable include immunity to noise, extremely high capacity, low attenuation, and very high reliability. It suffers from two disadvantages: cost and lack of power-carrying capability. Fiber-optic cable may cost up to 25 times more than UTP and is currently 20 percent more expensive than the best coaxial cable. In addition, the cost of connectors, couplers, and the labor to terminate fiber-optic cable is three to five times higher than that for copper media. Also, fiber-optic cable is not an electrical conduit, so it cannot carry electrical power on the cable itself. Additional copper cables are required if power must be supplied to inline devices such as amplifiers.

The advantages of fiber-optic cable usually outweigh the cost and power disadvantages. One of the more time-consuming and costly features associated with fiber-optic cabling is the termination and connection of the cables. This is also where the greatest potential for disaster lies. Some of the more common coupling problems are angular mismatch (*Figure 23*), lateral misalignment (*Figure 24*), and gapping (*Figure 25*).

Air-blown fiber (ABF) systems are also available. In the ABF system, a microduct is first installed, then individual or multiple optical fibers are blown in using compressed air. This method is effective for safely installing fiber in ducts with many bends. When the network needs to be changed, the installer blows out the old fiber, then blows in a new one.

In all of these situations, some of the rays of light escape the pipe, introducing losses into the system. If enough losses accumulate through improper termination, the LAN system may not function properly.

To help alleviate these problems and reduce attenuation, a number of coupling techniques have been developed. The highest-quality splices or connections are made by fusion. By literally melting the glass fibers together, the problems of gaps and misalignment are virtually eliminated. However, fusion splicing is expensive, requires special equipment, and uses high temperatures.

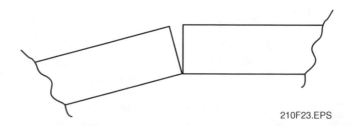

Figure 23 ◆ Angular mismatch.

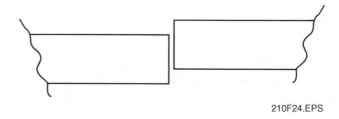

Figure 24 ◆ Lateral misalignment.

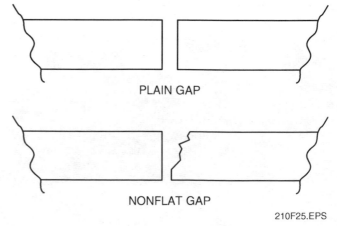

PLAIN GAP

NONFLAT GAP

Figure 25 ◆ Gapping.

Also, fusion splicers are difficult to operate. Epoxy splicing is a less expensive and generally acceptable alternative. In this technique, a filler (epoxy) is used to bond the fibers in the correct alignment and to fill any gaps, as shown in *Figure 26*. The least expensive technique is a simple mechanical connection. Alignment and fiber end preparation are critical in this technique, making the connection somewhat fragile. Because of this, mechanical connections are not recommended for use in industrial facilities.

7.3.3 Composite Cables

One of the newest wiring techniques uses composite cables. These custom-made cables can contain UTP, coaxial cable, and fiber-optic cable in a single jacket. The obvious advantage in using such a cable is that installation costs are lowered since there is only one cable to handle. Additionally, the fiber conductors will occupy spaces

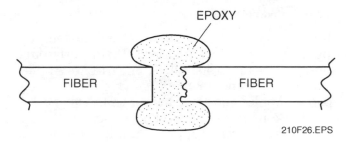

Figure 26 ◆ Using a filler to connect optical fibers.

between the metallic conductors, which normally would have been taken up by filler material.

Composite cables allow a network designer to run cables and implement each technology as the client's needs grow. Of course, using a composite cable when there is no indication that some additional cabling technology will be needed in the future does not make sense, because composite cables are expensive. It is also important to remember that not all cables are suited to all topologies.

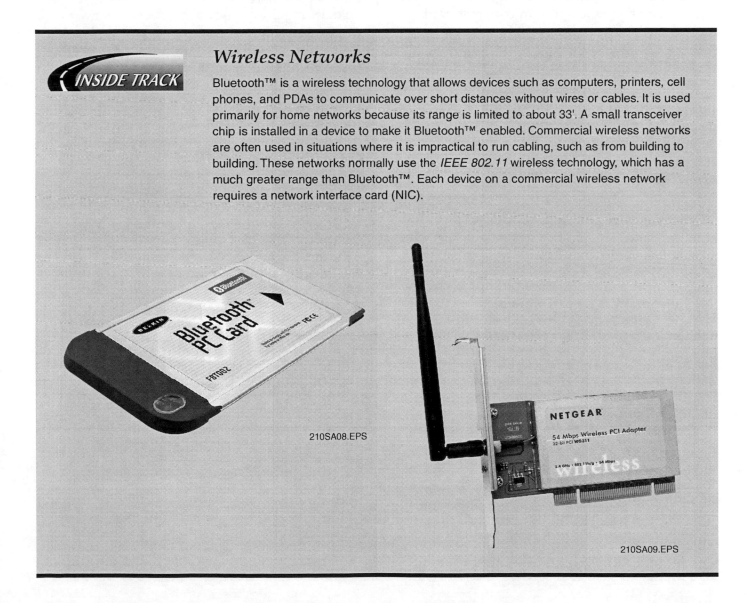

Wireless Networks

INSIDE TRACK

Bluetooth™ is a wireless technology that allows devices such as computers, printers, cell phones, and PDAs to communicate over short distances without wires or cables. It is used primarily for home networks because its range is limited to about 33'. A small transceiver chip is installed in a device to make it Bluetooth™ enabled. Commercial wireless networks are often used in situations where it is impractical to run cabling, such as from building to building. These networks normally use the *IEEE 802.11* wireless technology, which has a much greater range than Bluetooth™. Each device on a commercial wireless network requires a network interface card (NIC).

7.3.4 FireWire®

FireWire® is a very fast external bus developed under the *IEEE 1394* standard. It supports data transfer rates up to 400 Mbps. FireWire® is Apple® Computer's name for the bus, which was developed by Apple® and Texas Instruments. Other companies use names such as i.Link® and Lynx®. The *IEEE 1394* standard supports multiple (up to 63) external devices from a single port.

7.3.5 Wireless Media

As radios, television, and telephone microwave systems have demonstrated, wires or cables are not really needed for data communication. The air is an excellent medium for transmitting data signals. In fact, light traveling through the air has been used as a means of communication for centuries.

When data from wireless sources is to be incorporated into a real-time network, several options are available. One uses line-of-sight microwave, and the other uses satellite microwave. Both of these systems can be extremely expensive. Even when using satellite microwave, the interface with the satellite transceiver will still require transmission media. Transmission through free space is the most insecure form of data transmission and is subject to the greatest interference. If security of the data is critical, wireless media is probably not the best alternative unless it is the only one.

7.3.6 Optically Encoded Transmission

Optically encoded data transmission through the air offers better security and has been used to communicate between buildings when the installation of cabling is not technically or commercially feasible. Most systems use a modulated beam of infrared light. Optical communication is not affected by EMI, but a heavy layer of fog can render it useless. *Table 4* is a summary of the characteristics of each medium.

7.4.0 Physical Connections

Once the medium is in place, it must be attached to the devices that need to communicate with each other. Physical connection involves more than just the connector alone. In order to be transmitted, the data must be in a format that allows it to be transferred and interpreted correctly. This format may be individual digital pulses, referred to as baseband transmission, or signals that have had the data encoded on a carrier, referred to as a modulated signal. A certain amount of synchronization must be accomplished to ensure that the receiver is in a condition to accept the information. In some networks, a time-based form of medium sharing known as multiplexing is used to improve throughput and performance.

7.4.1 Connection to the Medium

Different coupling methods are used for each medium (copper, fiber, or air). Most copper-based technologies require a physical connection between the current-carrying network conductor and the attached device; optical systems and microwave or radio technologies do not. Because they are not current-carrying, optical systems rely on different coupling means than copper-based technologies. Because microwave and radio systems are outside the central focus of networking in the process industries, they will not be discussed further here.

One of the greatest problems facing the connection of networks in an industrial manufacturing setting is the environment. Industrial facilities inherently suffer from a variety of potential contaminants, with the type of contaminant depending on the industry. When using a copper-based medium, the integrity of the connection is vital to the integrity of the overall network. The number one cause of hardware-related network failures is poor connections. A poor connection can result in intermittent operation, which is an extremely difficult problem to isolate and repair.

Table 4 LAN Media Capabilities

Characteristics	Twisted Wire	Fiber Optics
Bandwidth	100 MHz	5 GHZ
Capacity (# of data channels)	1 or 2 per pair	1 or 2 per pair
Number of devices connected	Limited by physical space	Virtually unlimited
Video supported	Slow scan	Yes
Ease or speed of physical examination	Difficult to moderate	Easy
Port speed	Kbps to 100 Mbps	5 Gbps/channel
Error rates	Very low	Very low
Vulnerability to single point of failure	Moderate	Very low
Ease of relocation	Difficult to moderate	Difficult

Also present in many industrial settings are noxious or corrosive vapors. These vapors can contribute to the development of corrosion, which acts as an insulating layer, further reducing the conductivity of the connection. Another form of contaminant is electromagnetic interference or EMI, which is generally most severe where insufficient shielding or poor grounding occurs.

When a fiber-optic medium is selected, there is no need to provide for shielding or for ground continuity at the connection. Again, EMI is not a concern for the fiber-optic medium. Connection corrosion is a concern only to the extent that the connections remain light-tight.

The selection of the right connector for a given application, such as one that is secure, easily connected, and provides protection against gas and/or moisture intrusion, can be difficult and expensive. *Table 5* lists typical connector styles associated with data communication and classifies them according to industrial application standards.

7.5.0 Transmission Techniques

Once the connection has been made, the next step in moving the data through the network is the impression of the signal on the medium. Whether dealing with light waves or electrical signals, the two basic transmission techniques that are used are baseband and modulated signaling.

7.5.1 Baseband Transmission

Baseband is the most basic of all transmission techniques. It uses individual digital pulses to communicate information. These digital signals are generally on the order of 5V and have a very low current, so some form of amplification is nor-

mally required. There are two basic forms of baseband signaling:

- One that uses the presence or absence of voltage or light to distinguish between a binary one and a binary zero (*Figure 27*)
- One that uses the polarity of the signal to signify the binary value (*Figure 28*)

Since light cannot show polarity, fiber-optic systems employing baseband signaling only use the voltage method.

Using polarity to signify binary states has advantages, especially when transferring data over long distances. Polarity shifts generate by default (a third state—no signal present). This non-data state simplifies error checking and signal conditioning, and because the magnitude of the signal is not as important as the polarity, any signaling that uses this method is less sensitive to interference and therefore less prone to errors.

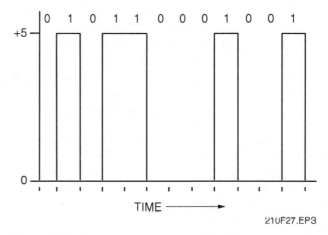

Figure 27 ◆ Binary state represented by the presence of voltage or light.

Table 5 Data Transmission Connectors

Connector Type	Physical Security	EMI Shielding	Gas Tightness	Typical Networks
RJ/MMJ	Poor	None	Poor	Token ring; Ethernet; serial
D-shell	Screwed – good Side clamp – fair	Can be shielded	Can be good, often fair	Token ring; Ethernet; serial
BNC	Very good	Very good	Good	Ethernet
N	Very good	Very good	Very good	Ethernet
Ethernet piercing tap	Fair to good	Good until removed	Poor to fair	Ethernet
IBM data	Good	Very good	Fair	Token ring
F	Very good	Very good	Poor to very good	Token ring
Twinaxial	Moderate to very good	Good to very good	Good	Various
Screw terminal	Fair to good	None	None	Various

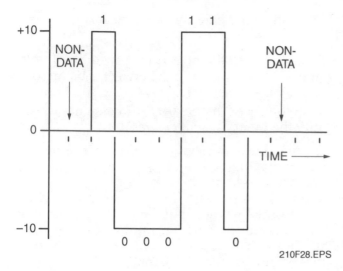

Figure 28 ◆ Binary state represented by polarity.

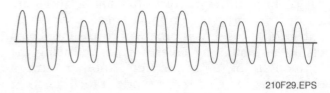

Figure 29 ◆ Amplitude modulation.

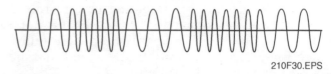

Figure 30 ◆ Frequency modulation.

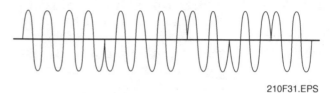

Figure 31 ◆ Phase modulation.

7.5.2 Modulated Signaling

An alternative to baseband signaling is modulated signaling, which encodes specific information onto a carrier frequency. Modulation is the basis for radio and television, where either video or audio signals are encoded into radio waves that carry the signals to the receiver where they are decoded. In digital data communication, this process is known as modulation and demodulation. Devices that perform these functions simultaneously to support bi-directional communication are known as modems.

Modulation can take several forms, from the simplest amplitude modulation (AM), through frequency modulation (FM), to the most complex **phase shift keying (PSK)**. An intricate explanation of these modulation techniques is beyond the scope of this module, but familiarity with the concept is required. *Figure 29* shows an amplitude modulated signal. *Figure 30* shows a frequency modulated signal.

There are two ways of performing frequency modulation, also known as **frequency shift keying (FSK)**. These are phase-coherent and phase-continuous frequency shift keying. These two modulation styles are not compatible. That is, if one modem is using phase-continuous FSK and another is using phase-coherent FSK, the two cannot communicate with each other.

The other widely used form of modulation for digital data communication is phase modulation (PM), or phase shift keying (PSK). In this instance, a simple change of phase can be used to signify a change in binary state. *Figure 31* shows how this is accomplished.

7.5.3 Carrier Band Transmission

A single digitally modulated carrier signal placed on a single conductor for transmission is called carrier band transmission. Carrier band transmission can travel over moderate grades of cable for reasonably long distances. It uses common coaxial cable with F-style connectors, which exhibit good resistance to environmental contamination and are easy to install and maintain. F-style connectors are generally used because they are weather-resistant.

7.5.4 Broadband Transmission

If more than one channel of information is required, there are several options. One is to string additional cables next to each other. A more practical solution is to use a broadband modulation scheme. In broadband modulation, multiple channels share a common cable by being assigned different carrier frequencies and then being modulated about that frequency. Broadband communication systems are one of the most common types in use today. Cable television is a one-way broadband cable communication system. Information is encoded onto several channels, and if one wants to access different information, the channel is simply changed. This model also works well in digital data communication. Multiple channels of data can exist on a single cable. If a person or device needs access to that information, a modem is tuned to the proper frequency and the channel can be shared.

Figure 32 shows how a broadband system shares the cable among multiple users. Note that data can be located on several channels, voice on other channels, and video signals for security, training, or other purposes on the remaining channels.

The IEEE has issued a standard for broadband data communication—*IEEE 802.7*. This standard provides an excellent primer on broadband cable usage and has sections devoted to documentation, testing, installation, and maintenance, as well as the expected specifications.

Broadband systems are not simple. They are true broadcast systems in which signal leakage can cause not only reliability problems, but can also result in a citation from federal authorities if it interferes with public forms of communication. Because these broadband systems are operating at the same frequencies as local community antenna television (CATV) systems and broadcast television and radio (FM bands), leakage can easily be detected by homeowners in the immediate vicinity. However, because broadband has the capability to operate over many kilometers, with many channels of data, voice, and video being transmitted concurrently, these systems are still popular, particularly in large facilities.

7.6.0 Synchronization

A means of controlling the timing between the sender and the receiver is essential. This is called synchronization. Two different types of synchronization are commonly used—asynchronous mode and synchronous mode. For lower-speed terminal to computer or computer to serial communication, asynchronous mode is used. For high-speed communication, synchronous mode is the better choice.

In asynchronous mode, the sending and receiving units have independent internal clocks, and the message contains information to facilitate proper processing. This information provides details about when the message starts, how many bits it contains, and when it ends.

To provide higher data rates, networks primarily use the synchronous mode of data transfer. In this mode, the data signal itself (through its transitions) becomes the receiver's clock. *Figure 33* illustrates how the data signal is used to generate a clocking signal in the receiver.

7.7.0 Multiplexing

Cable sharing was previously discussed using broadband transmission; however, there is

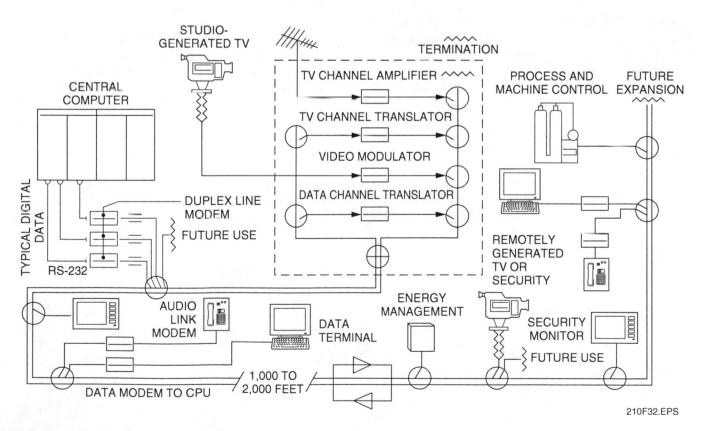

Figure 32 ◆ Broadband usage among multiple users.

210F32.EPS

- TIMING IS RECOVERED AT THE RECEIVER BY THE TRANSITIONS IN THE DATA.
- EXAMPLE: RECEIVING DATA

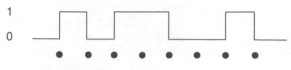

EXPECTED TRANSITIONS

- RECEIVER CLOCKING ERRORS CAN BE CORRECTED BY NOTING THE DIFFERENCE BETWEEN EXPECTED AND ACTUAL TIMING OF SIGNAL TRANSITIONS.

210F33.EPS

Figure 33 ◆ Synchronous clocking.

another form of cable sharing known as time division multiplexing (TDM). In time division multiplexing, a frame of data at a lower speed is compressed and inserted into a portion of a frame of a higher-speed transmission. *Figure 34* illustrates how three packets of data at 10 Mbps are multiplexed into a 50 Mbps message.

7.8.0 Physical Connection Summary

The physical connection to the medium is one of the most important and most often overlooked aspects of networking. The environment plays an important part in the decision of which medium is selected to maintain reliability. The choices of baseband, carrier band, or broadband and optical versus electrical are vitally important to the overall success of the network.

8.0.0 ◆ NETWORK ACCESS

Now that the transmission medium has been selected, the access to that medium must be managed. In typical networks, numerous devices are connected to a wire or cable. There must be a

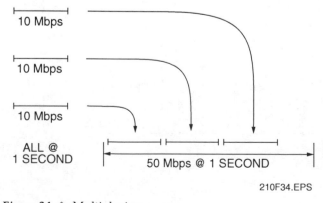

210F34.EPS

Figure 34 ◆ Multiplexing.

mechanism to control access to that cable so the devices do not all try to transmit at the same time. Some methods of wiring the devices together make access control management easier, but in doing so may create their own problems. The way these interconnections with the medium are made is known as topology. Each topology lends itself to the use of certain sharing mechanisms, each with strengths and weaknesses.

8.1.0 Topologies

One of the most basic topologies is the point-to-point or peer-to-peer topology. *Figure 35* is an example of a two-**node** point-to-point topology. If a third node is added, only a couple of additional links are needed. The problem arises when many nodes exist, all of which need to talk to each other. For simplicity and cost effectiveness, another means had to be developed to connect all of these devices.

8.1.1 Star Topology

One of the most prevalent topologies is the star topology. *Figure 36* is a simple seven-node system in which the seventh node serves as the **hub** of the star. This topology was first put to use by the telephone system, and it is the basis for many effective network systems as well as most existing industrial control systems. The major strength of star topology is the existence of a central node that simplifies management and **media access control (MAC)**. The biggest weakness of star topology is its reliance on a central hub, which becomes a single point of failure.

Cost is also a drawback to a star configuration. This is due to the large amount of wiring involved. For example, if there is a central point on the first floor of a 20-story building and every wire must be connected to that point, 100 terminals on the 20th floor will require running 100 sets of wires the depth of 20 floors to connect to that central hub.

8.1.2 Ring Topology

The use of ring topology is an attempt to circumvent the problems of star topology; namely, a central point of failure and excessive wiring lengths.

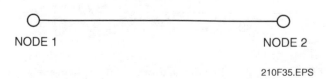

210F35.EPS

Figure 35 ◆ Two-node point-to-point network.

How Fast?

In a residential environment, there are three common methods used to access the Internet: cable modem, DSL, and dialup modem. The dialup modem uses a standard phone line and is the slowest method. The typical phone modem will download about 56 kilobits of data per second (Kbps). Uploading is slower. Besides being slow, the dialup modem ties up a phone line while you are on line. If it's connected to your only phone line, no one can call in or out while you are online. A digital subscriber line (DSL) also uses a phone line, but it does not interfere with regular voice traffic, so only one phone line is needed to accommodate both. Besides this advantage, DSL is blazingly fast in comparison to dialup—about 1.5 Mbps—because data is transmitted in digital form rather than the analog mode traditionally used for voice. A cable modem, such as the one shown here, moves data at about the same speed as DSL, but it uses the cable TV system rather than the phone system. Both of the faster methods require a special modem which is generally obtained from the provider as part of the installation. While each of the high-speed methods has its advantages and disadvantages, they are generally not noticeable to the average user. Here are two significant concerns:

- DSL becomes less effective as the distance between the head end and the subscriber increases. For that reason, DSL service is limited to users within a specified radius of the provider.
- Cable is not noticeably affected by distance, but is likely to degrade as the number of users on a channel increases. Adding a channel usually solves this problem.

Sizable commercial operations may subscribe to a T-1 line. T-1 service, like DSL, is offered by the phone company. A T-1 line operates at the same speed as DSL and cable, but is able to support hundreds of simultaneous users without significant degradation.

210SA10.EPS

In a ring, each node is connected to two adjacent nodes and the message is passed in a circle, as shown in *Figure 37*. If a node fails, the ring is broken and message flow is impeded. While it is possible for traffic to flow counter to the break, only sophisticated management can ensure that this happens correctly. Trying to constantly reverse the message flow could prove to be an insurmountable problem.

8.1.3 Bus Topology

Bus topology is becoming more prevalent, especially in process control networks. As shown in *Figure 38*, a bus topology merely snakes a common conductor among all of the nodes and connects them with a tap or drop. This parallel wiring system promotes easy connection and low susceptibility to complete network failure caused by the failure of a single device. The main weakness of bus topology stems from the difficulty in controlling access to the media.

8.1.4 Hybrid Topologies

Because no single wiring topology meets all needs, the most common approach is to use hybrid wiring schemes. Two of most common are the ring-wired star network and tree topology, which is a combination of busses and branching busses or stars.

A ring-wired star network is illustrated in *Figure 39*. It uses a backbone ring to connect localized rings, which are star-wired to local rings using centralized hubs. This minimizes wiring distances by keeping the star-connected segments in a local area, yet provides overall reliability to the rings.

Figure 40 represents the use of a bus to serve as the backbone or trunk of a network. It provides the long distance connections between the branches that serve the local areas. The branches may consist of another bus segment, as shown on the left, or star segments, as shown on the right.

There is another reason why hybrid topologies have become increasingly popular. Some physical media lend themselves to point-to-point wiring and some are more easily tapped; that is, better suited to bus topologies. Coaxial cable is the easiest medium to tap in its midpoint, so it is often used in bus topologies. Fiber-optic cabling and twisted-pair wiring are far more suitable for point-to-point connections and are generally used in ring or star topologies because these have point-to-point wiring schemes.

8.2.0 Hubs and Routers

Hubs and **routers** are network devices. Hubs are used to connect computers to a network. Routers are used to connect different networks. Hubs and routers both translate the network information from one network cable to another network cable.

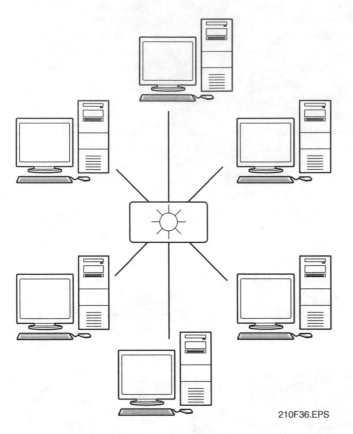

210F36.EPS

Figure 36 ◆ Star network.

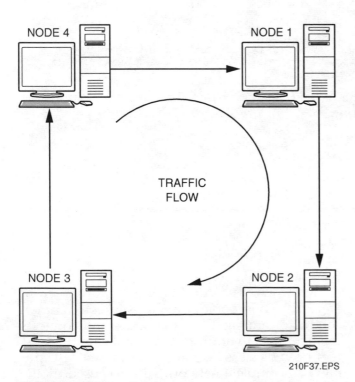

210F37.EPS

Figure 37 ◆ Ring network.

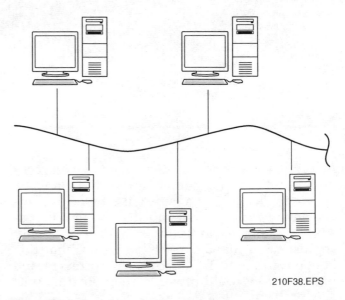

210F38.EPS

Figure 38 ◆ Bus network.

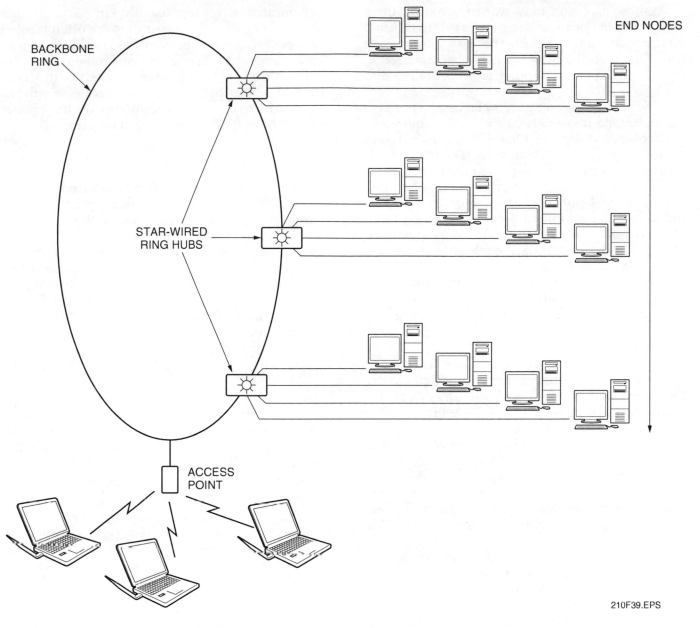

BACKBONE
RING

STAR-WIRED
RING HUBS

ACCESS
POINT

END NODES

210F39.EPS

Figure 39 ◆ Ring-wired star network.

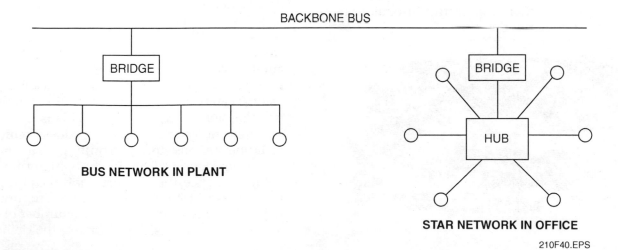

BACKBONE BUS

BRIDGE

BUS NETWORK IN PLANT

BRIDGE

HUB

STAR NETWORK IN OFFICE

210F40.EPS

Figure 40 ◆ Backbone bus.

Assume a person may want to connect three computers in one room to a network, but the room has only one network port. The person must use a hub. One end of a network cable is connected to the network port. The other end is connected to the hub.

Now each of the computers can be connected to the hub. One more network cable is required to connect each computer. One end of each network cable is connected to a computer. The other end is connected to an available port in the hub.

Hubs have several network ports. The most common are 4-port, 8-port, 16-port, and 32-port hubs. In our example, a 4-port hub is required. One port is connected to the network port. The remaining three ports are each connected to a computer.

For most hubs, any port can be used to connect to the network port. Some hubs have a port labeled uplink. If there is an uplink port on the hub, it must be used to connect to the network port in the room.

Switches are also used as hubs. Switches move the information between ports more efficiently than hubs, and are therefore becoming more common than hubs. A switch can be used in any situation where a hub is needed.

Routers (*Figure 41*) are used to connect networks together. For example, cable TV can be used to provide Internet access. A cable modem reads the information off the cable. A router connects to the cable TV network through the modem. It then translates the information to another port. This other port is a new network. Through the use of a hub, many computers can be connected to this new network. The computers on the new network can only communicate with other computers through the router.

The router listens to all the information on both networks. It looks for information on one network that is intended for a computer on the other network. The router then copies the information between the two networks.

210F41.EPS

Figure 41 ◆ Small office router.

Commercial routers usually have 2 ports, one for each network. Home routers are often labeled as 4-port routers. These devices combine a router and a 4-port hub in one device. Home routers are sold as cable/DSL routers. Cable TV and DSL (digital subscriber line service over a telephone wire) are the two most common ways for a home user to have high-speed Internet access.

8.3.0 Media Access

After a wiring topology has been selected, a means of controlling how that topology moves data is required. The granting of access to the network is referred to as media access control (MAC). Each of the transmission control techniques falls into one of two central categories: distributed control or centralized control. In distributed control, access to the medium is managed at each station. In centralized control, a single hub manages access to the medium for all stations.

8.3.1 Distributed Access Control

With **distributed control systems**, each station connected to the network manages its own access to the medium using a predetermined set of rules. Those rules will result in either a **stochastic** (randomly variable) or deterministic (predictable) response characteristic.

- *Stochastic access control* – Stochastic control is a random access method that has served as the basis for a number of popular networking technologies. The most commonly used random access control technique is the carrier sense multiple access with collision detection (CSMA/CD) protocol. CSMA/CD is the basis of networks based on both the Ethernet and *IEEE 802.3* standards. The protocol is simple. A station that wishes to transmit listens to see if any other station is transmitting and, if not, begins to transmit. This is the carrier sense multiple access part. While the station is transmitting, it continues to monitor the network to ensure that another station does not transmit while it is still transmitting. Such multiple transmissions (collisions) can be detected by the resulting garbled transmission on the network. This is the collision detection part. If a collision does occur, the colliding stations stop transmitting and wait for a random period of time before attempting to try again; hence, the stochastic nature of the protocol. It is not hard to imagine a situation in which very heavy traffic causes a number of collisions to occur, possibly delaying the delivery of a critical message. For this reason, few control system suppliers use CSMA/CD protocols for real-time process control.

- *Deterministic access control* – In real-time networks, predictable response times are often considered essential, as is the capability to resolve a priority scheme. Such priority issues include the ability to preempt routine message traffic to improve responses to emergency conditions. This determinism or predictability is a characteristic that is used to distinguish a class of access protocols. This class is further subdivided by a controlling method, such as central control or distributed control.

8.3.2 Centralized Access Control

Polling and circuit switching are the two primary types of centralized access control.

- *Polling* – This access method uses control by a single entity or **master** device connected to the medium, which queries its **slaves** by a technique known as polling. The master station sends out a message to each device on the network, inquiring if the device has a message. As soon as a device indicates that it has a message, it can transmit the message. When that transmission ends, the master station begins polling again. Polling can diminish overall access because one device may take a long time to transmit a large message; therefore, maximum transmission times are often enforced. Many control systems use polling as the basis for their communication services.
- *Circuit switching* – In circuit switching, direct peer-to-peer communication, which is not available when using master-slave techniques, is permitted. *Figure 42* shows that a central hub establishes physical connections between stations that need to communicate. Circuit switching is best implemented in a physical star topology.

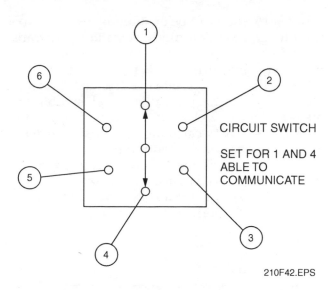

210F42.EPS

Figure 42 ◆ Circuit switching access control.

8.3.3 Token Passing

The most common form of local access control is **token** passing. In a token passing arrangement, each station is permitted to access the medium for transmission only when it possesses a token.

A token is a data message that signifies that a data transmission is not taking place and thus allows access to the transmission medium. This token is placed on the line and passed to each station.

When a station has a message to send, it takes the token and puts a message in its place. This message is then passed from station to station until it reaches its destination. The receiving station places an acknowledgment on the line, which is passed along to the message originator. When this acknowledgement is received, the token is placed back on the line for use by the next station that has a message to send.

Figure 43 illustrates token rotation in a physical bus topology and shows how the token rotates in a ring.

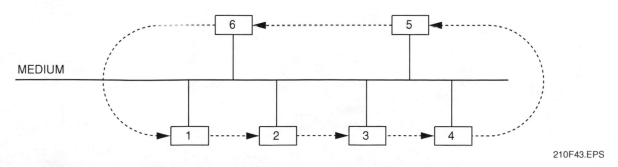

210F43.EPS

Figure 43 ◆ Physical bus with logical ring token passing.

Token rotation can be ensured by setting maximum token retention times. A station may transmit information for only a set time period and must then relinquish the token for the next station in line. If the time to rotate between the stations is known and the maximum time each station may talk is known, as well as the number of nodes connected, it is easy to calculate the worst-case delivery time under normal conditions.

The token passing technique is very adaptable and is therefore well suited to the use of a priority scheme. It is simple for a station with emergency traffic to attach a priority flag to the token. In this manner, only stations with a higher priority may claim the token and transmit. Therefore, while normal net cycle times are very predictable, they may be shortened to provide exceptional response in emergency situations. For this reason, several control system vendors have implemented token passing schemes in their proprietary networks.

The key to a successful use of a real-time control network is the selection of suitable topologies and associated media access control. *Table 6* indicates some of the pros and cons of the various topologies. *Table 7* indicates some of the pros and cons of the various access techniques.

Table 6 LAN Topologies

Topology	Pros	Cons
Star	Very easy to manage Well-suited to all access methods Supports fiber-optic cabling	Media intensive
Bus	Lowest cost routing of media Easiest to tap	Difficult access control Difficult to manage
Ring	Higher reliability Suitable for fiber optics	Star wiring needed to improve reliability and management Difficult to manage (true ring)
Point-to-point	Simplest design Can sometimes be converted to star	Often exists as an artifact of evolution Very medium intensive Extremely difficult to manage

Table 7 LAN Access Techniques

Access Technique	Pros	Cons
Dedicated channel	Contention-free uninterrupted service	Bandwidth waste Poor flexibility/expandability
Polling	Collision-free Centralized control simplifies network management, such as recovery from erroneous conditions	Net cycle time (polling message transmission time) Overhead increases with number of users Potential reliability problems; a single failure at master creates a problem; complex controls needed to solve problem Rigid under user population changes
Random access	Simple Accommodates large population of high-burst traffic nodes Low delay under limited throughput Suitable for broadcast environment	Limited throughput (collisions) May involve instabilities Not predictable
Token passing	Collision-free Predictable	Higher overhead More complex

7	APPLICATION
6	PRESENTATION
5	SESSION
4	TRANSPORT
3	NETWORK
2	DATA LINK
1	PHYSICAL

210F44.EPS

Figure 44 ◆ OSI reference model.

9.0.0 ◆ OSI REFERENCE MODEL

In 1977, the International Standards Organization (ISO) created the **open systems interconnection (OSI)** reference model. The OSI model is a functional guideline for communication tasks and consists of seven layers, as shown in *Figure 44*.

Each layer of the OSI model performs a specific function. The layers and functions were chosen based on natural subtask divisions. Each layer communicates with the same layer in other computers through layers in its own computer. Upper layers use the services of lower layers and provide services to higher layers.

See *Figure 45*. In this example, the transport layer (Layer 4) of Computer A wants to communicate with Layer 4 of Computer B. To do so, Layer 4 requests a service provided by the network (Layer 3) of Computer A. Layer 3 performs this service and in order to talk to Computer B, it requests a service of the data link (Layer 2). This process sends the request over the network medium. Once the message arrives at the destination computer, it ascends to Layer 4, where it is processed.

The seven functional layers of the OSI model are as follows:

- *Application layer* – The application layer (Layer 7) is the upper layer in the model and is concerned with the information in the message and how well it serves the user. This is where application programs (those computer programs that do the work) call upon the communication services. If this is not handled properly, the entire system is useless because the goal of the system is to serve the user when and how the user wants to be served. Typical protocols at

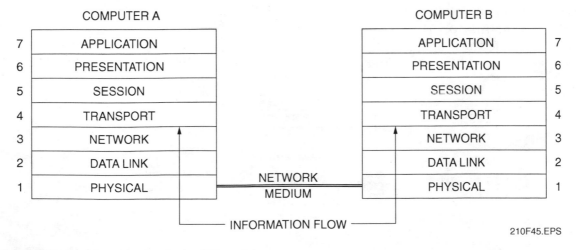

Figure 45 ◆ Network communication in the OSI model.

this layer include file transfer, access, and management (FTAM) and virtual terminal (VT).

- *Presentation layer* – The presentation layer (Layer 6) prepares the information for the application (*Figure 46*). An example of this function is the conversion of a file received from a computer using the **American Standard Code for Information Interchange (ASCII)** into the proper format for display on a system using **rich text format (RTF)**. Each system uses different codes to represent a letter in the alphabet. The presentation layer must know the differences and provide for them.

- *Session layer* – The session layer (Layer 5) is a coordinating function. It establishes the logical communication link between units and gradually feeds (buffers) the information to the device or program that performs the presentation function. The session layer also performs identification and authentication functions. It recognizes users and acknowledges both their arrival and departure. In some systems, the session layer can be a driving factor in system design; in others, it is a very small consideration.

- *Transport layer* – The transport layer (Layer 4) provides a common interface to the communication network. It translates whatever unique requirements the other higher layers might have into something the network can understand. It detects and corrects errors in transmission and provides for the expedited delivery of priority messages. It checks the data, puts it into the proper order (if necessary), and usually sends an acknowledgment back to the originating transport layer. It attempts to reestablish contact in the event of a network failure. Several industry and governmental standards exist for a transport function in data communication devices. The most common OSI protocol is called Transport Protocol Class 4.

- *Network layer* – The network layer (Layer 3) sets up a logical transmission path through a switched or dedicated network (*Figure 47*). In local networks, the path may only be theoretical, since the individual units are almost always electrically connected and the paths are defined by the network topology. In large systems, however, several transmission paths, and even alternative media, such as dialed telephone service versus leased service, may exist. The transmission path may be temporary, or it may provide a continuous connection for two users of the network. In a local network, the network control function can exist in one place (star topology) or be distributed (bus or ring topology).

- *Data link layer* – The data link layer (Layer 2) performs the accounting and traffic control functions that are necessary to transfer information on an electrical link, as shown in *Figure 48*. It forms the information to be moved into strings (long lines), or blocks (packages) of characters. The data link layer functions like a railroad yard supervisor who is making up a train. It puts every piece of information into the right place and checks it out before releasing it.

  Similarly, incoming information is broken down and properly routed within the receiving device.

- *Physical layer* – The physical layer (Layer 1) describes the electrical and physical connection between the communicating units, as shown in *Figure 49*. Often the most visible, it is sometimes the most troublesome part of the system.

9.1.0 Protocols

It is important to understand that the OSI model itself does not cause network communication to occur. Network communication requires a protocol calling for a particular implementation of one or more layers of the OSI model. Protocols are like blueprints for building a house. NetWare®, ARCnet, Ethernet, and transmission control protocol/internet protocol (TCP/IP) are examples of protocol blueprints.

Some protocols specify implementing several OSI layers, while others specify only one layer or

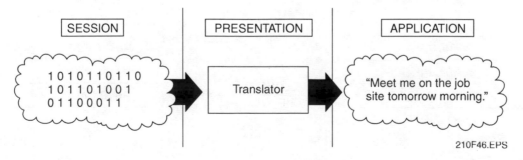

Figure 46 ◆ Presentation layer.

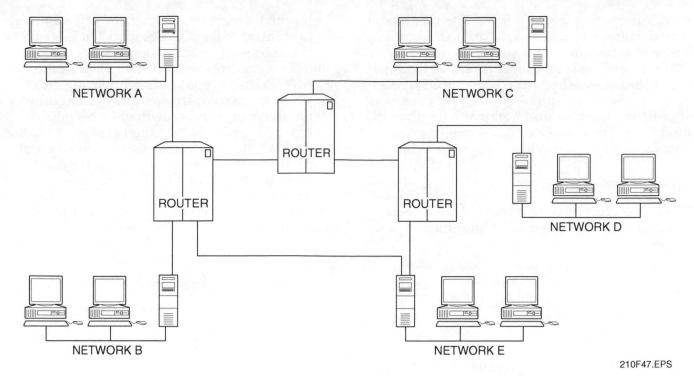

Figure 47 ◆ Network layer.

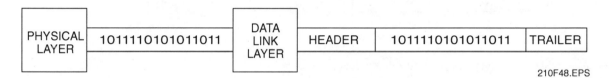

Figure 48 ◆ Data link layer.

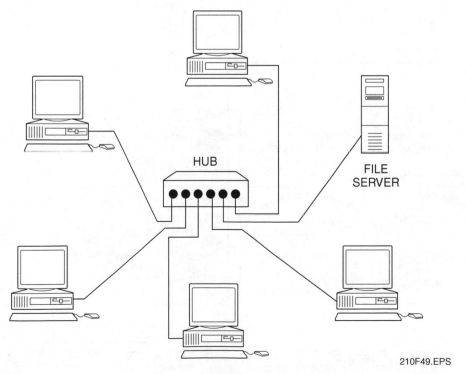

Figure 49 ◆ Physical layer.

even a portion of a single layer. The OSI model was designed to direct network development towards a standard. Many protocols were already in use when it was developed. Some of the existing protocols were back-fitted into the OSI model. To help the world understand how protocols should conform to and comply with the OSI model, ISO has developed a series of protocol specifications (*Figure 50*).

10.0.0 ◆ ETHERNET

Ethernet is one of the most successful networking protocols ever implemented. While Ethernet deals only with the physical and data link functionality in the context of the ISO-OSI model, it is the basis for numerous minicomputer, personal computer, and even real-time control networks. It is based on CSMA/CD access control and, originally, on a bus topology using coaxial cable. Ethernet led the way to the acceptance and growth of LANs.

Figure 51 compares the Ethernet standard to the *IEEE 802.3* standard. It is easy to see where the roots of the IEEE network lie. *Figure 52* is a comparison of Ethernet and IEEE frame formats. Note that the major differences are the use of a type field in Ethernet and the lack of both pad and length frames in the IEEE standard. The lack of a pad field

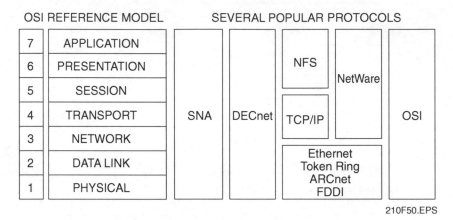

210F50.EPS

Figure 50 ◆ OSI model layers used by different protocols.

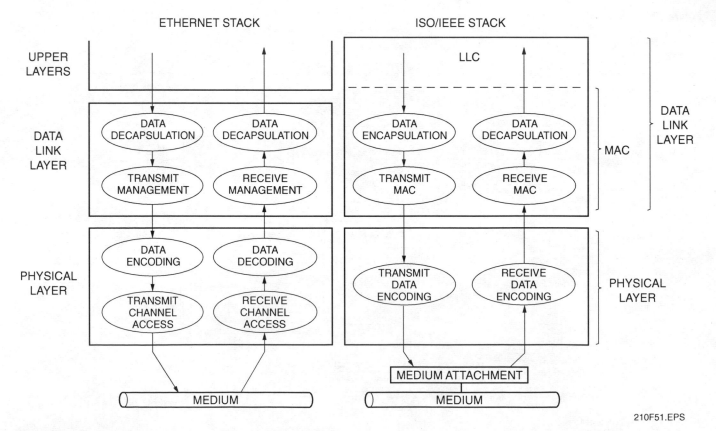

210F51.EPS

Figure 51 ◆ Ethernet standard versus *IEEE 802.3* standard.

ETHERNET FRAME

PREAMBLE	DESTINATION ADDRESS	SOURCE ADDRESS	TYPE FIELD	DATA FIELD	FRAME CHECK SEQUENCE
8 OCTETS	6 OCTETS	6 OCTETS	2 OCTETS	46–1,500 OCTETS	4 OCTETS

IEEE 802.3 FRAME

PREAMBLE	START DELIMITER	DESTINATION ADDRESS	SOURCE ADDRESS	LENGTH	DATA FIELD	PAD	FRAME CHECK SEQUENCE
7 OCTETS	1 OCTET	2 OR 6 OCTETS	2 OR 6 OCTETS	2 OCTETS	0–n OCTETS	0–P OCTETS	4 OCTETS

210F52.EPS

Figure 52 ◆ Frame format comparison.

is simple to explain: the IEEE frame uses a pad to allow 0 data bytes and maintain minimum frame length, while Ethernet uses a minimum data field length of 46 bytes. The type field was reserved for use by higher network layers. The similarities permit the two protocols to share a common cable, but the differences keep them from being able to talk to each other.

Because of the presence of such a large amount of Ethernet equipment and its use as a protocol for several control systems, the design of real-time control networks must take Ethernet into consideration. One of the most important considerations was previously noted—that while Ethernet and *IEEE 802.3* devices can share the same cable and coexist, they cannot communicate with each other. Given the similarities of the protocols, a number of network equipment manufacturers have developed interface cards that can operate over either protocol, but not simultaneously. It is important to know whether the control network has the correct interface for the type of CSMA/CD network installed.

Another issue is the life expectancy of the equipment. Any new device purchased should support *IEEE 802.3* protocols or both Ethernet and *IEEE 802.3*. Since a number of early control system implementations used Ethernet, it is probably best to specify compatibility with both to ensure that a useful life can be maintained.

Noise immunity is another critical issue. Baseband Ethernet has been criticized as being overly sensitive to the EMI present in industrial environments. Actual emission studies indicate that EMI is not as critical as once thought.

11.0.0 ◆ BASIC COMPUTER TROUBLESHOOTING

Due to the wide variety of computer systems and software available, there is no single method for troubleshooting computer problems. What is offered here are some basic steps to follow when trying to track down a system fault.

11.1.0 Temporary Problems

Some software errors on a computer occur when the system has been running too long, or when several different programs have been run either at the same time or in succession. Every program uses some of the computer's resources, such as memory or file handles. A file handle is a specific internal memory location that is used to read or write information from or to a file. A properly written program will reserve the resources that it needs when it starts, and release those resources for other programs to use when it is terminated. Unfortunately, there are many improperly written programs that do not release all of their resources. If many programs like this are run, or even if the same program is run many times, the system's pool of resources will run low, causing other, seemingly unrelated programs to fail. Problems of this sort can be fixed by rebooting the computer.

Resources and temporary files can also be a problem when a program has crashed, or if the user terminates the program through some means other than the normal method for exiting the program. Again, rebooting the computer will free up these resources and remove the temporary files.

11.2.0 Hardware Versus Software

Once a problem is detected, usually the first determination that must be made is whether the problem resides in the hardware or the software. Software problems usually leave the rest of your computer running, although you may have to reboot the system. Software problems are usually identified by error messages telling you that the computer cannot execute a part of the program, or a cryptic message such as *Invalid Function Call*,

Missing Library File, or any other message that does not specifically identify a piece of hardware that is malfunctioning.

A problem with a hard disk drive can be the hardest to define. It may seem like a software problem, stating that a part of the program cannot be found, or that a data file cannot be read. This could be due to a problem with the software, or it could occur because there is a problem with the disk on which the program or data is stored. Most computers will have some type of disk testing tool, often called a disk check program, or a disk scanner, to diagnose these problems.

If you determine that you do have a problem with a specific program, try removing that program from the computer and reinstalling it.

11.2.1 Hardware Problems

Hardware problems will typically leave some component of your system inoperable. For example, your printer may not operate, or you might not be able to read information from a diskette or CD-ROM. When experiencing hardware problems, check for loose connections and for component failures.

When a device that is external to the computer itself stops working properly, the first thing to do is to check all connections between the device and the computer. A loose cable will stop the flow of information between devices. These cables can come loose if the computer or device is moved. If the cable hangs from the edge of a table, its own weight may pull one end of it loose, or it may be jostled by anything passing nearby.

Connections for devices inside the computer case are less likely to come loose, but it does sometimes happen. Shut down the computer and unplug it. Wearing a properly grounded static discharge strap, open the computer case and make sure that all cables in the computer are firmly seated to their connections. If there is a device that does not appear to have a cable connected to it, refer the problem to a professional computer repair person.

Also check the power supply connections for devices such as disk drives and CD-ROM drives. These will usually have a small plug that is connected to the system's power supply.

After the system is closed again, start the computer. If the problem remains, then it probably is not a loose connection.

Once you have ruled out a software fault and a loose connection, you must determine which component has failed. For example, if there appears to be a problem with the printer, the error could be in the printer itself, the printer cable, or in the port on the computer to which the printer is connected. In cases like this, the simplest method is to swap out individual components until you eliminate all of those that work. Using the example of the printer, try using a different printer cable. If the printer still does not work, either try another printer, if one is available, or connect the printer to another computer. This allows you to determine whether you must replace the cable, repair or replace the printer, or repair the computer.

Monitors are particularly easy to diagnose this way. Usually, when a monitor fails, it simply does not display an image. Connect the monitor to another computer. If it is still blank, chances are the monitor has failed. Connect a monitor known to be working to the computer in question. If there is no image on this monitor, the video card or chipset of the computer has probably failed, or the computer is not booting up at all.

Use your senses – How can you tell if a computer is booting up if the monitor does not show an image? Most computers make distinctive sounds when they start up, which should still be heard. You can also usually tell that the computer is starting up by watching the disk drive activity light, which will light up when the computer is reading from the disk drive. If these signals indicate a normal startup sequence, and you know that the monitor is working properly, then there is probably a problem with the computer's video card.

Most hard drives will show signs that they are getting ready to fail before they stop operating completely. A high-pitched whine or squeal is usually hard from a drive that is going to stop working soon. You may also hear a rattling sound, which could be the bearings for either a disk drive or for an internal fan. On the other hand, no system should be completely silent. You should always hear the sound of at least one fan in the computer case. If no fan is running, the system is likely to overheat soon, which will cause a major system crash.

If the computer case or components feel excessively hot to the touch, this is a danger sign. Most computer components do not work well at high temperatures, and hot spots are unusual. Look for signs that a cooling fan is not operating, or for exposed wires.

The smell of burnt rubber is a sign that the insulation on one or more wires has been burnt off. This will usually be followed by a short circuit within the computer.

A smell of ozone (like you might smell after a nearby lightning strike) is a sign that an electrical arc or short has occurred. If the system is still operating after this has happened, there is still a high likelihood that one or more components within the system is not working properly.

Finally, the best way to know when something is going wrong with a computer is to know how it behaves when working properly. This makes it easier to determine what is not working when something goes wrong.

Most computer systems and peripheral devices come with manuals that contain their own troubleshooting guides. Refer to the manuals when trying to track down a problem.

Summary

The need for information and the requirement for compatibility among different systems to allow for information sharing has driven networks to their present-day sophistication.

This module has looked at the ISO-OSI reference model for data communication and it has shown the role communication plays in system connectivity, with each layer receiving information from the level below and providing service to the layer above.

The different types of transmission media have been identified, along with their strengths and weaknesses. This module has shown how the transmission media is physically connected to data transmission devices, and how access to that media is granted.

The different available network topologies were examined, along with their principles of operation. This module discussed how the different IEEE network standards function, as well as how they relate to the OSI model and the Ethernet standard. Internet function within the network context was also explained. This module explained the role that microcomputers play in the scheme of office and industrial LANs, and how they function in a network environment.

As more and more business, residential, and industrial sites are employing some type of digital communication network, all installation electricians will soon be required to demonstrate knowledge of these systems.

Review Questions

1. A computer's BIOS is normally located _____.
   a. on the hard drive
   b. in RAM
   c. on the back of the unit
   d. in ROM

2. Of the following types of transmission lines, _____ provides the highest bandwidth.
   a. 56 Kbps modem
   b. ISDN line
   c. T-1 line
   d. T-3 line

3. Of the following monitor standards, _____ provides the highest screen resolution.
   a. CGA
   b. VGA
   c. EGA
   d. VESA-compliant

4. The factors that most affect the operating efficiency of a hard drive are _____.
   a. rotational speed and access time
   b. size and shape
   c. access time and storage media thickness
   d. rotational speed and disk diameter

5. The process in which unnecessary data bits are removed from a computer file in order to reduce the size of the file is known as _____.
   a. disk fragmentation
   b. data compression
   c. baseband transmission
   d. frequency modulation

6. The type of program code that is permanently embedded in a memory chip by a manufacturing process is known as _____.
   a. RAM
   b. operating system software
   c. firmware
   d. shareware

7. The maximum distance that a data network can transfer information is _____.
   a. locally, within a building
   b. locally, within a small geographic area such as a city
   c. nationally
   d. globally

8. Screened twisted pair (ScTP) cable is often used in an industrial environment because _____.

    a. it is less susceptible to EMI
    b. it is cheaper than UTP cable
    c. it is more readily available than UTP
    d. all ScTP cable meets Category 5 standards

9. In an area where protection against gas penetration is required, a(n) _____ cable connector should be used.

    a. screw terminal
    b. Ethernet piercing tap
    c. N-type
    d. RJ/MMJ

10. What transmission technique sends and receives individual 1s and 0s?

    a. Broadband
    b. Carrier band
    c. Baseband
    d. Sideband

11. The type of network topology in which devices are connected to a common hub is known as a _____ topology.

    a. token ring
    b. bus
    c. star
    d. point-to-point

12. What media access technique has an inherently long net cycle time as one of its major drawbacks?

    a. Dedicated channel
    b. Random access
    c. Polling
    d. Token passing

13. The _____ layer of the OSI model performs traffic control functions.

    a. physical
    b. network
    c. transport
    d. data link

14. What would be the effect on network performance if an Ethernet system were connected to the same cable run as an IEEE CSMA/CD network?

    a. All nodes would stop communication.
    b. No effect would be noticed.
    c. An Ethernet node could only talk to other Ethernet nodes.
    d. Both groups of nodes could communicate with each other.

15. If it is necessary to open the computer to check internal connections, you should wear _____.

    a. a dust mask
    b. gloves
    c. a grounding wrist strap
    d. safety glasses

Trade Terms Introduced in This Module

Access control (AC): The means by which a communication standard grants access to the transmission media.

American Standard Code for Information Interchange (ASCII): A digital code for expressing the alphabet and numbers.

Basic input/output system (BIOS): The basic method by which a computer exchanges information.

Distributed control system: A manufacturing control system in which individual operations and processes are distributed throughout the system, but can be monitored by a central station.

Ethernet: A network based on bus topology and using the theory of stochastic actions of carrier sense multiple access/collision detection.

Firmware: Computer programs that are permanently burned into memory during a manufacturing process.

Frequency shift keying (FSK): A method of performing phase modulation in which a change in frequency is used to signify a change in binary state.

Hub: A common connection point for devices in a network.

Local area network (LAN): A network consisting of two or more computers connected together in a limited area, such as an office or a factory.

Master: A device that controls other devices.

Media access control (MAC): The means by which a networking system grants access to its transmission medium.

Microprocessor: An integrated circuit chip designed to perform computing functions. The microprocessor is the heart of a personal computer.

Monochrome: Able to display a single color.

Node: An element on the network that has a network interface card installed (for example, a computer workstation, network server, printer, modem, or other device).

Open systems interconnection (OSI) reference model: A seven-layer model developed by the International Standards Organization to describe how to connect any combination of devices for the purposes of communication.

Phase shift keying (PSK): A method of performing phase modulation in which a change in phase is used to signify a change in binary state.

Pixel: Abbreviation for picture element. A single dot in a graphic image on a computer screen.

Programmable logic controller (PLC): A device that can be programmed to accomplish specific tasks based on solving logic conditions. PLCs are often tied into a PC network.

Protocol: A set of predetermined rules that govern how two or more digital devices interact to exchange data. A protocol provides the foundation for network communication.

Real-time: A system used to control a function where response time is critical.

Rich text format (RTF): A text protocol intended to allow text to be exchanged with limited formatting, such as bold, italic, or bullets, between otherwise incompatible platforms.

Router: A device that forwards data within a network.

Semiconductor: Materials such as germanium and silicon in which the flow of current can be controlled by varying the applied voltage.

Slave: A device that is controlled by another device.

Stochastic: Information exchange involving a randomly variable response characteristic. Stochastic control is the basis for digital communication in the Ethernet network.

Time-domain reflectometer (TDR): A piece of test equipment that can measure cable characteristics.

Token: A data packet that is passed from node to node in a specific sequence. The node that has the token is allowed to transmit data on the network.

Token ring: A network architecture that is based on *IEEE 802.5* specifications and uses a token to control access to a ring-type network. Token rings were first developed by IBM® and are widely used in office environments.

Transistor: A semiconductor device that acts as a switch or amplifier. Transistors are the basic building blocks of integrated circuits.

Wide area network (WAN): A network whose elements are separated by distances great enough to require the use of telephone lines.

Computer and Network-Related Acronyms

AC	Access control
AM	Amplitude modulation
ANSI	American National Standards Institute
CAD	Computer-aided design
CAE	Computer-aided engineering
CAM	Computer-aided manufacturing
CD-ROM	Compact disk-read-only memory
CLNS	Connectionless network service
CONS	Connection-oriented network service
CPU	Central processing unit
CRC	Cyclical redundancy checksum
CSMA/CA	Carrier sense multiple access with collision avoidance
CSMA/CD	Data communication equipment

Additional Resources

This module is intended to present thorough resources for task training. The following reference works are suggested for both instructors and motivated participants interested in further study. These are optional materials for continued education rather than for task training.

How Networks Work, 2002. Frank Derfler, Les Freed. Indianapolis, IN: Que Corporation.

Network Design Reference Manual, Latest Edition. Tampa, FL: BICSI, www.bicsi.org.

Microsoft Encyclopedia of Networking, 2002. Mitch Tulloch, Ingrid Tulloch. Redmond, WA: Microsoft Press.

ACKNOWLEDGMENTS

Figure Credits

DPA Components International	210F02
Topaz Publications, Inc.	210F04, 210F05, 210F08, 210F09, 210SA02, 210SA03, 210F11, 210F12, 210F13, 210F17, 210SA05, 210SA06, 210SA07, 210F21, 210F22, 210SA10
Sun Microsystems, Inc.	210F06
Dukane Communication Systems	210SA01
Western Digital Corporation	210SA04
Advanced Computer and Network Corporation	210F14
Hewlett-Packard Company	210F18
Belkin Corporation	210SA08
Netgear, Inc.	210SA09
Linksys®, a division of Cisco Systems, Inc.	210F41

CONTREN® LEARNING SERIES — USER UPDATE

NCCER makes every effort to keep these textbooks up-to-date and free of technical errors. We appreciate your help in this process. If you have an idea for improving this textbook, or if you find an error, a typographical mistake, or an inaccuracy in NCCER's Contren® textbooks, please write us, using this form or a photocopy. Be sure to include the exact module number, page number, a detailed description, and the correction, if applicable. Your input will be brought to the attention of the Technical Review Committee. Thank you for your assistance.

Instructors – If you found that additional materials were necessary in order to teach this module effectively, please let us know so that we may include them in the Equipment/Materials list in the Annotated Instructor's Guide.

Write: Product Development and Revision
National Center for Construction Education and Research
P.O. Box 141104, Gainesville, FL 32614-1104

Fax: 352-334-0932

E-mail: curriculum@nccer.org

Craft _____ Module Name _____

Copyright Date _____ Module Number _____ Page Number(s) _____

Description _____

_____

_____

_____

_____

(Optional) Correction _____

_____

_____

(Optional) Your Name and Address _____

_____

_____

Module 33211-05

Advanced Test Equipment

COURSE MAP

This course map shows all of the modules in the second level of the *Electronic Systems Technician* curriculum. The suggested training order begins at the bottom and proceeds up. Skill levels increase as you advance on the course map. The local Training Program Sponsor may adjust the training order.

ELECTRONIC SYSTEMS TECHNICIAN LEVEL TWO

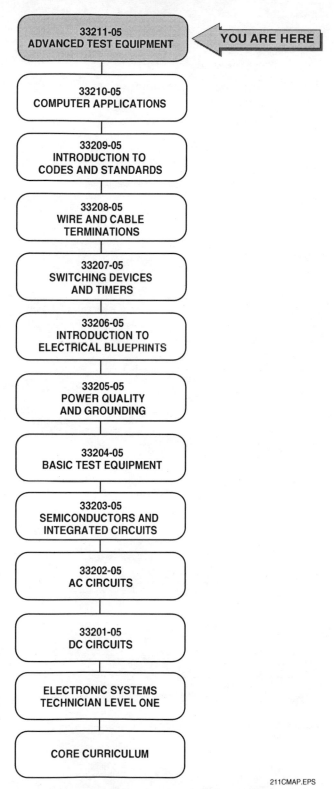

33211-05
ADVANCED TEST EQUIPMENT

YOU ARE HERE

33210-05
COMPUTER APPLICATIONS

33209-05
INTRODUCTION TO
CODES AND STANDARDS

33208-05
WIRE AND CABLE
TERMINATIONS

33207-05
SWITCHING DEVICES
AND TIMERS

33206-05
INTRODUCTION TO
ELECTRICAL BLUEPRINTS

33205-05
POWER QUALITY
AND GROUNDING

33204-05
BASIC TEST EQUIPMENT

33203-05
SEMICONDUCTORS AND
INTEGRATED CIRCUITS

33202-05
AC CIRCUITS

33201-05
DC CIRCUITS

ELECTRONIC SYSTEMS
TECHNICIAN LEVEL ONE

CORE CURRICULUM

211CMAP.EPS

Figures

Tables

Advanced Test Equipment

Objectives

When you have completed this module, you will be able to do the following:

1. Explain the operation and use of specialized test equipment used in the checkout and troubleshooting of electronic equipment, cables, and cabling systems.
2. Select the correct item of test equipment to be used in specific situations.
3. Use an oscilloscope to measure various waveforms.
4. Set up and use selected cable testers to check out cables and evaluate the performance of copper and optical fiber cable.
5. Use a signal generator to generate various waveforms.
6. Use a frequency meter to determine the frequency of a specific circuit.

Prerequisites

Before you begin this module, it is recommended that you successfully complete *Core Curriculum*; *Electronic Systems Technician Level One*; and *Electronic Systems Technician Level Two*, Modules 33201-05 through 33210-05.

Required Trainee Materials

1. Pencil and paper
2. Appropriate personal protective equipment

1.0.0 ◆ INTRODUCTION

The requirements for testing and troubleshooting electronic systems and devices go far beyond the capabilities of the basic test instruments covered previously. For example, it is often necessary to observe waveforms using an oscilloscope. In order to make precision measurements, it may be necessary to input a known signal using a signal generator. Because of the wide range of frequencies and the importance of frequency management, it may be necessary to use frequency counters and meters.

Some equipment is designed for use in testing specific technologies such as audio, video, and broadband. In addition, cables used in high-frequency applications require a variety of test instruments. Some simply verify cable continuity and the ability of the cable to pass a signal. Other, more sophisticated and expensive instruments are sometimes needed to provide multi-function cable testing to ensure that system cables meet the standards for the category of cable.

In this module, you will become familiar with this specialized test equipment that will eventually become an important part of your job.

2.0.0 ◆ OSCILLOSCOPES

The oscilloscope (*Figure 1*) is one of the most widely used and versatile test instruments. Its screen displays the actual shape of voltages that are changing with time so that waveform measurements can be made and one waveform can be compared to another. Oscilloscopes are also used to measure frequency, duration or time of occurrence of one or more cycles, phase relationships between voltage waveforms appearing at different points in the circuit, shapes of the waveform, and amplitude of the waveform. There are two basic types of oscilloscopes: analog and digital storage. Analog oscilloscopes display signals in real time. This is advantageous when adjusting and observing an electronic circuit since the results on the screen respond immediately to any changes. Analog scopes perform best when displaying waveforms with a high repetition rate. Digital storage oscilloscopes (*Figure 2*) sample the waveform, store the samples in memory, and display the signal on the screen. The remainder of this section will focus on the more commonly used analog (real-time) oscilloscope.

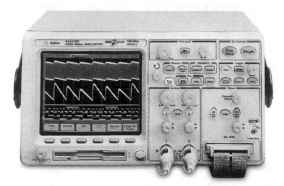

211F01.EPS

Figure 1 ◆ Typical oscilloscope.

211F02.EPS

Figure 2 ◆ Handheld digital oscilloscope.

2.1.0 Operation of the Analog Oscilloscope

The operation of an oscilloscope can vary depending on the manufacturer and model. The intent of this section is to familiarize you with some important characteristics of oscilloscopes and the function of the controls that apply to most general-purpose oscilloscopes.

2.1.1 Bandwidth

An oscilloscope's bandwidth is specified in megahertz (MHz). The bandwidth refers to the maximum frequency that can be observed and measured. Signals containing frequencies higher than the bandwidth will be restricted (filtered). The observation of complex signals requires an oscilloscope bandwidth of at least two or three times the highest frequency to be measured.

2.1.2 Number of Channels

Most oscilloscopes used today are at least dual-trace (two-channel) scopes that enable signals from two different inputs to be displayed on the CRT simultaneously. Many models are capable of displaying four or more channels. The number of signals that need to be observed simultaneously determines the number of channels required.

2.1.3 Controls

Controls and connectors can vary greatly in number. Usually, the more controls and connectors there are, the more versatile the instrument. Regardless of the number, all oscilloscopes have similar controls and connectors. Occasionally, identical controls will have different names from one model to another, but most controls are logically grouped and the names are suggestive of their function. The function of the controls common to most oscilloscopes are as follows:

- *Power switch* – Turns the power to the oscilloscope on and off. A light indicates when the power is on.
- *Intensity* – Varies the brightness of the trace(s) on the screen. It should be adjusted to provide an easily discernable signal. Caution should be used so that the intensity is not left too high for an extended period of time. Damage to the screen can result from excessive intensity.
- *Focus* – Focuses the beam so that the sweep (trace) is a fine, sharp line on the screen. An out-of-focus condition results in a fuzzy trace and waveforms.
- *Horizontal position* – Provides coarse and fine adjustment of the neutral (no signal applied) horizontal position of the trace on the screen. It is also used to reposition a waveform horizontally on the screen for more convenient viewing or measurement.
- *Vertical position* – Moves the trace up or down for easier measurement and observation of a signal. There is a vertical position control associated with each input channel.

- *AC-ground-DC switch* – Allows the input signal to be AC coupled, DC coupled, or grounded. There is a AC-ground-DC switch associated with each input channel. The AC coupling eliminates any DC component on the input signal. The DC coupling permits DC values to be displayed. The ground position allows a zero-volt reference to be established on the screen.
- *Volts/division control* – Selects the number of volts or millivolts to be represented by each major division on the vertical scale of the display. It indicates the range of voltage signal levels (sensitivity) that can be observed and measured. The higher the sensitivity, the lower the amplitude of the input signal the oscilloscope can detect. The lower the sensitivity, the higher the amplitude of the input signal the oscilloscope can detect. There is a volts/division control for each input channel.
- *Time/division control* – Specifies the limits of how fast or slow (sweep speed) the oscilloscope beam can sweep across the screen. A faster sweep speed provides more detail of a signal, and a slower sweep speed provides a greater amount of time to observe the signal. A sweep speed that can display a couple of cycles of the signal to be measured is adequate. Sweep speed can be specified in seconds per division (s/div), milliseconds per division (ms/div) and nanoseconds per division (ns/div).
- *Mode switches* – Provide for displaying either or both channel inputs, inverting channel signals, adding waveforms, and selecting between alternate and chop modes of sweep. The terms alternate and chop relate to the method by which an oscilloscope will display two independent signals on the same trace. In the alternate mode, the oscilloscope will display one full trace from one channel and then display the trace from the second channel. This allows the operator to view both traces at the same time because the phosphor-coated screen retains the display of the first trace while the second is being drawn. Due to the slow drawing times of low-frequency signals, the first trace may disappear before the second trace can be drawn. In order to eliminate this problem, the chop mode can be selected. In this mode, the display will switch from channel to channel at a predetermined rate during a sweep and only display small portions of each trace. The oscilloscope will display these traces so rapidly that your eyes can fill in the gaps in the trace. The result provides for a fairly good view of both waveforms.
- *Trigger controls* – Provide for synchronization of the horizontal sweep waveform and the input

signal waveform. As a result, the display of the input signal is stable on the screen, rather than appearing to drift across the screen. The trigger controls allow the beam to be triggered from various selected sources. The triggering of the beam causes it to begin its sweep across the screen. It can be triggered from an internally generated signal derived from an input signal, from line voltage, or from an externally applied trigger signal. An associated slope switch allows the triggering to occur on either the positive-going slope or the negative-going slope of the trigger waveform. The level control selects the voltage level on the trigger signal at which triggering occurs.

2.2.0 Use of Oscilloscope Probes

When measuring a signal with an oscilloscope, always use the appropriate probe. The use of a probe minimizes the effect of stray signals. The use of a plain test cable or lead rather than a probe will normally load the circuit under test and act as an antenna, picking up unwanted stray signals that will be displayed on the screen along with the signal of interest. The circuit can also be loaded by using the wrong probe to make a measurement. Circuit loading modifies the environment of the signals in the circuit you want to measure; it changes the signals in the circuit under test either a little or a lot, depending on the degree of loading.

Circuit loading is resistive, capacitive, and inductive. For signal frequencies under 5kHz, the most important component of loading is resistance. To avoid significant circuit loading here, all you need is a probe with a resistance of at least two orders of magnitude or 100 times greater than the circuit impedance, such as 100 megohm [MΩ] probes for 1MΩ sources; 1MΩ probes for 10kW sources, and so on.

When you are making measurements on a circuit that contains high-frequency signals, inductance and capacitance become important. You cannot avoid adding capacitance when you make connections, but you can avoid adding more capacitance or interference than necessary.

One way to avoid adding interference is to use an attenuator probe. Its design prevents the pickup of electromagnetic interference and offers a high-input impedance that minimizes circuit loading. The penalty is the reduction in signal amplitude from the 10:1 **attenuation**. These probes are adjustable to compensate for variations in oscilloscope input capacitance, and oscilloscopes contain a reference signal available at the front panel. Making this adjustment is called probe compensation. When you are measuring high frequencies, remember that the probe's impedance changes with frequency. Also remember to ground your probe securely with the shortest ground clip possible. In some very high-frequency applications, a special socket is provided in the circuit, and the probe is plugged into that.

Generally, you can divide probes by function into voltage-sensing and current-sensing types. Voltage probes can be further divided into passive and active types.

For most applications, the probes that are supplied with an oscilloscope are the ones that should be used. These will usually be attenuator probes. To make sure that the probe can faithfully reproduce the signal for your oscilloscope, the compensation of the probe should be adjustable. If you are not going to use the probes that came with your oscilloscope, pick your probe based on the voltage you intend to measure. For example, if you are going to be looking at a 50V signal and your largest vertical sensitivity is 5V, then that signal will take up ten major divisions of the screen. This is a situation where you need attenuation; a 10× probe would reduce the amplitude of your signal to reasonable proportions. *Table 1* lists the various types of common probes and their characteristics.

Proper termination is important to avoid unwanted reflections of the signal you want to measure within the cable. Probe/cable combinations designed to drive 1MΩ inputs are engineered to suppress these reflections. But for 50Ω scopes, 50Ω probes should be used. The proper termination is also necessary when you use a coaxial cable instead of a probe. If you use a 50Ω cable and a 1MΩ oscilloscope, be sure you also use a 50Ω terminator at the oscilloscope input.

The probe's ruggedness, flexibility, and the length of the cable can also be important, but remember that the longer the cable, the greater the capacitance at the probe tip. Check the specifications to see if the bandwidth of the probe is sufficient, and make sure you have the adapters and tips you will need. Most modern probes feature interchangeable tips and adapters for many applications. Retractable hook tips let you attach the probe to most circuit components. Other adapters connect probe leads to coaxial connectors or slip over square pins. Alligator clips for contacting large-diameter test points are another possibility.

Because of circuit loading and terminations, the best way to ensure that your oscilloscope and probe measurement system has the least effect on your measurements is to use the probe recommended for your oscilloscope and always make sure that it is compensated.

2.3.0 Waveform Characteristics and Terminology

In order to interpret correctly the waveforms displayed on an oscilloscope, it is important to understand the characteristics and terminology associated with waveforms.

2.3.1 Waveshape

As previously described, the waveform displayed on the oscilloscope's screen is created by the movement of an electron over time. The changes in the waveform over time form the waveshape, the most readily identifiable characteristic of a waveform. *Figure 3* illustrates some common waveshapes.

Basic waveshapes include sine waves and various non-sinusoidal waves such as triangle waves, square waves, and sawtooth waves. A square wave has equal amounts of time for its two states.

Table 1 Oscilloscope Probes

Probe Types	Characteristics
1× passive voltage-sensing	No signal reduction which allows the maximum sensitivity at the probe tip; limited bandwidths (4–30MHz); high capacitance (32–112pF); signal handling to 500V
10×/100×/1,000× passive voltage-sensing attenuator	Attenuate signals; bandwidths to 300MHz; adjustable capacitance; signal handling to 500V (10×); 1.5kV (100×) or 20kV (1,000×)
Active voltage-sensing field effect transistor (FET)	Switchable attenuation; capacitance as low as 1.5pF; more expensive, less rugged than other types; limited dynamic range, but bandwidths to 900MHz; minimum circuit loading
Current-sensing	Measure currents from 1mA to 1,000A; DC to 50MHz; very low loading
High-voltage	Signal handling to 40kV

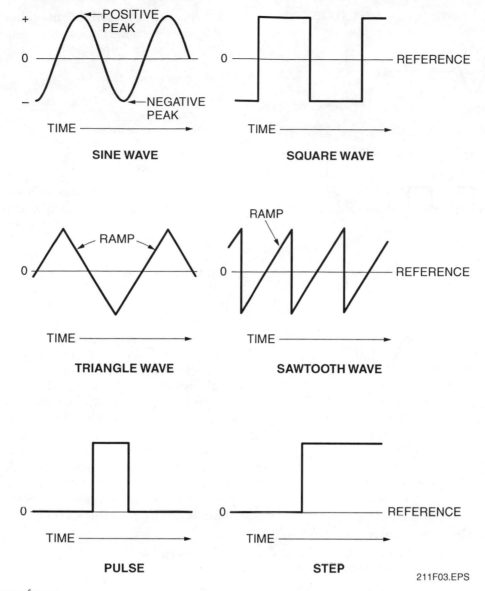

Figure 3 ◆ Basic waveforms.

Triangle and sawtooth waves are usually the result of circuits designed to control voltage with respect to time like the sweep of an oscilloscope.

In these waveforms, one or both transitions from state to state are made with a steady variation at a constant rate or ramp. Changes from one state to another on all waveforms except sine waves are called transitions. The last two drawings in *Figure 3* represent periodic single-shot waveforms. The first is a pulse. All pulses are marked by a rise, a finite duration, and a decay. The second one is a step, which is actually a single transition.

Waveshapes tell you a great deal about the signal. Whenever you see a change in the vertical dimension of a signal, you know that this amplitude change represents a voltage change. Whenever there is a flat horizontal line, there is no change for that length of time. Straight diagonal lines mean a

linear change (equal rise or fall of voltage for equal amounts of time). Sharp angles on a waveform mean a sudden change. But waveshapes alone are not the whole story. When you want to completely describe a waveform, you will want to find the parameters of that particular waveform. Depending on the signal, these parameters might include the signal amplitude, period, frequency, width, rise time, or phase.

2.3.2 Amplitude

Amplitude is a characteristic of all waveforms. It is the amount of displacement from the reference at a particular point in time, without regard to the direction of the change. In the waveforms shown in *Figure 4*, the amplitude is the same even though the sine wave is larger from peak to peak. In the

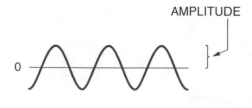

AMPLITUDE

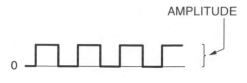

AMPLITUDE

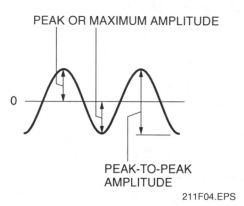

PEAK OR MAXIMUM AMPLITUDE

PEAK-TO-PEAK
AMPLITUDE

211F04.EPS

Figure 4 ◆ Waveform amplitude.

third drawing, an alternating current waveform is shown with peak or maximum amplitude and peak-to-peak amplitude parameters annotated. In oscilloscope measurements, amplitude usually means peak-to-peak amplitude.

2.3.3 Period

Period is the time required for one cycle of a signal if the signal repeats itself. See *Figure 5*. Period is a parameter whether the signal is symmetrically shaped like the sine and square waves shown or whether it has a more complex and asymmetrical shape like the rectangular and damped sine waves shown. Period is always expressed in units of time. Naturally, one-time signals (like the step) or uncorrelated signals without a time relation (like noise) have no period.

2.3.4 Frequency

If a signal is periodic, it has a frequency. Frequency is the number of times a signal repeats itself in a second and is measured in hertz as follows: 1Hz = 1 cycle per second, 1kHz (kilohertz) = 1,000 cycles

per second, and 1MHz (megahertz) = 1,000,000 cycles per second. Period and frequency are reciprocal, as shown here (see *Figure 6*):

$$Frequency = \frac{1}{period}$$

$$Period = \frac{1}{frequency}$$

2.3.5 Pulses

The parameters of a pulse can be important in a number of different applications. Pulse specifications are shown in *Figure 7* and include transition times measured on the leading edge of a positive-going transition; this is the rise time. The fall time is the transition time on a negative-going trailing edge. The pulsewidth is measured at the 50 percent points.

The **duty cycle**, **duty factor**, and repetition rate are all parameters of rectangular waves, as shown in *Figure 8*. They are particularly important in digital circuits. The duty cycle is the ratio of pulsewidth to signal period and is expressed as a percentage. For square waves, it is always 50 percent. For the pulse wave in the second drawing in *Figure 8*, it is 30 percent. The duty factor is the same thing as the duty cycle except it is expressed as a decimal, not a percentage. A repetition rate describes how often a pulse train occurs and is used instead of frequency to describe waveforms like that shown in the second drawing.

2.3.6 Phase

Phase is best explained with a sine wave. As shown in *Figure 9*, the plot changes as follows: 0 at 0 degrees, +1 at 90 degrees, 0 again at 180 degrees, −1 at 270 degrees, and finally 0 again at 360 degrees. Consequently, it is useful to refer to the phase angle (or simply phase) when you want to describe how much of the period has elapsed.

A phase comparison may be used to describe a relationship between two signals. Picture two clocks with their second hands sweeping the dial every 60 seconds. If the second hands touch the 12 at the same time, the clocks are in phase and if they do not, then they are out of phase. The phase shift (in degrees) is used to express the extent to which the signals are out of phase.

To illustrate, the waveform labeled current in *Figure 9* is said to be 90 degrees out of phase with the voltage waveform. Other ways of reporting the same information are that the current waveform has a 90 degrees phase angle with respect to the voltage waveform or that the current waveform lags the voltage waveform by 90 degrees.

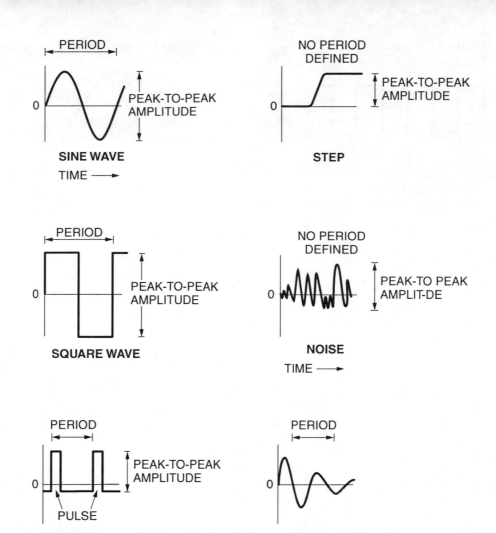

Figure 5 ◆ Waveform period.

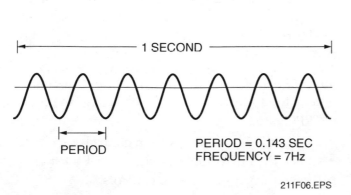

PERIOD = 0.143 SEC
FREQUENCY = 7Hz

211F06.EPS

Figure 6 ◆ Frequency measurement.

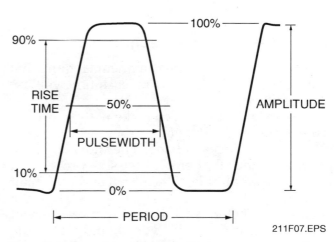

211F07.EPS

Figure 7 ◆ Parameters of a pulse.

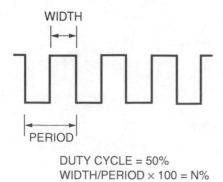

DUTY CYCLE = 50%
WIDTH/PERIOD × 100 = N%

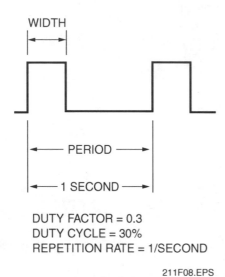

DUTY FACTOR = 0.3
DUTY CYCLE = 30%
REPETITION RATE = 1/SECOND

211F08.EPS

Figure 8 ◆ Parameters of rectangular waves.

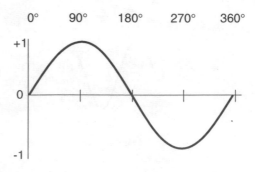

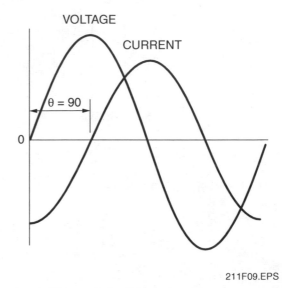

211F09.EPS

Figure 9 ◆ Phase relationships.

Note that there is always a reference to another waveform, in this case, between the voltage and current waveforms.

2.4.0 Measurement Techniques

Accurate oscilloscope measurements require that your system is properly set up each time you use your oscilloscope.

Most measurements you make with an oscilloscope require an attenuator probe, which is any probe that reduces voltage. The most common probes are the 10× (tenfold) passive probes that reduce the amplitude of the signal and the circuit loading by 10:1.

However, before you make any measurement with an attenuator probe, you should make sure the probe is compensated. *Figure 10* illustrates what can happen to waveforms when the probe is not properly compensated. Improperly compensated probes can distort the waveforms displayed on the oscilloscope.

Compensate the probe with the accessory tip.

NOTE

Do not compensate the probe in one vertical channel and then use it in another.

The most common mistake in making oscilloscope measurements is forgetting to compensate the probe. The second most common mistake is forgetting to check the control settings. Remember to take the following steps before using the oscilloscope:

- Check all the vertical system controls.
- Check the horizontal system control settings.
- Check the trigger system controls.

Always refer to the appropriate technical manual for initializing the oscilloscope.

Before you probe a circuit, make sure you have the right probe tips and adapters for the circuits you will be working on. Then check to see that the ground in the circuit under test is the same as the

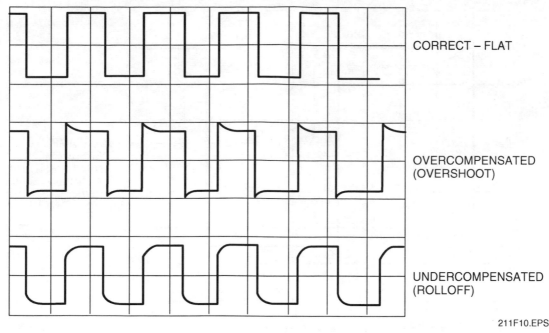

CORRECT – FLAT

OVERCOMPENSATED
(OVERSHOOT)

UNDERCOMPENSATED
(ROLLOFF)

211F10.EPS

Figure 10 ◆ Probe compensation.

oscilloscope ground; never assume that it is. The oscilloscope ground will always be the earth ground as long as you are using the proper power cord and plug. Check the circuit ground by touching the probe tip to the point you think is the ground before you make a hard ground by attaching the ground strap of your probe.

If you are going to be probing a lot of different points in the same circuit and measuring frequencies of less than 5MHz, you can ground that circuit to your oscilloscope once instead of each time you move the probe. To do this, connect the circuit ground to the jack marked GND on the front panel.

Rather than attempt to describe how to make every possible measurement with an oscilloscope, the following four sections describe common measurement techniques you can use in many applications.

2.4.1 Amplitude and Time Measurements

The two most basic measurements are amplitude and time. Almost every measurement is based on one of these two fundamental techniques.

Since the oscilloscope is a voltage-measuring device, voltage is shown as amplitude on the oscilloscope screen.

Amplitude measurements are best made with a signal that covers most of the screen vertically. Amplitude measurements are made by counting the major and minor divisions on the vertical graticule (grid) and multiplying by the setting of the volts/division switch to determine the signal

voltage, as shown in *Figure 11*. Each minor division is equal to two-tenths of a major division.

Time measurements are also more accurate when the signal covers a large area of the screen, as shown in *Figure 11*. Time measurements are made by counting the number of major and minor divisions on the horizontal graticule and multiplying this by the seconds/division switch to determine the signal's period.

2.4.2 Frequency and Other Derived Measurements

Voltage and time measurements are two examples of direct measurements. Once you have made a direct measurement, there are derived measurements you can calculate. Frequency is one example; it is derived from period measurements. While period is the length of time required to complete one cycle of a periodic waveform, frequency is the number of cycles that take place in a second. The measurement unit is a hertz, or one cycle per second, and it is the reciprocal of the period. Therefore, a period of 0.0025 second or 2.5 milliseconds means a frequency of 400Hz.

Additional examples of derived measurements are the alternating current measurements illustrated in *Figure 12*.

Derived measurements are the result of calculations made after direct measurements. For example, alternating current measurements require an amplitude measurement first. The easiest place to start is with a peak-to-peak amplitude measurement of the

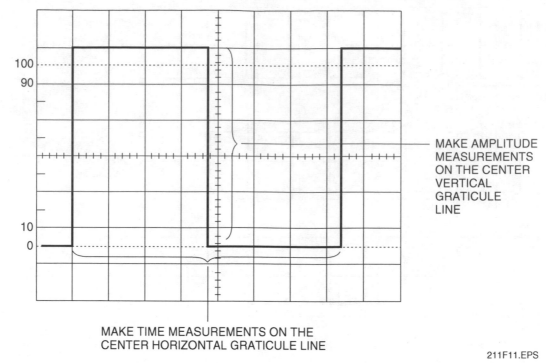

100 ──
90 ──

10 ──
0 ──

MAKE AMPLITUDE
MEASUREMENTS
ON THE CENTER
VERTICAL
GRATICULE
LINE

MAKE TIME MEASUREMENTS ON THE
CENTER HORIZONTAL GRATICULE LINE

211F11.EPS

Figure 11 ◆ Amplitude/time measurements.

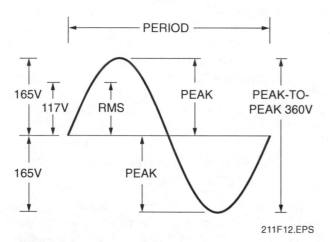

PERIOD

165V
117V RMS PEAK PEAK-TO-PEAK 360V
165V PEAK

211F12.EPS

Figure 12 ◆ Derived measurements.

voltage. The peak-to-peak voltage is 330V because peak-to-peak measurements ignore positive and negative signs. The peak voltage is one-half that (165V) when there is no DC offset. The root mean square (rms) voltage for this sine wave is equal to the peak value divided by the square root of two:

$$165 \div 1.414 = 117V$$

2.4.3 Pulse Measurements

Pulse measurements are important when you work with digital equipment and data communication devices. Some of the signal parameters of a pulse were discussed earlier, but that was an

example of an ideal pulse, not one that exists in the real world. The most important parameters of a real pulse are shown in *Figure 13*.

Real-pulse measurements include a few more parameters than those for an ideal pulse. Preshooting is a change of amplitude in the opposite direction that precedes the pulse. Overshooting and rounding are changes that occur after the initial transition. Ringing is a set of amplitude changes, usually a damped sinusoid, that follows overshooting. All are expressed as percentages of amplitude. Settling time expresses how long it takes the pulse to reach a maximum amplitude. Droop is a decrease in the maximum amplitude with time; nonlinearity is any variation from a straight line drawn through the 10 percent and 90 percent points of a transition.

2.4.4 Phase Measurements

The amount of time that has passed since the cycle began is called the phase and it is measured in degrees. Also, there can be a phase relationship between two or more waveforms, which is called the phase shift. There are two ways to measure the phase shift between two waveforms. One is by putting one waveform on each channel of a dual-channel oscilloscope and viewing them directly in the chop or alternate vertical mode with the trigger on either channel. Adjust the trigger level control for a stable display, and measure the period of the waveforms. Then measure the horizontal distance

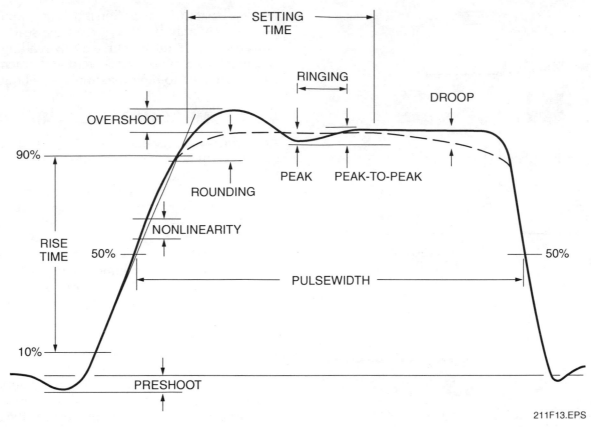

Figure 13 ◆ Real-pulse parameters.

between the same points on the two waveforms. The phase shift is the difference in time divided by the period and multiplied by 360 to give you degrees.

3.0.0 ◆ WATTMETER

Rather than performing two measurements and then calculating true power, a power-measuring meter called a wattmeter can be connected into a circuit to measure true power. The power can be read directly from the scale of this meter. Not only does a wattmeter simplify power measurements, but it has two other advantages.

First, voltage and current in an AC circuit are not always in phase; current sometimes either leads or lags the voltage (this is known as the power factor). When this happens, multiplying the voltage by the current results in apparent power, not true power. Therefore, in an AC circuit, measuring the voltage and current and then multiplying them can often result in an incorrect value of power dissipation by the circuit. However, the wattmeter takes the power factor into account and always indicates true power.

Second, voltmeters and ammeters consume power. The amount consumed depends on the levels of the voltage and the current in the circuit, and it cannot be accurately predicted. Therefore, very accurate power measurements cannot be made by measuring voltage and current and then calculating power. However, some wattmeters compensate for their own power losses so that only the power dissipated in the circuit is measured. If the wattmeter is not compensated, the power that is dissipated is sometimes marked on the meter, or it can easily be determined so that a very accurate measurement can be made. Typically, the accuracy of a wattmeter is within one percent.

The basic wattmeter consists of two stationary coils connected in series and one movable coil (*Figure 14*). The moving coil, wound with many turns of fine wire, has a high resistance.

The stationary coils, wound with a few turns of a larger wire, have a low resistance. The interaction of the magnetic fields around the different coils will cause the movable coil and its pointer to rotate in proportion to the voltage across the load and the current through the load. Thus, the meter indicates E times I, or power.

The two circuits in the wattmeter will be damaged if too much current passes through them. This fact is of special importance because the reading on the meter does not tell the user that the coils are being overheated. If an ammeter or voltmeter is overloaded, the pointer will indicate beyond full-scale deflection. In a wattmeter, both the current and potential (voltage) circuits may be carrying such an

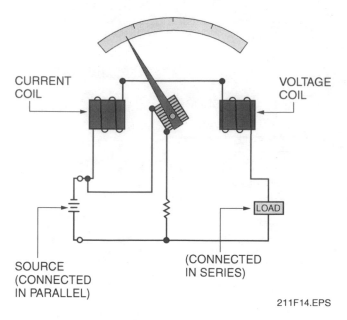

Figure 14 ◆ Wattmeter schematic.

overload that their insulation is burning, and yet the pointer may only be partway up the scale. A low-power factor circuit will give a low reading on the wattmeter even when the current and potential circuits are loaded to their maximum safe limits.

4.0.0 ◆ MEGOHMMETER (MEGGER)

An ordinary ohmmeter cannot be used for measuring resistances of several million ohms, such as those found in conductor insulation or between motor or transformer windings, and so on. To test these types of very high resistances adequately, it is necessary to use a much higher potential than is furnished by the battery of an ohmmeter. For this purpose, a megger is used (*Figure 15*). There are three types of meggers: hand, battery, and electric.

The megger is similar to a moving-coil meter except that it has two windings (coils). See *Figure 16*. Coil A is in series with resistor R_2 across the output of the generator.

This coil is wound so it causes the pointer to move toward the high-resistance end of the scale when the generator is in operation. Winding B is in series with R_1 and R_X (the unknown resistance to be measured). This winding is wound so that it causes the pointer to move toward the low or zero-resistance end of the scale when the generator is in operation.

When an extremely high resistance appears across the input terminals of the megger, the current through coil A causes the pointer to read infinity. Conversely, when a relatively low resistance appears across the input terminals, the current through coil B causes the pointer to deflect toward zero. The pointer stops at a point on the scale determined by the current through coil B, which is controlled by R_X.

Digital meggers use the same operational principles. Instead of having a scaled meter movement, these meters give the value of resistance in a digital readout display. The digital readout makes reading the measurement much easier and helps to eliminate errors.

To avoid excessive test voltages, most hand meggers are equipped with friction clutches. When the generator is cranked faster than its rated speed, the clutch slips, and the generator speed and output voltage are maintained at their rated values.

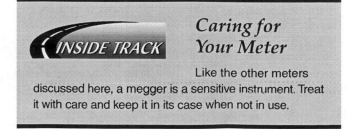

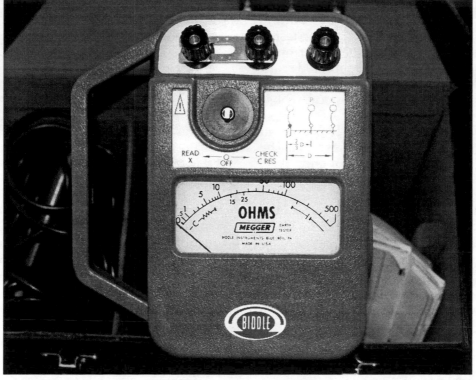

211F15.EPS

Figure 15 ◆ Typical megger.

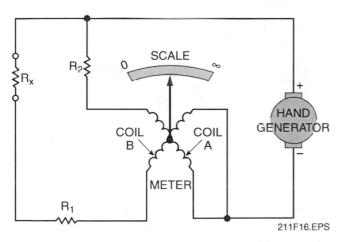

211F16.EPS

Figure 16 ◆ Megger schematic.

4.1.0 Safety Precautions

WARNING!

When a megger is used, the generator voltage is present on the test leads. This voltage could be hazardous to you or the equipment you are testing. *NEVER TOUCH THE TEST LEADS WHILE THE TESTER IS BEING USED.* Isolate the item you are testing from the circuit before using the megger. Protect all parts of the test subject from contact by others.

When using a megger, you could be injured or cause damage to the equipment if you do not observe the following minimum safety precautions:

- Use meggers on high-resistance measurements only (such as insulation measurements or to check two separate conductors in a cable).
- Never touch the test leads while the handle is being cranked.
- De-energize and verify the de-energization of the circuit completely before connecting the meter.
- Disconnect the item being checked from other circuitry, if possible, before using the meter.
- After the test, ground the tested circuit to discharge any energy that may be left in the circuit.

5.0.0 ◆ LINE FREQUENCY METER

Frequency is the number of cycles completed each second by a given AC voltage, and it is usually expressed in hertz (one hertz = one cycle per second). The frequency meter is used in AC power-producing devices such as generators to ensure that the correct frequency is produced. Failure to produce the correct frequency will result in excess heat and component damage.

There are two common types of analog frequency meters. One operates with a set of reeds having natural vibration frequencies that respond in the range being tested. The reed with a natural frequency closest to that of the current being tested will vibrate most strongly when the meter operates. The frequency is read from a calibrated scale.

A moving-disk frequency meter works with two coils, one of which is a magnetizing coil whose current varies inversely with the frequency. A disk with a pointer mounted between the coils turns in the direction determined by the stronger coil. The meter shown in *Figure 17* is a multi-function digital multimeter with the ability to measure frequency. Different meters are calibrated for different frequency ranges. This meter reads in the range of 0.001 Hz to 10 MHz. Others are designed for the audio frequency range, while still others are designed for high-frequency work.

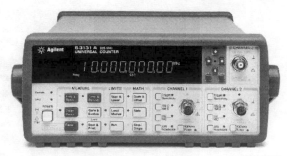

211F18.EPS

Figure 18 ◆ Frequency counter.

A harmonic is a sine wave whose frequency is a whole number multiple of the original base frequency. For example, a standard 60Hz sine wave may have second and third harmonics at 120Hz, 180Hz, and 240Hz, respectively. Harmonics may be caused by various circuit loads such as fluorescent lights and by certain three-phase transformer connections.

The results of harmonics are heating in wiring and a voltage or current that cannot be detected by most digital meters.

6.0.0 ◆ POWER FACTOR METER

The **power factor** is the ratio of true (actual) power to apparent power. The power factor of a circuit or piece of equipment may be found by using an ammeter, wattmeter, and voltmeter. To calculate the power factor, divide the wattmeter reading (true power) by the product of the ammeter and voltmeter readings (apparent power or EI). The ideal power factor is 1.

$$\text{Power factor} = \frac{\text{true power (wattmeter reading)}}{\text{apparent power (EI)}}$$

It is not necessary to calculate these readings if a power factor meter is available to read the power factor directly (*Figure 19*). This meter indicates the equivalent of pure resistance or unit power factor, which is a one-to-one ratio.

7.0.0 ◆ RECORDING INSTRUMENTS

The term recording instrument describes many instruments that make a permanent record of measured quantities over a period of time. Recording instruments can be divided into three general groups:

- Instruments that record electrical quantities, including potential difference, current, power, resistance, and frequency.

- Instruments that record nonelectrical quantities by electrical means (such as a temperature recorder that uses a potentiometer system to record thermocouple output).

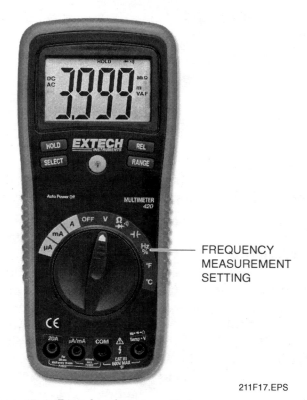

FREQUENCY MEASUREMENT SETTING

211F17.EPS

Figure 17 ◆ Digital multimeter with frequency measurement capability.

A frequency counter (*Figure 18*) is used to measure and/or adjust the frequency of signals generated by various electronic equipment. Most are microprocessor-controlled instruments. Frequency counters typically have automatic triggering, a high-stability time base, dual-channel inputs, and input voltage protection. They normally have multiple measuring functions such as frequency, period, period average, frequency ratio, and time interval. Typical instruments have a frequency measurement range between 5Hz and 175MHz.

Figure 19 ◆ Power factor meter.

- Instruments that record nonelectrical quantities by mechanical means (such as a temperature recorder that uses a bimetallic element to move a pen across an advancing strip of paper).

It is often necessary to know the conditions that exist in an electrical circuit over a period of time to determine such things as peak loads, voltage fluctuations, etc. It may be neither practical nor economical to assign a worker to watch an indicating instrument and record its readings. An automatic recording instrument can be connected to take continuous readings, and the record can be collected for review and analysis.

Recording instruments are basically the same as the indicating meters already covered, but they have recording mechanisms attached to them. They are generally made of the same parts, use the same electrical mechanisms, and are connected in the same way. The only basic difference is the permanent record.

7.1.0 Chart Recorders

Chart recorders are the most widely used recording instruments for electrical measurement. Their name comes from the fact that the record is made on a strip of paper or paper disk, usually four to six inches wide and up to 60 feet long. These can be used to record either voltage or current. A recording ammeter is shown in *Figure 20*.

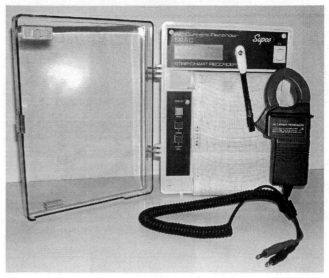

211F20.EPS

Figure 20 ◆ Recording ammeter.

Chart recorders offer several advantages in electrical measurement. The long charts allow the recording to cover a considerable length of time without operator attention, and chart recorders can be operated at a relatively high speed to provide very detailed records.

8.0.0 ◆ LINEMAN'S TEST SET

The lineman's test set (*Figure 21*) is also known as a butt set. It is used primarily to check for a dial tone on analog phone lines. The clip leads are attached to a loose cable pair or to a pair terminated at a punch block. The unit emulates a standard analog telephone with the added features of monitoring calls in progress and line polarity testing. Some sets also include speaker phone, mute, and speed dialing, and many are touchtone/pulse switchable.

211F21.EPS

Figure 21 ◆ Lineman's test set.

9.0.0 ◆ CABLE TONER

The technical name of this tester is inductive amplifier/tone generator, but it is often called a Fox and Hounds, which is the name used by a particular manufacturer (*Figure 22*). This device consists of two components. One is a generator that sends a signal through the cable. The other is a probe, or wand, that generates a tone when it receives the signal at the other end of the cable. Some of these devices can be used to trace a cable, and some of them will even locate the cable behind a sheetrock wall. Some sets also provide a talk path that allows line workers in different locations to communicate via their butt sets on a cable pair.

211F22.EPS

Figure 22 ◆ Cable toner.

10.0.0 ◆ CABLE CERTIFICATION TESTERS

Certification test sets go beyond the capabilities of the cable testers previously discussed. Certification testers can be used to verify that a cable meets the performance requirements of the applicable ANSI/TIA/EIA standard. Different tests have different features. The test set shown in *Figure 23A* is a basic certification tester for UTP and ScTP cable. The test set shown in *Figure 23B* can be used to certify single-mode and dual-mode optical fiber cable as well. Some tests can be interfaced to a computer using a USB port. Some certification test sets now on the market are capable of testing to 300MHz. A variety of tests can be performed, including insertion loss, cable length, crosstalk, return loss, and propagation delay.

11.0.0 ◆ SOUND PRESSURE LEVEL METERS

A sound pressure level (SPL) meter, also called an audio dB meter, is a device used to measure sound levels. It consists of a special calibrated microphone, a small amplifier, and a meter. It presents sound level measurements in decibels.

To manage different frequency requirements, an SPL meter (*Figure 24*) provides several weighting curves that contour the responsiveness of the meter to the frequencies being measured. This is necessary, as frequency plays an important role in perceived

---

Pair Testers

INSIDE TRACK

The devices shown here are known as wire map testers, or simply pair testers. They are used to test for opens, shorts, and other problems in UTP or ScTP cables. Not all pair testers are the same; the one on the left is a very basic test set. As you progress to the right, the testers provide additional features, but at a higher price.

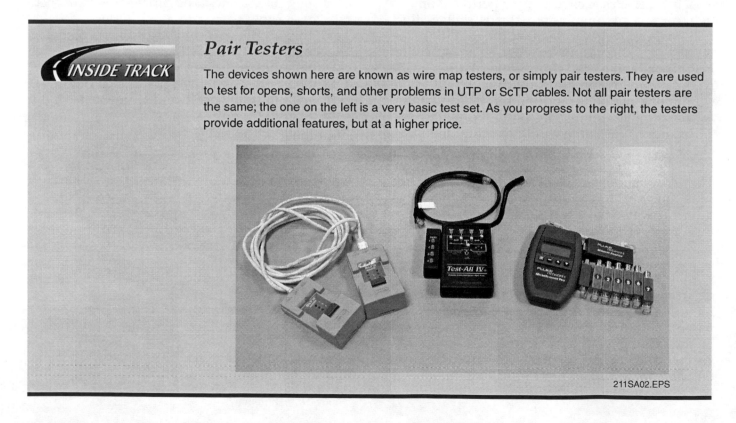

211SA02.EPS

---

loudness. A 1kHz sine wave at 90dB is rather loud, but a 30kHz sine wave at 80dB is inaudible.

A sound pressure meter should be held at arm's length. Typically, you should point it at 90 degrees to the sound source. Outdoor readings are usually taken with the microphone pointing up. These actions serve to minimize the effects of surface reflections.

12.0.0 ◆ RF POWER METER

Radio frequency (RF) power meters (*Figure 25*) are used in conjunction with the appropriate power sensor to measure the power of various RF signals found in wireless systems. Different model instruments cover different frequency ranges. A typical power meter may be capable of measuring power levels ranging from –70dBm to +45dBm and cover

a frequency range of 10kHz to 100GHz. In newer instruments, all major functions are menu driven and/or selected by front panel touch-sensitive buttons. The power level of the measured RF signal expressed in dBm is typically displayed on a 4½- or 5½-digit LCD display.

211F24.EPS

Figure 24 ◆ Sound pressure level meter.

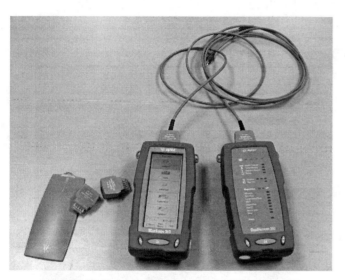

(A) AGILENT CABLE CERTIFICATION TESTER

(B) OMNI CABLE CERTIFICATION TESTER

211F23.EPS

Figure 23 ◆ Cable certification test sets.

211F25.EPS

Figure 25 ◆ RF power meter.

13.0.0 ◆ SIGNAL LEVEL METER

A signal level meter (SLM) is the most used item of test equipment in coaxial cable systems for maintaining and monitoring the signal levels. System measurements commonly made with an SLM include the following:

- Audio and video carrier levels on each channel
- Audio-to-video signal level ratio
- Carrier-to-noise ratio

SLMs detect RF energy and display the measurement in dBmV and/or dBµV. They are made by a number of manufacturers and in several models, ranging from relatively simple, manually tuned models to microprocessor-controlled programmable models (*Figure 26*). Depending on the model, the SLM can be an analog or digital instrument. Digital SLMs are more common. Digital SLMs usually show the signal level and other measured parameters as text on a liquid crystal display (LCD). Typically, they have the capability to tune automatically by frequency and/or by channel, and automatically switch in signal level attenuation as needed to give an accurate reading. The SLM should be able to measure all the frequencies within the broadband CATV range of 55 to 890MHz. Most SLMs have a built-in speaker that provides demodulated audio for the channel under test. Sophisticated models have the capability to interface with a printer and/or computer, allowing the measured signal data to be downloaded to these devices.

211F26.EPS

Figure 26 ◆ Signal level meter.

14.0.0 ◆ RADIO FREQUENCY (RF) ANALYZER METERS

The radio frequency (RF) analyzer meter (*Figure 27*) is a portable, microprocessor-controlled RF measuring instrument that transmits a low-power RF signal that is used to check and adjust antennas and related feedlines and RF networks. It can be used to measure antenna standing wave ratio (SWR) and impedance. It can also be used to measure transmission line parameters such as loss, impedance, and electrical length. Some models also measure capacitance and inductance, R and X components of impedance, parallel R and X, and impedance at the far end of a feedline. Higher-quality instruments can be used to check feedlines with impedances ranging from 25Ω to 450Ω. Some can automatically determine if a load is inductive or reactive and display the value of a series coil or capacitor to add in order to eliminate a series reactance and yield the lowest SWR. The operation of a radio frequency analyzer depends on the specific manufacturer and model and should be done in accordance with the manufacturer's instructions.

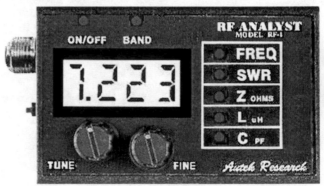

211F27.EPS

Figure 27 ◆ RF analyzer meter.

15.0.0 ◆ IMPEDANCE BRIDGES

An impedance bridge, or impedance meter, measures resistance, conductance, inductance, and capacitance (*Figure 28*). In audio, it is used to troubleshoot the impedance of a circuit. For example, it can be used to calculate the impedance of speaker circuits.

This device is useful when troubleshooting a sound reinforcement system. It can help identify and locate opens, shorts, and overloaded circuits.

16.0.0 ◆ TIME-DOMAIN REFLECTOMETER

The time-domain reflectometer (TDR), commonly called a cable-length meter or cable fault locator, can

211F28.EPS

Figure 28 ◆ An impedance meter.

211F29.EPS

Figure 29 ◆ Handheld time-domain reflectometer (TDR).

be used to locate faults on almost all types of metallic cables such as power cables, data cables, CATV/CCTV cables, etc. It can also be used to measure the length of cable stored on a reel or in a cable run. A TDR like the one shown in *Figure 29* is a handheld, battery-operated, microprocessor-controlled unit. It generates a test pulse that is transmitted down the cable or wire. This pulse is reflected back on the cable or wire toward the meter when it encounters a fault such as an open or short circuit. The time between the transmission and the reflection of the pulse is proportional to the distance to the fault or the open end of the wire on a reel. The manufacturer's instructions normally show how the polarity and amplitude of the returned reflected signal can be interpreted to identify the type of fault encountered; for example, short, open, or a transition from a low impedance to high impedance or vice versa.

An optical time-domain reflectometer (OTDR) is a very specialized instrument used to analyze fiber-optic cables (see *Figure 30*). It can be used to identify signal loss and disruptions on multimode fiber and to identify reflective breaks. As its name implies, an OTDR allows evaluation of an optical fiber in the time domain.

An OTDR requires access to only one end of the fiber in cable segments that may be tens of miles long. Timing how long it takes light to travel from the instrument to a point in the fiber and back can locate flaws and junctions in the fiber. In this way, an OTDR is like optical radar. The OTDR is normally used in conjunction with another device, such as a cable tester.

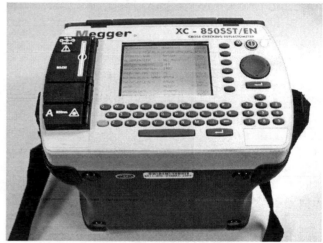

211F30.EPS

Figure 30 ◆ Optical time-domain reflectometer.

17.0.0 ◆ SIGNAL GENERATORS

A large variety of signal generators are available with signal source outputs ranging in various increments from DC to 110 GHz. They cover frequency applications from audio through cellular phone, digital phone, mobile radio, and television to millimeter-wave satellite systems. *Figure 31* shows three general types of signal generators. The function/arbitrary waveform generator shown can supply sine and square waves from 1 micro-Hz to 80 MHz. It can also produce triangular, pulse, ramp, noise, and DC waveforms. The output can be internally modulated using AM, FM, or other modulation methods. The output can also be swept using linear or logarithmic modes.

FUNCTION/ARBITRARY
SIGNAL GENERATOR

ANALOG SIGNAL
GENERATOR

DIGITAL SIGNAL
GENERATOR

211F31.EPS

Figure 31 ◆ Typical signal generators.

The arbitrary waveform generation capability can be used to create signals that mimic noise, vibration, control pulse strings, and any other complex signal needed for realistic tests. These types of generators are used for a wide variety of applications including audio, communications receiver, control circuit, and sensor circuit testing.

An example of a typical analog signal generator is also shown. This particular unit covers a CW frequency range from 250 kHz to 67 GHz with digital-step or ramp sweep to cover testing of certain RF and microwave systems, communications transceivers, control systems, and components. It also provides square, triangular, ramp, and noise waveforms. With available options, it can provide AM, FM, and other internal modulation modes.

Figure 31 also shows a typical digital signal generator. This particular generator provides a signal from 250 kHz to 6 GHz with a wide range of analog and digital modulation capabilities. Software-driven modulated signals are also provided for a variety of wireless communications formats.

18.0.0 ◆ TESTING AND TROUBLESHOOTING

Cable testing following installation is imperative to ensure proper operation of the low-voltage system. This testing can detect problems that may prevent operation or damage sensitive equipment.

Different testing requirements exist for commercial and residential wiring. Residential testing is primarily concerned with ensuring that the system is wired properly and that all stations are operational. Commercial testing also ensures this, but much more emphasis is placed on overall system performance.

Two stages of testing are performed: wiring integrity and operational testing. The wiring integrity is tested after installation and prior to connection to the service cable or operational equipment. Operational testing is conducted after the service provider cable or operational equipment is connected to the premises cabling.

18.1.0 Test Parameters

Test parameters are the values achieved during testing. These values are compared against standard minimum or maximum values to determine if the installation meets minimum specifications to be considered operational and effective. A number of parameters are usually checked, but the two basic parameters measured are attenuation and NEXT.

Attenuation is the measurement of signal strength loss over the length of a cable. It is accomplished for twisted pair, coaxial, and fiber-optic cable. To measure this, a signal of known frequency and power is injected into one end of the cable. At the other end, a measurement is taken using a power meter or spectrum analyzer. The difference in power levels is the amount of attenuation of the cable. The acceptable attenuation loss values for each type of cable at specific frequencies are as follows:

- *Coaxial cable* – As an example, standard 50Ω RG-58 coaxial cable has 1.5dB of attenuation per 100 feet at 10MHz and 5dB per 100 feet at 100MHz.

- *Twisted pair cabling* – There are acceptable attenuation loss values for each type of twisted pair cable governed by ANSI/TIA/EIA standards. All values for twisted pair cable attenuation are for lengths of 328 feet (100 meters).

NEXT is the measurement of near-end crosstalk for twisted pair cabling, or the amount of signal that bleeds into an adjacent conductor through transformer action. To measure this, a signal is injected into a wire pair. A measurement is then taken of the signal that is passed into an adjacent wire pair. This passed signal must be below a certain level. For example, in Category 5 cable at 10MHz, the passed signal must be a minimum of 47dB below the original signal level. NEXT is not a concern in coaxial and fiber optic cables.

18.2.0 Troubleshooting

Most problems with electronic systems can be traced to cabling and interconnections, with the

Basic Cable Tester

A handheld device like the one shown here can be used to detect and locate shorts and opens in copper cable and to measure the length of a copper cable. While such devices are not as sophisticated as other types of testers, they are much less expensive. Their relatively low cost makes them suitable for day-to-day use by technicians and installers.

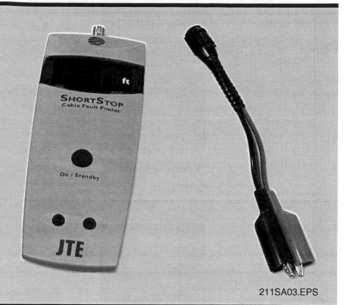

211SA03.EPS

most likely causes being grounds and opens. Opens, short circuits, or crossed conductors are easy to detect using a cable tester. The cable tester is placed on both ends of the cable and turned on. The tester will indicate the type of fault, which will in turn determine if the cable requires replacement. When testing for an open, physically move the cable around to eliminate the possibility of an intermittent problem.

A TDR can also be used to detect and locate opens, shorts, and grounds in a cable. The fault will cause a reflection to be displayed on a TDR connected to one end of the cable. The position of the reflection on the display is directly related to the location of the fault in the cable.

Grounds or short circuits are also easily detected using a multimeter set up to check for continuity. To test for a short or ground, disconnect both ends of the cable. To check for shorts, place the test leads of the multimeter across a pair of the conductors. If continuity exists, the conductors are shorted together. To check for grounds, place one test lead of the multimeter on a conductor and the other on a common ground. If continuity exists, the conductor is grounded. Finding the ground, however, can sometimes be time consuming because the entire cable length may have to be physically examined to find where the contact with ground is taking place. However, a TDR can be used to locate the exact position of a ground in any transmission line.

Another possible problem is excessive noise. If proper cable routing guidelines were followed, this should not be a problem. However, if a noise problem suddenly appears after installation, this can usually be traced to the addition of a piece of equipment near a data or voice transmission line. When equipment is to be added, always consider the placement carefully. A spectrum analyzer can be used to observe the nature and frequency of the noise and to help identify its source.

Other network problems may involve faulty operation of one of the connected devices. These problems may be detected using a network analyzer and will require the troubleshooting skills of a trained network administrator or engineer.

Normally, the length of a balanced pair of conductors (twisted pair) in a cable or a coax cable or the distance to a short circuit in a cable may be accurately and easily found using the equipment discussed above. However, in the event that specialized equipment is not immediately available, the DC loop resistance of a pair of conductors or a coax cable can be measured using a multimeter set up to measure resistance, and the cable length or distance to a short can be approximated as described in the following sections.

18.2.1 DC Loop Resistance Test for a Balanced Pair of Conductors

The following are guidelines for performing a DC loop resistance test for a balanced pair of conductors:

Step 1   If testing in order to verify the total length of a balanced pair of conductors, short out one end of the pair of conductors.

Step 2   Set the multimeter to the lowest ohm scale, and connect the leads of the meter together. Zero out any meter reading.

ADVANCED TEST EQUIPMENT

Step 3 Connect the multimeter leads to the non-shorted end of the pair of conductors.

Step 4 Read and record the DC loop resistance of the pair of conductors.

Step 5 Using applicable tables, determine the DC resistance per foot of the single AWG wire size being checked, and multiply the value by two to obtain the nominal DC loop resistance of a pair of the conductors for one foot.

NOTE
The nominal DC loop resistance of a pair of 24 AWG wires is 0.0572Ω/foot (300 mm). The nominal DC loop resistance of a pair of 22 AWG wires is 0.036Ω/foot (300 mm).

Step 6 Divide the DC loop resistance reading recorded from the meter in Step 4 by the nominal DC loop resistance per foot of the pair of conductors determined in Step 5 to obtain the approximate length of the pair of conductors or the distance to a short.

18.2.2 DC Loop Resistance Test for Coaxial Cable

The following are guidelines for performing a DC loop resistance test for coaxial cable:

Step 1 If testing in order to verify the total length of a coaxial cable, fabricate a shorting jack by connecting the center conductor of an appropriate jack to the outer body of the jack with a very short length of wire. Connect the shorting jack to one end of the cable.

Step 2 Set the multimeter to the lowest ohm scale, and connect the leads of the meter together. Zero out any meter reading.

Step 3 Connect the multimeter leads between the center conductor and the shield of the non-shorted end of the coaxial cable.

Step 4 Read and record the DC loop resistance of the center conductor plus the shield.

Step 5 From the cable manufacturer's data, determine the nominal DC resistance of the center conductor per foot and the nominal DC resistance of the shield per foot. Add the two values together to obtain the nominal DC loop resistance per foot of the cable.

Step 6 Divide the DC loop resistance reading recorded from the meter in Step 4 by the nominal DC loop resistance per foot of the pair of conductors determined in Step 5 to obtain the approximate length of the coaxial cable or the distance to a short.

19.0.0 ◆ SAFETY

Safety must be the primary responsibility of all personnel. The use of test equipment and testing of electrical equipment enforces a stern safety code because carelessness on the part of the technician can result in serious injury or death due to electrical shock, falls, or burns. After an accident has occurred, investigation usually shows that it could have been prevented by the exercise of simple safety precautions and procedures. Each person concerned with electrical equipment is responsible for reading and becoming thoroughly familiar with the safety practices and procedures contained in all safety codes and equipment technical manuals before performing work on electrical equipment. It is your personal responsibility to identify and eliminate unsafe conditions and unsafe acts which cause accidents.

Summary

The equipment covered in this module goes far beyond the basic meters used to check voltage, current, and resistance. In addition to those items, the electronic systems technician must know how to select and use a variety of specialized test equipment in order to test, certify, and troubleshoot the complex systems that he or she will encounter. For example, the EST will sometimes find it necessary to observe waveforms using an oscilloscope because slight distortions of a waveform may be the only clues to the source of a problem.

When working with high-performance twisted pair, coaxial, and optical fiber cable, it is not enough to simply verify that the cable will pass a signal. Instead, it will be necessary to verify that a cabling system meets the applicable specifications and standards. This will require highly sophisticated test instruments and the ability to use them properly.

An EST working with audio, video, or RF systems will find it necessary to use frequency measuring equipment and signal strength analyzers to check out and troubleshoot those systems.

This module is intended to familiarize you with the various instruments an EST might use in the field. You will learn more about the use of these instruments in subsequent modules.

Review Questions

1. The peak value of the waveform shown in *Figure 1* is _____.
   a. 0.6V
   b. 1.5V
   c. 3.0V
   d. 4.5V

2. The period for the waveform shown in *Figure 1* is _____ milliseconds.
   a. 10
   b. 20
   c. 25
   d. 30

3. The frequency of the waveform shown in *Figure 1* is _____.
   a. 40Hz
   b. 50Hz
   c. 60Hz
   d. 65Hz

4. Wattmeters are used to measure _____.
   a. power
   b. voltage
   c. resistance
   d. impedance

5. To check the resistance between an insulated conductor and ground or between two insulated conductors, use _____.
   a. the resistance function of a multimeter
   b. a continuity tester
   c. a megohmmeter
   d. a time-domain reflectometer

6. A cable toner is commonly used to check cable continuity.
   a. True
   b. False

7. A test set that can be used to perform insertion loss, cable length, crosstalk, return loss, and propagation delay for UTP, ScTP, and optical fiber cable is the _____.
   a. pair tester
   b. cable toner
   c. certification test set
   d. OTDR

8. The signal level meter is primarily used to test _____.
   a. coaxial cable
   b. UTP cable
   c. optical fiber cable
   d. wireless devices

9. An instrument that can be used to determine the impedance of a transmission line is a(n) _____.
   a. RF analyzer meter
   b. chart recorder
   c. time-domain reflectometer
   d. oscilloscope

10. A TDR or OTDR can detect a fault, but is not capable of identifying the type of fault.
    a. True
    b. False

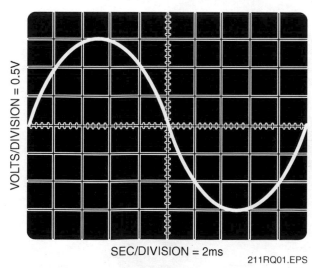

VOLTS/DIVISION = 0.5V

SEC/DIVISION = 2ms

211RQ01.EPS

Figure 1

GLOSSARY

Trade Terms Introduced in This Module

Attenuation: A decrease in the magnitude of a signal. It represents a loss of signal power between two points and is measured in decibels (dB).

Duty cycle: The ratio of pulsewidth to signal period expressed as a percentage.

Duty factor: Same as duty cycle, but expressed as a percentage.

Power factor: The ratio of true power to apparent power.

Additional Resources

This module is intended to be a thorough resource for task training. The following reference works are suggested for further study. These are optional materials for continued education rather than for task training.

Electronics Fundamentals: Circuits, Devices, and Applications, 2004. Thomas L. Floyd. New York, NY: Prentice Hall.

Principles of Electric Circuits: Conventional Current Version, 2003. Thomas L. Floyd. New York, NY: Prentice Hall.

Figure Credits

AEMC Instruments	211F19
Agilent Technologies, Inc.	211F01, 211F18, 211F25, 211F31
Autek Research	211F27
Blonder Tongue Laboratories, Inc.	211F26
Extech Instruments, Inc.	211F17, 211F24
Fluke Corporation **Reproduced with permission**	211F02
Gold Line	211F28
Ideal Industries, Inc.	211F21
Sealed Unit Parts Co., Inc.	211F20
Telewave, Inc.	211SA01
Topaz Publications, Inc.	211F15, 211F22, 211SA02, 211F23, 211F29, 211F30, 211SA03

NCCER makes every effort to keep these textbooks up-to-date and free of technical errors. We appreciate your help in this process. If you have an idea for improving this textbook, or if you find an error, a typographical mistake, or an inaccuracy in NCCER's Contren® textbooks, please write us, using this form or a photocopy. Be sure to include the exact module number, page number, a detailed description, and the correction, if applicable. Your input will be brought to the attention of the Technical Review Committee. Thank you for your assistance.

Instructors – If you found that additional materials were necessary in order to teach this module effectively, please let us know so that we may include them in the Equipment/Materials list in the Annotated Instructor's Guide.

Write: Product Development and Revision
National Center for Construction Education and Research
P.O. Box 141104, Gainesville, FL 32614-1104

Fax: 352-334-0932

E-mail: curriculum@nccer.org

Craft _____ Module Name _____

Copyright Date _____ Module Number _____ Page Number(s) _____

Description _____
_____
_____
_____
_____

(Optional) Correction _____
_____
_____

(Optional) Your Name and Address _____
_____
_____

Index

Index

A

Abbreviations and acronyms, 6.16, 6.17, 10.1–10.3, 10.40
ABF. *See* Air-blown fiber system
Above the finished floor (AFF), 6.27
Absolute zero, 7.12, 7.33
AC. *See* Access control; Alternating current circuit
Access control (AC), 10.1, 10.32, 10.36–10.38, 10.47
Accidents. *See* Electric shock
Actuator, 7.2, 7.7
Adapter, graphic, 10.8
AFF. *See* Above the finished floor
AHJ. *See* Authority Having Jurisdiction
AIA. *See* American Institute of Architects
Air-blown fiber system (ABF), 10.26
Air conditioner, 2.35, 5.21
Air conditions and static electricity, 5.46
Air gap, in lightning arrester, 5.26
Air terminal, 5.19, 5.20
Alarm systems. *See* Fire alarm system; Security system
Alternating current circuit (AC)
    average value, 2.5
    capacitance in, 2.13, 2.14–2.19
    conversion to DC, 5.31, 5.33, 5.34. *See also* Rectifier
    effective value. *See* Root-mean-square
    inductance in, 2.11–2.14
    lightning or surge protection device placement, 5.26
    overview, 2.1, 2.45, 5.4–5.5
    phase relationships, 2.6–2.8, 2.10, 2.18, 2.19, 2.20, 2.22, 5.5
    power in, 2.32–2.36
    resistance in, 2.10–2.11, 5.2
    safety, 2.44
    transformers, 2.36–2.45, 5.2, 5.3, 5.4
    waveform. *See* Waveforms
Aluminum, 1.6, 3.3, 5.12, 5.20
American Institute of Architects (AIA), 6.17
American National Standards Institute (ANSI), 9.19. *See also*
    ANSI/TIA/EIA standards
American Standard Code for Information Interchange
    (ASCII), 10.40, 10.47
Ammeter. *See also* Multimeter
    clamp-on, 4.4, 4.6, 4.15
    overview, 1.12, 1.27, 4.2–4.6
    power loss from, 11.11
    recording, 11.15
    symbol, 1.8

Amperage capacity, 8.48, 8.66
Ampere (amp), 1.4, 1.6, 1.13, 1.27. *See also* Ohm's law
Ampere-hour, 1.5
Amplifier, 3.10, 3.18–3.19, 3.21
Amplitude of a waveform
    on oscilloscope display, 11.5–11.6, 11.7, 11.9, 11.10
    peak-to-peak, 2.4, 2.10, 3.5, 5.5, 11.6, 11.7, 11.9–11.10
    relationship with heat, 2.6
AMP Special Industries, 8.34, 8.35
Anode
    battery, 2
    definition, 7.33
    diode, 3.4, 3.5, 3.7, 10.3
    silicon-controlled rectifier, 3.13, 7.10
ANSI. *See* American National Standards Institute
ANSI/TIA/EIA standards
    coaxial cable, 8.3
    electrical symbols, 6.15, 6.16
    equalizing conductor, 5.16
    fiber-optic cable connectors, 8.60
    overview, 9.13
    patch-cord wiring color codes, 8.29, 8.30
    residential telecommunications infrastructure, 9.15
    telecommunications cabling, 9.13–9.14
    telecommunications grounding and bonding, 9.15–9.16
    telecommunications pathways and spaces, 9.14–9.15
    UTP cable, 8.9, 8.20, 10.25
Antenna, 5.23, 11.18
Antimony, 3.3, 5.40, 5.41, 5.42
Apple®, 10.5, 10.28
Approval block, 6.11
ARCnet®, 10.22, 10.40
Argon, 3.3
Armature, 7.18, 7.19, 7.23, 7.33
Arrester, 5.17, 5.24, 5.25–5.27
Arsenic, 3.3
ASCII. *See* American Standard Code for Information
    Interchange
Atom, 1.1–1.3, 1.27, 3.2–3.3
Attenuation, 8.33, 11.3, 11.4, 11.20, 11.25
Audio dB meter, 11.16–11.17
Audio signals, 4.10, 4.13–4.14
Authority Having Jurisdiction (AHJ), 5.15, 9.3–9.4
Autoranging capability, 4.7, 4.16
Autotransformer, 2.42–2.43, 5.29, 5.63

Software
    applications, 10.22
    CAD, 6.9, 6.17, 6.22, 10.12
    overview, 10.20
    vs. hardware, 10.43–10.45
Soil conditions and ground testing, 5.49
Solar cells (photovoltaics), 5.41, 5.42, 5.63, 7.12, 7.13
Solder, 8.53–8.54, 8.56, 8.58. *See also under* Connectors
Soldering gun, 8.57
Soldering iron, 8.54–8.56
Soldering procedure, 8.55–8.59
Soldering station, 8.55
Solder sucker, 8.57–8.58
Solenoid, 3.21
Solid-state electronics. *See also* Diode; Semiconductor
    definition and overview, 3.1, 3.24, 3.27
    digital meter, 4.2
    relays, 7.15, 7.23–7.26
    surge protector, 5.25
    timers, 7.28–7.29
    uninterruptible power supply, 5.33
Sound card, 10.9, 10.12
Sound pressure level meter (SPL), 11.16–11.17
Sound system, 6.2, 6.31–6.32, 11.18
Southern Building Code Congress International (SBCCI),
    9.13
Spark gap, in lightning arrester, 5.27–5.28
Speakers, computer, 10.7, 10.12
Specifications, written, 6.17, 6.42, 6.45–6.46, 6.50, 9.2. *See also*
    MasterFormat™
Spike, 5.20–5.21
SPL. *See* Sound pressure level meter
Splice, 5.12, 5.13, 5.15, 8.35–8.36, 10.26–10.27
Spreadsheet software, 10.22
Sprinkler system, 6.2, 9.12
SSR. *See* Relay, solid-state
Standard network interface unit, 5.17–5.18
Standards. *See* Codes and standards
Standing wave ratio (SWR), 11.18
Static electricity, 3.18, 5.45–5.46, 8.55, 10.44
Stochastic, 10.48
Stochastic access control, 10.36–10.37
Stripper, cable or wire, 8.3–8.4, 8.21, 8.33, 8.44
Substation, 2.2, 5.2, 5.3
Sulfuric acid, 5.40
Surge, power, 5.2, 5.6, 5.7, 5.20–5.21. *See also* Lightning
Surge protection devices, 5.24–5.28, 5.31
Surge strip, 5.28
Survey, property and topographic, 6.23
Swell, voltage, 5.21
Switch(es)
    automatic (actuator), 7.2, 7.7
    classification, 7.2–7.4, 7.30
    definition, 7.1
    DIP, 10.2
    electronic. *See* Rectifier, silicon-controlled
    float-level, 7.4–7.5
    function, 4.10, 4.14
    installation, 7.1, 7.9
    limit, 7.8–7.9, 7.10
    microprocessor, 3.10
    mode, 11.3
    motion detector controller, 7.13
    as a network hub, 10.36
    oscilloscope, 11.2, 11.3
    panel-mounted, 7.4, 7.5
    photocell, 7.10–7.12

    pressure, 7.5–7.8, 7.9
    programmable electronic time, 7.29–7.30
    range
        ammeter, 4.4–4.6
        multimeter, 4.10, 4.12, 4.14, 4.15, 4.16
        voltmeter, 4.7, 4.8
    reed, 7.16–7.18
    solid-state timer, 7.28, 7.29
    symbols, 1.8, 6.15, 6.16, 6.17, 6.18, 6.27, 6.41
    tap, 5.29
    time, 7.26–7.28
    uninterruptible power supply, 5.34, 5.35
SWR. *See* Standing wave ratio
Symbols
    amplifier, 3.21
    in circuit diagrams, 1.6, 1.7, 1.8, 1.10, 2.10
    contacts, 6.42
    diac, 3.14
    diode, 3.8, 3.10
    electrical, 6.14–6.18, 6.27
    inverter, 3.21
    list (legend), 6.2, 6.18, 6.26–6.27
    logic gates, 3.20, 3.21, 3.22, 3.23, 3.24
    operational amplifier, 3.19
    silicon-controlled rectifier, 3.13
    transistor, 3.11, 3.13
    triac, 3.14
    varistor, 5.28
Synchronization, 10.28, 10.31, 10.32, 11.3
Synthesizer, magnetic, 5.30–5.31

T

Tap, in voltage regulator, 5.29, 5.34, 5.35
TCP/IP. *See* Transmission control protocol/internet
    protocol
TDR. *See* Reflectometer, time-domain
Telecommunications bonding backbone (TBB), 5.15–5.16
Telecommunications Industry Association (TIA), 9.13, 9.19.
    See also ANSI/TIA/EIA standards
Telecommunications system, 5.15–5.18, 5.23, 6.25, 9.13–9.16.
    See also Telephone system
Telephone system
    lineman's test set to check for dial tone, 11.15
    line used by dialup modem, 10.33
    old-style, connections for, 8.60, 8.62–8.63
    residential, 5.17–5.18, 5.28, 8.60
Television
    broadband transmission, 10.30
    community antenna, 5.17–5.19, 8.3, 8.6, 10.31, 10.36, 11.18
    in electrical system plan, 6.31–6.32
    electromagnetic interference from, 5.23
    remote control, 7.12
    resonant circuits in, 2.33
    surge protection, 5.28
Temperature
    absolute zero, 7.12, 7.33
    battery, 5.40
    and capacitor voltage rating, 2.16
    computer, 10.44
    diode tolerance, 3.7
    effects on multimeter, 4.12
    effects on solid-state relay, 7.25–7.26
    and infrared detection, 7.12
    and power rating of resistor, 1.14
Temperature recorder, 11.14, 11.15
Terminal, "dumb", 10.5
Terminal block, 8.50–8.52

Vibrator, in ground testers, 5.50
Video Electronic Standards Association (VESA), 10.8
Video graphics array (VGA), 10.8
Virtual memory, 10.3
Virus, computer, 10.14
Volt(s). *See also* Ohm's law
    bits in binary data, 10.19, 10.23, 10.30
    definition, 1.28
    from generating station, 5.2, 5.3
    oscilloscope operation, 11.3
    relationship with angle to lines of flux, 2.3–2.4
    static electricity, 5.46
    theory, 1.4, 1.5, 1.6, 1.13
Voltage. *See also* Ohm's law
    average value, 2.5
    breakover, 3.14
    and capacitors, 2.14, 2.16, 2.17–2.18
    closed path, 1.23–1.24
    commonly used, 5.2
    control with relays, 7.15
    conversion to different, 5.31
    DC power supply, 5.38
    definition, 1.28
    interruptions, 5.21–5.22, 5.31, 5.33
    Kirchhoff's law, 1.22–1.24, 2.26
    measurement, 1.11, 4.12–4.13, 4.17–4.18. *See also*
        Multimeter; Voltage detector; Voltmeter
    over and under, 5.21, 5.31
    peak, 2.5, 2.10, 5.5, 5.20–5.21
    peak inverse, 3.5, 3.7, 3.8
    phase relationships, 2.6–2.8, 2.10, 2.18, 2.19, 2.20, 5.5
    power line, 5.48–5.49
    and power quality, 5.2, 5.20–5.22
    regulation, 5.2, 5.28–5.30
    relationship with current, 2.2, 2.10, 2.11, 2.12–2.13,
        2.17–2.18, 2.20
    ripple, 5.38, 5.39, 5.43
    RLC circuit, 2.30–2.31
    sag, 5.21, 5.31
    step up or step down, 2.37, 2.39–2.40, 5.2, 5.3, 5.8
    storage battery, 5.39
    theory, 1.4, 1.5, 1.6
    and transformers, 2.37, 2.38–2.44
    transients and surges, 5.2, 5.6, 5.7, 5.20–5.21, 5.23, 5.31,
        5.46. *See also* Lightning
Voltage detector, 4.9
Voltage drop
    ammeter, 4.2–4.3
    definition, 1.28
    as effect of harmonics, 5.23
    LC circuit, 2.26, 2.27
    multimeter, 4.11, 4.14, 4.20
    ohmmeter, 4.8–4.9
    RC circuit, 2.25
    relays, 7.25
    and resistors, 1.9, 1.23
    RLC circuit, 2.28
    RL circuit, 2.21, 2.22, 2.23, 2.24
    series circuit, 1.19
    voltmeter, 4.6, 4.7
Voltage ratio, 2.39–2.40
Volt-amperes (VA), 2.33, 2.44, 5.2
Volt-amperes-reactive (VAR), 2.34
Voltmeter. *See also* Multimeter
    circuit, 5.50–5.51
    definition, 1.28
    overview, 1.11–1.12, 4.6–4.8

symbol, 1.8
    use in ground testing, 5.49
    use prior to an ohmmeter, 4.9
Volt-ohm-milliammeter (VOM). *See* Multimeter

W
WAN. *See* Networks, wide area
Water cooler, 6.30
Watt, 1.6, 1.13–1.14, 1.28, 5.2
Wattage (power rating), 1.14, 8.55
Wattmeter, 1.8, 11.11–11.12
Waveforms
    AC phase relationships, 2.6–2.8, 2.10, 2.18, 2.19, 2.20, 2.22,
        5.5
    amplitude. *See* Amplitude of a waveform
    analysis for troubleshooting. *See* Oscilloscope
    from DC power supply, 5.38
    generation, 11.19–11.20
    harmonic distortion, 5.22
    nonsinusoidal, 2.9–2.10, 11.5
    on oscilloscope display, 11.4–11.8
    pulse, 11.5, 11.6, 11.7, 11.10, 11.11
    sawtooth, 2.9, 11.5
    sinusoidal, 2.1–2.8, 5.4, 5.5, 11.5
    square or rectangular, 2.9, 11.5, 11.6, 11.8
    step, 11.5
    triangle, 11.5
    from uninterruptible AC power supply, 5.33
Wavelength, 2.5
Welding, exothermic, 5.12, 5.13
Windings. *See* Coil
Windows®, 10.8, 10.14, 10.18, 10.20–10.21
Windows®Server, 10.21
Wire and wiring. *See also* Cable; Terminations
    AWG size and color codes, 8.34
    continuity testing, 4.22
    cutting procedure, 8.44
    DC loop resistance, 11.22
    desoldering procedure, 8.57–8.58
    diagrams, 6.6, 6.36, 6.38–6.40
    drain, 8.52, 8.66, 10.25, 10.26
    electrical drafting lines, 6.14
    high-voltage. *See* Power transmission lines
    jumper, 8.38
    local area network, 10.23–10.24
    neutral, 5.8, 5.13, 5.22, 5.23
    premises, 5.3–5.4
    soldering procedure, 8.57
    stripping procedure, 8.44, 8.45, 8.57
    switch, 7.3–7.4
    symbols, 1.8, 6.15, 6.17
    time clock, 7.27
    USOC scheme, 8.29, 8.31
Wireless technology, 10.27, 10.28
Wire map tester, 8.32–8.33, 11.16
Wire running list, 6.40
Word processing, 10.22
Work, in electrical theory, 1.13
Workstation, 8.27–8.28
Wrist strap for grounding, 5.46, 10.44, 11.9

X
Xerox®, 10.22

Z
Zinc, 1.6

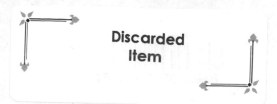